M. Markussen R. Buse
H. Garrelts M.A. Máñez Costa
S. Menzel R. Marggraf

Valuation and Conservation of Biodiversity

Interdisciplinary Perspectives on the Convention on Biological Diversity

Michael Markussen Ralph Buse
Heiko Garrelts María A. Máñez Costa
Susanne Menzel Rainer Marggraf

Valuation and Conservation of Biodiversity

Interdisciplinary Perspectives on the Convention on Biological Diversity

with 57 Figures and 34 Tables

Springer

Dr. Michael Markussen
Friedrich-Ebert-Str. 21
33102 Paderborn, Germany

Heiko Garrelts
Institute for Forest Policy,
Forest History
and Nature Protection
University of Göttingen
Büsgenweg 3
37077 Göttingen, Germany

Dr. Susanne Menzel
Department of Agricultural
Economics
University of Göttingen
Platz der Göttinger Sieben 5
37073 Göttingen, Germany

Ralph Buse
Institute for Production
and Investment
Department of Corporate
Planning
University of Göttingen
Platz der Göttinger Sieben 3
37073 Göttingen, Germany

Dr. María A. Máñez Costa
Alfredo Culla 4-15
46018 Valencia, Spain

Professor Dr. Rainer Marggraf
Department of Agricultural
Economics
University of Göttingen
Platz der Göttinger Sieben 5
37073 Göttingen, Germany

Cover photo: Petra Maass

Library of Congress Control Number: 2004115722

ISBN 3-540-24022-5 Springer Berlin Heidelberg New York

Springer is a part of Springer Science+Business Media
springeronline.com

Printed in Germany

Cover design: E. Kirchner, Heidelberg
Production: Almas Schimmel
Typesetting: Ralph Buse, Göttingen
Printing: Mercedes-Druck, Berlin
Binding: Stein + Lehmann, Berlin

Printed on acid-free paper 30/3141/as 5 4 3 2 1 0

Preface

The continued loss of global diversity brings into question the approaches used to conserve it. This volume brings together stimulating research and findings from Germany and Guatemala asking fundamental questions. For example, one paper questions the economic rationality of current biodiversity conservation methods, while another one attempts to answer this by pointing out that companies are increasingly investing in conservation because they expect this to become a social norm – and therefore respecting and following a social norm is a rational strategy.

It is more evident now than ever that biodiversity conservation is successful only in societies where basic socio-economic needs have been met. This means that poor societies will conserve the flora and fauna around them if given the opportunity to be effective practitioners by including their indigenous knowledge, having effective access and benefit sharing regimes, making national parks more equitable, and tackling poverty and particularly land hunger which is devastating in its effects through deforestation.

Economic instruments are increasingly recognised as having an important role to play in the protection of biological diversity. Well-defined property and use rights can promote the conservation and sustainable use of biodiversity. Tradable catch and export permits can encourage the protection of endangered species, and incentives such as charges and taxes can help to maintain critical habitats, including wetlands. The use of economic instruments has expanded significantly over the last few decades. This reflects a growing understanding that economic instruments can increase the efficiency and effectiveness of environmental management, generate financial resources and incentives for investment, and expand the involvement of private agents in environmental protection.

An areas of great interest to UNEP, among others is the Great Apes Survival Project Partnership (GRASP), which has been faced with the immediate challenge of lifting the threat of imminent extinction faced by bonobos and gorillas, and serious threats to chimpanzees and orangutans. GRASP in its endeavour to conserve viable, wild populations of every kind of great ape and to make sure that their interactions with humans are mutually positive and sustainable, seeks to exemplify and relieve the threats faced by other kinds of animals, birds and plants sharing the forests where apes survive and to illustrate what can be achieved through a genuine partnership between all the stakeholders in these fragile Eco-systems.

All of these perspectives are highly valuable contributions to an increasingly complex problem of stemming the loss of biodiversity globally. The findings from Germany and Guatemala represent the spectrum of issues arising in biodiversity conservation both in rich and poorer countries, and strengthen the discourse on how to effectively implement the 1992 Conservation on Biological Diversity for the benefit of all.

I trust that this publication will stimulate critical thinking and most importantly promote sound and effective conservation actions that will significantly reduce the loss of biodiversity.

Klaus Töpfer

Executive Director,
United Nations Environmental Programme (UNEP)

General introduction

The German Advisory Council for Global Change (WBGU) recognises biodiversity damage and loss as one of the key problems of global change, which, in turn, is rated by leading scientists as the greatest global ecological danger. The global relevance of this problem is emphasised by the fact that one of the three international agreements signed at the United Nations Conference on Environment and Development held in Rio de Janeiro in 1992 was exclusively targeting at the conservation and the sustainable use of biodiversity. So far, the Convention on Biological Diversity (CBD) is the most comprehensive regime of regulations for the protection of biological diversity under international law. Like other environmentally related agreements of international scope, the CBD is regarded as an initial milestone towards a new ecological world order and is seen as a substantial element of a new world order. The concept of sustainability, in which the CBD is explicitly integrated, liberates the ecological question from its isolation and makes it an integral part of the global societal development. This poses a considerable challenge to scientific research; after all, the point is to identify determinants of vastly complex systems – ecological, economic and social systems alike – and to develop suggestions how to network them both locally and internationally in a justifiable way.

Science concerned with the global problem of the loss of biological diversity faces the following demands: it needs to be internationally oriented and needs to emphasise the interconnection of disciplines of social, economic and natural sciences. The latter can be achieved by the disciplines being concertedly geared to a real-life problem which requires an interdisciplinary endeavour. Results provided by natural sciences, e.g., conservation biology, ecology, agricultural and forest sciences, have to be interlinked with those provided by social sciences, e.g., law, economics, political science, ethics and sociology.

In October 2000 the University of Göttingen, Germany, launched an interdisciplinary graduate research programme entitled »Valuation and conservation of biodiversity: implementation of nature conservation strategies within the framework of the Convention on Biological Diversity.« This programme was designed to explicitly overcome the deficiencies of the German internationally oriented biodiversity related research criticised by the WBGU. Not only did the programme take care of an interconnectedness of the various disciplines engaged in the programme but it explicitly took a perspective centred on social sciences. This approach was different to those of projects which had been funded by the German Research Foundation (DFG) in the frame of the expired biodiversity related priority programme. An analysis of the problems of and the perspectives on the conservation of biological diversity connected to the implementation of the CBD was at the heart of the graduate research programme. As the area of regulation of the CBD is markedly large, the subject of the programme had to be confined in a way appropriate for interdisciplinary problem-oriented research. It is against this background that the subject was limited to the »in-situ conservation« codified in article 8 of the CBD, which promotes the strategy of establishing a system of protected areas.

Conserving biodiversity by establishing a system of protected areas is an element of national nature protection strategies and takes place under the conditions given by the CBD. Thus, the scientific questions of the graduate programme were related to two layers:

In the global layer general issues – issues that are concerned with the CBD as a whole – were dealt with. These issues are independent of concrete considerations about national implementation strategies and point beyond national borders. They link national implementation strategies to each other and are concerned with the background information and the framework for their development.

In the national layer problems of and perspectives on the implementation of the CBD were to be analysed in regards to national conditions and frameworks found in selected nations. As developing and developed nations differ in various aspects – their area of unspoilt nature and general ecologic conditions as well as their economic, social and political-institutional potentials for implementation – one developing (Guatemala) and one developed country (Germany) were included into the analysis.

The editors

Michael Markussen
Ralph Buse
Heiko Garrelts
María A. Máñez Costa
Susanne Menzel
Rainer Marggraf

Introduction to the structure and contents of this volume

The 20 contributions to this volume present the most important research results of the interdisciplinary graduate research programme »Valuation and conservation of biodiversity: implementation of nature conservation strategies within the framework of the Convention on Biological Diversity«. Scientists from 14 disciplines were involved.

Part I of this volume takes a global and general perspective on the Convention on Biological Diversity.

Part II takes local, regional and nationwide perspectives on the Convention on Biological Diversity by investigations in Germany. The local and regional study areas in Germany are primarily located in the Lower Oder Valley National Park at the border to Poland. The River Oder is a typical low land river in Central Europe with a three to four km wide floodplain adjacent to the river.

Part III takes local, regional and nationwide perspectives on the Convention on Biological Diversity by investigations in Guatemala. The local and regional study areas in Guatemala are located in the central highlands of Guatemala and in the lowlands in the surroundings of the National Park Laguna Lachuá. All investigation sites in Guatemala are part of the department of Alta Verapaz.

Part IV takes a general perspective on the evaluation of interdisciplinary research.

In the following the contribution's perspectives on the evaluation and conservation of biodiversity in chronological order are given.

Rainer Marggraf Marggraf analyses the loss of biodiversity and the measures to protect biodiversity theoretically. From this economic point of view the central arguments for or against protection are benefits and costs. To take this into account means to act rationally in the economic sense. After giving a thorough yet intelligible introduction into the term of economic rationality Marggraf outlines the need for organised, collective international action to internalise border crossing external benefits of biodiversity protection. For this internalisation there are different opportunities: biodiversity can be protected by trade restrictions or money can be transferred international indirectly by ›debt for nature swaps‹ or directly by ›transfer payments‹ like those provided through the GEF. A policy implication is that biodiversity protection should be orientated on its social costs and benefits. Neither trade restrictions implemented in GATT, the GEF mechanism nor ›debt for nature swaps‹ succeed in this task but payments through the GEF are a better alternative than ›dept for nature swaps‹ to transfer money for biodiversity protection in developing countries.

Susanne Menzel This contribution deals with the costs of biodiversity conservation and the financial contributions made so far by industrialised countries for biodiversity conservation in developing countries within the framework of CBD. Menzel shows that the payments of industrialised countries don't cover the costs of a complete conservation of biodiversity. She finds that the current GEF mechanism under the CBD leads to an under-supply of global biodiversity protection as industrialised countries are made to pay less for the protection of biodiversity than they benefit from it. As an example Menzel analyses whether the German GEF contribution corresponds to the economically determined valuation of biodiversity by the German population. An empirical study is presented: Menzel derives a sample benefit for an industrialised country from an empirical study of German citizens' willingness to pay using the contingent valuation method.

Alexander Behrens The transfer, handling and use of living modified organisms (LMOs) resulting from modern biotechnology leads to a threat to biodiversity and is therefore posed under international rules. The objective of the Protocol of Carthagena is to contribute to ensuring an adequate level of protection and focuses the regulation of trade. This raises the question of the Protocol's relation to other international trade regulations, specifically to world trade law as embodied in WTO law. Apart from this tension field the question has to be addressed in how far the focus on trade constitutes efficient mechanisms to regulate biotechnology. And are the provided instruments adequate to this regulation? Behrens analyses this aspects from an environment perspective.

Heiko Garrelts, Regina Birner and Heidi Wittmer In general, measures for biodiversity conservation don't have a high political priority. Nonetheless biodiversity conservation policies sometimes are surprisingly successful, for example the formation of National Parks in East Germany and Guatemala. Following the *Policy Window Approach* Garrelts, Wittmer and Birner explain this policy as as a result of international influence on conservation policy, regime change, well organised conservation groups and political entrepreneurship. They clearly distinguish there own analyses from economic analyses, where rational choices and preferences are assumed.

Ralph Buse Buse specifies a model of the problem space ›the development of the role and importance of biodiversity for industrial companies‹. To do so, an appropriate approach to identify the *value* of biodiversity for industrial companies is the key. Buse argues that if the companies have the motivation to conserve biodiversity they do not refer to the scientifically based definition of terms but it is more the public perception and understanding of ›biodiversity‹ that is crucial. He assumes that the so called ›diffusion of associations‹ mechanism is decisive for both the number of companies motivated to engage in the protection of biodiversity and the quality of their actions. This mechanism has the public, the legislature, NGOs and pro-biodiversely acting companies as its determining players. Finally Buse's approach suggests another alternative evaluation method for biodiversity using multiagent based social simulation techniques.

Matthias Schaefer Schaefer shows an overview of the problems and perspectives of conservation of biodiversity in Germany. He analyses the positive and negative sides of the situation for conservation of biodiversity. Numerous species exist in a diverse landscape with much anthropogenic disturbance of different kinds. Schaefer presents a »comprehensive« approach with the focus on the maintenance and enhancement of the diversity of the (natural and cultural) landscape. Typical dangers to biodiversity in Germany are discussed. Protective measures have to take into consideration these specific risk situations, i.e., besides a protection of species and a protection of habitats, a protection of the landscapes is of particular importance.

Tobias Hellenbroich Hellenbroich analyses the juridical perspectives of the biodiversity-protection on the basis of National Parks. The following issues are raised: Does the German Nature protection law correspond and coincide with the standards of the International Union of the Conservation of Nature (IUCN)? In this context Hellenbroich picks the question out in how far anthropogenically influenced areas are declarable as a National Park – the famous case of the legally failed National Park *Elbtalauen* marks exactly the lack of untouched landscapes in central Europe. Behind this background the contribution describes the learning process of the German government which created within the amended Nature Protection Law the category of a National Park in process (*Entwicklungs-Nationalpark*) which makes further declarations possible. In the second part Hellenbroich describes the proceeding of a National-Park-declaration.

Judith Rothenbücher, Kai Bentlage and Peter Just On the basis of the National Park *Lower Oder Valley* conservation management and process conservation as different protection strategies are critically analysed. Above all the contribution deals with the target-species approach and its aim to preserve and to manage not only the selected target species but also a larger area (»windfall gain«). Three case studies serve as a research base, among them the corncrake (*Crex crex* L.) as a highly threatened bird species. To combine conservation of rare endangered species as well as of biological diversity the combination of both strategies seems to be a successful approach. At the same time the contribution shows that there are still some open questions concerning the valuation and conservation of biodiversity in Germany.

Ingo Bräuer Ingo Bräuer argues that a proper quantification of the value of nature conservation projects needs to be based on an adequate economic evaluation of ecosystems services. He presents an innovative approach proposing an adequate way of measuring the economic value of this ecosystem services in an appropriate user friendly procedure. In case of doubt Bräuer deliberately has chosen low values for the benefits of an nature conservation project in order not to overestimate its economic value. He shows that a major part of the projects costs could be counterbalanced if all beneficial ecosystem services provided by the project were adequately accounted for in a cost benefit analysis. The contribution of Bräuer to the overall target of conservation is given through the results obtained in his calculations. With his contribution, Bräuer is building a bridge between the results of scientific research and its policy implications.

Anke Fischer In the stress field between land use and conservation, payments for environmental services may – according to the Convention of Biological Diversity (CBD) Art. 10;11 – be an appropriate instrument to encourage sustainable land use. The concept of payments for environmental services has several advantages that make it a particularly suitable incentive measure. If payment schemes could be designed carefully and were compatible with ecological, economic and social aspects of sustainability, they could be quite a powerful instrument used to promote sustainability. Fischer presents such a concept of sustainable protection of agricultural biodiversity in Germany. This concept takes into consideration the preferences of the population for regional biodiversity. The author has made an empirical study which proves the following: a) the population is able to express its valuation of biodiversity as it is able to express its valuation of market goods. b) The population is prepared to make a financial contribution to the conservation of regional biodiversity.

Anne Holl When talking about local knowledge and conservation of biodiversity the reader tends to expect a location different from German home gardens. Holl presents a brilliant way of including people's perception of biodiversity and biodiversity conservation. The contribution of Anne Holl pleads for a higher weight of social sciences within the biodiversity research. But this research should be more oriented to the everyday knowledge and should refer to the individual nature perception. Thus people should be allowed to develop their own personal meanings to the issues; it is not the researcher's role to decide which knowledge is useful or of value. Within her home garden studies, which base on episodic interviews with a strong narrative and biographic components and are supported by participant's observations and group discussions, Holl shows that the knowledge on nature and biodiversity is part of the people's identity and thus depends on time and space. This fact has to be a point of departure of political campaigns on biological diversity instead of simply giving information of the loss of genes and species. Developing their personal view on the subjects in question so that informants can give their own personal meanings to the world. Since the CBD – in its article 8 – encourages the in situ conservation and the introduction of local knowledge for the overall protection of biodiversity, the contribution of Holl reveals important aspects of the failures of science and gives us insights in how significant local knowledge is for the protection and preservation of biodiversity.

Johannes Klose Human valuation of biodiversity has changed in history. It is determined by the respective prevailing social conditions. Since ›Nature‹ is a social construct to take into account the development of the value of birds along the last centuries is a bright contribution for the development of future conservation plans. In his contribution Klose examines the importance attributed to the protection of birds in Germany between the 16th and 20th century. By doing so it becomes clear how manifold the motives are which determine the attitude to songbirds. Important is not only to notice the changes on human population's behaviour and the intrinsic changes in animal abundance due to changes in behaviour, but also how socio-economic and political changes can affect fauna and flora and how preferences change along centuries.

Regina Birner, Heidi Wittmer, Augustin Berghöfer and Michael Mühlenberg The development of biodiversity conservation policies in Guatemala is outlined. Guatemala belongs to those developing countries which make genuine efforts to protect biodiversity. Based on the existing social and natural conditions in Guatemala potentials of and challenges to a successful protection of biodiversity can be determined. If one wants to make use of them a co-operation of GOs and NGOs is of particular importance.

This contribution explains how such a co-operation in Guatemala was achieved and to which results it led regarding protection of biodiversity.

Maria Máñez Costa and Manfred Zeller Usually the integration of strategies, which promote natural protection into peasants' production systems, result in increased costs at the farm-household level. Máñez Costa and Zeller investigate direct payments for environmental services as an opportunity to conserve resources important for freshwater supply, biodiversity protection, and poverty reduction on the farm-household level. Whereas Fischer in this volume deals with the quantification of benefits of biodiversity protection in Germany, Máñez Costa and Zeller focus on the costs of biodiversity protecting activities in Guatemala. Using data from the household level they calculate payments within a linear programming model to asses reasonable payments, which would encourage farmers to produce environmental services instead of crops. These payments can provide incentives for more biodiversity friendly management of agricultural used land. Additionally, they can contribute to the protection of primary forest, as the sustainable way of cultivation reduces the need for additional land.

Petra Maass This contribution introduces the concept of co-management to contribute to the conservation management plan of a national park in Guatemala. This concept implicates the joint effort of conservationists and local population introducing local knowledge systems in policy decision-making processes. It is seen a way of reaching the targets of chapter 8j of the CBD.

The contribution about the cultural context of biodiversity conservation includes an ethnographic case study on indigenous knowledge and its applicability in conservation strategies among Q'eqchi' communities in Alta Verapaz. Natural and cultural aspects of biodiversity are linked in this contribution. The connection of social and spiritual dimensions of biodiversity are elucidated.

Maria Máñez Costa and Swen Renner Máñez Costa and Renner create a meaningful combination of socio-economic and ecological perspectives of the conservation of biodiversity for the highlands of Guatemala. They generate different scenarios and illustrate the future trend of land use dynamics in the region of tropical montane cloud forests in Central Guatemala. Even the still stand of further deforestation can protect the endemic birds. Máñez Costa and Renner arrive at the conclusion, that direct payments could be a usable instrument to protect the natural resources and with that the habitats of flora and fauna.

Michael Markussen and Swen Renner Markussen and Renner emphasise the dimension of soil quality for the protection of biodiversity. On the one hand they investigate deforestation with satellite images on a large scale. On the other hand they study soil quality and micro climate of different land-use systems on a micro scale. Whereas they can surprisingly not identify loss of forest between 1986 and 2000, they detect soil degradation (loss of soil fertility) as a result of anthropogenic influence (agricultural use). Markussen and Renner found that degradation was less harmful when farmers used soil protection strategies. As the authors assume a relationship between soil quality and species diversity, they deduce a positive impact from soil protection strategies for biodiversity protection. As policy implication they argue for the implementation of soil protection strategies not only to care for soil quality, but to protect species biodiversity.

Swen Renner and Michael Markussen Renner and Markussen use birds as keystone species to measure the impact of deforestation and fragmentation on biodiversity. Their results show higher diversity of birds in more human influenced secondary forests than in primary forests. However, they do not interpret this as a ›good sign‹, as the mean body mass per species is higher in primary (more natural) forests and specialists were not caught in secondary forests. Especially species like *Penelopina nigra* Highland Guan or *Pharomachrus mocinno* Resplendent Quetzal are absolutely dependent on natural forests. These species will be extinct at the latest when the last natural forests are gone. The contribution of Renner and Markussen shows that focusing on special species like the Quetzal leads to different protection strategies than concentration on number of species. If the protection of specialists is an aim primary forest have to be protected.

Ludger-Josef Löning and Hermann Sautter Subject of this contribution are the reasons which determine the loss of tropical rainforests in Guatemala. The contribution from Löning and Sautter shows with the comprehensive evaluation of literature a good overview about the fundamental causes of deforestation and habitat loss in Guatemala. The loss of tropical rainforests in Guatemala is analysed in a socio-historical examination. By doing so it becomes clear that there are a variety of reasons which have an impact ranging from poverty, population growth, migration, insufficient property rights to the level of education. It will be worked out that a lesser pressure on the tropical rainforest requires an increase in the welfare of the population.

Philip W. Balsiger and Rudolf Kötter Several approaches exist to describe the co-operation of several scientific disciplines: supra-, inter- and transdiciplinary. In order to avoid misunderstandings Balsiger and Kötter clarify this terms, asking in a next step which form of research in the field of biodiversity is actually done and which form of research may be considered a suitable form. Finally, their contribution focuses on the problem of evaluation of research projects and its need in the field of biodiversity.

The editors

Michael Markussen
Ralph Buse
Heiko Garrelts
María A. Máñez Costa
Susanne Menzel
Rainer Marggraf

Brief Contents

Part I Global and general perspectives on the Convention on Biological Diversity

Global Conservation of Biodiversity from an Economic Point of View
Rainer Marggraf ... 3

Financial support for biodiversity protection in developing countries – does the CBD mechanism lead to an appropriate level of biodiversity protection?
Susanne Menzel ... 23

The Cartagena Protocol: trade related measures as a means to protect biological diversity from risks deriving from genetically modified organisms
Alexander Behrens ... 43

Policy-Windows for the Declaration of Protected Areas – A Comparative Case Study of East Germany and Guatemala
Heiko Garrelts, Regina Birner, Heidi Wittmer ... 65

Will companies engage in the conservation of biodiversity? A prototypical model of aggregated pro-biodiverse actions of industrial companies
Ralph Buse ... 85

Part II Local, regional and nationwide perspectives on the Convention on Biological Diversity: Examples from Germany

Problems and Prospects of the Conservation of Biodiversity in Germany
Matthias Schaefer ... 107

The Designation of National Parks in German Nature Conservation Law
Tobias Hellenbroich ... 133

Conservation management of target species or conservation of processes – Winners and losers of two different conservation strategies
Judith Rothenbücher, Kai Bentlage, Peter Just 155

Valuation of ecosystem services provided by biodiversity conservation: an integrated hydrological and economic model to value the enhanced nitrogen retention in renaturated streams
Ingo Bräuer 193

Towards sustainable land use: Public demand for plant diversity in agricultural landscapes of central Germany
Anke Fischer 205

Narrating diversity: Plants, personal knowledge and life stories in German home gardens
Anne Holl 221

Aspects of bird valuation in Brandenburg-Prussia: Towards the significance of socio-economic conditions for biodiversity perception between the 16th and 20th century
Johannes Klose 249

Part III Local, regional and nationwide perspectives on the Convention on Biological Diversity: Examples from Guatemala

---

Prospects and Challenges for Biodiversity Conservation in Guatemala
Regina Birner, Heidi Wittmer, Augustin Berghöfer, Michael Mühlenberg 285

Calculating Incentives for Watershed Protection. A Case Study In Guatemala
María A. Máñez Costa, Manfred Zeller 297

The cultural context of biodiversity conservation
Petra Maass 315

Direct payments for conservation – the importance of environmental measures in farming systems for bird populations in a fragmented landscape. A case study in Guatemala
María A. Máñez Costa, Swen C. Renner 343

Land use changes and abiotic aspects as basic conditions for conservation of biodiversity in a tropical montane cloud forest (Guatemala)
Michael Markussen, Swen C. Renner 357

Human impact on bird diversity and community structure in a tropical montane cloud forest in Alta Verapaz, Guatemala, with special reference to the Quetzal (*Pharomachrus mocinno*)
Swen C. Renner, Michael Markussen 373

What drives biodiversity loss in the land of trees? A review of the economic and historical parameters causing deforestation in Guatemala
Ludger-Josef Loening, Hermann Sautter 391

Part IV Epilogue: A general perspective on the evaluation of interdisciplinary research

Permanent Evaluation: An important tool for a quality assurance in interdisciplinary research
Philip W. Balsiger, Rudolf Kötter 421

Contents

Part I Global and general perspectives on the Convention on Biological Diversity

Global Conservation of Biodiversity from an Economic Point of View
Rainer Marggraf ... 3
1 Introduction ... 3
2 What characterises an economically rational way of »dealing«with biodiversity? ... 4
3 The economic value of biodiversity ... 5
4 The loss of biodiversity as an economic problem ... 7
5 Biodiversity conservation measures ... 11
6 The protection of biodiversity by trade restrictions ... 12
7 International transfers for biodiversity conservation ... 15
8 Concluding remarks ... 19
References ... 20

Financial support for biodiversity protection in developing countries – does the CBD mechanism lead to an appropriate level of biodiversity protection?
Susanne Menzel ... 23
1 Introduction ... 23
2 Economic analysis and economic perspective ... 24
3 Contribution of CBD to provision of biodiversity protection in developing countries ... 25
4 Financial resources from the GEF and costs of worldwide biodiversity protection ... 27
4.1 Financial resources provided by the GEF ... 27
4.2 Estimated costs of global biodiversity protection ... 28
4.3 Comparison of biodiversity protection costs and GEF spending ... 29
5 Negotiations for the replenishment of the GEF fund ... 30
5.1 Procedure of replenishment ... 30
5.2 Results of negotiations ... 30

5.3 Short analysis of negotiations 31
6 Undersupply thesis 32
7 Case study – willingness to pay for the conservation of biodiversity 32
7.1 Study design 33
7.2 Study results 34
7.3 Discussion of results 37
8 Conclusion 37
References 38

The Cartagena Protocol: trade related measures as a means to protect biological diversity from risks deriving from genetically modified organisms
Alexander Behrens 43
1 Introduction 44
2 Risks and benefits of genetically modified organisms 44
3 Historical development of the Cartagena Protocol 46
4 Central provisions of the Protocol 47
4.1 Scope of application 47
4.2 Trade between parties 48
4.3 Trade with non-parties 55
4.4 Domestic management 56
4.5 Unintentional transboundary movements 56
4.6 Supportive measures 56
5 Environmental protection by trade regulation? 57
5.1 Background to the focus on the regulation of trade with LMOs 58
5.2 The functioning of the trade related measures 60
5.3 Assessment of the trade related measures from an environmental point of view 62
References 63

Policy-Windows for the Declaration of Protected Areas – A Comparative Case Study of East Germany and Guatemala
Heiko Garrelts, Regina Birner, Heidi Wittmer 65
1 Introduction 65
2 Theoretical framework: Kingdon's policy window approach 66
2.1 The garbage-can model as a basis for the policy window approach . 66
2.2 The Policy-Window Approach 67
3 Case Studies 69
3.1 Case study East Germany 69
3.2 Case Study Guatemala 73
3.3 Discussion 77
4 Conclusions 80
References 81

Will companies engage in the conservation of biodiversity? A prototypical model of aggregated pro-biodiverse actions of industrial companies
Ralph Buse ... 85
1 Introduction: motivation, goal and operationalising questions ... 85
2 Question I: Why do companies act pro-biodiversely? ... 86
2.1 General motivation for companies to take pro-biodiverse action ... 86
2.2 The biodiversity related social environment: stakeholders ... 88
2.3 How aware are stakeholders of biodiversity? ... 91
3 Question II: Which companies have a motivation to act pro-biodiversely? 94
4 Question III: What are companies doing when they act pro-biodiversely? 97
5 A prototypical model specification: summary of answers ... 100
6 Concluding remarks ... 102
7 Outlook ... 103
References ... 103

Part II Local, regional and nationwide perspectives on the Convention on Biological Diversity: Examples from Germany

Problems and Prospects of the Conservation of Biodiversity in Germany
Matthias Schaefer ... 107
1 Introduction ... 107
2 Basic patterns: from natural to cultural habitats ... 108
2.1 History ... 109
2.2 The cultural landscapes ... 109
2.3 General aspects of biodiversity in Germany ... 111
2.4 Biodiversity in natural and non-natural habitats ... 111
2.5 Recent patterns of change in species diversity ... 113
3 Problems: from the threats to the biota to the design of management strategies ... 115
3.1 Population ecology: small populations and spatial dynamics ... 116
3.2 Anthropogenic stress factors on a macro- and mesoscale ... 117
3.3 Criteria for conservation ... 118
3.4 Necessity of managing habitats ... 119
4 Prospects: from the species to the landscape approach ... 120
4.1 Basis for conservation: the ecology of the species ... 120
4.2 Programs and strategies for conservation ... 121
4.3 The species approach: preservation of species; the problem of Red data books ... 122
4.4 The habitat approach: preservation of rare habitat, foundation of larger Protected Areas ... 124
4.5 The comprehensive or landscape/habitat/species approach: preservation, restitution and/or creation of a diverse landscape ... 125
5 Synthesis and conclusions ... 127
References ... 128

The Designation of National Parks in German Nature Conservation Law
Tobias Hellenbroich ... 133
1 Introduction ... 133
2 The legal status in Germany ... 134
2.1 Before the 2002 amendment ... 135
2.2 Since the 2002 amendment ... 140
2.3 Juridical requirements for the configuration of national parks ... 143
3 Conformity of German law and international requirements ... 149
3.1 International requirements ... 149
3.2 Comparison with German law ... 150
4 Conclusion ... 151
References ... 152

Conservation management of target species or conservation of processes – Winners and losers of two different conservation strategies
Judith Rothenbücher, Kai Bentlage, Peter Just ... 155
1 Introduction ... 155
2 Two conservation strategies ... 156
3 Study area ... 158
4 Three case studies from the Lower Oder Valley National Park, Germany ... 161
4.1 Impact of mowing on the suitability of grassland as habitat of corncrakes (*Crex crex*) in the Lower Oder Valley National Park ... 161
4.2 Impact of flooding on the amphibian community ... 167
4.3 Impact of land use and flooding on the diversity of insects ... 174
5 Winners and losers of the two conservation strategies ... 185
6 Acknowledgements ... 188
References ... 188

Valuation of ecosystem services provided by biodiversity conservation: an integrated hydrological and economic model to value the enhanced nitrogen retention in renaturated streams
Ingo Bräuer ... 193
1 Introduction ... 193
2 Investigated conservation programme and study site ... 195
3 Materials and Methods ... 195
4 Results ... 197
4.1 Quantification of the ecosystem function ... 198
4.2 Definition of the substitute and its marginal costs ... 198
4.3 Valuation of the ecosystem service ... 200
5 Discussion ... 201
6 Conclusions ... 202
References ... 203

Towards sustainable land use: Public demand for plant diversity in agricultural landscapes of central Germany
Anke Fischer ... 205
1 Introduction ... 205
2 Payments for environmental services of agriculture: a pilot scheme in central Germany ... 206
3 Study objectives ... 209
4 Methods ... 209
5 Results ... 211
5.1 Stated willingness to pay ... 211
5.2 Information processing ... 212
5.3 Interpretation ... 213
6 Discussion: participation of the local populace in agri-environmental decision making ... 214
7 Summary ... 218
References ... 219

Narrating diversity: Plants, personal knowledge and life stories in German home gardens
Anne Holl ... 221
1 Introduction ... 221
2 Misreading the social in biodiversity research ... 222
3 Knowledge, action and the world ›out there‹ ... 227
4 Getting close to people's knowledge ... 232
5 Stories of difference ... 238
6 Towards a new culture of knowledge ... 243
References ... 245

Aspects of bird valuation in Brandenburg-Prussia: Towards the significance of socio-economic conditions for biodiversity perception between the 16th and 20th century
Johannes Klose ... 249
1 Introduction ... 250
2 Research approach, historical background and hypotheses ... 251
3 Birds as game, trading goods, food and delicacy ... 259
3.1 The significance of small birds for nutrition ... 260
3.2 Some characteristics of bird markets ... 261
3.3 Highly valued in cuisine: The field larch and the fieldfare ... 262
4 Birds as vermin ... 264
4.1 The control of the sparrow: Sensible from an economic perspective? 264
4.2 The control of the crane: Who encroached upon whose habitat? ... 266
4.3 The control of the great bustard: What made the crop-devastator proliferate? ... 268
4.4 Concluding remarks ... 270

5 The development of bird conservation for functional, ethical and aesthetic reasons ... 270
5.1 The conservation of the nightingale and the thrush nightingale (since 1686) ... 270
5.2 Means to overcome a caterpillar calamity: the conservation of insect-eating birds (1792–1802) ... 271
5.3 First approaches to permanent bird conservation: town decrees and school curricula (1813–1858) ... 272
5.4 Turning the tide: the bird conservation decrees from 1867 ... 272
5.5 The German Bird Conservation Act of 1888 ... 273
5.6 Progress and relapse: the aims of conserving the fieldfare ... 274
6 Discussion and summary ... 275
6.1 There is a general trend in the historical perception of birds ... 275
6.2 There are two turning points in bird perception ... 276
6.3 The general trend in bird perception follows socio-economic development. Does this apply to the perception of biodiversity, too? 277
References ... 279

Part III Local, regional and nationwide perspectives on the Convention on Biological Diversity: Examples from Guatemala

Prospects and Challenges for Biodiversity Conservation in Guatemala
Regina Birner, Heidi Wittmer, Augustin Berghöfer, Michael Mühlenberg ... 285
1 Introduction ... 285
2 Socio-Economic and Political Situation ... 285
3 Biodiversity in Guatemala ... 287
3.1 Bio-Geography ... 287
3.2 Cloud Forests ... 288
4 Policy Approaches to Conserve Biodiversity ... 289
5 Prospects and Challenges for Biodiversity Conservation ... 291
References ... 294

Calculating Incentives for Watershed Protection. A Case Study In Guatemala
María A. Máñez Costa, Manfred Zeller ... 297
1 Introduction ... 297
1.1 Legal and policy frameworks for sustainable production systems in Guatemala ... 297
1.2 Research area ... 299
1.3 Research objectives ... 300
2 Conceptual framework ... 301
3 Data collection and the categorization of farm households ... 302
3.1 Categorization of farm households ... 302
3.2 Environmental Services ... 305

3.3 Recent behaviour of the farmers after the coffee crisis 306
3.4 Measuring the demand of water 307
4 Model simulations 308
5 Findings and Conclusions 311
References 312

The cultural context of biodiversity conservation
Petra Maass 315
1 The global context 315
1.1 Indigenous communities and biodiversity conservation 316
1.2 The Convention on Biological Diversity 317
2 The discursive context 318
2.1 Anthropological perspectives 319
2.2 Cultural concepts of nature 321
2.3 Indigenous knowledge 321
2.4 The landscape approach 324
3 The local context 325
3.1 The ethnographic experience 327
3.2 Protected area management and local livelihoods 328
3.3 Expressions of indigenous knowledge 331
3.4 The sacred landscape 335
4 Concluding remarks 339
4.1 From local to global 339
4.2 Rethinking scientific assumptions 340
References 341

Direct payments for conservation – the importance of environmental measures in farming systems for bird populations in a fragmented landscape. A case study in Guatemala
María A. Máñez Costa, Swen C. Renner 343
1 Introduction 344
2 Background of the problem 344
3 Predicting the future of the region 347
4 The needs of environmental conservation 350
5 Policy Implications 352
References 354

Land use changes and abiotic aspects as basic conditions for conservation of biodiversity in a tropical montane cloud forest (Guatemala)
Michael Markussen, Swen C. Renner 357
1 Introduction 357
2 Study area 358
3 Methods and Material 360
4 Results 360
4.1 Forest cover and forest conversion 360
4.2 Soil properties and its degradation 362

4.3 Micro climate changes ... 365
5 Discussion ... 368
6 Conclusions ... 369
References ... 370

Human impact on bird diversity and community structure in a tropical montane cloud forest in Alta Verapaz, Guatemala, with special reference to the Quetzal (*Pharomachrus mocinno*)
Swen C. Renner, Michael Markussen ... 373
1 Introduction ... 374
2 Study area ... 374
3 Methods ... 375
4 Diversity pattern and community structure of the birds in the Sierra Yalijux ... 376
4.1 Species set and composition ... 376
4.2 Estimation of species richness ... 377
4.3 Habitat selection ... 377
4.4 Recaptures ... 379
4.5 Body mass distribution ... 380
4.6 Pharomachrus mocinno and its regional population ... 382
5 Discussion ... 384
6 Conclusions ... 386
References ... 387

What drives biodiversity loss in the land of trees? A review of the economic and historical parameters causing deforestation in Guatemala
Ludger-Josef Loening, Hermann Sautter ... 391
1 Introduction ... 391
2 From past to present ... 393
2.1 Mayan civilisation and Spanish colonial period ... 393
2.2 Production of export crops ... 394
2.3 Directed and spontaneous colonisation ... 397
2.4 Protected areas and civil war refugees ... 400
3 Magnitude and location of forest cover loss ... 401
3.1 Magnitude of deforestation ... 401
3.2 Location of deforestation ... 403
3.3 Environmental diversity ... 404
4 Direct sources of deforestation ... 404
5 Underlying determinants of deforestation ... 408
5.1 Poverty, population growth and migration ... 409
5.2 Soil conditions and road construction ... 410
5.3 Agricultural productivity ... 411
5.4 Property rights ... 412
5.5 Ethnicity ... 412
5.6 Rural non-farm employment and education ... 413
6 Concluding summary ... 413
References ... 414

Part IV Epilogue: A general perspective on the evaluation of interdisciplinary research

Permanent Evaluation: An important tool for a quality assurance in interdisciplinary research
Philip W. Balsiger, Rudolf Kötter 421
1 Introduction 421
2 On terminological inconveniences 422
2.1 Problem-oriented vs. topic-oriented research 422
2.2 Interdisciplinarity and the question of how to guide the research process 423
2.3 Transdisciplinarity 424
2.4 Biodiversity and its real relationship with supradisciplinary research 425
3 Evaluation 426
References 428

Part I

Global and general perspectives on the Convention on Biological Diversity

Global Conservation of Biodiversity from an Economic Point of View

Rainer Marggraf

Institute of Agroeconomics, Georg-August University of Göttingen, Platz der Göttinger Sieben 5, 37073 Göttingen, Germany, Tel 0049-551-394829, Fax 0049-551-394812, email: rmarggr@gwdg.de

Summary. This paper demonstrates from an economic point of view what characterises rational biodiversity conservation efforts, why the present methods for the conservation of biodiversity are not rational, and what measures should be taken to change this state. This economical approach is demonstrated by analysing international policy mechanisms used for the avoidance of one of the internationally greatest ecological risks – the continuing loss of global biodiversity. The international environmental policy is formed by sovereign states which cannot order each other to take certain measures. Rather, they must choose different ways to influence global biodiversity conservation. The states must choose to realise this conservation either by unilateral or international agreement. In Sect. 6 and 7 each of these possibilities is analysed as to the extent to which it is used in international biodiversity conservation policy. Further, it is explained from an economical point of view how well the central task of international policy measure, e.g., the internalisation of international external benefits and costs, is fulfilled.

Key words: international biodiversity conservation policy, economic analysis of biodiversity loss and conservation, economic value of nature, convention on biological diversity

1 Introduction

This contribution demonstrates from an economic point of view what characterises rational biodiversity conservation efforts, why the present methods for the conservation of biodiversity are not rational, and what measures should be taken to change this state. This economical approach is demonstrated by analysing international policy mechanisms used for the avoidance of one of the internationally greatest ecological risks – the continuing loss of global biodiversity.

Sect. 2 explains which economical criterion decides whether a biodiversity conservation effort is justified or not. The economic value of nature is important for this criterion and this value concept is described in Sect. 3. Sect. 4 demonstrates the present overexploitation of biodiversity and Sect. 5 shows the possibilities for avoiding this.

The international environmental policy is formed by sovereign states which cannot order each other to take certain measures. Rather, they must choose different ways to influence global biodiversity conservation. The states must choose to realise this conservation either by unilateral or international agreement. In Sect. 6 and 7 each of these possibilities is analysed as to the extent to which it is used in international biodiversity conservation policy. Further, it is explained from an economical point of view how well it fulfils the central task of international policy measure, e.g., the internalisation of international external benefits and costs. The last section contains some concluding remarks.

2 What characterises an economically rational way of »dealing« with biodiversity?

Economic evaluations and the corresponding derived recommendations for action raise an ethical awareness. These are not merely uncommitted personal value judgments made by particular economists without obligations, but statements which demand an utilitarian ethical quality. The economic theory of valuation is applied utilitarian ethics. The ethical basis of economic valuation is the ethical theory of social subjectivism[1].

This ethical theory is determined by three components. The first component is comprised of the utilitarian value theory. According to this theory only the benefit (= the fulfilment of the interests) of the affected members of the society are relevant when evaluating particular actions. Interests of the state, enterprise objectives and the welfare of animals or other entities are not considered. The only basis for the evaluation is human interests. These are considered according to the so-called principle of consumer sovereignty in economics. It is assumed that every individual knows what is best for her or him. The ideas of societies' members are accepted as they are, that is the interests are not differentiated as ›important‹ or ›unimportant‹ and ›fair‹ or ›unfair‹. The second component involves the aggregation of individual interests. Different proposals exist as to how the interests should be aggregated. The utilitarian proposal plays a predominant role in the ethical discussion. In this case the individuals' interests in the action to be evaluated are simply added up. This proposal implies particular ideas as to the measurability and comparability of individual interests. Most economists do not agree with this method. Therefore, they prefer a different procedure for aggregation by which the different individual interests in a particular action are transformed to costs and benefits. This occurs in three steps. First, all of the aspects to be evaluated are divided into two categories: those which the members of society would consider positive and those which they would consider negative. This can be explained with the help of the following example:

[1] For an overview of the various subjective and objective ethics theories see Kutschera (1982).

A street which bypasses a community is to be built in order to reduce the amount of traffic in the centre of the city. The planners on the transportation committee have determined that only a particular route can be financed by the available funds. The route for the bypass however, has the consequence that it would pass directly through and destroy a natural biotope in which rare animal and plant species live. If the street is not built, then the money will be used to expand a preschool. There are at least two positive and two negative consequences tied to this street construction project.

Positive consequences are that for those living on the currently used streets a reduction in noise pollution will result, and those who drive daily to work will save time and fuel. The negative results include the destruction of the natural biotope and the fact that the preschool will not be expanded.

In the second step, the social benefits and social costs of the consequences are determined. A project's consequence leads to social benefits or social costs if it is determined to be positive or negative, respectively. In our example the consequences ›quieter streets‹ and ›new highway‹ are considered social benefits while the project consequences ›destruction of a biotope‹ and ›no pre-school expansion‹ cause social costs. Social benefits and costs are measured in monetary units. They reflect the respective social gain and loss which would result from the project consequences and are defined as follows:

The social benefits of a project consequence are equal to the sum of the maximum amount of money that can be taken away from the individuals in order to leave them just as well off as they were before the consequence happened. The social costs correspond to the minimum amount of money necessary to compensate the individuals for the consequence.

The final step consists of the summation of the economic benefits and costs of all consequences to calculate the total social benefits and the total social costs of the project.

The third component of the ethical theory of social subjectivism is the concept of the consequential criterion of correct action. An action is economically justified when the social benefits are greater than the social costs. Thus, it is necessary to determine the corresponding economic increase or loss in value for each of the resulting consequences in order to decide if the planned action should be done. If any of these consequences has an effect on biodiversity, then this effect must also be economically assessed.

3 The economic value of biodiversity

Biodiversity is highly significant for the satisfactory securing of very different human needs (Pearce and Moran 1994). It is the basis for food, it is a natural pharmaceutical laboratory and it provides storage of raw materials. Biodiversity is important for fundamental scientific research, it serves as a model for technical development, and it has indicator and monitor functions for mineral resources and harmful chemicals. The adaptability and exploitation abilities of biological diversity are used. And finally, the usage of biological diversity as a source for the satisfying aesthetic and emotional needs is appraised. All of these contributions to the human benefit must be considered in economic valuations of biodiversity.

Many methods for systematising the economic value of biodiversity can be found in literature. The fundamental valuation categories are formed from the production and consumption values (Marggraf and Streb 1997). Individual components of biological diversity (in economic terminology: biodiversity goods) are often in the form of natural mineral resources and are thus entered as production factors in the manufacturing process of market goods. Changes in quantity and quality of these resources affect production costs, prices and income, thus influencing the supply of these market goods for members of society. In these cases, in which biodiversity goods are significant because they are implemented in the production of market goods, these goods are said to have a production value. Tropical rainforest for example, possesses a production value. The destruction of tropical rainforest contributes to the greenhouse effect which influences agricultural production and the supply of agricultural products.

As a rule, biodiversity goods influence human well-being not only through the ›detour‹ of market goods, but also directly. Biodiversity goods and market goods are equally important factors for determining people's utility. A change in the condition of biodiversity goods also has an effect on human well-being when the supply of market goods remains unchanged. The economic value of such changes is known as the consumption value of biodiversity. A tropical rainforest, for example, is not only valued for its production value. One would like it to be preserved in order to visit it or because one believes it is not right to destroy the ecosystem with the greatest number of species in the world. Thus, the rainforest also possesses a consumption value.

The consumption value can be use-dependent or non-use-dependent (Marggraf and Streb 1997). When people go for a walk in an undisturbed countryside, hike to natural wonders or biotopes or visit near and far resort areas, then the biodiversity goods concerned contribute a *use* value.

Many individuals however, also find pleasure in the mere existence of certain natural and landscape goods without ever planning to use them. They are concerned about preserving whales, bearded vultures or bog lands, even if they never get to see these animals or landscapes. In all these cases the concerned biodiversity goods also possess a *non-use* value. Non-use values can be divided into existence values and bequest values. When benefit is gained by merely knowing that certain animal species and biotopes exist, then these biodiversity goods possess an existence value. If the individuals are interested in preserving the environment for future generations, they are assigned a bequest value.

The production value of biodiversity is of interest commercially since profit and income can be achieved with these value components. Biological resources are of particular importance as inputs in the production process of market goods in agriculture and the pharmaceutical industry. The demand in the agricultural realm is primarily concerned with the genetic material of plants living in the wild, which when crossed with domestic plants can improve resistance to pests, drought and other threats and can secure greater productivity and steady growth. Recently, a number of studies, for example, those in the European BIODEPTH-Project (Hector et al. 1999), have been completed. They have proven that the diversity of nature on cultivated land and pastures has a positive effect on agricultural yields.

The biological diversity is of interest to the pharmaceutical industry because the genetic code of living things contains the »recipe« for the manufacture of commercially utilised chemical compounds. Pharmaceutical research of natural substances is more interested in developing lead structures than identifying natural substances which can be utilised unchanged. Lead structures are promising molecules or structures of compounds that can be drawn upon for the development of new pharmaceutical products.

Commercial benefit can also be derived from the use component of consumption value. Here, the market demand on nature is for rest and relaxation purposes. This is sustained by the preservation of refuge areas nearby and through eco-tourism opportunities. The tourism industry is one of the most significant areas of benefit for biological diversity in many developing countries. The expenditure of tourists from overseas to observe animals in national parks comprises a high portion of foreign income (McNeely et al. 1990).

The entire non-use (consumption) value of nature is useless commercially. No expenditure is tied to the non-use values allowing for profit or an increase in income to be acquired. However, from an economic point of view, the non-use values are as relevant as the use values which do lead to commercial values.

Some results of studies about non-use values of rare and threatened animal species are summarised in Table 1.

Most of the studies were conducted in the USA. The values show how much an American person is willing to pay on average per year for nature conservation measures to secure the respective animal species existence. The values were determined during single unique independent studies. Therefore, the values cannot be summed to determine a non-use value for all the above listed animal species.

The values in Table 1 illustrate that non-use values for biodiversity cannot be neglected. As a result, it was stated in an OECD study that the economic and commercial values of biodiversity are »light years« apart (Brown 1997).

4 The loss of biodiversity as an economic problem

From an economic point of view biodiversity losses are too high. In the following section, this fact will be demonstrated by an example of land conversion in which high diversity land use such as tropical rainforest is altered to low diversity land use such as intensive agriculture. First, the reflections of an individual farmer owning an area of tropical rainforest and who must decide how to use this area will be considered. Secondly, this decision is considered from society's point of view.

Assuming the farmer has the choice between two competing options: (i) sustainable use of the forest, e. g., by marketing non timber products or eco-tourism

Table 1. Willingness to pay for some rare and threatened animal species (Average value 1990 in US $ per person per year) [Source: Geisendorf et al. 1999 and Perrings 1995]

Northern spotted owl	21
Pacific salmon	7.6
Grizzly bear	18.5
Whooping crane	6.5
Sea otter	8.1
Gray whale	9.3
Bald eagle	19.1
Bighorn sheep	8.6
Minnow	4
Wild turkeys	11.4
Sea eagle, white-tailed eagle	19
Brown bear, wolverine and wolf (together)	15
Humpback whale	50
Coyote	5.1

and (ii) conversion of forest land into agricultural use. Economic theory suggests that the farmer behaves ›rationally‹ and chooses the option of land conversion if the net benefits of this option are higher than the net benefits from sustainable use. The net benefits of an option can be calculated as its benefits minus its direct costs. It is reasonable to assume that the individual will take into consideration only the benefits he receives and the costs he has to bear. In economic theory these are called *private* benefits and costs. The net benefits which are forgone because an option is not chosen are also considered as so-called *opportunity costs* in economic theory. If there are two options for using a particular resource, the benefits forgone of the option which is not chosen have to be added as opportunity costs to the other costs of the option which is chosen. Hence, the condition for choosing the land conversion option can be written as

$$\mathrm{PB}(\mathrm{AGR}) - \mathrm{PC}(\mathrm{AGR}) > 0 \tag{1}$$

where $\mathrm{PB}(\mathrm{AGR})$ = private benefits of agricultural use of the land (= market value of the agricultural products) and $\mathrm{PC}(\mathrm{AGR})$ = private direct and opportunity costs of agricultural use of the land (= costs of the agricultural production factors and profits forgone for the sustainable use of the forest)

In contrast, the economic (= social) decision rule is determined by the benefit-cost-criterion. Applied to the rainforest example, the benefit-cost-criterion implies that the decision for the land conversion of the tropical rainforest is economically justified or economically rational if, and only if

$$SB(AGR) - SC(AGR) > 0 \qquad (2)$$

where $SB(AGR)$ = social benefits of agricultural use of the land and $SC(AGR)$ = social costs of agricultural use of the land.

Comparing decision rules 1 and 2 we see: In order to judge whether the farmer's individually rational choice is economically rational we have to examine whether there is a difference between social and private benefits and costs. If $SC(AGR) > PC(AGR)$ or $SB(AGR) < PB(AGR)$ applies, the farmer has not taken into consideration all negative consequences for his fellow human beings. Economists consider these differences as so-called *external costs*. In the opposite case, that is if $SC(AGR) < PC(AGR)$ or $SB(AGR) > PB(AGR)$ applies, economists speak of external benefits.

The land conversion from tropical rainforest to agricultural use typically leads to external costs. The loss of biodiversity which results from land conversion is considered to be negative by many people. However, the farmer does not feel the social costs of this consequence (that is, the sum of the individual compensating variations of the biodiversity loss) if he decides in favour of the land conversion. Therefore, these costs are not a component of the private costs of agricultural use of the land. Accordingly, the social costs of agricultural use are higher than the private costs ($SC(AGR) > PC(AGR)$). Economists distinguish between two major causes leading to external costs: market failures and policy failures. A market failure occurs if there are no market forces which transform the willingness to pay for biological resources into an income source for the farmer. In some developing countries, e. g. Brazil, Costa Rica, Panama, it is a policy of the government to subsidise forest conversion into agricultural land. This is an example of a policy failure. The subsidies reduce the private direct costs of land conversion and therefore, increase the gap between social and private costs even further. Up to now, we have shown that the private decision rule 1 does not correspond to the economic decision rule 2. However, this does not necessarily mean that the farmer's decision is to be criticised from an economic point of view. Fig. 1 presents a diagrammatic exposition of this issue. The horizontal axis measures the amount of land converted from forest to agricultural land and the vertical axis measures monetary units, for example US$. The line MPB represents the marginal private benefits of the converted land. »Marginal« means that the line maps the private benefits of a very little change of forest into agricultural land. Assuming that the market value of the agricultural products reflects the willingness to pay for these commodities, marginal private and social benefits coincide ($MPB = MSB$). The line is downward sloping which indicates that the converted land is ordered according to the height of its benefits. The line MPC represents the marginal private costs of land conversion, i.e., the additional costs which arise, if an additional unit of land is converted. This line is upward sloping because the prices of labour and capital resources, which are used for land conversion, rise if more land is converted.

It is assumed that L_1 marks the actual area of converted land and that the question is, whether this area should be augmented to L_2. For the individual, the answer

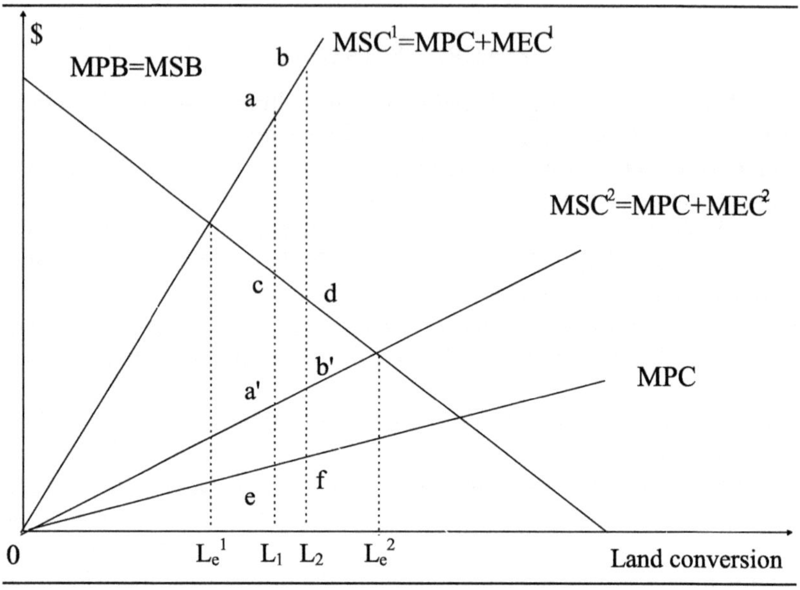

Fig. 1. Land conversion from an individual and an economic point of view

is yes because the private benefits (cdL_1L_2) exceed the private costs (efL_2L_1). From an economic point of view, the answer depends on what the marginal social cost curve (MSC) looks like. To derive MSC we have to add the marginal external costs (MEC) to MPC. The point of intersection of MSC and MSB determines the efficient area of converted land. In Figure 1 we have made two different assumptions about the amount of the marginal external costs (MEC^1 and MEC^2). Therefore, there are two different MSC curves (MSC^1 and MSC^2) and two different efficient areas of converted land (L_e^1 and L_e^2). If MSC^1 indicates the true social costs, the augmentation of the converted land from L_1 to L_2 is economically not justified because its social costs (abL_1L_2) exceed its social benefits (cdL_2L_1). The conversion is justified if MSC^2 indicates the true social costs. In this case the social benefits suffice to compensate for the social costs ($a'b'L_2L_1$).

The undervaluation of biodiversity is only to be considered as an economic problem if it results in an economically unjustified (= inefficient) decision or action. In order to determine whether or not this is the case, one has to calculate the external effect – the external costs of a biodiversity loss or the external benefits of biodiversity conservation.

So far, there has been a wealth of studies in which these effects were evaluated (for a survey see UNEP 1995 : 865). The absolute figures are interesting, but the following fact is more important: In almost all cases it was proven that the social costs of an action leading to a biodiversity loss exceeded its social benefits and that the social benefits of a biodiversity conservation project exceeded its social costs. This implies that how we deal with biodiversity is economically not justified. In terms of Fig. 1: The true marginal social costs of land conversion are indicated by MSC^1, and we are already on the right side of the economic optimum L_e^1. Therefore, economists are interested in instruments and institutions which contribute to a conservation of biodiversity.

5 Biodiversity conservation measures

In contrast to market goods, biodiversity goods often benefit several people simultaneously. The number of beneficiaries at locally significant biotopes is manageable; however, other biodiversity goods such as tropical rainforest are valued worldwide. Normally, it is either impossible or very expensive to control the number of beneficiaries. In addition, a single person cannot guarantee the sustainability of nature alone whether it is by a change in behaviour with regard to his use of nature or by financial contributions.

In order to ensure that the entire economic value of biodiversity is considered when decisions regarding the conservation and the use of biodiversity are made (= that the external benefit of biodiversity conservation and the external costs of biodiversity use are internalised), *organised, collective action* is imperative. This necessity is dealt with at the governmental level by passing legislation (making policies) on biodiversity conservation measures. Biodiversity conservation policies can implement methods which directly and indirectly affect behaviour. *Direct measures* for regulating behaviour are sanctioned regulations and bans which have proven effective. *Indirect methods for influencing behaviour* include informing, appealing and offering financial incentives. Let us consider the example illustrated above. The farmer can be forbidden to clear an area of forest $L_2 - L_1$ by penalisation. He could also be informed of the consequences of his actions if he clears land and be asked to refrain from clearing more land to expand the amount he could plant. Financial incentives are aimed at increasing the private costs for making the change, such that the farmer will calculate with the correct social costs. Two possibilities exist. Either the farmer must pay a tax for clearing the land or he receives payment for maintaining the tropical rainforest. In the first case the costs for clearing are increased and in the second case the opportunity costs are increased. When the tax amount and payment amount correspond to the economic value of biodiversity, then with both alternatives the private costs achieve the level of the social costs. The (previously) external costs are totally *internalised.*

The indirect measures for influencing behaviour (i.e., by taxes or duties) have more benefits compared to the direct influence of behaviour with regulations and bans. Indirect measures are not based on force and thus offer more autonomy in decision making. More latitude can be used to make biodiversity conservation as inexpensive as possible to implement. It also enables the individual to take more initiative and responsibility in searching for more protective measures of behaviour with regard to biodiversity. The direct measures for influencing behaviour do not offer any incentive for realising a potential reduction in costs. Regulations and bans usually determine behaviour after a certain threshold level. In contrast, the indirect methods make clear, with respect to all of biodiversity, that it is a scarce resource which must not be used wastefully. If possible, the indirect measures of influencing behaviour possibilities for determining the value of biological diversity should be implemented. From an economic point of view, internalisation of the external costs through positive and negative financial incentives is necessary in most cases.

The fact that biodiversity has an economic value just as market goods have is most obvious in the payment of money for the conservation of biodiversity. In addition, rewards are usually much more effective in influencing actions than penalties which frequently cause evasion behaviour and require a great effort of control. With the correct design of the payment/reward for the conservation of biodiversity, none of society's members should suffer a loss in benefit. Those exploiting biodiversity decide freely for a different use. The benefit increase of those individuals who welcome a more protective treatment of biodiversity is so high that they can finance the monetary reward for biodiversity preservation without reducing their level of utility.

However, it must be considered that payments for the conservation of biodiversity may violate the generally accepted principle that the individual who causes damage must bear the cost (Kirchgässner 1990). Thus, it is necessary to examine each case when deciding whether payment should be made for the protection of biodiversity. The result will depend in particular upon the involved individuals. In the example above the payment for the protection of biodiversity can surely be a substantiated alternative if the farmer is a subsistence one who cannot support his family unless he achieves a higher income. With respect to international biodiversity conservation measures one has to consider the fact that sovereign nations cannot ›give orders‹ concerning certain measures to one another. They have to choose other ways to influence biodiversity conservation in the respective other countries. They have the fundamental option to do so either against the will of these countries or with their consent. In the next two sections of this contribution it will be shown how far both options apply to biodiversity policies. Moreover, it will be explained how well they comply with the central economic task, i.e., the internalisation of border crossing external benefits and costs.

6 The protection of biodiversity by trade restrictions

The regulation of existing markets by setting environmental standards for international trade is a traditional instrument implemented by international environmental policy (Knorr 1995). Biodiversity conservation oriented trade policy measures are components of international agreements. Individual countries however, have moved to do the same. The USA, in particular, attempts to influence biodiversity conservation in other countries by restricting trade. For example, they do not allow the import of tuna fish and shrimp which have been caught in developing countries by methods in which dolphins and sea turtles are killed, respectively.

The community of states has created its own set of rules for international trade. Thus, the question can be asked whether the environment related trade restricting measures (of a single country or an international environmental treaty signed by several countries) are compatible with the set of rules of the General Agreement of Tariffs and Trade (GATT). In reality this question has only been discussed in disputed cases in which the question of reconciliation of the GATT-rules and the environmental trade restrictions were initiated by a single country (Knorr 1995).

Possible conflicts are addressed twice in the GATT-contract. Article XI allows export bans on important goods and on natural resources in certain special cases. Article XX opens the possibility of trade limitations necessary to protect human, animal or plant life or health (Art. XXb) and »for the conservation of exhaustible natural resources« (Art. XXg) to be made.

All of these regulations enable countries to set environmental standards and to stipulate how the standards will be obeyed for imported goods. The regulations must be applied independent of where the goods originated, thus, they are equally binding for goods produced inland. In addition, the trade related measures for the protection of the natural environment must be related to the particular product. Measures for trade restriction for the influence of undesired production standards abroad are generally forbidden.

The World Trade Organisation (WTO) increasingly applies Article XX in an ›environmentally friendly‹ way. This can be seen in its appeals committee which has the authority to clarify disputed cases. For example, in 1994 in the disputed case where the USA decided to ban all tuna fish from Mexico, the WTO appeals committee determined that tuna fish caught in such a way that dolphins are spared is not the same as methods where more than a specific number of dolphins are killed (Senti 1998 : 54).

The increased significance of the protection of global biodiversity goods is also illustrated in the so-called sea turtle-shrimp case of 1998. Sea turtles are threatened by extinction amongst other reasons because they become tangled in the nets of shrimp fishermen and are killed after the nets are brought in. Therefore, since the end of the 1980s American shrimp fishermen must use a specific protective device. Under pressure from environmental protection organisations, in 1986 the American government ordered that shrimp can only be imported from countries whose fishermen can prove that they use this protective device for sea turtles. India, Malaysia, Pakistan and Thailand claimed that this import ban was a violation of the GATT-rules. In October 1998, the WTO appeals committee dealt with this case. The committee held this import restriction to be unjustified and an arbitrary discrimination of undesired competition as there was no serious effort made to find a unanimous solution. In addition, the foreign producers were only given a short amount of time to change their methods and no transparent procedure existed for controlling whether the importers were really using the desired production methods. Another problem was that out of several environmentally protective production methods, one particular method was stipulated. The shrimp import ban imposed by the USA was thus determined to be a violation of the WTO agreement.

The fact that the appeals committee recognised the following important aspects is more important than the decisions made in the above individual cases (Pfahl 2000):

- Animals and plants are included in the exhaustible natural resources mentioned in Article XXg. It is not significant that they have the ability to reproduce, but rather despite this ability they can as a matter of fact become extinct.
- The natural resources that can become extinct that are supposed to be protected by one-sided trade restrictions can be found outside the sovereignty of the affected country. It is critical that the trade restrictions function in the same way as those that are protected domestically.

This made clear that trade restrictions created to influence the use of the natural environment in other countries are not in general a violation of the GATT-contract.

Trade restrictions only affect the private benefits and costs with respect to the use of nature in other countries indirectly, and are therefore often not an appropriate means of influence. Trade restrictions should be examined for their effectiveness before they are implemented. It would be little effective, for example, to protect tropical rainforests through trade restrictions on tropical hardwood as has been demanded by many environmental conservation organisations (Amelung 1997). Various measures have been proposed to address this such as general import bans, quantitative restrictions on imports, import bans on tropical hardwoods that are not produced in a sustainable way and high import taxes. All of these measures would only make sense if the private benefit of tropical wood resulted primarily from trade. Globally however, this is not the case. Calculations have shown that only 5 % of tropical hardwoods are exported. A complete relinquishment of international trade with tropical hardwoods would only lead to a reduction in the private benefit for a small percentage of users of tropical rainforests. Thus, it is merely a small contribution towards saving the tropical rainforests in the world.

Trade restrictions related to biodiversity protection cannot only be ineffective, they can also be counter productive thus contributing to the origin of international external costs (Swanson 1994). In some African countries such effects could happen while efforts are made to protect elephants from extinction by implementing trade restrictions. The commercial trade ban on elephants is designed to raise the private economic benefit for using the resource ›elephant‹ beyond the natural regeneration rate to approach zero. This ban makes sense when the elephant is an open-access-resource to which every person has free access.

It is different if there is a legal system which secures property rights for elephants. In this event, users are interested in a long-term use of the resource. If the private benefit of elephants solely or to a major part only derives from trade, this interest will get lost with a forced loss in demand. A trade ban will also have counterproductive effects in countries in which elephants are threatened by extinction because an increasing part of the area in which they live is used for agriculture. That is why countries such as South Africa, Zimbabwe, Namibia, and Botswana have pronounced themselves against a trade ban for elephants (Douglas-Hamilton

1992). In all of these countries, the legal system is sufficient to secure the continued existence of elephants. In these countries, as in all developing countries, the increase in human population will lead to an increased demand for agriculturally productive land. This threat to elephants cannot be met by a trade boycott. On the contrary, such a boycott makes the conversion to agriculturally productive land even more attractive, and elephants will be even more threatened by extinction as would be the case without a boycott.

7 International transfers for biodiversity conservation

An internalisation of the international external benefits and costs of biodiversity conservation is also possible with the help of payments from industrialised nations transferred to developing countries for their changes in biodiversity usage which have been agreed upon. A possibility for the institutional design of the compensation payment is to relate it to the country's (where the change in biodiversity use is to occur) credit relationship.

Non-governmental organisations in industrialised countries are especially attempting to persuade governments in developing countries to implement measures for biodiversity conservation through the so-called *debt-for-nature-swaps* (Amelung 1997). At least three parties are involved in a *debt-for-nature-swap*: The government (and the national bank) of the indebted country, the current credit organisation and a non-governmental organisation which would like to affect a change in the use of biodiversity in the indebted country. First, the non-governmental organisation must obtain a debt instrument for the indebted nations. As a rule, this usually occurs by a purchase thereof on the secondary market where governmental debts of developing countries are dealt with by private banks. Sometimes non-governmental organisations obtain such instruments from credit banks as a donation. The market price which is paid for the debt instrument upon purchase can be significantly lower (up to 70 %) than the nominal value. Then the new owners of the debt instrument, the national bank, and the government of the indebted nation agree to cash in the debt instrument for domestic currency or to convert it to a new debt in the country's currency. Finally, a government environmental protection agency or a nature protection organisation acting in the public interest according to the countries laws in the respective indebted countries receives the nominal value in domestic currency. This is paid out at the time of redemption or the new interest from the domestic debt instrument in order to implement the biodiversity conservation project. The foreign debt of the indebted countries has at least been reduced through such a *debt-for-nature-swap*. The entire debt will be purchased when the debt instrument is traded for domestic currency. If they are merely exchanged for domestic liabilities, then the *debt-for-nature-swaps* affect debt relief when the non-governmental organisation as the buyer of the debt instrument receives a higher discount price than the indebted nation and passes the higher discount to the developing country.

The first *debt-for-nature-swap* took place in 1987 between the American non-governmental organisation Conservation International and the Bolivian government (Laser 1997). Conservation International purchased the Bolivian debt having a nominal value of US$ 650,000 for US$ 100,000. In return, the Bolivian government committed to expand an environmentally protected area by 1.2 million hectares in which 13 of the 18 threatened species in this country live. By 1999, non-governmental organisations had initiated and paid almost US$ 200 million for *debt-for-nature-swaps*. When looking at the total debt of developing countries in 1999 (US$ 2.554 trillion), it becomes apparent that this instrument is of little significance. The interest payments and repayments alone were higher in 1999 at US$ 350 billion than the transaction volume of the *debt-for-nature-swaps* where non-governmental organisations were involved.

The significance of *debt-for-nature-swaps* with the involvement of non-governmental organisations is limited because the secondary market for developing countries' government debt is an oligopolistic structured market in which the debt instrument of a country is owned by a few banks (Amelung 1997). Therefore, a relatively high increase in price is the result of an increase in demand of the debt instruments of a developed country. The *debt-for-nature-swaps* thus become increasingly more expensive. In addition, the debt instruments of only approximately 20 developing countries (primarily Central and South American) are traded frequently. Non-government organisations can therefore only influence the environmental quality in a few countries with the *debt-for-nature-swaps*.

Both limitations do not hold true for *debt-for-nature-swaps* in which governments of the creditor countries are involved. In this variation, the creditors' governments are willing to cancel the loan if the funds in domestic money, which are then made available, are applied to concrete environmental and resource protection measures. For example, such an agreement was made between Germany and Kenya in 1989 (Krieg 1992). The USA has to date remitted public debt of over US$ 1 billion for specific causes (Mutlak 2001). The transaction volume of *debt-for-nature-swaps* where governments are involved is now higher than those ›swaps‹ initiated by non-government organisations.

It should be considered that if these instruments are implemented more often, inflationary effects could occur. The danger is great both if (a) it is planned to trade the debts in domestic funds and (b) the money is immediately spent on nature conservation projects. Therefore, it is suggested that the debt instrument should not be remitted immediately, but rather to finance the nature protection project through interest payments and repayment. This procedure has its advantages if the developing country does not follow through the conditions put forth in the *debt-for-nature-swap* agreement. The developing country can be sanctioned by demanding the repayment of the remaining debt.

In the last years, *debt-for-nature-swaps* have often been used to establish environmental funds. Environmental funds serve to protect nature by using the interest for agreed upon nature protection projects. In this way, Germany has been able to relieve Ecuador of approximately US$ 11 million debt to date. In return, Ecuador has made 30 % of this sum available as starting capital for an environmental protection fund whose interest will be available to maintain and expand a national park. To date, approximately US$ 1 billion have been invested in environmental funds, and US$ 200 million in interest have been spent on nature protection programmes. The transfer of international financial contributions for the protection of biodiversity must not be tied to credit transactions. The industrialised countries might grant financial compensation for the protection of biodiversity to the developing countries regardless of the indebtedness. This solution has been chosen in the convention on biodiversity. According to articles 20 and 21, industrialised countries are committed to providing new and additional financial resources to support developing countries. These financial means should enable developing countries to bear the full incremental costs which are necessary for the implementation of protective measures. If these financial means are not provided, the developing countries are not committed to taking protective measures. The implementation of the financial transactions was assigned to the financial mechanism of the Global Environmental Facility (GEF) (see Menzel this vol.).

Decisions concerning the distribution of funds are made in a leading body which consists of 16 representatives from developing countries, 14 representatives from industrialised countries, and two representatives from countries in transition. If an agreement cannot be reached, a decision has to be taken on the basis of a qualified majority which at the same time represents 60 % of the overall participants of the GEF as well as 60 % of all membership payments. As the developing countries hold 50 % of the votes in the leading body, they can block decisions but are unable to carry them through on their own.

As mentioned, financing by the GEF is designed to cover the incremental costs which result from biodiversity protection projects in developing countries. The concept of incremental costs has multiple meanings. First, financial and social costs associated with an environmental protection project do not, as a rule, coincide. Secondly, the incremental costs of a project can be interpreted according to the gross or net principle, that is, with or without deducting the project's benefit which remains within the country. No explanations in the conventional rhetoric are able to clarify these points, but the guidelines set forth by the GEF for determining the incremental costs do (GEF 1996). According to these guidelines the incremental costs are to reflect the burden on the total economy induced by the project. Thus the social cost concept is the relevant one. The governmental expenditure for the project (= the financial costs) forms a starting point that makes sense for calculating the incremental costs. However, they should not be set equal to one another.

Given the choice between gross and net principle, the GEF chose a solution somewhere in the middle. When calculating the incremental costs, the social costs are not reduced by the total national social benefits from the project, but rather only by the savings in costs which result because the biodiversity conservation project renders some other planned activities superfluous. Since the savings in costs usually lie below the national benefit, this is a solution somewhere in between the gross and net principle. If the policies orient their goals and objectives exclusively towards the benefit-cost concept, then this is always the case.

Positive effects of the biodiversity conservation project in the regarded country only lead to a decrease in the incremental costs when the country's policies are laid out to realise these effects. In this case, not the social benefits of these positive effects are deducted, but the costs of the measures no longer needed. The following two examples attempt to clarify this regulation:

1. It can be expected that an increase in foreign tourism and thus an increase in foreign money annually, will be a result of a nature conservation project. This then leads to a decrease in the reimbursable incremental costs if the increase in foreign money income is an official goal of the economic policy. If this is the case, then the social costs of the project are not reduced by the social benefits from the increase in foreign money (which coincide with the increase in foreign money), but by the costs of an alternative measure from which a similar increase in annual foreign income can be expected.
2. A nature conservation project leads to an improvement in water quality. If the country has planned measures to improve the water quality that have become unnecessary due to the nature conservation project, then the costs of these measures should be deducted from the social project costs. If the country believes the water quality to be good enough and is not planning any measures for its improvement, then no savings in costs result from the nature conservation project for the protection of the water supply. In this case the reimbursable project costs do not need to be corrected (downwards).

How should the GEF's calculation method be evaluated? When social benefits (SB) associated with the project is greater than the social costs (SC), then a project should be implemented according to the logic of the benefit-cost analysis.

$$\mathrm{SB} > \mathrm{SC} \tag{3}$$

International benefit and not only national benefit is associated with biodiversity conservation projects ($\mathrm{SB} = nat.\mathrm{SB} + int.\mathrm{SB}$). The social costs of such projects are usually only incurred in the country in which the project is conducted ($\mathrm{SC} = nat.\mathrm{SC}$). Decision rule 3 can thus be transformed to

$$int.\mathrm{SB} > nat.\mathrm{SC} - nat.\mathrm{SB} \tag{4}$$

Equation 4 makes clear that the national benefit should not be disregarded when deciding whether a biodiversity conservation project makes sense economically. In other words, according to equation 4 the incremental costs in the sense of the net principle are relevant for decision making.

If one applies the net principle when calculating the incremental costs according to the GEF, then the developing countries achieve neither national economic advantages nor disadvantages. Thus, there is no incentive for them to implement a nature conservation project. It is understandable that the GEF avoids the net principle and reimburses more than only the national social net costs. A result of this deviation is that with the given financing too few economically sound nature conservations projects are realised. In order to make these types of projects interesting for developing countries, this is the ›price‹ to be paid.

The way in which the GEF determines which are the reimbursable incremental costs is to be criticised. As explained, these incremental costs primarily depend upon what the political goals of the country in which the project is to be conducted are and which activities are planned to attain the goals. It is therefore possible, as shown in the second example above, for a country whose government is not interested in environmental protection to receive more money for a nature conservation project than a country making efforts towards environmental protection. This can occur even when the consequences of the financial regulations in both cases are equally high. The consequences of these financial regulations are naturally undesirable from the GEF's point of view. Therefore, part of the GEF's work is continuing to find a satisfying concept for the calculation of incremental costs.

8 Concluding remarks

As we have shown, the contributions of economists to biodiversity conservation may be classified into three different categories: the investigation of the causes of biodiversity loss, the economic evaluation of biodiversity and the design of instruments and institutions for biodiversity conservation.

Global biodiversity policy has the responsibility from an economic point of view to internalise external benefits and costs for the use of the environment internationally. This is to guarantee that the natural environment is used in a way in which the interests of people living in other countries are considered. This problem can be solved conjointly. The internalisation of the external benefits and costs can be strived for, even against the will of the country in which the external costs originate.

The latter way is being pursued by international biodiversity policy with the help of environmentally related trade restrictions. Trade policy measures only influence the international external benefits and costs of biodiversity use indirectly. Therefore, they do not always attain the goal and sometimes they even affect an undesired change in the use of the environment.

Global biodiversity policy measures which were mutually agreed upon by the countries involved, implicitly or explicitly include financial transfers from industrialised countries to developing countries. Apart from the debt-for-nature-swaps, the Global Environmental Facility in particular should be mentioned. Debt-for-nature- swaps have the big disadvantage that they can only be used for biodiversity conservation in certain developing countries. The design of the ›spending policies‹ of the GEF gives wrong incentives. But in principle, all measures adopted for global biodiversity conservation are justified from an economic point of view.

References

Amelung T (1997) Globaler Umweltschutz als Verteilungsproblem im Nord-Süd-Konflikt. Peter Lang, Frankfurt am Main

Brown G (1997) Benefits of Biodiversity Evaluation: Why Biodiversity Valuation is Imperative. In: OECD Proceedings Investing in Biological Diversity. The Cairns Conference, OECD, Paris, pp 57-86

Douglas-Hamilton I (1992) Battle for the elephant. Wiley, New York

Geisendorf S, Gronemann S, Hampicke U, Immler H (1999) Die Bedeutung des Naturvermögens und der Biodiversität für eine nachhaltige Wirtschaftsweise. Möglichkeiten und Grenzen ihrer Erfaßbarkeit und Wertmessung. Berichte des Umweltbundesamtes 6/1998, Berlin et al.

Global Environmental Facility (GEF) (1996) Incremental Cost Discussion Paper. World Bank, Washington

Hector A, Schmid B, Beierkuhnlein C, Caldeira MC, Diemer M, Dimitrakopoulos PG, Finn JA, Freitas H, Giller PS, Good J, Harris R, Högberg P, Huss-Danell K, Joshi J, Jumpponen A, Körner C, Leadley PW, Loreau M, Minns A, Mulder CPH, O'Donovan G, Otway SJ, Pereira JS, Prinz A, Read DJ, Scherer-Lorenzen M, Schulze E-D, Siamantziouras A-SD, Spehn EM, Terry AC, Troumbis AY, Woodward FI, Yachi S, Lawton JH (1999) Plant Diversity and Productivity Experiments in European Grasslands. Science 286: 1123-1127

Kirchgässner G (1990) Das Verursacherprinzip: Leere Formel oder regulative Idee? Juristenzeitung 22: 1042-1046

Knorr A (1995) Welthandelsordnung und Umweltschutz. Ordo-Jahrbuch für Wirtschaft und Gesellschaft 46: 203-254

Krieg H-H (1992) Der Tausch Schulden gegen Umwelt: Ein Beitrag zur Lösung der Schuldenkrise und Umweltkrise der Wirtschaft. Metropolis, Marburg

Kutschera F von (1982) Grundlagen der Ethik. De Gruyter, Berlin, New York

Laser J (1997) Dept-for-nature-swaps: Ein Instrument zur Bekämpfung der Schulden- und Umweltkrise? Zeitschrift für Angewandte Umweltforschung 10: 552-560

Marggraf R, Streb S (1997) Ökonomische Bewertung der natürlichen Umwelt: Theorie, politische Bedeutung, ethische Diskussion. Spektrum, Heidelberg, Berlin

McNeely JA, Miller KR, Reid WR, Mittermeier RA, Werner TB (1990) Conserving the World's Biological Diversity. IUCN, Gland

Mutlak N (2001) Schuldenerlass gegen Umweltschutz. Ökologisches Wirtschaften 6/2001: 8-9

Pearce DW, Moran D (1994) The Economic Value of Biodiversity. Earthscan, London

Perrings C (1995) Economic Values of Biodiversity. Heywood VH, Watson RT (eds) Global Biodiversity Assessment (published for UNEP). Cambridge University Press, Cambridge

Pfahl S (2000) Internationaler Handel und Umweltschutz. Springer, Berlin

Senti R (1998) Handel und Umweltschutz. GAIA 7: 53-54

Swanson TM (1994) The Economics of Extinction Revisited and Revised: A Generalised Framework for the Analysis of the Problems of Endangered Species and Biodiversity Losses. Oxford Economic Papers 46: 800-821

United Nations Environment Programme (UNEP) (1995) Global Biodiversity Assessment. Cambridge University Press, Cambridge

Financial support for biodiversity protection in developing countries – does the CBD mechanism lead to an appropriate level of biodiversity protection?

Susanne Menzel

Environmental and Resource Economics, Department of Agricultural Economics, Georg-August University of Göttingen, Platz der Göttinger Sieben 5, 37073 Göttingen, Germany, email: `smenzel@gwdg.de`

Summary. The contribution concerns the Convention on Biodiversity (CBD) and the provision of money for biodiversity protection in developing countries. The CBD proposes a solution to solve the problem of insufficient biodiversity protection in developing countries: the contracting developed country parties have committed themselves to provide »new and additional financial resources« to developing countries in order to enable them to protect their own biodiversity (Article 20). This contribution describes what is understood as an economic analysis and an economic perspective. Biodiversity services are considered as global public goods. The problem of providing these goods is explained. Articles 20 and 21 are presented as a solution to the market failure problem of providing global public goods. The following questions are examined: (1) Do the new and additional financial resources provided through GEF lead to an appropriate level of biodiversity protection? (2) Will the negotiations lead to an appropriate level of biodiversity protection? To answer the first question, estimated costs for biodiversity protection and actual spending of the GEF are compared. With regard to the second question, the replenishment regulations are analysed. Both of these considerations lead to the hypothesis that the actual regulations lead to an undersupply of global biodiversity protection. It is argued that the economic criterion for the amounts of money provided by the industrialised countries should equal the benefits derived from biodiversity protection. To corroborate the hypothesis a contingent valuation study and its results are presented. The results of the survey are compared to the actual spending of the GEF. They support the assumption that the request for global biodiversity protection is definitely higher than the actual spending of the GEF.

Key words: financing biodiversity protection, Global Environmental Facility, contingent valuation **JEL Codes:** Q26, Q28, D62, E62, F39

1 Introduction

This paper considers the provision of biodiversity protection in developing countries in connection with the Convention on Biodiversity (CBD). The CBD proposes a solution to solve the problem of insufficient biodiversity protection in developing countries. A number of major developed countries have committed to provide »new and additional financial resources« to developing countries to enable them to protect their own biodiversity (Article 20). Using article 21 of the Conferences of the Parties the Global Environmental Facility (GEF) was appointed to operate the financial mechanism under the CBD.

The overall question now is whether the regulations concerning the financing of biodiversity protection lead to efficient provision of biodiversity protection in developing countries.

Two specific questions are:

- Do the new and additional financial resources provided through the GEF lead to an appropriate level of biodiversity protection?
- Will the financial and political negotiations lead to an appropriate level of biodiversity protection?

This paper will address these two questions and is organised as follows: Firstly, there is a description of what is understood as economic analysis and an economic perspective. In this section biodiversity services are described as public goods and global public goods, respectively. In the third section articles 20 and 21 are presented as one possible solution of the market failure problem in the context of biodiversity protection. Here it is argued that the economic criterion for the amounts of money provided by the industrialised countries should be their benefits from protection of biodiversity. At the end of this section we consider the question, if the theoretical ›right‹ solution of articles 20 and 21 to the problem of worldwide efficient biodiversity protection does solve or can solve the problem.

To answer this question the two aforementioned questions are examined with a comparison of estimated costs for biodiversity protection and actual spending of the GEF (Sect. 4) and an analysis of the replenishment regulations (Sect. 5). Both of these considerations lead to the thesis that the current regulations contribute to an undersupply of global biodiversity protection (Sect. 6). To verify the thesis a contingent valuation study, its results and validity are discussed and compared to the actual spending of the GEF (Sect. 7). The results of this study support the assumption that the request for global biodiversity protection is significantly higher than the actual spending of the GEF. This further supports the argument for greater attention from political decision makers on preferences of the public for biodiversity protection spending. This contribution results in a conclusion.

2 Economic analysis and economic perspective

In this paper economic analysis is understood as questioning whether the use of resources leads to an efficient production of a private or public good (Zimmermann and Henke 1994; Hanley et al. 1997). The economic perspective includes the assumption of self-interested individuals maximising their personal benefits. With biodiversity protection and services we are confronted with provision problems as the good in question is a (global) public good.

In a complete market, once several conditions are achieved, the activities of self-interested individuals lead to a socially optimal production of goods. The individual marginal costs correspond to the social marginal benefit. If not all the conditions are met, markets are incomplete and externalities occur. Public goods are a case of externalities - no one can be excluded from the consumption of these benefits and additionally, no rivalry in consumption occurs (Hanley et al. 1997; Cansier 1993). There are few incentives for companies to produce goods with positive external effects. As the users cannot be excluded from the consumption of the goods, they tend not to reveal their demand. They can consume the good without paying for it and consequently the producer gains no revenue from (a part of) the benefits he produces and therefore his production is lower than the actual demand for the good. The classic solution for this case of market failure is the provision of the public good by the government (Hanley et al. 1997).

Because biodiversity services are regarded as public goods, it is easy to see how the market failure problem described above can occur. This leads to the implication that biodiversity services could be undersupplied. If this would be the case, and additional provision of biodiversity protection would produce more benefit than costs, it should be provided by the government. But it is not easy for the government to provide the good, unlike in the case of ›normal‹ public goods. Some of the positive external effects of biodiversity protection occur globally thus, biodiversity services are referred to as global public goods (Kindleberger 1986, Kaul et al. 1999, Perrings and Gadgil 2003).

The provision of global public goods produces special challenges. There is no government which can provide the public good (Kaul et al. 1999). In the case of biodiversity, there is no supranational institution or »world government« which would have the authority to internalise the external effects of biodiversity conservation (Hanley et al. 1997 : 163). From an idealised economic point of view, this institution would have to organise the protection of biodiversity in developing countries at a level where the global marginal cost of biodiversity protection corresponds to its global marginal benefit.

As participating nations began to recognise the importance of biodiversity protection in developing countries and the problem of undersupply of the public good, they tried to find solutions to the problem. Subsequently, parts of the CBD can be regarded simply as regulations to ensure the provision of biodiversity services in developing countries.

3 Contribution of CBD to provision of biodiversity protection in developing countries

With articles 20 and 21 of CBD, a multinational regulation was agreed upon to contribute to overcome the undersupply of biodiversity protection in developing countries. The incremental cost approach, which leads to the sharing of expenses for biodiversity protection projects, was also agreed. Developed countries are committed to providing new and additional financial resources for biodiversity conservation. Referring to the principle of equivalence they decided to pay only for the *global environmental benefit* of measures to protect biodiversity in developing countries[1]. The national or local benefit that occurs from these projects in developing countries has to be financed by national governments or co-financers. And Article 20(2) states, that »implementation of these commitments shall take into account the need for adequacy«.

From an economic point of view this *need for adequacy* can be interpreted in several ways. In most cases, the discussion about adequacy concerns sharing of burdens between developing and developed countries. Developing countries interpret ›adequacy‹ by estimating that the contributions should be as high as the individual costs incurred when protection measures are implemented. The developed nations viewed that the contributions should be adequate to finance only the social costs of these protection measures. Beyond the discussion of cost sharing between developed and developing country parties, the need for adequacy can be alternatively interpreted: as a level of contributions from donor countries that allows for a global efficient level of protection of biodiversity in developing countries.

An optimal global level of biodiversity conservation in developing countries can be achieved when the global social marginal costs of protection are as high as the social marginal benefit of the protection. The question is thus: are the commitments of donor countries adequate in terms of global efficiency? From a welfare economic point of view, an adequate amount of money should represent the benefits of the industrialised countries from protection of biodiversity.

With Article 21 and the second Conference of the Parties (COP 2), the Global Environmental Facility (GEF) was appointed to operate the financial mechanism under the CBD. Thus, the GEF is the institution that organises the governmental provision of financial resources for biodiversity protection in developing countries.

Does the regulation laid down by the CBD and achieved by the GEF lead to a level of biodiversity protection where global social costs and benefit are equal?

[1] »The developed country Parties shall provide new and additional financial resources to enable developing country Parties to meet the agreed full incremental costs to them of implementing measures which fulfil the obligations of this Convention and to benefit from its provisions and which costs are agreed between a developing country Party and the institutional structure referred to in Article 21, in accordance with policy, strategy, programme priorities and eligibility criteria and an indicative list of incremental costs established by the Conference of the Parties.« Article 20 (2) CBD.

4 Financial resources from the GEF and costs of worldwide biodiversity protection

4.1 Financial resources provided by the GEF

When analysing GEF finances, it is important to distinguish between payments to and from the fund. As well as biodiversity protection projects, the GEF fund also finances projects subject to climate change, international waters, ozone, land degradation and persistent organic pollutants projects.

Table 1. Payments in the GEF (in US$ million with exception of 1)

	GEF-1 (1994–1997)	GEF-2 (1998–2002)	GEF-3 (2003–2007)[a]
(1) commitments in US$ billion[b,c]	approx. 2	approx. 2	2,92 (2,21)[d]
(2) percentages of contributions for biodiversity[e,f]	41.7 %	39 %	32 %
(3) absolute payments for biodiversity[g]	834	676	960
(4) calculated annual payments for biodiversity[h]	208	169	240
(5) Germany's contributions (absolute and as percentages of whole GEF)[i]	238 (11.9 %)	228 (11.4 %)	264 (11 %)
(6) average annual contributions of Germans for biodiversity to GEF[j]	approx. 25	approx. 22	approx. 21

Source: see Footnotes (own assembly)
[a] planned
[b] (GEF 2002b)
[c] (GEF 2002c)
[d] US$ 2.21 million are new commitments (Horta, Round and Young 2002). The difference results from carryover of GEF-2 resources and investment income.
[e] (Bundesregierung 1999)
[f] (GEF 2002c)
[g] calculated from (1) and (2)
[h] calculated (3) /(4)
[i] (GEF 2002b; GEF 2002c)
[j] calculated from (2) and (5)

Every four years the donor countries decide on the payments for the following four years. After the pilot phase of the GEF (1991-1993) the fund was replenished three times: in 1993 (GEF-1), 1997 (GEF-2) and 2002 (GEF-3) (Streck 2001; Horta et al. 2002; GEF 2002b; GEF 2002c). The intended ratio for the focal area biodiversity was 32–41 % of the whole GEF fund (Bundesregierung 1999, GEF 2002c). For GEF-1 approximately US$ 834 million was agreed on to finance biodiversity conservation (GEF 2002b). For 2003–2007, US$ 960 million is planned for biodiversity protection in developing countries (GEF 2002c). From these figures, we can deduce a computed value of US$ 240 million annually planned payments into the GEF fund.

Germany contributes 11–12 % of the total amount of the donor countries (GEF 2002b; GEF 2002c). Thus, via the GEF, Germany spends a calculated annual amount of approximately US$ 25 million on biodiversity protection in developing countries (see Table 1).

Table 2. Payments from the GEF for biodiversity protecting projects

Source	Period	Spending	Annual spending
(GEF 2002a)	1992–2002	US$ 1.4 billion	US$ 140 million
GEF (gefweb.org)	1991–1999	US$ 991 million (grants) plus USD 1.5 million (co financing)	US$ 310 million

Source: own assembly

Figures concerning payments from the GEF projects are as follows: »from 1992 to 2002, GEF had allocated nearly US$ 1.4 billion for 470 biodiversity projects in 160 countries« (GEF 2002a). Over the ten year period 1992–2002, an average of US$ 140 million was spent annually on biodiversity projects. According to the GEF website US$ 991 million in grant payments was made between 1991 and 1999. An additional US$ 1.5 billion in co-financing for biological diversity projects was mobilized (GEF). In total, the annual contribution from industrialised countries for biodiversity protection in developing countries from 1991–1999 was US$ 310 million (see Table 2).

4.2 Estimated costs of global biodiversity protection

There are only a few estimates of the costs of global biodiversity protection and biodiversity protection in developing countries. The annual estimates range from US$ 170 million for traditional protection of tropical rain forest (Reid and Miller 1989) to over US$ 500 million for the protection of 25 biodiversity hot spots, mostly in developing countries. This is equal to the protection of 1.4 % of the earth's land surface or 44 % of the vascular plant species and 35 % of the species in four vertebrate groups (Myers et al. 2000). Another estimate of US$ 27.5 billion is for a representative global network of protected areas (James et al. 1999) (see Table 3).

These estimations refer to different items of protection and have been reached using different methods. Thus, a ›definite‹ assessment of costs does not exist and the appraisals should be considered as rough approximations.

Table 3. Annual cost of biodiversity protection

Subject of protection	Annual costs (US$ million)	Author
Traditional protection of tropical rain forests	170	Reid and Miller (1989)
2000 species (500 individuals per species)	1250	Reid and Miller (1989)
Activities outlined in agenda 21 for biodiversity protection	440	Agenda 21 (1992)
Comprehensive global conservation programme	300,000	James et al. (1999)
Representative global network of protected areas	27,500	James et al. (1999)
1.4 % of the land surface of the earth. (44 % species of vascular plant species, 35 % species of vertebrate species)	500	Myers et al. (2000)

Source: different objects of protection, authors, years and estimations (own assembly)

4.3 Comparison of biodiversity protection costs and GEF spending

It is evident that all biodiversity protection cost estimations are higher than the actual GEF spending. The rough estimations vary substantially. If we assume the estimates of US$ 27.5–500 million are realistic and compare them to the estimated payments of US$ 140–310 million, the factors between spending of the GEF and estimated costs are 1.5 to 200.

However, if we do use this figures we have to bear in mind that biologists see this problem from a natural scientific viewpoint. They do not deal with the needs of people. From a biological perspective the dimension of a necessary biodiversity protection cannot be defined, as biologists do not consider the ›objective‹ necessity of biodiversity protection. They can estimate the costs of protecting 25 of the ›hottest hot spots‹ which cover 1.4 % of earth's land surface, or the possible costs of strictly protecting 10 % of the land surface to get a representative global network of protected areas. Previously, natural scientists defined levels of protection whilst often ignoring the needs of the population. However, over the last decade social needs have been taken into account and the target of protecting 1.4–10 % of earth's surface is now, to some degree, concerned with the needs of people and can therefore be used as a benchmark for biodiversity protection. Their cost estimations are however, based on the actual individual costs of protection projects.

Often the policies of developing countries are ›environmentally unfriendly‹ or ignore the scarcity of the environment. This makes nature consumptive activities relatively cheap and nature ›sparing‹ activities expensive. If governments of developing countries introduced more environmentally friendly policies, environmental protection would be cheaper and therefore new costs would be lower than the calculations mentioned above.

Thus, the estimated costs should not be perceived as economically necessary or reasonable, but rather as a frank indication that the present payments are too low in terms of an optimal economic level of protection.

5 Negotiations for the replenishment of the GEF fund

5.1 Procedure of replenishment

Every four years the GEF fund has to be replenished. There is little information publicly available on how the replenishment negotiations actually take place, but it roughly follows this procedure. The council of the CBD asks the World Bank for renewal of the GEF fund. The World Bank convenes the donor countries and negotiations begin to restock the fund. Representatives of donor countries negotiate their individual contributions and how the total donation amount should be divided. This is based on a mechanism developed by the International Development Association of the World Bank for the shares of development assistance (Horta et al. 2002; Hanley et al. 1997), which is based on the economic strength of each nation. »In practice, however, contributions reflect political will more than agreed formulas.« (Horta et al. 2002, chap. 2). Furthermore, national representatives act under strong restrictions of (national) ministers of finance (Kaiser 2003).

5.2 Results of negotiations

The replenishment negotiations of the GEF are recorded in the »Summary of Negotiations« (see Appendix I). The following is an excerpt concerning the levels of overall commitment from donor countries.

The biggest amount is provided by the USA (21 % [of the whole GEF fund spending]) followed by Japan (18 %), Germany (11 %), and the UK (11 %) (GEF 2002c). The order in which contributions to the fund are made corresponds to the order in which more developed countries would appear on a national gross domestic products (GDP) list. However, the donor countries contribute different percentages of their GDP to the fund. Nations which we could call ›leaders‹ committed a total of 0.01 % of their GDP to the GEF over four years (Japan, France, Germany, UK, and Canada). Some smaller countries, we could call them ›front-runners‹, spent more than 0.02 % of their GDP e. g. Sweden and Denmark. The ›taillights‹ of the more developed donor countries committed 0.005 % or less of their GDPs: e. g. USA or Spain (see Table 4) (GEF 2002c).

Table 4. Country contributions to the GEF-3 (extract)

Nation	Percentage of GDP for GEF (for four years)	Contribution in US$ million
Sweden	0.03	72
Denmark	0.023	36
UK	0.012	190
Germany	0.012	264
Japan	0.011	423
France	0.011	163
Canada	0.011	103
USA	0.0023	500
Spain	0.0023	19

Source: commitments to the GEF 3 according to the Summary of Negotiations (GEF 2002c), GDPs according to the CIA world fact book (`http://www.cia.gov/cia/publications/factbook/fields/2001.html`) [own calculations]

5.3 Short analysis of negotiations

If we assume, according to public choice theory, that representatives of governments try to maximise their personal benefits and that these personal benefits depend on (national) re-election, then we have unfavourable requirements to solve the problem characterised as follows:

The task to protect biodiversity of global interest demands international collective action. At the same time the probability of this approach to the problem being successful is relatively low.

Thus, the requirements for the negotiations on contributions for biodiversity protection in developing countries are as follows: There are representatives of (developed) nations who are expected to solve the problem of biodiversity loss, who, however, also have national interests. At the same time very little is known about the benefits of biodiversity protection. Thus every invested amount is at risk of becoming a misdirected investment. Furthermore, every nation has little influence on overall protection activity and the level of biodiversity protection. National governments can trust or hope that the good (biodiversity services) will be provided even if their own contribution is low. Particularly if they assume that the importance of global biodiversity protection is relatively low for voters in their home country, national governments have low incentives to contribute a large amount of money. Moreover, the level of difference the contribution has made in terms of biodiversity protection is not easy to communicate in the home country. If governments act in national (election) interests, because the sources for (potential) votes for re-election are the citizens of these nations, it is in their personal interest to invest national funds on projects that have a direct impact on national problems. And biodiversity protection in developing countries does not seem to directly influence the short term well-being of citizens in developed countries.

These considerations lead to a conclusion that the national governments have high incentives to act as a ›free rider‹ and have low incentives to spend money on global biodiversity protection.

On the other hand nearly all of the developed nations want to be seen as participating in a ›worthy cause‹ and there are higher expectations on some developed countries. Not to lose international credibility as a country which cares about the global environment explains in my opinion why nations contribute at all.

6 Undersupply thesis

We have seen, that the GEF provides only US$ 150 to 300 annual to protect the biodiversity in developing countries. We also have seen, that biologists estimate costs in the range of US$ 170 to 300,000 million for the protection of biodiversity on an international level. Additional there are low or not existing incentives for the national representatives to spend a lot for the protection of biodiversity in developing countries.

The analysis of GEF spending and negotiation leads to the following thesis: the actual amount and procedure resulting in commitments from donor countries to the GEF do not lead to a globally adequate level of protection.

From what we have seen so far it seems worthwhile to estimate the benefit of biodiversity protection in developing countries in order to verify the thesis.

7 Case study – willingness to pay for the conservation of biodiversity

There are several economic methods to estimate the benefits of items or issues. Some are based on revealed preferences (or market data) and are called indirect methods (like the travel cost method or the replacement cost method). Others are based on stated preferences collected through direct questioning on willingness to pay for described goods or through observed choices of goods with slightly different attributes or parameter values, respectively (=direct methods).

In this study the direct Contingent Valuation Method (CVM) was chosen to estimate the benefit which arises in developed countries from the protection of biodiversity.

CVM is very popular in environmental economics to estimate values of non-market goods (Hanley et al. 1997). In Contingent Valuation (CV) studies respondents are asked how much they are willing to pay for a non-market good or whether they are willing to pay a specific amount for a good[2]. From these answers the median and mean of payments are computed and the economic benefit of providing the valued good can be estimated with respect to the basic population. Many studies evaluate the benefit of conservation of a single species or specific conservation projects like national parks (Loomis and White 1996; Hanley et al. 1998). The aim of this study was to survey the benefit of biodiversity protection in developing countries.

7.1 Study design

The population of the survey consists of German residents (native and foreign) aged 18 and older[3,4]. Because of the huge basic population of 66.4 million, a minimum sample of 1,000 people was interviewed to ensure a representative result. A telephone survey was selected as the interview technique, primarily because of restricted financial resources. Telephone numbers were generated using the »random digit dialling method«[5], and the »last-birthday-method«[6] was applied. The object of valuation was the protection of half of the endangered species expected to become extinct in the next 10 years[7]. A tax increase was used as the payment vehicle[8] because it was felt that respondents take this form of payment seriously (compared with making a donation to a nature conservation organisation). A nature tax (comparable to a visitor's tax) was considered implausible for respondents, particularly because of the non-excludability from the benefits of biodiversity conservation (Bateman et al. 2002). The dichotomous choice format[9] was chosen for the willingness to pay (wtp) question. This format is favoured because it is most similar to purchase decisions (Spash 1999). It is also particularly suitable when respondents are unfamiliar with the good they are valuing. The wtp question for the

---

[2] First question format is called »open-ended« the second »dichotomous choice«.

[3] One part of the basic population is its eligible voters. In Germany people are eligible to vote when they turn 18. 1998: 60.8 million (1998). (Federal Statistical Office, Bundeswahlleiter).

[4] The other part of the basic population are the foreigners, who are 5.775 million (2001), 5.561million (1998) people aged 18 and older (Federal Statistical Office).

[5] With this method random telephone numbers are generated.

[6] When contact is established, the person recently had birthday is asked to participate.

[7] At first glance the protection of surface area seems to be the appropriate measure. However, focus group interviews determined people do not imagine areas in acres or hectares. Also the goal of protection of, e.g., tropical rain forests is not only to protect biodiversity, but also concerns the climate. As we wanted to assess the value of biodiversity protection, the protection of ›natural areas‹ or tropical rain forests was deemed inappropriate.

[8] To decide on an appropriate payment vehicle the popular pro and con arguments were considered (Mitchell 1988; Bateman 2002).

[9] A question requiring a »yes« or »no« answer.

protection of half of the endangered species in developing countries was evaluated as an unfamiliar question, demanding attention and thought from respondents. Thus, the dichotomous format should facilitate an answer to the wtp question. Results of CV studies are difficult to validate and the object in question here is particularly abstract. Furthermore, the results of this wtp question are difficult to compare with data of other studies. To test the validity of the wtp result, additional variables were collected. These were derived from a socio-psychological theory, the protection motivation theory developed by Rogers (1977, 1983). It is assumed that if the answers to the wtp question can be explained by the answers to the questions derived from the psychological theory, the responses to the wtp question are valid.

7.2 Study results

In April and May 2001 a total of 12,000 numbers were dialled, 3,675 persons were contacted and read the screening text. 58 % refused to participate in an interview, but only 1.5 % dropped out and a total of 1,017 people were interviewed (see Table 5).

54.7 % of respondents were female, 45.3 % were male. The age group ranging from 25–45 years was over-represented and people older than 65 were under-represented (see Table 6).

The sample is more or less evenly distributed over different income categories and this seems comparable to the basic population, however, it is not easy to evaluate the overall representation of the sample (see Appendix II).

The respondents were asked randomly if they were willing to pay a specific amount ranging from € 1 to € 40. The bid levels were 1, 3, 4, 5, 8, 10, 17, 26, and 40 Euros. The acceptance and rejection rate respectively range from 39 % to 82 % depending on the bid level. If a respondent had to decide about the € 1 amount, the probability of a »yes« answer was 82 %. If the bid level was € 40 the probability of a »yes« answer was 39 %.

A logistic regression with variables from the protection motivation theory and socio-demographic variables was performed to test the validity of the wtp answers and to calculate average wtp resulting from the model.

Explanatory variables for the acceptance or rejection of the payment are:

- self-efficacy (= the belief in the effect of the own payment for the protection of biodiversity in developing countries)
- bid level (financial costs of contribution to protection)
- responsibility (of the respondent for the protection of species in developing countries)
- age
- threat appraisal (perceived threat as consequence of loss of biodiversity)
- opinion about whether the industrialised countries have the right to interfere in biodiversity protection affairs of developing countries

Table 5. Sample report

	Cases	Percentages
Telephone-Number Total	12,000	100.0 %
neutral outfalls	5,177	43.1 %
No connection	4,537	37.8 %
wrong connection / number has changed	83	0.7 %
business telephone number	557	4.6 %
Revised Gross I	6,823	100.0 %
other outfalls	3,148	46.1 %
no connection tone, no contact	1,701	24.9 %
busy	86	1.3 %
answering machine / mailbox	601	8.8 %
fax machine/ modem (whistle)	541	7.9 %
strong communication problems	219	3.2 %
Revised Gross II	3,675	100.0 %
not neutral outfalls	2,658	72.3 %
cancelled appointments	41	1.1 %
person not available in given time period (10 contact attempts)	427	11.6 %
refusals	2,135	58.1 %
drop outs	55	1.5 %
Realised Interviews	1,017	27.7 %

Source: own research and own calculations

- whether the respondent had visited a developing country[10]

From the median values of the explanatory variables and their regression coefficients, the average wtp can be calculated as follows (Backhaus et al. 2000):

$$x_1 = \frac{\beta_0 + \beta_0 * \bar{x_2} + \ldots + \beta_j * \bar{x_j} + \ldots + \beta_k * \bar{x_k}}{-\beta_1} \quad (1)$$

x_1 = median of the wtp = bid level where an average respondent is indifferent
β_0 = constant
β_2 = regression coefficient of e. g. age

[10] The income is also a significant explanatory variable. However, 16.9 % of the respondents who agreed to pay and 28.6 % of the respondents who did not agree to pay did not provide information about their income. If income is included as an explanatory variable the percentage of explainable model cases increases, while the number of analysed cases decreases. Furthermore, biases in the mean and median wtp result are generated. Thus, the model was calculated without consideration of income.

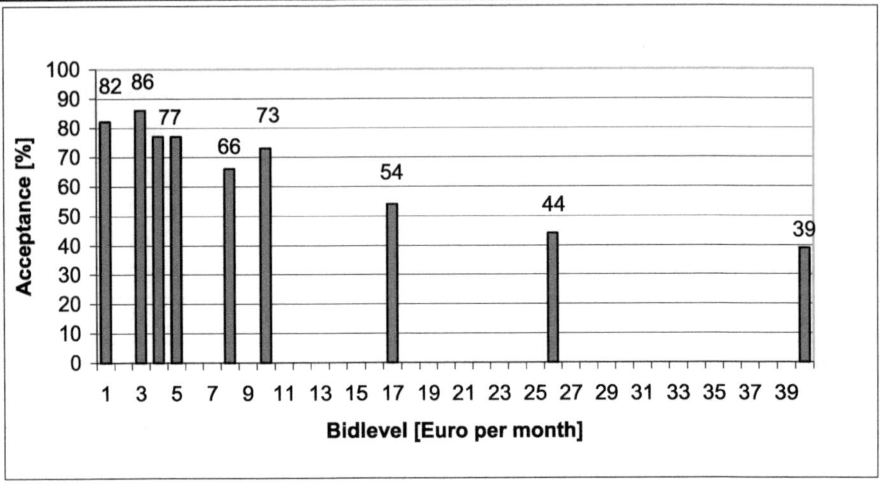

Fig. 1. Acceptance rate at different bid levels (own calculations)

$\bar{x}_2$ = average value of an explanatory variable e. g age x_2
β_j = regression coefficient of the variable
$\bar{x}_j$ = average parameter value of the explanatory variable
β_1 = regression coefficient of the bid level

The median of wtp (x_1) is € 22. This can be deduced from the model as an average monthly wtp of the respondents. In terms of conservative average wtp estimation, we assume that people who refused to participate in the survey or dropped out (total = 59.6 %) have a wtp of zero Euros. We also assume that people who cancelled their appointments or who were not available at the time the survey was taken have the same wtp as people who took part in the interviews (total = 40.4 %). The multiplication of the calculated wtp of € 22 with the sample population (40.4 %) results in an average wtp of approximately € 9.

This can be interpreted as an expression of (monthly) benefits for an ›average‹ German resident resulting from the protection of 25,000 species in developing countries over the next 10 years. We calculated a basic population of 66.4 million people (see Sect. 7.1). A multiplication of the average wtp and the basic population results in approximately € 600 million monthly and 7 billion annual wtp, respectively. This can be taken as a potential (annual) benefit for Germany if 25,000 species in developing countries are prevented from extinction in the next 10 years.

With the acquired value of € 9 we can attempt to roughly estimate the benefits that occur in the ›main‹ donor countries[11]. Taking a comparable percentage of beneficiaries in these countries (80 % of the population), a similar average benefit

[11] USA, UK, Japan, France, Germany, Australia, Belgium, Canada, Finland, Italy, Netherlands, Norway, Sweden and Switzerland (CIA 2002).

per capita and the population of the main donor countries, we can calculate a benefit resulting from protection of 25,000 species in developing countries to be € 4.4 billion.

7.3 Discussion of results

Every endeavour was made to get a representative sample but the final breakdowns indicate that this objective may not have been achieved.

Some people might have refused to participate because the expression »nature conservation« was used in the screening text or because they were unwilling to take part in a telephone interview or any other form of survey. The results may be biased if respondents are affected by specific expressions, e.g., »nature conservation«. The sample is clearly not representative in terms of age and this may bias the results as older people have lower wtps than younger people. As gender is not an explanatory variable, it does not matter that the sample is not gender representative.

The wtp answers can be seen as valid. The pseudo r^2(Nagelkerke) of the regression model is 0.339. This figure indicates a moderate to good model (Backhaus et al. 2000). In order to make a conservative calculation of the average wtp estimation (Arrow et al. 1993), a wtp of zero is assumed for people who refused to participate.

The size of the basic population has a strong influence on the result and we cannot be certain that some people did not reply on behalf of their household rather than just for themselves (as was requested). If that occurred frequently, the previous calculation would have led to an over-estimation of Germany's wtp. If we calculate with the number of households (= 34,777[12]), Germany's annual benefit would be approximately € 3.8 billion.

8 Conclusion

GEF spending on biodiversity protection and the estimated costs of global biodiversity protection together with the analysis of regulations concerning the replenishment of the GEF fund, lead to the hypothesis that payments from developed countries into the fund are lower than economically adequate in regard to a *global economic optimum*.

The result of a contingent valuation study can be interpreted as verification of this thesis. Germany currently contributes € 25 million annually into the GEF fund for the protection of biodiversity. However, the wtp of people living in Germany can be estimated to be as high as € 3.8 billion. Thus, the benefit and wtp of people is higher than actual national contributions. If we assume a similar wtp in other main donor countries, the benefit of the protection of 25,000 species over the next 10 years can be appraised to € 4.4 billion annually.

[12] Indication for the year 2001 according to Federal Statistical Office, Germany `http://www.destatis.de/basis/d/evs/budtab2.htm`

This value may be condemned as unrealistic, but it should be understood within the dimension of the estimated costs for a representative global network of protected areas and a comprehensive global conservation programme, respectively.

The results of the study show that the commitment from donor countries should be higher. Furthermore, politicians should be more considerate of the preferences of the population for biodiversity protection in developing countries. The results of this study can be used to emphasise the need for greater attention to preferences of the public in decision making on biodiversity protection activity and spending.

References

Arrow K, Solow R, Portney P, Leamer E, Radner R, Schuman H (1993) Report of the NOAA - Panel on Contingent Valuation. Federal Register 58: 4601-4614

Backhaus K Erichson B Plinke W Weiber R (2000) Multivariate Analysemethoden. Springer, Berlin Heidelberg New York

Bateman I, Day B, Hanemann WM, Hanley N, Hett T, Jones-Lee M, Loomes G, Mourato S, Özdemiroglu E, Pearce D, Sugden R, Swanson J (2002) Economic Valuation with Stated Preference Techniques: A Manual. Edward Elgar, Northhampton

Bundesregierung (1999) Schutz und Bewirtschaftung der Tropenwälder - Tropenwaldbericht der Bundesregierung 6. Bericht Juni 1999. available at: http://www.verbraucherministerium.de/wald_forst/6_tropenwaldbericht/inhalt.htm

Cansier D (1993) Umweltökonomie. UTB, Stuttgart

CIA (2002) World Fact Book. Capital Intelligence Agency. Available at: http://www.cia.gov/cia/publications/factbook

Federal Statistical Office Germany: data available at: http://www.destatis.de

Federal Statistical Office (2002): Datenreport 2002. Bonn: Available at: http://www.destatis.de/datenreport/d_daten.htm

Frey BS, Kirchgässner G (1994) Demokratische Wirtschaftspolitik. Vahlen, München

GEF (2002a) Biodiversity Matters: GEF´s Contribution to Preserving and Sustaining the Natural Systems that Shape our Lives. Washington DC. Available at: http://www.gefweb.org/Outreach/outreach-PUblications/GEF_Biodiversity_CRA.pdf

GEF (2002b) GEF-2 Current and Projected Funding Status and Estimated Carryover and Projected Investment Income for the GEF-3 Replenishment Period. Available at: http://www.gefweb.org/Replenishment/Reple_Documents/R324.pdf

GEF (2002c) Summary of Negotiations on the Third Replenishment of the GEF Trust Fund. Available at: http://gefweb.org/Replenishment/Summary_of_negotiations_-_ENGLISH_Revised_11-5.doc

Hanley N, Shogren JF, White B (1997) Environmental Economics in Theory and Practice. Macmillan, Houndmilles and others

Hanley N, Wright RE, Adamowicz V (1998) Using Choice Experiments to Value the Environment. Environmental and Resource Economics 11: 413-428

Horta K, Round R, Young Z (2002) The Global Environment Facility - The First Ten Years – Growing Pains or Inherent Flaws? Halifaxinitiative. Available at: `http://www.halifaxinitiative.org/hiphp/WB/333`

James AN, Gaston KJ, Balmford A (1999) Balancing the Earth´s accounts. Nature 401: 323-324

Kaul I, Grunberg I, Stern M A (eds) (1999) Global Public Goods - International Cooperation in the 21st Century. Oxford University Press, New York Oxford

Kindleberger C P (1986) International Public Goods without International Government. The American Economic Review 76: 1-13

Loomis JB, White DS (1996) Economic benefits of rare and endangered species: summary and meta-analysis. Ecological Economics 18: 197-206

Mitchell R, Carson R (1988) Using Surveys to Value Public Goods: The Contingent Valuation Method. Resources for the Future, Washington DC

Myers N, Mittermeier RA, Mittermeier CG, da Fonseca GAB, Kent J (2000) Biodiversity Hotspots for Conservation Priorities. Nature 403: 853-858

Perrings C, Gadgil M (2003) Conserving Biodiversity: Reconciling Local and Global Public Benefits. In: Kaul I, CONCEIÇÃO P, Goulven, K, Mendoza, R U (eds) Providing Global Public Goods: Managing Globalization, Oxford University Press, New York Oxford, pp 532-555

Reid WV, Miller KR (1989) Keeping Options Alive: The Scientific Basis for Conserving Biodiversity. World Resources Institute, Washington DC

Spash CL (1999) Lexicographic Preferences and the Contingent Valuation of Coral Reef Biodiversity in Curaçao and Jamaica. In: Gustavson K, Huber RM, Ruitenbeek, JR (eds) Integrated Coastal Zone Management for Coral Reefs: Decision Support Modeling. Word Bank, Washington DC, pp 97-117

Streck C (2001) The Global Environment Facility – a Role Model for International Governance? Global Environmental Politics 1: 71-94

Zimmermann H, Henke K-D (1994) Finanzwissenschaft. Vahlen, München

Appendix I

GLOBAL ENVIRONMENT FACILITY TRUST FUND
THIRD REPLENISHMENT OF RESOURCES

Attachmen

CONTRIBUTIONS
(in millions)

	Calculated Basic Contributions		Supplemental Contributions	GEF-3 Total Contributions			
Contributing Participants	(%)	SDR	SDR	%	SDR	National Currency	g/
Australia	1.46%	27.60		1.46%	27.60	68.16	
Austria	0.90%	17.01	0.69 a/	0.94%	17.70	24.38	
Belgium	1.55%	29.30	3.67 a/	1.74%	32.97	41.98	
Canada	4.28%	80.91		4.28%	80.91	158.94	
China	-	4.00 b/	4.44 a/ d/	0.45%	8.44	78.71	
Cote d'Ivoire	-	4.00 b/		0.21%	4.00	3,758.86	
Czech Republic	-	4.00 b/	0.50 a/	0.24%	4.50	194.36	
Denmark	1.30%	24.58	3.37	1.48%	27.95	298.18	a
Finland	1.00%	18.91	2.03	1.11%	20.94	30.00	a
France	6.81%	128.84 a/		6.81%	128.84	164.00	
Germany	11.00%	207.96		11.00%	207.96	297.92	
Greece	0.05%	0.95	3.55 a/ c/	0.24%	4.50	5.73	
India	-	4.00 b/	3.99 a/ d/	0.42%	7.99	426.39	
Ireland	0.11%	2.08	2.42 a/ c/	0.24%	4.50	5.73	
Italy	4.39%	82.99	h/	4.39%	82.99	118.90	
Japan	17.63%	333.41 a/		17.63%	333.41	48,754.33	
Korea	0.23%	4.35		0.23%	4.35	7,142.95	a
Luxembourg	0.05%	0.95	3.05 c/	0.21%	4.00	5.73	
Mexico	-	4.00 b/		0.21%	4.00	4.00	a
Netherlands	3.30%	62.39		3.30%	62.39	89.38	k
New Zealand	0.12%	2.27	1.73 c/	0.21%	4.00	12.13	
Nigeria	-	4.00 b/	0.50 a/	0.24%	4.50	4.00	
Norway	1.06%	19.96		1.06%	19.96	228.32	
Pakistan	-	4.00 b/		0.21%	4.00	320.63	
Portugal	0.12%	2.27	1.73 c/	0.21%	4.00	5.73	
Slovenia	-	1.00	0.13 a/	0.06%	1.13	313.94	
Spain	0.80%	15.12		0.80%	15.12	21.67	
Sweden	2.62%	49.53	7.45	3.01%	56.98	764.67	
Switzerland	2.43%	45.94		2.43%	45.94	99.07	
Turkey	-	4.00 b/		0.21%	4.00	4.00	
United Kingdom	6.92%	130.82 a/	19.09 a/	7.93%	149.91	117.83	
United States	20.86%	394.36		20.86%	394.36	500.00	f
1. New Funding from Donors	88.99% **	1,715.50	58.34	93.82%	1,773.84		
2. Supplemental Contributions including Credits			12.50 a/ k/	0.66%	12.50		
3. Investment Income e/					105.00		
4. Carryover of GEF Resources j/					450.00		
5. Total Projected Resources to Cover GEF-3 Work Program (1 + 2 + 3 + 4)					**2,341.34**		

** GEF basic shares, which are originally derived from the GEF-1 and were largely maintained in the GEF-2, do not add up to 100%.

a/ Contributing Participants have the option of taking a discount or credit for acceleration of encashment and; (i) including such credit as part of their basic share; (ii) counting such credit as a supplemental contribution; or (iii) taking such discount against the national currency contribution. France and Japan have opted to include the credit for accelerated encashment in their basic share. The United Kingdom has chosen to accelerate encashment of its basic and supplemental contributions. A credit for accelerated encashment is thus included in its basic share and its supplemental contribution. Austria, Belgium, China, Czech Republic, Greece, India, Ireland, Nigeria, and Slovenia have opted to include the credit for accelerated encashment as a supplemental contribution. Denmark, Finland, Korea, and Mexico have opted to take a discount against their national currency contribution. Canada chose to acclerate encashment of its contribution but not to take either a discount or a credit.

b/ Represents the agreed minimum contribution level to the GEF-3.

c/ These Contributing Participants have agreed to adjust their contributions upward to the agreed minimum contribution level of SDR 4 million.

d/ China and India have indicated that they would contribute more than the agreed minimum contribution level of SDR 4 million.

e/ Represents projected investment income expected to be earned on resources projected to be held in the GEF Trust Fund over the GEF-3 commitment period (FY03 through FY06).

f/ In addition to four annual installments of USD 107.5 million, the United States will provide USD 70 million in the final year of the replenishment upon achievement of the performance measures outlined in Schedule 1 to this Table. The achievement of such measures will be determined by the Council on the basis of verification by the Independent Monitoring and Evaluation Unit, and taking into account any unforeseen events or circumstances that may prevent their achievement.

g/ Calculated by converting the SDR amount to the national currency using an average daily exchange rate over the period from May 15, 2001 to Nov 15, 2001, as agreed by the Contributing Participants at the May 7, 2001, GEF-3 replenishment meeting.

h/ For this Contributing Participant, acceleration of encashment is under consideration.

j/ Represents the amount carried over to the GEF-3 pursuant to paragraph 9 of Resolution No. ____, valued on the basis of June 30, 2002 exchange rates.

k/ Represents a credit from acceleration from Canada in the amount of SDR 10.13 million and a supplemental contribution from The Netherlands in the amount of SDR 2.37 million, bringing The Netherlands' t national currency contribution to EUR 92.76 million.

Fig. 2. Summary of negotiations on the third replenishment of the GEF Trust Fund (GEF 2002c)

Appendix II

Table 6. Percentages of people in age groups in the sample and in the basic population

Age group	Percentage of sample	Percentage of basic population
15(18)-25	15	13
25-45	45	36
45-65	29	31
65+	12	20

Source: own research and own calculations.
Data for basic population: Federal Statistical Office (Germany), 2002

Table 7. Payments in the GEF (in US$ million with exception of 1)

	Study sample			Basis population	
	freq.	[%]	Valid percentages	Income-categories[a]	Percentages basis population
Less than € 900	127	12	16	<920	16.7
€ 900–1,250	107	11	30	920–1,534	27.7
€ 1,251–1,600	138	14			
€ 1,601–2,000	124	12	30	1,534–2,556	32.5
€ 2,001–2,500	107	11			
More than € 2,500	197	19	25	>2,556	22.9
total	800	77	100		
Do not know / unspec.	217	21			

Own research data
[a] Federal Statistical Office, Datenreport 2002 : 112

The Cartagena Protocol: trade related measures as a means to protect biological diversity from risks deriving from genetically modified organisms

Alexander Behrens

Institute of International Law, Georg-August University of Göttingen, Platz der Göttinger Sieben 6, 37073 Göttingen, Germany

Summary. The rise of modern biotechnology during the 1980s has been associated with various hopes in different fields. At the same time however, this development has also raised serious concerns. To control the risks which this technology might pose to the environment, the international community addressed the issue of biosafety in the Convention of Biological Diversity. As the Convention's general obligations were considered insufficient, states agreed to regulate the field of biosafety in a specific treaty, the Cartagena Protocol which was signed in January 2000 and came into force in September 2003. The Cartagena Protocol focuses on the regulation of trade in LMOs. This has recently triggerd a heated debated in international law about the question of the Protocol's relation to other international trade regulations, specifically to WTO law. While this question considers the Cartagena Protocol essentially from a trade perspective, this contribution approaches the Protocol from a more environmental perspective and asks whether the focus on trade constitutes an efficient mechanism to regulate biotechnology and whether the instruments which the Protocol provides for this regulation are adequate. To answer these questions, the contribution describes the risks and benefits associated with biotechnology to clarify the regulatory object of the Protocol and undertakes a detailed analysis of the central provisions of the Protocol and their functions. Based on this this contribution comes to two conclusions: first, if risks to the biological diversity should occur, these will not be limited to the country in which the LMO was released as an unintentional transboundary movement of these organisms will be hardly avoidable. As a result, the focus on the transboundary pollution caused by trade is insufficient. Second, if one considers the efficacy of the trade measures as such, one can note that the flow of information especially from developed to developing countries will strengthen the decision-making capacities of developing countries and will thus contribute to a heightened ability of these countries to exercise, in practice, their sovereignty with regard to imports of LMOs. Summing up, one can therefore state that the Cartagena Protocol represents some amelioration with regard to the status quo ante, that the concept of the focus on trade as well as the way in which the AIA regulates this trade, however, remains behind what would have been desirable from an environmental point of view.

Key words: biotechnology, living modified organisms, trade related measures, MEAs

1 Introduction

The rise of modern biotechnology during the 1980s has been associated with various hopes in different fields. At the same time however, this development has also raised serious concerns. These include, among others, potential risks to biological diversity. Scientists fear for example, that genetically modified crops or fish might become invasive, unintended transfer of the inserted genetic material could occur or non-target species might be adversely affected by crops that were designed to be resistant to insect pests. To control these risks, the international community addressed the issue of biosafety in the Convention of Biological Diversity. As the Convention's general obligations were considered insufficient, states agreed to regulate the field of biosafety in a specific treaty, a so called Protocol. After almost six years of heated negotiations, the Cartagena Protocol on Biosafety (Cartagena Protocol) was signed in January 2000 and came into force in September 2003. As of December 2003, 78 states have ratified the Protocol. The Cartagena Protocol focuses on the regulation of trade in LMOs. This raises the question of the Protocol's relation to other international trade regulations, specifically to world trade law as embodied in WTO law, which has attracted the attention of many authors. While this question considers the Cartagena Protocol essentially from a trade perspective, this contribution approaches the Protocol from a more environmental perspective: does the focus on trade constitute an efficient mechanism to regulate biotechnology and are the instruments which the Protocol provides for this regulation adequate? To answer these questions, Sect. 2 describes the risks and benefits associated with biotechnology to clarify the regulatory object of the Protocol. Sect. 3 explores the historical development which led to the adoption of the Protocol. Thereafter, a detailed analysis of the central provisions of the Protocol will be undertaken. Based on an exploration of the reasons of the Protocol's focus on trade and an analysis of the aims which the trade related provisions intend to pursue, Sect. 5 comes to an assessment of the adequacy of the Protocol's approach from an environmental perspective.

2 Risks and benefits of genetically modified organisms

Biological processes have been used for centuries to modify food in order to improve taste, palatability and safety, especially by using the technique of selective breeding. This technique relied on the genetic variation already available in the population or naturally generated spontaneous mutations. In the 1950s, scientists discovered the structure of the DNA. This discovery paved the way for a new technique by which the genetic coding of organisms could be altered to give them new characteristics that natural evolution or selective breeding could not produce[1]. Practical applications deriving from this discovery became possible two decades

[1] Teel J (2000) Regulating Genetically Modified Products and Processes: An Overview of Approaches, New York University Environmental Law Journal 8, p 649.

later, when science reached a new breakthrough by finding out how to isolate individual genes, refashion them and copy them in cells. First, commercial applications were developed in the field of medicine, soon followed by technologies using so called genetically modified organisms (GMOs) to produce new fine chemicals. However, not until 1994, when the Flav'r Sav'r tomatoes came on the market in the US, had whole food been genetically altered[2]. The technique of genetic modification differs from selective breeding mainly in that it removes genes directly from one organism and inserts them into the DNA of the cells of another. Thus, it is faster and more exact than the generally random based approach applied in selective breeding. Importantly, it also opens the possibility of transferring genes across natural borders between different organisms[3]. In the field of agriculture, this has meant the opportunity to add genes to transgenic crops that can express such traits as pesticide resistance, herbicide resistance, increased viability in harsh environments and added nutritional content[4]. Thus, agricultural outputs could be immensely raised in terms of quantity and quality, for example, in the form of more nutritious foods, which offered wider economic margins. In addition, by avoiding or minimising the use of pesticides, environmental and human health purposes could be served. For some, research, development and marketing of these new crops are therefore hailed as great advances for the environment and public health. As a consequence, genetically modified products in the field of agriculture have mushroomed since 1994. For example, the global area devoted worldwide to grow transgenic crops has increased by 44 % to 39.9 million hectares in 1999 alone[5]. It should however be noted, that in 2001 99 % of all GMO crop area world-wide was grown in four countries: 68 % of the crop area planted with GMOs was in the USA, 22 % in Argentina, 6 % in Canada, and 3 % in China[6]. However, this development has also raised serious concerns. These range from ethical and religious considerations about mankind's role in the world, to various socio-economic issues (for example disruption of small scale farming systems especially in developing countries). Concerns also encompass potential risks for human health and the environment[7]. The extent to which GMOs pose those risks however, is controversial and remains uncertain. With regard to human health, problems like enhanced allergies and resistance to antibiotics are discussed, but to date no adverse effects of GM foods on human health have been reported in the peer-reviewed scientific lit-

[2] Royal Commission on Genetic Modification - New Zealand (2001) Report of the Royal Commission on Genetic Modification - New Zealand, Auckland, p 362.

[3] Katz D (2001) The Mismatch Between the Biosafety Protocol and the Precautionary Principle, Georgetown International Environmental Law Review 13, p 940, 949.

[4] Katz (cp. Footnote 3), p 953.

[5] Hagen PE, Weiner JB (2000) The Cartagena Protocol on Biosafety: New Rules for International Trade in Living Modified Organisms, Georgetown International Environmental Law Review 12, p 697, 698.

[6] IUCN (2003) Explanatory Guide to the Cartagena Protocol on Biosafety, Environmental Policy and Law Paper No. 46, Cambridge, p 7.

[7] For both these aspects see the detailed analysis of Katz (cp. Footnote 3), pp 967.

erature[8]. At the moment, concerns therefore seem to centre on risks to biological diversity, especially if genetically modified organisms are released into the environment, for example for the growing of genetically modified crops or the growing of fish in aquaculture projects, as these organisms might become invasive[9]. Moreover, transfer of the inserted genetic material could occur or non-target species might be adversely affected by crops that were designed to be resistant to insect pests. Given the relatively small amount of experience with the applications of biotechnology to date, the potential of these effects is still however, a heated issue of scientific debate[10].

3 Historical development of the Cartagena Protocol

The regulation of biotechnology with a view to preventing adverse effects is often referred to as biosafety or biosecurity. On the regional level, international attempts in this regard were firstly undertaken by the EU[11]. On the global level, the non-legally binding Agenda 21, adopted at the 1992 United Nations Conference on Environment and Development in Rio de Janeiro (Rio Summit), addressed the issue of biotechnology in Chapter 16. It recognises the promise of biotechnology and calls upon the international community to ensure that biotechnology is developed and applied in an ecologically sound manner. Later global efforts include the Voluntary Code of Conduct for Environmental Release of Genetically Modified Organisms, elaborated in the framework of the UNIDO, and the UNEP International Technical Guidelines for Safety in Biotechnology, which are both of non-legally binding nature.

The issue of biosafety emerged for the first time in the context of a global legally binding instrument during the negotiations of the Convention on Biological Diversity (CBD). The CBD was adopted in 1992 with the aim of conserving biological diversity, the sustainable use of its components and the fair and equitable sharing of benefits arising out of utilisation of genetic resources[12]. As GMOs constitute a potential risk to the biological diversity[13], negotiators agreed, after lengthy debate, to include language addressing this aspect, but only in very vague terms[14].

---

[8] Eggers B, Mackenzie R (2000) The Cartagena Protocol on Biosafety. In: Journal of International Economic Law, p 525, 526.

[9] The definition of biological diversity is still contentious. Art. 2 of the Convention on Biological Diversity defines it as the »variability among living organisms from all sources [...] and the ecological complexes of which they are part; this includes diversity within species, between species and of ecosystems«.

[10] Teel (cp. Footnote 1), p 650.

[11] For an intensive discussion see MacKenzie R, Francesconi S (2000) The Regulation of Genetically Modified Foods in the European Union: An Overview, New York University School of Law Environmental Law Journal 8, pp 530.

[12] Art. 1 CBD.

[13] As described above.

[14] According to Art. 8 (g) CBD, every party shall, as far as possible and as appropriate, »establish or maintain means to regulate, manage or control the risks associated with the use

In addition, the CBD calls upon parties to consider the need for and modalities of a protocol setting out appropriate procedures, including, in particular, advance informed agreement, in the field of GMOs[15].

In 1994, at the first meeting of the Conference of the Parties, preparatory works were authorised in this regard. Based on these findings, the second COP set up the Open-ended Ad Hoc Working Group on Biosafety with a view to draft a Protocol particularly focussing on the transboundary movement of GMOs[16]. The Working Group was scheduled to forward a draft to the First Extraordinary Meeting of the Conference of the Parties to the CBD (ExCOP) immediately after its sixth meeting, held in 1999 in Cartagena. However, negotiators failed to reach a compromise at this meeting, thus, the meeting had to be suspended. After informal meetings the ExCOP was eventually resumed in January 2000 in Montreal where the Protocol could be adopted. The Protocol came into force in September 2003, which is ninety days after the date of deposit of the fiftieth instrument of ratification. As of December 2003, 78 countries have ratified the Protocol.

4 Central provisions of the Protocol

4.1 Scope of application

The scope of the Cartagena Protocol is established in Art. 4 which has to be read in conjunction with Art. 3 and 5. According to Art. 4, the Cartagena Protocol shall apply to the transboundary movement, transit, handling and use of all living modified organisms (LMOs) that may have adverse effects on the biological diversity, also taking into account risks to human health. The terms »modified« and »living« are further specified by Art. 3. According to Art. 3 (g), a living *modified* organism means »any living organism that possesses a novel combination of genetic material obtained through the use of modern biotechnology«[17]. As a consequence of the reference to modern biotechnology, the term LMO effectively equals to what is generally known as a GMO. Consequently, the first drafts of the Protocol still contained the term »genetically modified organism«. The substitution with »living modified organisms« was established at the end of the negotiations in response to US claims that the latter expression was more neutral than the slightly negatively perceived term of GMOs[18].

A *living* organism means any biological entity capable of replicating genetical material[19]. The intention of this qualification was particularly to exclude the

and release of living modified organisms resulting from biotechnology which are likely to have adverse environmental impacts that could affect the conservation and sustainable use of biological diversity, taking also into account the risks to human health«.

15 Art. 19 (3) CBD.

16 COP Decision II/5, which establishes the so-called Jakarta Mandate.

17 Modern biotechnology, in turn, is defined in Art. 3 (i).

18 Gupta A (1999) Framing »Biosafety« in an International Context: The Biosafety Protocol Negotiations, New York, p 5.

19 Art. 3 (h).

so-called ›products thereof‹, that are processed materials of LMO origin containing detectable novel combinations of genetic material, for example cereals deriving from genetically modified crops, from the scope of application. Moreover, Art. 5 exempts, under certain conditions, one specific category of LMOs, pharmaceuticals for humans, from the applicability of the Protocol[20].

4.2 Trade between parties

As mentioned above, the various trade related measures constitute the core of the Protocol.

Regulations for LMOs intended for »intentional introduction into the environment«: the Advance Informed Agreement (AIA)

The key mechanism of the Cartagena Protocol is the advance informed agreement procedure. Before describing the functioning of this procedure, its scope as defined by Art. 6 and 7 shall be outlined as it differs from the scope of the Protocol as such. This is the result of a compromise between developing countries, which favoured an inclusion of all LMOs in the scope of the Protocol, and developed countries, which generally pushed for a more limited applicability[21].

The scope of the AIA procedure is regulated in Art. 6 and 7 in quite a complicated way that seems to reflect the underlying political controversy. The general principle is laid down in Art. 7 (1) and (2). Under Art. 7 (1), the AIA procedure shall apply prior to the first intentional transboundary movement of LMOs for »intentional introduction into the environment«. This phrase is however, not defined in the Protocol. Instead, Art. 7 (2) specifies - negatively - that the term »intentional introduction« does not refer to LMOs intended for direct use as food or feed, or for processing, like for example grains from modified crops. Art. 6 and Art. 7 (4) then contain different kinds of exceptions from this principle regarding LMOs in transit, LMOs destined for contained use and LMOs identified in a decision of the Conference of the Parties as being not likely to have adverse effects. In practical terms, the AIA procedure will therefore apply particularly to the growing of agricultural crops, release of fish and of modified micro-organisms[22].

The obligations which are applicable to those LMOs covered by the scope of the AIA are mainly contained in Art. 8, 9, 10, 12 and 15 and Annex III. According to Art. 8 (1), the procedure starts by a notification by the exporting country or the exporter to the importing country prior to the first intentional transboundary movement of

---

[20] Note that, strictly speaking, Art. 5 (1) refers only to the *transboundary movement* of LMOs which are pharmaceuticals for humans. Therefore, one could argue that general provisions, for example with regard to capacity-building, also apply to this category of LMOs.

[21] Gupta A (2001) Governing Trade in Genetically Modified Organisms: The Cartagena Protocol on Biosafety, Environment 42/4, p 23, 25.

[22] IUCN (cp. Footnote 6), p 68.

an LMO that falls within the scope of the procedure[23]. The notification shall, at a minimum, contain the information as specified in Annex I. Annex I covers a wide range of information, which may be loosely grouped into three categories: information concerning the LMO itself aimed at providing the importing country with factual information[24]; regulatory status in the exporting party intended to inform the country of import about the cost-benefit assessment of the state of origin; and suggested methods for safe handling and use.

As a second step of the AIA procedure, Art. 9 requires the importing state to acknowledge receipt of the information to the notifier within ninety days of its receipt. The acknowledgement shall include information whether to proceed according to the procedure specified in Art. 10, or according to the domestic regulatory framework of the party of import that shall be consistent with the Protocol. This latter aspect may be interpreted in such a way that it allows those parties which have already established regulatory systems in place to continue using them, thus not being forced to enact new laws in order to comply with the Protocol, as long as they serve the purpose of the Protocol. Finally, Art. 9 (4) explicitly states that a failure by the party of import to acknowledge receipt shall not imply its consent.

Art. 10, then, is concerned with the decision-making stage. It contains procedural and substantive requirements and also provides some specific supportive measures for the importing countries? decision-making process. With regard to the procedure to be followed, Art. 10 (2) provides that the importing state shall, together with the acknowledgement of receipt, notify to the exporting state as to whether the transaction shall proceed only after the party of import has given written consent or after no less than 90 days without written consent. That means that it is up to every importing country to decide, on the merits of each single case, whether it wishes the application of the consent requirement. Furthermore, Art. 13 provides the possibility for importing countries to specify in advance to the Biosafety Clearing-House (a) cases in which the transboundary movement may take place at the same time as the movement is notified to the party of import and (b) imports of LMOs to be exempted from the AIA procedure[25].

If the importing state opts for written consent, Art. 10 (3) provides that the importing country has to communicate within 270 days whether or not consent is granted, whether the period of 270 days shall be extended, or whether additional information is required. Except in a case in which consent is unconditional, the importing country shall also disclose the reasons on which its decision is based, as per Art. 10 (4). Negotiators of the Cartagena Protocol too were forced to face

[23] The limitation of the procedure to the *first* intentional transboundary movement is not explicitly stated in Art. 8 (1), but results from Art. 7 (1).

[24] This information includes, inter alia, the name and identity of the LMO in question, centres of origin of the recipient organisms, description of the modification and intended use.

[25] Art. 14 offers an additional possibility for parties to deviate from the requirements of the AIA procedure by entering into bilateral, regional and multilateral agreements »consistent with the objective of this Protocol and provided that such agreements do not result in a lower level of protection than that provided for by the Protocol«.

the question of which rules to apply, if the importing party fails to communicate its decision. Art. 10 (5) states that such a failure »shall not imply its consent« to the transboundary movement. Thus, Art. 10 (5) states negatively how the case of no-answer may not be interpreted. However, it does not specify which rules will apply in this case. An interpretation of Art. 10 (5) should firstly consider the above-mentioned Art. 10 (2). It seems to provide that in any case in which a country does not communicate that written consent is unnecessary; this consent is a precondition of every transboundary movement. This interpretation is secondly supported by a comparative analysis of Art. 10 (5) with Art. 11 (7). This provision is relevant in the case that an importing country fails to communicate its decision with regard to LMOs destined for food, feed or processing. It provides that such a failure shall not imply its »consent *or refusal* to the import«. From the different wording of this provision, one can conclude that Art. 10 (5) must be read as requiring explicit consent before any transboundary movement may commence.

Another important aspect is equally not explicitly provided in the Protocol; that is the question of whether exporting countries are bound to enforce the consent requirement, i.e., to allow the export only in case that written consent has been notified by the importing country. The Protocol does not contain an explicit obligation in this regard. In addition, no further hints which could assist in interpreting this aspect can be found in other provisions of the Protocol. Therefore, clarification on this essential aspect seems necessary, and desirable on behalf of the Conference of the Parties.

While the above mentioned provisions stipulate essentially the formal process of interaction between the exporting and the importing state, much attention of the Cartagena Protocol is also devoted to laying down procedural and substantive criteria which the importing state has to take into consideration when making its import decision. For this sake, the Cartagena Protocol distinguishes basically between the procedural requirement of risk assessment and the broader notion of risk management containing substantive conditions. Art. 10 (1) merely states that decisions taken by the importing party »shall be in accordance with Art. 15«. The main requirement of Art. 15 (1), in turn, is that risk assessments be carried out in a scientifically sound manner and must be aimed at evaluating the possible adverse effects of LMOs. Moreover, the risk assessment must be in accordance with Annex III and take into account recognised risk assessment technologies. Annex III further specifies the modalities of this risk assessment. Accordingly, risks associated with LMOs should be considered in the context of the risks posed by non-modified recipients. Annex III furthermore, sets out the methodology for the risk assessment as well as aspects which have to be taken into account. Thus, Annex III does not only point to the requirement of scientific soundness, but also provides a point of reference for the risk assessment technique.

One important and hotly debated question in the context of risk assessment was what kind of adverse effects should be taken into account: should the assessment be limited to adverse effects on the biological diversity or should those on human health also be considered. The origin of this debate can be traced to the text of the Convention on Biological Diversity. Art. 19 (3) of the Convention calls upon parties to consider the need for a Protocol, but makes no reference to human health. In contrast, Art. 8 (g) of this Convention requires parties to control risks resulting from adverse effects of LMOs to the biological diversity, »taking into account the risks to human health«. By way of compromise, negotiators agreed on language stating that human health issues are to be »taken into account«. However, the very meaning of this phrase is as contentious as its inclusion, and reflects the fact that the compromise was more a linguistic, than a substantive one.

While the above mentioned risk assessment aims at the gathering of information that is to be considered, the actual process of deciding as such can be characterised as a weighing of policy alternatives (in concrete terms with regard to the import decision) in light of considering the risk assessment and possibly other factors[26]. In the Protocol, this process is referred to in Art. 16 as risk management[27]. While the Protocol does not comprehensively regulate this process, it lays down, however, three substantive criteria in this context, i.e., scientific uncertainty, »socio-economic considerations« and »necessity«.

A scientifically operated risk assessment can easily reveal scientific uncertainty with regard to specific adverse effects, or disagreement at a scientific level can occur[28], in particular in an emerging technique like biotechnology. This raises the question of how to cope with these situations in the context of the (import) decision - which hence, was amongst the most contentious issues during the negotiations[29]. The Protocol touches on this issue at various points. With specific regard to decision-making in the context of the AIA procedure, Art. 10 (6) establishes that lack of scientific certainty due to insufficient relevant scientific information and knowledge shall not prevent parties from taking a decision in order to avoid or minimise such potential adverse effects. Hence, the Protocol explicitly grants parties protective measures beyond clear scientific justification in reaching their decisions on import. Thus, it explicitly introduces the so-called »precautionary approach«[30].

---

[26] See for this definition, among many others, the »Codex General Standard for Contaminants and Toxins in Foods« of the Codex Alimentarius Commission, available at: `www.who.int/fsf/Codexreview/GENERALSTANDARDCONTAMINANTSANDTOXINSInFOODS.pdf`. (accessed on 10 December 2003). Note however, that the Protocol itself does not contain a definition either of risk assessment or of risk management.

[27] While Art. 16 is entitled as »Risk management«, respective criteria are also embodied in other provisions as will be argued below. On the other hand, Art. 16 does not only incorporate certain criteria to be considered in the risk management, but also contains the obligation for parties to undertake such a risk management. For further details see below.

[28] For example, as referred to at the outset of the article, there is still disagreement about the probability that antibiotic resistance will be enhanced by the consumption of food based on LMOs.

[29] Adler J (2000) More Sorry Than Safe: Assessing the Precautionary Principle and the Proposed International Biosafety Protocol, Texas International Law Journal 35, p 173, 194.

[30] Stoll T (1999) Controlling the Risks of Genetically Modified Organisms: The Cartagena Protocol on Biosafety and the SPS Agreement, Yearbook of International Environmental

Whereas the issue of precautionary measures remains within the overall context of decision-making according to scientific criteria, the question arises whether countries should, in taking a decision, only consider (scientifically based) risks to biodiversity and human health or whether other aspects could also be taken into account. During the negotiations, this issue was heavily debated with regard to the so-called ›socio-economic considerations‹. Developing countries feared the socio-economic impact of LMOs and biotechnology, especially on indigenous communities, and therefore pushed for inclusion of these aspects in the decision-making process of the Protocol. Developed countries instead were afraid that this could open the door to mere protectionist measures[31]. Art. 26 now states that parties may take into account socio-economic considerations arising from the impact of LMOs on the biological diversity, but only in so far as consistent with their international obligations. The last qualification will likely limit the relevance of Art. 26 as it seems, for the time being, unclear how science-based WTO obligations and socio-economic aspects can be reconciled[32].

A third substantive criterion with regard to the decision-making process is incorporated into Art. 16 (2). Accordingly, measures »shall be imposed to the extent necessary to prevent adverse effects« within the territory of the importing party. In practical terms, this provision allows for a wide margin of discretion on behalf of each party and does not give specific guidance. Nevertheless, it is interesting from a structural point of view as the term ›necessary‹ could also be interpreted as not only to oblige parties, but also to limit their control measures to the extent to which they are necessary. This limitation would strive to prevent unnecessary hurdles to international trade and thus to prevent a conflict between the Cartagena Protocol and WTO law.

According to the rules of the Cartagena Protocol as referred to above, the risk assessment and the risk management with regard to each import decision are the sole responsibility of the importing country. To assist importing countries in this process the Protocol however, does not only provide for general supportive measures in form of generic capacity-building[33], but includes in Art. 15 regulations specifically designed to facilitate concrete transactions. While the first sentence of Art. 15 (2) confirms that the party of import shall ensure that risk assessments are carried out before a decision is taken, the second sentence stipulates that the importing state may require the exporter to carry out the risk assessment. Moreover, as regards the cost of risk assessment, Art. 15 (3) establishes that this shall be borne by the notifier if the importing state so requires. Put differently, the im-

Law 10, p 82, 98. Further references to the precautionary approach are contained in the preamble, in Art. 1, Art. 11 (8) and in Annex III. The term »precautionary approach« itself is however, used only once in the preamble of the Protocol. The reference to ›approach‹ rather than to ›principle‹ is due to opposition from some developed countries which, during the negotiations, denied the actual existence of a respective principle or considered existing expressions as too nebulous to name it a principle.

[31] Gupta (cp. Footnote 18), p 17.

[32] Stoll (cp. Footnote 30), p 97.

[33] For details in this regard see below.

porting state is responsible for the fact that a risk assessment will be conducted, but does not have to conduct it on its own and does not have to bear the costs. The latter two aspects seem to be an expression of the ›polluter pays‹ principle - well known to modern environmental law.

As mentioned before, risk assessment is only one step of the decision procedure. The decision procedure itself, or the risk management, equally requires related capacities. To assist parties that lack these capacities, Art. 10 (7) provides that the Conference of the Parties shall, at its first meeting, decide upon appropriate mechanisms to facilitate decision-making by importing parties. As this provision is very vague, the first and second Intergovernmental Committees on the Cartagena Protocol (ICCP I and II) have addressed the issue and have made recommendations of how to implement the article. Accordingly, the COP shall develop guidelines for the decision process as well as for procedures to seek external consultation[34].

Regulations regarding LMOs intended »for direct use as food or feed, or for processing« (LMO-FFPs): Art. 11

As mentioned above, exporting countries managed during negotiations to have the scope of the AIA procedure limited to LMOs destined for intentional introduction into the environment. However, this was only accepted by other countries under the condition that another procedure was established for those LMOs not intended for intentional introduction, i.e., LMOs intended for direct use as food or feed or for processing, for example genetically modified fruits for human consumption or genetically modified soya destined for processing into edible oils. These rules are contained in Art. 11.

According to Art. 11 (1), a party that makes a final decision regarding domestic use of such a LMO-FFP, i.e., the commercial growing or the placing on the market, that may be subject to transboundary movement shall inform the parties through the Biosafety Clearing-House[35]. This notification shall contain the information as set out in Annex II[36]. Thus, Art. 11 (1) functions as a means of information sharing with respect to domestic regulations in the field of LMOs.

While Art. 11 (1) regards national decisions related to the domestic use of LMO-FFPs, Art. 11 (4) refers to national decisions on imports. It does not contain an obligation, but asserts parties' right to make a decision for import of LMO-FFPs under its domestic regulatory framework that is »consistent with the objective of the Protocol«. As the discussion during the negotiations and the ongoing debate in the framework of the WTO evidence, the question of what kind of import restrictions may be imposed is very contentious with regard to LMOs, Art. 11 (4) and particularly its reference to the »objectives of the Protocol«, which include the

[34] UNEP/CBD/ICCP/2/15, Annex I, Recommendation 2/7.

[35] The Biosafety Clearing House is the key mechanism of the Cartagena Protocol for centralised information exchange.

[36] This information is similar to that required in Annex I for LMOs destined for deliberate release into the environment, but is not so extensive.

substantive criteria for import restrictions as discussed in the context of the AIA procedure, could gain considerable importance.

The related Art. 11 (5) requires parties to make such decisions, laws and regulations available to the Biosafety Clearing-House. While this information could also serve as a source of information for other countries that consider import decisions, its main purpose lies in promoting transparency and predictability for those who intend to export LMO-FFPs[37].

Art. 11 (6) addresses specific needs of developing countries or countries that have an economy in transition. If one of these countries lacks a domestic regulatory framework for imports of LMO-FFPs it can declare through the Biosafety Clearing-House that its decision prior to the first import of a LMO-FFP, on which information has been provided under Art. 11 (1), will be taken according to (a) a risk assessment in accordance with Annex III and (b) a decision made within a predictable timeframe, not exceeding 270 days. With regard to the question of non-response, Art. 11 (7) states that a failure of an importing country to communicate its decision whether to permit or not the import of a specific LLM-FFP in the context of a procedure under Art. 11 (6) shall not imply »its consent *or refusal*«. As already noted, Art. 10 (5), which addresses the situation of non response in the context of the AIA procedure, simply states that this does »not imply its consent«. The additional wording of »or refusal« can, therefore, only be interpreted in the sense that export may also take place in the absence of an explicit consent. As in the case of the AIA procedure, it remains unclear however, whether the state of export is obliged to enforce this international rule. Therefore, a clarifying interpretation of the Conference of the Parties seems necessary with regard also to Art. 11.

If one compares the procedure for LMO-FFPs under Art. 11 to that of LMOs destined for intentional introduction into the environment under the AIA, two main differences can be observed: Firstly, while the Protocol itself mandates a specific procedure to be followed for LMOs of the latter category (the AIA procedure), it does not do the same for LMO-FFPs, but only offers developing countries the possibility to declare a specific procedure as applicable under the Protocol. So, the importing party has to trigger off the procedure by a specific declaration and thus, carries the relevant burden. Secondly, while the AIA provides for the requirement of consent before a transaction may take place, under Art. 11 (6), exports may also take place without the explicit consent of the importing country. Summing up, it can easily be noticed that the procedural protection which is offered especially to developing countries by Art. 11 is far lower than that presented by the AIA procedure.

While the preceding analysis has revealed considerable differences between the procedures to be followed under Art. 11 (6) and the one of the AIA, it should be noted that the criteria to be applied by the importing state when taking the import decision are the same in both cases. In particular, Art. 11 (6.a) also requires a risk assessment to be undertaken in accordance with Annex III and Art. 11 (8) explicitly

[37] In this regard, the provision is quite similar to notification requirements as prescribed by WTO agreements.

allows parties to take a precautionary approach to decision-making on imports. In addition, the general rules concerning risk management and socio-economic considerations are also applicable.

Further obligations with regard to trade between parties: Handling, transport, packaging, identification and illegal transboundary movement

In addition to the procedures laid down above, the Protocol stipulates some further obligations for exporting states which in particular, deal with the modalities of a transfer which was permitted by the importing state. Art. 18 considers handling, transport, packaging and identification of LMOs. Art. 18 (1) deals with the first three. Accordingly, each party shall take necessary measures to require that LMOs destined for transboundary movement are handled, packaged and transported under conditions of safety, taking into consideration relevant international rules and standards. Art. 18 (2) provides for very vague identification requirements. In addition, Art. 18 (3) states that the Conference of the Parties shall consider the need for, and modalities of, developing standards with regard to identification, handling, transport and packaging. The vagueness of Art. 18 (1 and 2) is evidence of the fact that parties could not arrive at greater compromise during negotiations. Instead, they postponed further measures to the Conference of the Parties[38].

Finally, Art. 25 (1) calls upon parties to prevent illegal transboundary movements of LMOs. More importantly, Art. 25 (2) regulates the consequences of such an illegal transfer obliging exporting parties, upon request, to dispose of, at its own expense, the LMOs illegally transferred by repatriation or destruction.

4.3 Trade with non-parties

The Cartagena Protocol addresses not only trade between parties, but also trade with non-parties. Art. 24 (1) establishes that this trade shall be consistent with the objective of the Protocol[39]. Generally speaking, regulations concerning trade with non-parties are supposed to serve two different ends: firstly to ensure that parties of a treaty do not get involved in transactions which take place under lower environmental conditions than laid down in the treaty; and secondly to encourage access of non-parties. However, given the very soft approach of the Cartagena Protocol, simply requiring in Art. 24 (1) a trade consistent with the objectives, there is little (negative) incentive for non-parties to join the agreement. Therefore, a second paragraph was deemed to be necessary calling upon parties to encourage non-parties to adhere to the Protocol. Given its vagueness, the practical impact of the whole Art. 24 may however, be doubted.

---

[38] Goldman KA (2000) Labelling of Genetically Modified Foods: Legal and Scientific Issues, Georgetown International Environmental Law Review 12, p 717, 721.

[39] In addition, Art. 24 (1) allows parties to enter into agreements with non-parties.

4.4 Domestic management

The Cartagena Protocol also contains provisions concerning the domestic management of LMOs. According to Art. 2 (2)[40], parties shall ensure that the development, handling, transport, use, transfer and release of any LMO »are undertaken in a manner that prevents or reduces the risks to biological diversity, taking also into account risks to human health«. In addition, Art. 16 (risk management) contains some respective obligations. Art. 16 (1) states that parties shall establish and maintain appropriate measures to regulate and control risks associated with the use, handling and transboundary movement of LMOs. More specifically, Art. 16 (4) calls upon parties to »endeavour to ensure that any living modified organism, whether imported or locally developed, has undergone an appropriate period of observation that is commensurate with its life-cycle or generation time before it is put into use«. Thus, on one hand, the Cartagena Protocol also addresses the domestic management of LMOs. On the other hand, all respective obligations are very vague. This is evident for the first two mentioned provisions, but also Art. 16 (4) which could be seen as a specification of the general obligations is watered down by the use of the term »endeavour«. The relevance of the respective provisions is further more questionable, as they add little new substance compared to what was already provided by the Convention on Biological Diversity[41].

4.5 Unintentional transboundary movements

According to Art. 16 (3), each party shall take appropriate measures to prevent unintentional transboundary movements of LMOs. If such a kind of movement nevertheless occurs, Art. 17 specifies which notifications a party has to undertake to inform neighbouring states of this movement to enable them to take appropriate counter-measures. This set of obligations does not exactly fit into the categorisation of trade, domestic and supportive measures as it requires domestic measures, but only to the extent that neighbouring states might be affected.

4.6 Supportive measures

The trade measures described above, in particular the AIA procedure and the procedure according to Art. 11 require an active role of importing developing countries. Most importantly, they have to undertake an assessment whether to consent to the import or not. The implementation of this obligation – and of the domestic management measures – presupposes according capacities which developing countries often lack. Therefore, these countries sought to include mechanisms into the Protocol providing for technology and financial assistance, arguing that the above procedures would otherwise fail in its objective of protecting them against

---

[40] Art. 2 is entitled »general obligations«.

[41] Stoll (cp. Footnote 30), p 88. For the exact wording of Art. 8 (g) cp. Footnote 14.

unwanted and illegal imports[42]. As they do not themselves lay down obligations with regard to specific trade transactions, but support the implementation of other (trade) measures, they can be referred to as supportive measures.

The Cartagena Protocol contains two different kinds of supportive measures: first, provisions specifically related to an import decision within the context of the AIA procedure or within the context of Art. 11 which have already been discussed above[43]; and second, those related to general assistance. Provisions in this regard are contained in Art. 20 (Information Sharing and Biosafety Clearing-House), Art. 22 (Capacity-building) and Art. 28 (Financial Mechanisms and Resources). All of them make reference to the need of assistance for developing countries, often particularly highlighting least developed countries and small island states. None of them however, contains specific commitments. Nevertheless, it should be stressed that Art. 28 states that the financial mechanism established in Art. 21 of the Convention on Biodiversity - that is the Global Environmental Facility (GEF) - shall also be the financial mechanism for the Protocol. As the GEF seems, for the time being, to provide sufficient funds, this will ensure at least a certain level of financial resources. This hope is underlined by the fact that GEF has already provided financial resources for capacity-building in the form of a project on biosafety frameworks to be implemented by UNEP[44]. In addition, the Intergovernmental Committee on the Cartagena Protocol in its second meeting endorsed the »Action Plan for Building Capacities for the Effective Implementation of the Cartagena Protocol«, with a view to facilitate and support the development of capacity to effectively implement the Protocol[45].

5 Environmental protection by trade regulation?

If one considers the regulations described above, one can easily notice that the main focus of the Cartagena Protocol is on the transboundary movement of LMOs between parties to the Protocol whereas the aspect of domestic management of LMOs and trade with non-parties are only regulated in very broad terms. This conclusion is confirmed by the fact that the rules on domestic management in the field of biosafety as laid down by the Cartagena Protocol barely goes beyond what had already been prescribed by the Convention on Biological Diversity. This

42 Report of the Working Group on its 1st session, UNEP/WG.182/3; Report of the Working Group on its 2nd session, UNEP/WG.186/3.

43 See Art. 10 (7) and Art. 11 (9), respectively.

44 The UNEP-GEF Global Project on the Development of National Biosafety Frameworks began in June 2001. This three year project is designed to assist up to 100 countries to develop their National Biosafety Frameworks so that they can comply with the Cartagena Protocol on Biosafety. The project will also promote regional and sub-regional cooperation on biosafety. For details see http://www.unep.ch/biosafety/index.htm (accessed on 28 December 2003).

45 Report of the Intergovernmental Committee for the Cartagena Protocol, Second Meeting, Nairobi 1-5 October 2001, UNEP/CBD/ICCP/2/15.

raises three questions: (1) What were the motives for negotiators to regulate trade in LMOs? (2) Which aims pursue the respective provisions intended to regulate this trade? And finally, (3) does this focus represent an adequate mechanism to ensure global biosafety?

5.1 Background to the focus on the regulation of trade with LMOs

In order to understand the focus of the Cartagena Protocol on trade matters, one has to go back to its historical roots. The proposal for a provision on biosafety within the Convention on Biological Diversity came first from Malaysia, i.e., from a developing country[46]. Concerns of developing countries with regard to LMOs are linked to their respective national regulatory systems: the potential risks posed by genetically modified organisms led, in most OECD countries during the late 1980s and early 1990s, to the promulgation of respective national safety regulations (which however, differed widely from country to country causing trade restrictions). As far as developing countries are concerned, some of them – especially those which engaged themselves in the development and growing of genetically modified agricultural crops, i.e., Brazil and India – have regulations dating back to the beginning of the 1990s. The majority of them however, lacked experience in this technology and started only recently to develop domestic biosafety regulations[47]. As a result, those countries were afraid of being misused as testing grounds for LMOs by companies based in industrialised countries. That these fears were not merely hypothetical is evidenced by an incident dating back to 1986 when NGOs discovered that genetically altered rabies vaccine had been field tested by a US institute on an Argentinean farm without the knowledge or consent of the Argentinean government[48]. However, thereafter no major similar incidents seem to be reported. Does that mean that one single incident and the non-proven fear of subsequent ones were as such, sufficient to cause states to sign a global agreement?

This sounds barely convincing given the long tradition of typical unwillingness of most states to enter into environmental agreements restricting their national sovereignty. Therefore, the above mentioned aspect can be seen as offering an explanation for developing countries? desire to specifically regulate trade in LMOs. To understand why this could lead to the beginning of international negotiations, it seems however that another aspect also has to be taken into account: the international regulatory context of the Protocol, namely its predecessors. In the section »historical development«, the Voluntary Code of Conduct for Environmental Release of Genetically Modified Organisms, elaborated in the framework of the UNIDO, and the UNEP International Technical Guidelines for Safety in Biotechnology are described as non-legally binding predecessors of the Cartagena Protocol. This is however, only true in so far as the global regulation of biosafety as a matter of substance is concerned. In contrast, when analysing the Protocol's focus on

[46] Gupta (cp. Footnote 18), p 4.

[47] Hagen and Weiner (cp. Footnote 5), p 700.

[48] Rajan MG (1997) Global Environmental Politics: India and the North-South Politics of Global Environmental Issues, New Delhi, p 179.

trade between parties, it seems more adequate to see the Protocol in the context of two other legally binding agreements: the Basel Convention on the Control of Transboundary Movements of Hazardous Wastes and their Disposal (Basel Convention) which was signed in 1989, and the Rotterdam Convention on the Prior Informed Consent Procedure for Certain Hazardous Chemicals and Pesticides in International Trade (Rotterdam Convention), signed in 1998. Similar to the Cartagena Protocol, both these agreements focus on trade in the related substances and use similar regulatory mechanisms in that they subject trade in these substances to a procedure of »prior informed consent«[49] which is structurally very similar to the AIA procedure. In fact, the first drafts of the Cartagena Protocol even used the term »prior informed consent«; some industrialised countries were, however, concerned that the use of this term could induce the perception that LMOs were as dangerous as hazardous wastes or chemicals and, therefore, pushed for a replacement of this term by »advance informed agreement«[50]. Negotiations to both these agreements were triggered off by massive exports of hazardous wastes or chemicals respectively from developed countries to developing countries, often without explicit consent of these states, which sometimes resulted in disastrous incidents. To understand the atmosphere which surrounded trade in these substances two of these incidents may be recalled. One of the most prominent of the incidents in the field of hazardous wastes took place in 1986, when the ship Khian Sea left the US carrying twenty-eight million pounds of toxic incinerator ash. On the pretence that the ship contained fertiliser ash, the ship docked in Haiti, and then dumped 4000 tons of this waste before the Haitian government became aware of the misinformation and forced the Khian Sea to leave the port. Thereafter, she wandered the oceans for eighteen months, refused any import permission. Eventually, she anchored in Singapore without her toxic cargo and nobody knew exactly where the freight had ended up[51]. The problems particularly of the North-South trade in hazardous chemicals, become best tangible if one considers the incident of the pesti-

---

[49] It should be noted however, that the Basel Convention was amended in 1995 by Decision III/1 of the Conference of the Parties inserting a ban of hazardous wastes exports for final disposal and recycling from what are known as Annex VII countries (Basel Convention Parties that are members of the EU, OECD, Liechtenstein) to non-Annex VII countries (all other Parties to the Convention). The amendment has however, not yet entered into force which gives evidence of the political problems associated with mechanisms which are more trade restrictive than the prior informed consent procedure.

[50] Gupta (cp. Footnote 18), p 6. Thus, the rationale for using this wording is similar to what was said with regard to the term »living modified organisms«.

[51] Rublack S (1989) Fighting Transboundary Waste Streams: Will the Basel Convention Help?, Verfassung und Recht in Übersee 22, p 364, 369. An extensive compilation of information on illegal hazardous waste export schemes, including the case mentioned, is also given in Spalding H, Valette J (1990) The International Trade in Wastes: A Greenpeace Inventory, 5th ed, Washington D.C., pp 21.

cide »Lephtophos«[52] which was one of the first cases that raised concerns about the practices of international trade in certain hazardous chemicals. In 1975, Velsicol, a Texas based corporation, produced over 3 million pounds of the pesticide »Lephtophos«. The pesticide could not be sold in the United States because it had never been registered with the relevant authorities. Nevertheless, during 1975 alone, Velsicol shipped the pesticide to more than 30 different nations, including Egypt that at that time did not have any procedures for pesticides regulation. Later on, it was discovered that the use of Lephtophos had resulted in the death and illness of many rural farmers and in the killing of over a thousand water buffaloes. Despite these incidences, Velsicol continued to market Lephtophos abroad while proclaiming its safety. Those and similar cases led to the perception of developing countries to be misused as ›dumping grounds‹ for developed countries and fierce diplomatic eruptions. It is very probable that developed countries had these developments in minds and the desire to avoid their recurrence and the related political and diplomatic damage when they agreed to start negotiations on a biosafety agreement focussing on trade.

5.2 The functioning of the trade related measures

The functioning of the trade related measures of the Cartagena Protocol can best be understood if seen in the context of the above mentioned key reason for its conclusion; that is developing countries' fear to become ›dumping grounds‹. While this is a political term, the incidents referred to above give evidence of the factual ramifications of this term. Put in abstract legal terms, the problem can be described as follows: from an environmental perspective, trade in (potentially) hazardous substances, including LMOs, can be perceived as a form of transboundary pollution. This kind of pollution however, differs strongly from the situations which have traditionally led to the adoption of global environmental agreements, that is situations in which the source of pollution, for example an air-polluting factory, is situated in one state, while the adverse effects reach beyond this state's borders. In these cases, states are *legally* not able to protect their own territory because the source of pollution, be it industrial factories or products, is situated in the territory of another state. In contrast, the negotiations leading to the conclusion of the Cartagena Protocol concentrated on a different kind of transboundary pollution, which is specific to products, that is pollution caused by the intentional transfer of these products (i.e., trade). To prevent this form of pollution, it is not necessary to impose restrictions to the domestic management of the respective substances, rather it is sufficient to regulate the transboundary pollution as such – this means to regulate the trade in these substances[53]. In contrast to traditional forms of transboundary

[52] For more details of this incident see Bryan D (1980) International Consumer Protection: Export of Hazardous Products from the United States, A.S.I.L.S. International Law Journal 4, p 1.

[53] For the sake of clarity, it should be specified that the trade itself does not cause the pollution. However, it enables the pollution, by bringing possible sources of pollution into the country.

pollution, every state has the legal possibility of preventing pollution caused by trade by enacting import controls. However, the Cartagena Protocol responds to the fact that developing countries are often *factually* unable to protect themselves efficiently from this kind of transboundary pollution due to missing legislative and administrative capacities.

How does the advance informed agreement as the key trade related mechanism strive to address this problem? To answer this question, it seems advisable to reduce the complex procedure to its core elements which are basically three: first, information is provided by the exporting country to the importing country concerning the LMO to be transferred. This instrument aims at healing the factual inability of importing countries to properly exercise its sovereignty with regard to imports of LMOs resulting from a lack of information as a precondition of a free and conscious decision. This lack of information concerns both the administrative and the legislative bodies in many developing countries and is a consequence of the fact that regulatory knowledge in the field of biotechnology is often linked to domestic industrial capacities in certain fields – which is missing in the case of biotechnology in many developing countries. Thus, even if the kind of information required by Annex I varies considerably, they all aim at assisting the importing country to take an informed decision about the import of a specific LMO[54].

The second element of the advance informed agreement procedure consists in obliging importing countries to take a decision, which has to be communicated to the exporting country. In addition, the Protocol lays down certain criteria which have to be taken into account in the decision-making process, namely an adequate protection of biological diversity and human health. Thus, it becomes clear that in contrast to the first element of the AIA, the second one does not aim at strengthening importing countries' capacities, but rather attributes to them a certain procedural and substantive responsibility aiming at the protection of the biological diversity.

The third element, which is typical to the procedures of prior informed consent as contained in the Basel Convention and the Rotterdam Convention on whose model the AIA was designed, is that the exporting country has to prevent exports from taking place without the prior consent of the importing country, within certain limits. This element is related to the enforcement of import decision. This enforcement poses problems especially for developing countries as they provide only limited capacities to effectively oversee and control the transboundary trade which takes place between private entities. An obligation on behalf of exporting (developed) countries to enforce importing countries' decision serves as a de facto transfer of enforcement technology and resources. However, in light of a lack of an unambiguous provision in the Cartagena Protocol, it seems that this (potential) function of a procedure of prior informed consent is not granted by the AIA as it now stands.

[54] While the first category can be said to assist in the risk analysis, category two and three tend to address the stage of risk assessment.

Summing up, it can be stated that the information element of the AIA is aimed at strengthening the decision-making capacity of importing countries and thus its capacity to actually exercise its sovereignty with regard to imports. In an indirect way, this also serves environmental purposes in so far as the state of import's ability to take environmentally protective measures is strengthened. The more efficient enforcement of importing countries' decisions which is a major aspect of the similar procedures of the Basel and Rotterdam Convention is barely secured due to vague provisions. Finally, direct environmental protection is only provided for in form of the substantive criteria to be taken into account by importing states. As the procedure applicable to LMOs destined for food, feed, or processing in terms of Art. 11 is structurally very similar to the AIA, the same is true in this regard, whereby it has to be noted that the procedural protection of the factual sovereignty of the importing state is less accentuated in the context.

5.3 Assessment of the trade related measures from an environmental point of view

If one undertakes to assess the trade related measures, in particular the AIA procedure, from an environmental perspective, it seems necessary to strictly differentiate between two aspects: first, one has to ask whether the focus on trade as such can be considered an adequate approach, second does the AIA constitute an effective instrument to this aim?

With regard to the first aspect, one first has to set up criteria for the adequacy of a global environmental agreement. For the sake of this contribution, the author wants to refer to the traditional approach of international environmental law, that is to regulate transboundary pollution[55]. Resulting from this criteria, the key question seems to be whether transboundary pollution in the field of LMOs derives only from the more recent form of trade as such, in these substances or – at least to a considerable degree – also from the traditional form of transboundary pollution, that is domestic use of LMOs which may result in an unintended transfer of these organisms to neighbouring states. Even in light of the great amount of scientific uncertainty in the field of biotechnology it seems highly probable that if risks to the biological diversity should occur, these will not be limited to the country in which the LMO was released as an unintentional transboundary movement of these organisms is hardly controllable. As a result, one may conclude that the focus on the transboundary pollution caused by trade is insufficient. One may object to this assessment, however, that the Basel Convention which centres on trade in waste appears more than one decade after it came into force as a widely recognised efficient environmental instrument. But there is one difference between the underlying problems of waste and LMOs which has to be taken into account when

[55] It is admitted that there is a more recent tendency to address at the international level environmental pollution whose adverse effects are widely limited to the domestic environment. For the time being, most of these instruments are however, or of a non-legally binding character or are limited to general obligations.

coming to an assessment: in contrast to LMOs, wastes only pose a very small risk of unintentional transboundary movement; and this is exactly the reason which justifies the limitation of the Basel Convention to trade – and which is missing in the area of LMOs.

If one considers the efficacy of the trade measures as such, a differentiated conclusion can be drawn. On one hand, the flow of information especially from developed to developing countries will strengthen the decision-making capacities of developing countries and will thus contribute to a heightened ability of these countries to exercise, in practice, their sovereignty with regard to imports of LMOs. On the other hand, the lack of a clear-cut provision in the context of both the AIA procedure and the procedure as laid down in Art. 11 which would oblige exporting countries to allow exports only when consent has been granted by the competent authorities of the importing country, results in the fact that the Cartagena Protocol does not also strengthen the enforcement capacity of importing countries in the same way as granted by the Basel Convention. Summing up, one can therefore state that the Cartagena Protocol represents some amelioration with regard to the status quo ante, that the concept of the focus on trade as well as the way in which the AIA regulates this trade, however, remains behind what would have been desirable from an environmental point of view.

References

Adler J (2000) More Sorry Than Safe: Assessing the Precautionary Principle and the Proposed International Biosafety Protocol, Texas International Law Journal 35, p 173

Bryan D (1980) International Consumer Protection: Export of Hazardous Products from the United States, A.S.I.L.S. International Law Journal 4, p 1

Eggers B, Mackenzie R (2000) The Cartagena Protocol on Biosafety. In: Journal of International Economic Law, p 525

Goldman KA (2000) Labelling of Genetically Modified Foods: Legal and Scientific Issues, Georgetown International Environmental Law Review 12, p 717

Gupta A (1999) Framing »Biosafety« in an International Context: The Biosafety Protocol Negotiations, New York

Gupta A (2001) Governing Trade in Genetically Modified Organisms: The Cartagena Protocol on Biosafety, Environment 42/4, p 23

Hagen PE, Weiner JB (2000) The Cartagena Protocol on Biosafety: New Rules for International Trade in Living Modified Organisms, Georgetown International Environmental Law Review 12, p 697

IUCN (2003) Explanatory Guide to the Cartagena Protocol on Biosafety, Environmental Policy and Law Paper No. 46, Cambridge

Katz D (2001) The Mismatch Between the Biosafety Protocol and the Precautionary Principle, Georgetown International Environmental Law Review 13, p 940

MacKenzie R, Francesconi S (2000) The Regulation of Genetically Modified Foods in the European Union: An Overview, New York University School of Law Environmental Law Journal 8, p 530

Rajan MG (1997) Global Environmental Politics: India and the North-South Politics of Global Environmental Issues, New Delhi

Royal Commission on Genetic Modification - New Zealand (2001) Report of the Royal Commission on Genetic Modification. New Zealand, Auckland

Rublack S (1989) Fighting Transboundary Waste Streams: Will the Basel Convention Help?, Verfassung und Recht in Übersee 22, p 364

Spalding H, Valette J (1990) The International Trade in Wastes: A Greenpeace Inventory, 5th ed, Washington D.C., p 21

Stoll T (1999) Controlling the Risks of Genetically Modified Organisms: The Cartagena Protocol on Biosafety and the SPS Agreement, Yearbook of International Environmental Law 10, p 82

Teel J (2000) Regulating Genetically Modified Products and Processes: An Overview of Approaches, New York University Environmental Law Journal 8, p 649

Policy-Windows for the Declaration of Protected Areas – A Comparative Case Study of East Germany and Guatemala

Heiko Garrelts[1], Regina Birner[2], and Heidi Wittmer[3]

[1] Institute for Forest Policy, Forest History and Nature Protection, Georg-August University of Göttingen, Germany, email to hgarrel@gwdg.de
[2] Institute of Rural Development, University of Göttingen, email to rbirner@gwdg.de
[3] Centre for Environmental Research (UFZ), Leipzig, email to heidi.wittmer@ufz.de

Summary. The declaration of protected areas is the most important policy instrument in nature conservation worldwide. Using the policy window approach developed by John Kingdon (cp. Kingdon 1984), this contribution deals with the conditions under which it is politically feasible to place comparatively large areas of land under protection. We compare the case of East Germany, as example of an industrialized country, and Guatemala, as an example of a developing country. The policy window approach stipulates an analysis that distinguishes between problem stream, policy steam and politics stream, and places attention on the coupling of these streams by political entrepreneurs and their ability to make use of windows of opportunity. Our comparative study of East Germany and Guatemala shows that in spite of country-specific differences a fundamental change of the political regime, the formation of a conservation movement and the development of policy options under the prior regime made it possible in both cases to place a comparatively large proportion of the country under protection in a short period of time. Further conditions were political entrepreneurship of leaders in the conservation movement as well as international support. Our study shows that this analytical approach provides valuable insights for the analysis of important events in conservation policy as there is a growing need to better understand the political conditions under which protected areas are being implemented.

Key words: Nature protection policy, political transition, policy windows, protected areas, East germany, Guatemala

1 Introduction

The declaration of protected areas remains the major policy instrument for biodiversity conservation worldwide. According to Article 8 of the CBD concerning in-situ-conservation, each contracting party is required to establish a system of protected areas. Yet, the declaration of protected areas is often confronted with social conflict, both in industrial and developing countries. Therefore, there is a need to better understand the political conditions under which this policy instrument has the best chances to be implemented, especially if comparatively large areas are to be protected. This contribution analyses this question with a comparative perspective, considering Germany, as an example of an industrialised country, and Guatemala, as an example of a developing country. In each country, a comparatively large area of land was placed under protection at one particular opportunity. In former East Germany, within the »National Park Programme« an unexpectedly high number of protected areas was established – covering almost 10 percent of the country's surface – in the last very brief period prior to Germany's reunification, between 1989 and 1990. In Guatemala, almost 25 percent of the entire country was put under legal protection within a few years after the State Agency for Protected Areas (CONAP) had been established in 1989.

The goal of this contribution is to better understand this radical change in protected-area policy against the background of the swift transformation of an entire system – both in East Germany and Guatemala. We seek to answer four questions: How did this change occur? What were the special circumstances that made this change possible? Which differences and similarities can be observed in the two cases? And, what conclusions have to be drawn, what lessons can be learnt from these experiences?

As a theoretical foundation for the empirical analysis our contribution draws on the »policy-window« approach developed by Kingdon (1984). To explain policy changes, Kingdon distinguishes three streams – the problem stream, the policy stream, and the politics stream. He argues that the greatest policy changes occur, if these streams are coupled. A situation where this happens is considered a policy window and well-prepared policy entrepreneurs may be able to use such a window of opportunity and invest the necessary resources in order to achieve the adoption of policy concepts. This contribution focuses on a positive policy analysis explaining why these changes were possible. It is not concerned with an evaluation or a normative assessment or of these policy changes.

The contribution proceeds as follows: Sect. 2 explains Kingdon's policy window approach, which serves as an analytical framework. The third section presents the two case studies. Sect. 4 discusses the findings of the case studies with a comparative perspective. The final section draws some conclusions.

2 Theoretical framework: Kingdon's policy window approach

2.1 The garbage-can model as a basis for the policy window approach

Kingdon developed his »policy-window«-approach on the basis of the so-called »garbage-can« model of organisational choice (Cohen, March and Olsen 1972). This model analyses decision processes in »organised anarchies«, which are defined as organisations that lack a clearly defined set of preferences, both on the part of the members and the organisation. Cohen et al. (1972) considers universities as examples of such organised anarchies, in which actors often fail or refuse to define their goals. An organisation thus appears as »a loose collection of ideas rather than that it acts on the basis of preferences« (Kingdon 1984 : 89). Furthermore, such organisations are characterised by an unclear technology. The members of an organised anarchy do not understand the organisations' processes very well. They may know their own goals, and the organisation's as a whole, but »its members have only fragmentary and rudimentary understandings of why they are doing what they are doing and how their jobs fit into a more general picture of the organisation« (p 89). The »left hand doesn't know what the right hand is doing« (p 90). The third property of an organised anarchy is fluid participation: »Participants vary in the amount of time and effort devoted to different domains; involvement varies from one time to another« (Cohen, March and Olsen 1972 : 1). Consequently, the boundaries of the organisation are uncertain and changing. According to Cohen, March and Olsen, a decision can be seen as an outcome of four different streams within an organisation: problems, solutions, participants and choice opportunities. Although not completely independent of each other, each of the streams has a »life« of its own. The structure of decision making resembles »collections of choices looking for problems, issues and feelings looking for decision situations in which they might be aired, solutions looking for issues they might be an answer to, and decision makers looking for work« (p 2). A choice opportunity can be seen as »a garbage can into which various kinds of problems and solutions are dumped by participants as they are generated« (p 2). Within the context of a choice opportunity the various streams are coupled. If the problems, issues and decision makers fit with each other, a decision can be the result, depending on the content of the garbage can and its perception by the actors. If the coupling of streams within a choice opportunity fails, no decision will be made. The problem remains unsolved, and together with the solutions and participants it drifts to the next choice opportunity.

2.2 The Policy-Window Approach

Kingdon (1984) further developed the garbage-can model for the analysis of public policies. He is especially concerned with the question why within public policy, some issues and subjects emerge and are seriously considered while others are neglected. Kingdon defines public policy as »a set of processes, including at least (1) the setting of an agenda, (2) the specification of alternatives from which a choice is to be made, (3) an authoritative choice among those specified alternatives, as in a legislative vote or a presidential decision, and (4) the implementation of the decision« (Kingdon 1984 : 5). Both the garbage-can model and the policy window approach developed on this basis reject the central idea of rational choice approaches, which assume given problems and given preferences. They provide an alternative to the problem cycle approach, which assumes a target-oriented, coordinated process with well-defined stages.

As in the garbage-can model, Kingdon applies the metaphor of different streams as basis of his analysis. Using the Federal Government of the USA as an example, he differentiates between three major process streams, namely a problem stream, a policy stream, and a politics streams. He holds that each stream is largely independent of the other and each develops according to its own dynamics and rules (p 20). In the first stream, various problems capture the attention of people in and around the government. They define conditions as problems by comparing current conditions with their values concerning more ideal states of affairs. Often their own performance is compared with that of other countries (pp 20 and 95ff). The problem stream is also composed of events like crises and disasters or by feedback from the operation of current programs. Systematic indicators play a crucial role. These problems are recognised independently of whether or not there is a solution. Kingdon's focus is on »how and why one set of problems rather than another comes to occupy officials' attention« (p 92). The policy stream refers to instruments, conceptions, ideas, proposals, speeches etc. which float around in a »primeval soup« (pp 21 and 122ff). They are generated by a community of specialists like bureaucrats, politicians, interest groups, researchers or academics. Policies are designed whether or not they are solving a problem. The politics stream contains changes of the national mood, changes of public opinion, election results, a new administration, pressure group campaigns, or shifts in partisan or ideological distributions in Congress (p 93). Politics can also imply power or the lack of opposition. Again, political dynamics move along at their own pace. For instance, the politics stream may change whether or not political actors are ready and whether or not the problems a country is facing have changed.

The greatest policy changes occur when the three streams (problems, policies and politics) are joined through a choice opportunity or »coupled into a package« (p 21). This is what Kingdon calls a policy window – an »opportunity for advocates to push their pet solutions, or to push attention to their problems.« (pp 173ff) Policy windows – sometimes predictable, sometimes not – are opened by events in either the problems or the political streams (p 213). A new problem or even a catastrophe may create an opportunity to attach a solution to it. Events leading to policy windows that are relevant for the present study include dramatic shifts in the socio-political-economic context such as changes of a national mood, changes of the administration, the turnover of officials, and vigorous lobbying (p 176).

The linkage of the streams itself often relies on skilful policy entrepreneurs »for coupling solutions to problems and coupling both problems and solutions to politics« (p 21). Policy entrepreneurs can be scientists, politicians or citizens willing to invest resources in return for influencing policies (p 214). They have their proposals or their concerns about problems ready, and they push them at the propitious moments. They mobilise the public opinion and help set the decision-making agenda. They are also able to mobilise political support by other actors by building so-called advocacy-coalitions (Sabatier 1988) or epistemic communities (Haas 1992). Thus, the »appearance of the right entrepreneur at the right time« (Kingdon 1984 : 214) can be seen as a crucial factor. Of course dramatic policy changes rely on »considerable doses of messiness, accident, fortuitous coupling and dump luck« (p 216). It remains to underline the fact that policy windows are small and scarce: »Opportunities come, but they also pass. Windows do not stay open long« (p 213).

3 Case Studies

In the following section, we present the two case studies. We use the policy-window approach as a heuristic device to describe and analyze the main factors involved. Following Kingdon, we distinguish problems, policies and politics as explaining variables.

3.1 Case study East Germany

Between the fall of the Berlin Wall on November 9th 1989, and the unification of the two German states in October 1990, nature-protection actors succeeded in declaring six biosphere reserves, three nature parks and five national parks as protected areas. These areas were part of the German Democratic Republic's (GDR) National Park Programme and were adopted and fixed in the Unification Contract (*Einigungsvertrag*) that provided the legal basis for the reunification of Germany. Together with additional areas, which were temporarily protected, 9.6 % of GDR territory was declared protected areas.

This case study is based on key informant interviews and a review of documents.

Problems

In 1989 the GDR had – despite of the general perception of a catastrophic environmental situation focusing on industrial pollution – large and nearly untouched natural areas and vast intact cultural landscapes. In these areas nature could, uniquely in Middle-Europe, be conserved or regenerated (cp. Knapp 1991 : 5; Knapp 2001 : 35; Reichhoff and Böhnert 1991 : 195; Succow 1991 : 11). But this fact was by no means the result of a successful nature protection policy in the GDR. These results can rather be attributed to »exaggerated security needs and green privileges of the former leadership« (Succow 1991 : 11 [translation by the authors]). In the seclusion of the locked border areas along the coast and along the Iron Curtain, in the vast security zones of the military training areas, and in the state-used hunting areas altogether about 15 % of the total land area was excluded from an intensive use (Succow 1991 : 11). With the unification of the two German states, nature protection actors saw the problem of an acute threat by commercialisation and settlement leading to fragmentation through economic and tourist exploitation (Knapp 1991 : 5). It was necessary to preserve »the last rests of characteristic large landscapes in the GDR from further fragmentation and selling off« (p 5). Besides an »ecologisation of land-use« the idea of a »landscape free from utilization« came up.

Policies

Although after the GDR 1970 complied with international efforts of nature protection (for instance by joining the Washington Convention on International Trade of

Endangered Species (CITES), the Biosphere Programme and the Ramsar Convention) no national park was founded between 1949 and 1989. This was due to ideological reasons – the concept of the national park was clearly of American origin (Schurig 1991 : 370). Although national parks existed in other socialist countries like Hungary, Czechoslovakia, Poland and even in the Soviet Union, the GDR attached importance to an ideological delimitation against the Federal Republic of Germany (FRG), and the USA standing behind the FRG. This »aimed not only at the US-imperialism as the economic core value, but also at the less central American cultural values« (p 370) under which national parks had to be subsumed, as well. Accordingly, demanding the creation of national parks was regarded as a state critique. Another reason to reject national parks was the specific idea of socialist land-culture. In this view, it was necessary to optimise and intensify land use (p 368). Nevertheless, many discussions and initiatives to establish national parks took place (cp. Gilsenbach 1998; Hemke 1998; Knapp 2001; Rösler 1998a), last but not least due to the examples in the neighbouring socialist states. In 1989 / 1990, nature protection actors could refer to these initiatives, in the sense of existing ideas and proposals. Thus, proposals for four of the five national parks and four of the six Biosphere Reserves declared in 2000 had already been presented in 1976 (Rösler 1998a : 553).

It came to be an advantage that collecting scientific data on nature was regarded as »harmless« and therefore allowed by the socialist regime, while nature-protection activities that aimed at creating public awareness or criticising the state were not permitted. Many natural scientists were organised in zoological and botanical groups and conducted field work. Thus, large-scale and costly inventories were not necessary since basic ecological data concerning the selection of the territories and their boundaries and zonation were already available (Rösler 1998b : 587–588).

Finally, a legal foundation for the declaration of the protected areas was necessary, one which would keep the legal status after the unification[4]. One decisive aspect of the policy dimension can be seen in the so-called »Environment Union« (*Umweltunion*) which was implemented – together with the »Currency Union« (*Währungsunion*) on July 1st 1990. With this event, the West-German Nature Protection Law (cp. Hellenbroich this vol) directly entered into force in East Germany, and the national-park category became available.

Politics

Change of institutions

Members of the Socialist Party were dominant in all ministries of the GDR-Government and among employees. They had supported the system over years.

[4] Until that moment a temporary declaration was based on §27 of the GDR-Landeskulturgesetz and the First Executive Ordinance (*Erste Durchführungsverordnung*) of May 18th, 1989 that arranged the declaration of nature protection areas of central importance (cp. Bauerschmitt and Dräger 1990 : 80f).

Between January and April 1990 the heads of almost all ministries changed frequently. The Ministry for Environmental Affairs alone was directed by three different Ministers in only five months. With the background of a dramatic development in politics, personal uncertainties marked the scenery. The major question always was who would be superseded next, so that nobody was encouraged to take the initiative. Other actors lacked experience in government positions and in dealing with authorities (Rösler 1998b : 585–586). Against this background, and in view of a complete change of state authorities concerning environmental and nature protection affairs, it was not too difficult to detach the nature protection administration from the forestry administration, and to establish a separate ministry which was responsible for nature protection and regional planning (*Raumordnung*). This institutional change was achieved by political entrepreneurs like Prof. Michael Succow, who was later awarded the *Alternative Nobel Peace Price*[5]. He was appointed as the Second Minister for Environmental Affairs. Under his domain, he only employed people who had not worked for the Ministry before. This new beginning with a completely new staff had the »big advantage that there was no need to convince former staff nor to take into account former structures or habits.« (p 586).

Institutionalisation of nature protection and nationalization of networks

Whereas in all administration sectors jobs were reduced, the nature-protection administration was extended, following an order of the government (Knapp 2001 : 42). Until the Berlin Wall came down, only one person was responsible for nature protection on the regional level, and that person was also in charge of hunting affairs. Then, as indicated above, nature protection received an administrative structure of its own. Until April 1990, about 1,000 jobs were created within this structure. These positions were filled – like in the ministry – with actors from voluntary nature protection groups (p 42). Both the new staff working for the National Park Programme and the staff working at the regional level knew each other personally. Thus time-consuming coordination was not necessary. The usually existing differentiation between private and state actors concerning environmental affairs did not exist in this decisive period. This facilitated unbureaucratic action, without the frictions and time losses usually caused by coordination, forms, rules etc. (Rösler 1998b : 589).

Environmental protection as a central issue

Under the GDR regime, environmental issues had been placed under a taboo. This was one reason why pressure by environmentally concerned groups became popular, especially in view of the dramatised ecological problems that existed particularly in the industrial sector.

Activities by these groups were considered to be very important. Another important political aspect was the fact that the political change in East Germany was

[5] To name further persons who also acted as key actors in this period: Mathias Freude, Leberecht Jeschke, Hans-Dieter Knapp (in detail see Rösler 1998b).

forced – next to pacifist and church groups – by environmentalists. In the period after the wall came down »a bad conscience was the dominating feeling in the national mood towards the environment«[6]. The *Round Table* in East-Berlin and similar groups at the regional level supported the National Park Programme. The government was asked to make the necessary means available (Rösler 1998b : 566). Nature conservation and environmental protection issues had a big weight in the public opinion. Thus, the last GDR-government, consisting of members of the Socialist Party and former Block Parties, agreed on March 16th 1990 to the declaration of six biosphere reserves, five national parks and twelve nature parks and their temporary protection as »protected areas of central importance«. This was two days before the first free elections took place in the GDR (March 18th 1990). Later on, these protected areas were adopted in the »Environmental Law Providing Guidelines« (*Umweltrahmengesetz*) between GDR and FRG (July 1st 1990).

No formal participation and lengthy proceedings of declaring protected areas

Until October 3rd 1990, no federal states (*Länder*) existed in the GDR, and the centralistic role of the ministries in Berlin dominated policy making. This applies also to the National Park Programme that was coordinated in East-Berlin. This condition facilitated the activities of the local and regional environmental initiatives, which were able to take advantage of the still existing power structures. Thus, the declaration of the protected areas was characterized both by bottom-up and top-down characteristics at the same time. Regional activities coincided with existing and *used* liberties of action at the central level. The Minister of Environmental Affairs had free hand to declare the protected areas. Lengthy proceedings for the declaration with the participation of public groups, which nowadays can last several years, were not necessary in the main phase[7].

Lack of opposition and resistance

At the governmental level, the above mentioned insecurity due to frequent change of staff had the consequence that no organised resistance against the National Park Programme came up. This is the case especially for the forest administration, which was organised in a hierarchically, ideologically state-oriented and authoritarian way[8]. As the National Park Programme included mainly marshy, mountain and coastal areas, farming was not directly concerned. With regard to the restrictions of Nature Parks and Biosphere Reserves, farming actors wanted to wait for improved liberties of action. Above all, enterprise-oriented lobby-groups against nature protection were not yet organised. In that period before the re-privatization of almost the whole GDR-territory, private investors had no access to public-owned areas.

[6] Expert Interview October 2003.

[7] It has to be mentioned that after the »Environment Union« and the »Environmental Law Guidelines« such a formal participation was obligatory and took place, especially concerned actors in ministries, regions and on the local level (then-mayors) participated (for details cp. Müller-Helmbrecht 1998 : 599).

[8] Expert Interview September 2003.

Support by Western actors

GDR nature protection actors soon contacted Western environmental groups (for instance WWF, DBV/NABU, SEN) and persons responsible for the National Park Bavaria (*Bayrischer Wald*) in order to obtain support. Since February 1990, meetings of the two German nature protection administrations took place. The West-German Ministry of Environmental, Nature Protection and Nuclear Safety (BMU) offered its support, as well. Early in that period, it became clear that success depends less on scientific but rather on juridical issues. The creation of legally binding and valid orders was the main challenge in order to provide the entire programme with instruments that referred both to the still valid GDR-law and to the West-German laws expected to become valid in the future, as well as to the prepared Unification contract (Rösler 1998b : 574).

Surprising actor constellations

The idea of a non-utilized landscape was suddenly also supported by actors which had not exactly belonged to the nature protectionists before. Very surprisingly, members in the Berlin Ministry for Environmental Affairs were contacted by high-ranking militaries who proposed a conversion of 15–20 military training areas to National Parks[9].

3.2 Case Study Guatemala

When the first civil administration in Guatemala was established in 1986 after 16 years of military government, around 2.5 % of the country's land area was protected. In 1989, this administration passed a law, which established a public administration for protected areas (CONAP), confirmed the legal status of de facto existing protected areas and mandated to establish new ones. This law is the basis for 42 different protected areas. In the following year, the two largest protected areas, Sierra de las Minas and Biosfera Maya, were established by specific laws, accounting together for 17 % of the land area. After further declarations in 1990, approximately 25 % of the country's land area is under legal protection within 17 different categories ranging from strict conservation to multiple-use areas. In addition to this spectacular increase in land area under protection, the management for the Sierra de las Minas Biosphere Reserve, which harbours 70 % of the reptile, bird and mammal species reported for Guatemala and Belize, many of which are endemic, was delegated by law to a national NGO, Defensores de la Naturaleza. This was the first incidence of formally delegating the full management authority to an NGO in Latin America (Secaira et al. 2000).

Similar to the case study of East Germany, this case study is based on a series of interviews held between 2001 and 2003 with representatives of different governmental and non-governmental organisations in the field of nature conservation policy, and on a review of secondary literature.

[9] Expert Interview September 2003.

Problems

By the middle of the 1980s, the issue of tropical deforestation had been recognised as a major problem of nature conservation at the international level (Nations et al. 1989). In Guatemala, 68 % of the territory was still under forest cover in 1960. By the end of 1981, the forest cover had been reduced to 39 %. In the same time period, the population more than doubled from 4.0 to 8.6 million. (Leonard 1987, quoted in Secaira et al. 2000 : 5). Since Guatemala's biological diversity is one of the highest in Latin America and the country ranks among the 25 most plant-rich countries of the world, the rapid loss of forest cover was recognised as a serious problem for nature conservation.

As mentioned above, only about 2.5 % of Guatemala's territory had been placed under protection when the military rule ended in 1986. These areas had been created and managed for different purposes and by different institutions, notably the forest administration, the administration for archaeological sites, and the state University of San Carlos. Even though the civil war was terminated officially only with the Peace Accords in 1996, the new constitution of 1985 and the change to the first civilian government in 1986 constituted the hope for a new beginning. This fact was expected to increase the migration into the sparsely populated areas of the Petén, which was still characterised by large not fragmented forest cover. At the same time, the prospect of returning refugees in search of land constituted a further problem with regard to forest conservation. The migration into forest areas was further aggravated by the highly unequal land distribution in Guatemala, even though conservationists appeared to perceive population increase, rather than unequal land distribution, as the major cause of the migration and deforestation problem[10].

Policies

In the middle of the 1980s, the declaration of protected areas as the major policy instrument for nature conservation was well established both on the international agenda and in Guatemala. As in other countries, conservation ideas, promoted by writings like those of Rachel Carson, had been embraced by members of the educated elite in Guatemala since the 1960s. Already the UN Conference on the Human Environment in Stockholm in 1972 had been attended by Guatemalan environmentalists, who were later to formulate the relevant articles in the 1985 constitution[11]. Their background and motivation was not so much »anti-system« but philanthropic. At the Second World Parks Congress in Bali in 1980, IUCN declared the goal of placing 10 % of the earth's surface under protection, which created considerable incentives for conservationists around the world to pursue this goal in their countries. In the 1980s, conservationists in Guatemala, who had studied abroad and had international connections, clearly considered the establishment of protected areas for large, threatened and biologically valuable areas in their country

---

[10] Interview with representatives of conservation NGOs, Guatemala City, May 2002.

[11] Expert Interviews, Guatemala City, May 2002.

as a necessary policy instrument for nature conservation[12]. Supported by international funding, the conservationists, who were organised in non-governmental organisations (NGOs), and members of the Universities carried out biological surveys to identify suitable areas for protection. New conservation NGOs were formed during this period. The policy goal of forest protection was also promoted with the development of a National Forest Action Plan, which was supported in Guatemala at the end of the 1980s by a respective programme of the UN Food and Agriculture Organisation FAO[13].

In the Guatemalan context, it was, however, evident that the capacity of the state to effectively manage protected areas was seriously limited. Having suffered from civil war for decades (1960–1996), the per capita gross domestic product in Guatemala in the middle of the 1980s was significantly below the average of other Latin American countries. Between 1981 and 1995 an average of 53,3 % of the population was living below the poverty line, and the war-torn state budget did not seem to allow for the funding of nature conservation (World Bank 1997 : 214). However, conservation NGOs, supported by private contributions and considerable international funds, had been formed. In this situation, the policy option of legally delegating the management of protected areas to NGOs emerged as an attractive instrument, especially as these NGOs appeared to possess the know-how, connections and capacities the state was lacking.

Politics

Access of conservationists to the political system

In 1985, Vinicio Cerezo, candidate of the Christian Democrat Party won the elections that established, as mentioned above, the first civil government after 16 years of military rule. The Christian Democratic Party won a 51 % majority in the Parliament, and Alfonso Cabrera became President of the Parliament. This constellation created a unique opportunity for conservationists to get access to political decision-making. President Vinicio Cerezo was the father of Marco Cerezo, a leading conservationist, who was the director of National Forest Action Plan. Alfonso Cabrera, the President of the Parliament, was the brother of Jorge Cabrera, who also was a prominent figure in the conservation movement. These personal relations obviously provided excellent opportunities for conservationists to lobby for laws on nature conservation.

Creation of new institutions

The Constitution of 1985 mandated the creation of a legal and institutional framework for the environment, which already reflects the ability of the environmental movement to influence politics in the first half of the 1980s. In 1989, the Environmental Protection and Improvement Law, which established the National Environmental Commission (CONAMA) was passed. In the same year, the Protected Areas

[12] Interview with representatives of conservation NGOs, Guatemala City, May 2002.
[13] Interviews with experts on forestry, Guatemala City, May 2002 and August 2003.

Law was enacted, which established CONAP as the implementing agency for protected areas. Jorge Cabrera (see above) became the head of CONAMA and Andreas Lehnhoff, another leading figure in the conservation movement, was appointed as head of CONAP. Due to the support by leading politicians, the newly appointed administrators had considerable room for implementing the conservation policies.

CONAP was established as an agency that directly depends on the Presidency of the Republic, which indicates the comparatively high political priority of conservation that its proponents were able to create. Its governing council includes representatives of the national government and decentralized institutions, the national University, and those non-governmental organisations involved in the management of larger public protected areas (Ley de Areás Protegidas decreto no. 4 1989 Art. 59). This composition indicates that the conservation groups were able to create political space for themselves within this new state institution.

Declaration of protected areas and delegation of management authority

As mentioned above, the Protected Areas Law, that established CONAP, confirmed and created a total of 42 protected areas, accounting for almost 25 % of Guatemalan territory. The first areas established by separate laws were the Sierra de las Minas Biosphere Reserve and the Maya Biosphere Reserve, which together account for 17 % of the national territory (Secaira et al. 2000 : 6). Thus, during the Cerezo administration from 1986 to 1991, about 21 % of the territory was placed under protection (Castro and de Leon 2003). Under the subsequent two administrations, only minimal conservation policy measures were adopted in spite of a pro-environmental rhetoric (Lehnhoff and Nunez 1998, quoted in Secaira et al. 2000 : 6). The law establishing the Sierra de Las Minas Biosphere Reserve delegated the management to the NGO Defensores de la Naturaleza. The fact that the leadership of Defensores accepted the challenge of this task in view of considerable uncertainty concerning their prospects of success can be seen as a bold entrepreneurial decision. As indicated above, this was the first delegation of protected area management in Latin America. Delegation of management authority to NGOs has, since then, become a major strategy in protected area management in Guatemala. Today, 21 % of the total protected area is under some type of co-management or fully delegated management; either with NGOs, the University of San Carlos or with municipalities (UNEP 1996 : 35; Secaira et al. 2000 : 10).

Weak opposition

Since nature conservation was introduced as a relatively new theme on the political agenda, politicians and lobbyists favouring conservation were able to reach a »gentlemen's agreement« with opposition leaders not to use this topic for party politics. This agreement was probably influenced by the fact that conservation groups belonged to the »establishment« and were explicitly not opposed to the political system. Against the background of Guatemala's history of civil war, this clearly distinguished them from those political groups concerned with social issues such as the land question who had been engaged in the civil war. Against this background, there was no major political opposition at the national level for passing the laws mentioned above. Compared with others the topic was not controversial[14].

The establishment of large protected areas, however, was confronted with opposition from landowners and commercial forest users in the areas to be placed under protection. In the case of the Sierra de las Minas Biosphere Reserve, the NGO Defensores de la Naturaleza together with CONAP successfully used a negotiation strategy to win the support of the opponents, including the municipalities and the Chamber of Industry's Forestry Guild. The military also opposed the conservation efforts, but generally lost influence under the new civil government. A few weeks after the declaration of the Sierra de las Minas Biosphere Reserve, landowners, who could not be convinced by the negotiation approach, requested the Constitutional Court to repeal this law. However, the Court upheld the law, thus setting an important precedent for nature conservation in Guatemala (Secaira et al. 2000 : 7).

The declaration of protected areas also affected local and indigenous communities, who had customarily used parts of the land and forest areas to be placed under protection. However, against the background of the civil war, these groups had not been able to organise themselves politically in a way that would have allowed them to oppose the new conservation laws or to have their interests reflected in these laws.

International funding

International influence was not only important for the policy stream, as indicated above, but also for the politics stream. International funds were an important factor in allowing conservation organisations to become political actors. Due to funds from international donors, CONAP and conservation NGOs were able to work in the remote areas of the Petén, which was until then under military administration. Other (state) organisations lacked the funding to work in such regions. Therefore, the NGOs gained a unique political position in proposing protected areas in such regions and in becoming entrusted with their management. International funding was also essential in sponsoring the newly created CONAP (see below). The US-based Nature Conservancy (TNC) financed the first year of operation of CONAP. Likewise, international funds were essential for the operation of the newly established protected areas. In the case of Sierra de Las Minas Biosphere, which was, as outlined above, under delegated management, CONAP's contribution accounted for only 2 % of its budget (Secaira et al. 2000 : 8).

3.3 Discussion

A comparison of the two cases reveals interesting similarities as well as differences. In both cases, the fact that a far-reaching regime change took place was essential for the appearance of a policy window that made it possible to declare a comparatively high percentage of land as protected areas. In the conceptual framework adopted here, the regime change can be attributed to the politics stream. It had, however, important implications for the problem stream and the policy stream, as well.

[14] Expert Interview Guatemala City, May 2002.

Problem Stream

Both in East Germany and Guatemala, the regime change led to increasing concerns that however politically welcome the change was, it might cause threats with regard to nature conservation. In both countries, there were considerable areas of land that had been excluded from agricultural or other uses. In neither East Germany nor Guatemala this had been the effect of a planned conservation strategy. Rather it was a by-product of the existing political conditions. While in East Germany, security concerns and hunting privileges of the elite had the effect that large areas remained under natural conditions, in Guatemala, it was the civil war conditions that prevented large forest areas from being settled by migrants. In both cases, the regime change led to the concern that these unused areas might become subject to increased settlement, fragmentation, and conversion to agricultural land. This problem perception coincided in both cases with the perception that environmental issues had been neglected by the prior regime. Due to industrial pollution problems, this perception was particularly pronounced in the case of East Germany.

Policy Stream

The policy options for nature conservation observed in the policy stream were similar in both cases. Due to the wide-reaching international influence on conservation policy, the policy approach to declare protected areas as a major instrument of nature conservation dominated the policy stream in both countries. In the case of East Germany, a decidedly anti-Western political ideology had, however, prevented the establishment of National Parks, which were disapproved of on the basis of their US-American origin. In Guatemala, it was the situation of the civil war that had lasted for almost 30 years, which had left little political space for implementing conservation policies. In particular, protected area categories that involve a wider range of stakeholders such as biosphere reserves, had not been implemented. In both cases, it was the regime change that opened opportunities for the realisation of options that had already prevailed in the policy stream.

In spite of these similarities concerning the policy stream, there was a remarkable difference. In the Guatemalan case, the fact that the implementation capacity of the state did not match well with the policy option of state-managed protected areas gave rise to an important policy innovation – the delegation of protected area management to organisations of the civil society. Such a comparable mismatch between existing policy options and implementation possibilities did not exist in the case of East Germany, and no comparable policy innovation emerged in its policy stream.

Politics Stream

Both in East Germany and in Guatemala, the regime change created astonishing opportunities for conservationists to become political actors. In both cases, an important aspect of the politics stream was the fact that conservation groups had been able to organise themselves. One difference was that in East Germany, the movement had a broad basis and formed a dense social network while conservation groups in Guatemala were comparatively small groups dominated by the elite. In the German Democratic Republic, nature conservation activities practiced by these groups, such as biological surveys, were not considered to be a threat to the then prevailing regime. However, in this case, nature conservation also provided an umbrella for members that were politically opposed to the prevailing regime.

In both cases, one could observe considerable political entrepreneurship, which was essential for coupling the three streams and making use of the policy windows. Both in East Germany and in Guatemala, political entrepreneurs in the conservation movement managed to create new political and administrative institutions for conservation and to assume the key positions in these newly created institutions. This included the position of the Minister for Environment in East Germany, and the positions of the director of the state agency in charge of protected areas (CONAP), which directly belonged to the Presidency in Guatemala, and the state agency in charge of the environment (CONAMA). Thus, conservation leaders became part of the government and, therefore, had excellent opportunities to successfully implement the policies that had been developed in the policy stream. In the Guatemalan case, access of conservation groups to the political system was further promoted by the family relations between conservation leaders and the heads of both the executive and the legislative. In both cases, the political entrepreneurs acted as described in Kingdon's explanatory approach (cp. Sect. 2). They had their proposals or their concerns about problems ready, and they pushed them at the propitious moments. The »appearance of the right entrepreneur at the right time« (Kingdon 1984 : 214) took place in both cases.

An important difference between both cases can be seen in the role of the opposition. In East Germany, the conservationists managed to have the areas protected before any major opposition could emerge. An enterprise-oriented lobby had not yet been formed and the re-privatisation process had not yet started. In Guatemala, by contrast, the private sector timber lobby, municipalities and private landowners opposed declaration of protected areas. However, as the topic was comparatively new on the political agenda and a »gentlemen's agreement« not to use the topic for party politics had been reached, they lacked political support. Opposition groups then either agreed to the declaration of protected areas after negotiations, as in the case of the timber industry, or they challenged the decisions in court, but lost, as in the case of the landowners. The role of the military also differed between East Germany and Guatemala. In East Germany, high-ranking militaries – though not belonging to the conservation movement – proposed the conversion of military training areas to National Parks, probably to prevent other uses that they might have found objectionable for political reasons. In Guatemala, the military did not support the declaration of protected areas, but – under the new civil rule – they did not prevent the declaration of protected areas.

The role of international support for these political entrepreneurs constitutes another difference of the two cases. In the Guatemalan case, the support of international conservation organisations, especially the funding provided by them, was an essential precondition for the conservation group to become political actors and to make use of the policy window. In the East German case, it was support by environmental organisations and by employees of the Ministry for the Environment (BMU) in Western Germany especially in terms of juridical advice – that was crucial for putting the National Park Programme in concrete forms and for sustaining it in the frame of the unification. However, this concerns a *later* phase. Before, the activities of the GDR-protectionists have to be considered as one of the very few examples for »endogenous action and actor potentials« (Reissig 1996 : 248 [translation by the authors]). Although incorporated by Western Germany, in terms of nature protection the former GDR did not become a mere copy of the FRG. Own alternatives were realized.

A difference within the politics stream of the two cases can also be seen in the time frame of the policy window. In East Germany, the window was created by a specific constellation that lasted only for a few months. With the creation of the Environment Union on July 1st, 1990, the categories of protected areas of Western Germany, including the National Park Category, became available in East Germany. In the Guatemalan case, the policy window was not so narrow since it was mainly constrained by the increasing opposition and finally, the change of government in 1991.

4 Conclusions

The study has shown that the policy window approach provides a useful framework for a comparative analysis of similar events of policy adoption in nature conservation policy, in this case the protection of comparatively large areas at one point in time. The analysis showed that in spite of country-specific differences, a fundamental change of the politcal regime, the formation of a conservation movement and the development of policy options under the prior regime made it possible to place a comparatively large proportion of the country under protection in a short period of time. Further conditions were a remarkable political entrepreneurship of leaders in the conservation movement as well as international support. The policy window approach stipulates an analysis that distinguishes between problem stream, policy stream and politics stream, and places attention on the coupling of these streams by political entrepreneurs and their ability to make use of windows of opportunity. Our study shows that this analytical approach provides valuable insights for the analysis of important events in conservation policy. The study also shows that the policy window framework is useful for comparing similar events of policy adoption in countries that are otherwise rather different in terms of economic development, culture and political system.

Even though the protection of large areas is considered as a remarkable success by conservation organisations in both countries, one has to acknowledge that in Guatemala, the people at the local level affected by the declaration of these areas were largely not included in the political decision making. This raises normative questions concerning political participation and legitimacy. Moreover, not including affected groups in political decision-making may have negative repercussions on the possibilities of reaching conservation objectives in the long run. While participation in decision making prolongs political processes, thus making it more difficult to use narrow policy windows, participation is widely acknowledged as a means to increase legitimacy and, consequently, reduce conservation-related conflicts and lower the transaction costs of managing protected areas (Hanna 1995; Borrini-Feyerabend et al. 2000). These considerations are especially relevant for areas where conservation competes with agricultural land use.

The two cases analysed in this study can also be placed in a wider context. The protection of comparatively large areas in a comparatively short period of time is not confined to these two cases. Such »spectacular« events have also been observed in other cases. In 1972, motivated by WWF's »Operation Tiger« campaign, India established 23 Tiger Reserves (WWF 1999). At the World Parks Congress in Bali in 1982, Indonesia declared the establishment of 40 National Parks, covering 15 million hectares (Mappatoba 2004). At the World Parks Congress in Durban 2003, President Ravalo Manana of Madagascar announced that his government would increase the amount of protected areas in his country from 4.3 million acres to 15 million in the next five years (U.N. Wire 2003). At the same event, the leader of the Brazilian state of Amapá, declared plans for the creation of a 10-million hectare Biodiversity Corridor that covers 71 % of the state, blankets the world's largest tropical rainforest park, and safeguards hundreds of unique plant and animal species (CI 2003). Based on the insights of this study, it might be useful to further analyse (1) which specific features of conservation policy make such events possible, and (2) to which extent such events are conducive both for reaching conservation goals and, more generally, for finding a balance between the economic, social and ecological objectives of sustainable development.

References

Bauerschmitt J, Dräger E (1990) Gesetzliche Grundlagen. In: Rösler et al. (eds), pp 77–87

Borrini-Feyerabend G, Farvar MT, Nguinguiri JC, Ndangang V (2000) Co-management of Natural Resources: Organising, Negotiating and Learning-by-Doing. GTZ, Frankfurt and IUCN, Gland

Castro F, de León F (2003) Informe Nacional de Areas Protegidas de Guatemala. CONAP, Guatemala City

CI (2003) Brazil – New Corridor Links 12 Protected Areas in Amazon. Conservation International (CI) Press release, Sept. 16, 2003. www.conservation.org/xp/news/press_releases/2003/091603_amapa_eng.xml

Cohen M, March J, Olsen J (1972) A Garbage Can Model of Organizational Choice. Administrative Science Quarterly 17: 1–25

CONAP (Consejo Nacional de Areas Protegidas) (2002) Informe Institutional 2002. CONAP, Guatemala

Gilsenbach R (1998) Die größte DDR der Welt – ein Staat ohne Nationalparke. Des Merkens Würdiges aus meiner grünen Donquichotterie. In: Institut für Umweltgeschichte und Regionalentwicklung (ed), pp 533–546

Haas PM (1992) Introduction: Epistemic Communities and International Policy Coordination. International Organization 46, 1: 1–37

Hanna S (1995) Efficiencies of User Participation in Natural Resources Management. In: Hanna S, Munasinghe M (eds) Property Rights and the Environment: Social and Ecological Issues. The Beijer Institute of Ecological Economics and the World Bank, Washington, pp 59–67

Hemke E (1998) Der Müritz-Seen-Park – ein Schritt in Richtung Nationalpark. In: Institut für Umweltgeschichte und Regionalentwicklung (ed), pp 611–616

Institut für Umweltgeschichte und Regionalentwicklung (ed) (1998) Naturschutz in den neuen Bundesländern – ein Rückblick. Halbband II, Verlag des Bundes demokratischer Wissenschaftlerinnen und Wissenschaftler, Marburg

Kingdon J (1984) Agendas, Alternatives, and Public Policies. Little, Brown and Co., Boston and others

Knapp HD (1990) Nationalparke in der DDR. Bausteine für ein gemeinsames europäisches Haus. Nationalpark 67, 2: 4–9

Kollmorgen R, Reißig R, Weiß J (eds) (1996) Sozialer Wandel und Akteure in Ostdeutschland. Empirische Befunde und theoretische Aufsätze. Leske und Budrich, Opladen

Ley de Areás Protegidas decreto no. 4 (1989) In: CALAS. Programa de Información Estratégica 2003. Legislación Ambiental Guatemalteca I. CALAS, Guatemala: 43–66

Mappatoba M (2004) Co-management of Protected Areas: The Case of Community Agreements on Conservation in the Lore Lindu National Park, Central Sulawesi, Indonesia, Doctoral Dissertation, University of Göttingen

Müller-Helmbrecht A (1998) Endspurt – das Nationalparkprogramm im Wettlauf mit der Zeit. In: Institut für Umweltgeschichte und Regionalentwicklung (ed), pp 597–608

Nations J, Houseal B, et al. (1988) Biodiversity in Guatemala: Biological Diversity and Tropical Forests Assessment. World Directory of Country Environmental Studies. World Resources Institute / United States Agency for International Development. `www.wri.org/data/dces.html`

Reichhoff L, Böhnert W (1991) Das Nationalparkprogramm der ehemaligen DDR. Natur und Landschaft 66, 4: 195–202

Reissig R (1996) Perspektivenwechsel in der Transformationsforschung. In: Kollmorgen et al. (eds), pp 245–262

Rösler M (1998a) Nationalparkinitiativen in der DDR bis zur Wende 1989. In: Institut für Umweltgeschichte und Regionalentwicklung (ed), pp 547–560

Rösler M (1998b) Das Nationalparkprogramm der DDR. In: Institut für Umweltgeschichte und Regionalentwicklung (ed), pp 561–596

Rösler M, Schwab E, Lambrecht M (eds) (1990) Naturschutz in der DDR. Economica Verlag, Bonn

Sabatier P (1988) An Advocacy Coalition Framework of Policy Change and the Role of Policy Oriented Learning Therein. Policy Sciences, 21: 129–68

Secaira E, Lehnhoff A, Dix A, Rojas O (2000) Delegating protected area management to an NGO: The case of Guatemala's Sierra de las Minas biosphere reserve. Washington, D.C., Biodiversity Support Program

Schurig V (1991) Politischer Naturschutz: Warum wurde in der DDR (1949–1989) kein Nationalpark gegründet? Natur und Landschaft 66, 7/8: 363–371

Succow M (1991) Einmalige Chance. Die Wende hat dem Naturschutz unerwartete Möglichkeiten eröffnet. Politische Ökologie 24: 11–14

Succow M, Jeschke L, Knapp HD (2001) Die Krise als Chance – Naturschutz in neuer Dimension. Findling Verlag, Neuenhagen

UNEP-CEP (Caribbean Environment Programme) (1996) Technical Report No. 36: Status of Protected Area Systems in the Wider Caribbean Region: Guatemala. UNEP-CEP Regional Coordinating Unit, Kingston, Jamaica, www.cep.unep.org/pubs/techreports/tr36en/countries/guat.html, last update 8/1999

U.N. Wire (2003) World Parks Congress closes with Conservation Accord. United Nations Foundation National Journal, 17.9.2003, published online at: http://www.unwire.org/UNWire/20030917/449_8513.asp

World Bank (1997) The State in a Changing World. World Development Report 1997, New York

WWF (1999) http//:www.panda.org/resources/publications/species/tiger99

Expert Interviews

Dr. Hans Dieter Knapp, Bundesamt für Naturschutz

Prof. Dr. Mathias Freude, Landesumweltamt Brandenburg

Names of experts in Guatemala are not given since some of them requested anonymity.

[illegible] P. (1988) An Advocacy Coalition Framework of Policy Change and the Role of Policy-oriented Learning Therein. Policy Sciences, 21: 129-168

[illegible] A, [illegible] A, [illegible] O (2000) Delegating protected area management [illegible] de la [illegible] de los [illegible]

[illegible]

Will companies engage in the conservation of biodiversity? A prototypical model of aggregated pro-biodiverse actions of industrial companies

Ralph Buse

Institute for Production and Investment, Department of Corporate Planning, University of Göttingen, Platz der Göttinger Sieben 3, 37073 Göttingen, Germany, email to rbuse1@gwdg.de

Summary. It is an implicit goal of the Convention on Biological Diversity (CBD) and the hope of some non-governmental organisations (NGOs) that the conservation of biodiversity will become a major issue on the agenda of industrial companies. I will show that this development is not something one could only hope for but that these dynamics follow a mechanism, which in principle can be modelled, explored, explained and, last but not least, influenced. Here the first prototypical model of this mechanism is specified. The main feature of this specification is the so called ›diffusion of associations‹ mechanism. This mechanism has the public, the legislature, NGOs and pro-biodiversely acting companies as its determining players. Finally the approach presented here offers another alternative evaluation method for biodiversity using multiagent based social simulation techniques.

Key words: economic modelling, corporate environmental management, biodiversity, agent based social simulation, multiagent systems

1 Introduction: motivation, goal and operationalising questions

> »It is the hope of our three organisations that a wider cross-section of the private sector would accept the challenge, integrated biodiversity concerns into their management systems, and take action to conserve biodiversity that is at the core of our planet's ability to sustain future generations.« EARTHWATCH, IUCN and WBCSD (2002 : 1)

It is the goal of this contribution to address the question whether pro-biodiverse actions of companies today will remain an exception or whether they will successfully ›diffuse‹ in the industrial sector.

To do so, I will identify the factors that are likely to boost and hinder the take-up and the attributed importance of the concept of biodiversity by industrial companies. I will focus on three questions that will be addressed from the perspective of business economics:

1. *Why* do companies act pro-biodiversely? In section 2 I discuss why biodiversity can be or become important for industrial companies. I specify the frame for and the *range of motivation* to take corporate pro-biodiverse action.
2. *Which* companies can have a motivation to act pro-biodiversely? In section 3 the possibility for and the usefulness of intra-sectoral diversification, a *biodiversity specific typology of companies*, is discussed.
3. *What* are companies doing when they act pro-biodiversely? In section 4 the *range of pro-biodiverse activities* of companies is determined. The chances for a time, place or company type related classification is analysed.

To distill a prototypical model that is apt to be formally (computationally) represented in a later stage from the findings of sections 2 to 4 is the aim of section 5, which is followed by the obligatory concluding remarks (Sect. 6).

Methodological remarks

The sought factors need to be all and only those players, roles, rules, interactions etc. which qualify to explain the actions of companies in regards to biodiversity. As the problem-space is inherently connected to various disciplines[1], the array of factors (players, roles, rules etc.) to ›choose‹ from and thus to incorporate in the model's specification is vast. In the course of argumentation some simplifying assumptions have to be made to distinguish *problem-relevant* from *problem-irrelevant* factors. In terms of disciplines business economics and corporate environmental management theory are chosen as the ›simplifying‹ perspectives to start with, i.e., whenever possible the model's specification is based on corporate environmental management theory[2].

2 Question I: Why do companies act pro-biodiversely?

2.1 General motivation for companies to take pro-biodiverse action

Three categories of what I denote as *situation-goal-motivation triples*[3] are distinguished which are all assumed to be potentially leading to (possibly different kinds of) pro-biodiverse action. The format of these triples is given in *IF situation AND IF*

[1] Problem related approaches are inherently interdisciplinary – in contrast to topic-related ones, which are, at most, multidisciplinary (cp., e.g., Balsiger and Kötter in this volume). Here the disciplines of, e.g., economics, microeconomics, business economics, sociology, psychology, ecology and others can all contribute to the specification of the model.

[2] Yet, sometimes, due to a lack of information, my argumentation is based on rationality, plausibility or even on common sense.

[3] Goals are assumed to be persistent. In certain situations, which are transient, these goals have the potential to be met. This potential – the result of a certain transient situation-goal constellation – is denoted as the (transient) motivation (the incentive if you like). Whether this motivation is sufficient to take action is the result of a (rational) decision making ›inside‹ a company which is not discussed here.

goal THEN motivation style. These triples (or rules) are by no means complete but are given here to illustrate their characteristic differences (cp. Table 1):

1. IF a company explicitly uses or relies on the *diversity* of species, genes or ecosystems as the main input in their production process AND the company has the information that this diversity is or will be endangered AND IF a company has the goal to *maintain the operational business* (trivial) THEN this company has a motivation to protect or sustainably use biodiversity itself (or at least the part of biodiversity which they are interested in) AND it has a motivation to make other companies endangering the (diversity of) resources act in the same way.
   A company belongs to this category, i.e., it shares this motivation, by virtue of its type of production (which is – in most cases – constant over time) AND of its areas of production, i.e., the location of its production sites (which may be varying over time) AND of the extent of the danger for biodiversity perceived by the company[4].
2. IF a company poses a danger to biodiversity from the perspective of its *stakeholders*[5,6] AND these stakeholders ›care‹ for biodiversity and make this explicit in a way the company has access to this information AND IF a company has the goal / heuristics to ›get along‹ with the company's stakeholders, i.e., to be aware of the company's ›social situatedness‹ and to act in regards to the social environment (to, e.g., reduce transaction costs, simplify access to biodiversity rich regions etc) THEN this company has a motivation to *convince their stakeholders*, possibly – but not necessarily – by ›protecting or sustainably using biodiversity‹ in a sense that the stakeholders feel it should be done[7].
   A company belongs to this category (shares this motivation) depending on the overall perception level of biodiversity among the stakeholders AND the importance biodiversity conservation is given by them AND depending on how strongly stakeholders associate the company (e.g. the type and the site of production) with biodiversity.
3. IF the general attitude towards / awareness of biodiversity in the *public* is ›biodiversity-friendly‹ AND IF the company is generally eager on taking chances on current topics as it ›believes‹ to gain competitive advantage by this ›style‹ of acting THEN this company has the motivation of taking chances on the issue of biodiversity in general, e.g, by voluntarily entering commitments or expressing empathy, campaigning etc. and might do so in certain cost benefit constellations. Taking chances on an environmental friendly image can be

---

[4] This perception may, of course, be correlated to the natural scientific knowledge about the ecosystem-functions responsible for biodiversity and to other information from different sources.

[5] For a differentiated account of the notion ›stakeholders‹ cp. Sect. 2.2.

[6] Features of a company which can be made responsible for the stakeholders' linkage of the company with biodiversity are discussed in Sect. 3.

[7] An elaboration on the scope of what can be done to convince stakeholders regarding biodiversity conservation can be found in Sect. 4.

worth the effort if there is a public awareness for environment conservation, regardless whether there is a perceived connection of the company's production process with endangering the environment.

Of course, it is most interesting to be able to understand and explain the *actions* resulting from these motivations. However, it is easy to see that the motivations mentioned in Table 1 may well be recognised and evaluated differently by individual companies and thus may well result in different strategies or actions considered appropriate. Being highly individual, pro-biodiverse strategies (and actions) cannot be scientifically accessed at this stage of modelling. Given that the goals of companies don't vary over time (i.e. assuming that companies don't ›learn‹ on the level of goal formation) it becomes obvious that the *situations* a company has been, is or will be in is responsible for the company to have the motivation to take pro-biodiverse action in any of the ways described above. Thus, it is the situation a company is locally and socially situated in that is highly relevant for an explanation of a generalisable motivation, which then is the basis for an analysis of anything more fine-grained, e.g., actions resulting from individual ›style‹.

This makes the *situatedness* of companies the keystone of my analysis at this stage. From now on I will analyse triple 2 in greater detail and disregard triple 1 and triple 3 for reasons of effectiveness, simplicity and conciseness[8].

2.2 The biodiversity related social environment: stakeholders

›Stakeholders‹ is an umbrella term for individuals and organisations that are associated with a company in various respects. Customers, suppliers, creditors are examples of market related external stakeholders, where NGOs, the public, media, consumers' associations and the legislature are examples of non market related external stakeholders. They form the social environment of a company. Internal stakeholders are for example employees, shareholders etc. (classification is taken from Meffert and Kirchgeorg 1998 : 94–96).

It is a central assumption of corporate environmental management theory that companies' stakeholders' opinions play a dominant role to identify the risk a company needs to deal with resulting from the ecological impacts which the company (e.g. its production process) has on its natural environment. Different to triple 1, in triple 2 the *diversity* of species, genes, and ecosystems is an *externality*[9], i.e., the

[8] The disciplines, management techniques and relevant information to deal with triple 1 and 3 are fundamentally different to those of triple 2 so that a combined consideration is rather confusing. Moreover, triple 2 is one most likely to drive the diffusion process in question for two reasons: (i) obviously, only the second and the third situation-goal-motivation triple are effective for an *undetermined* group of companies and hence are the only candidates for driving a ›qualitative‹ diffusion process. (ii) It is likely that the motivation attached to triple 2 will be more powerful in terms of initiating a diffusion process than the one of triple 3 as ›risk‹ is generally considered a stronger driving force than ›chance‹.

[9] For an elaboration of the term ›externality‹ cp. Marggraf (in this volume).

Table 1. Starting point: assumed biodiversity related situation-goal-motivation triples from the perspective of companies

triple#	situation	superordinate goal (or motive) / static action-leading feature	primary motivation / possible temporary goal	major influential factors for motivation
1	operation relies on biodiversity AND danger for biodiversity is known	maintain operational business is goal	use biodiversity sustainably AND make others use biodiversity sustainably	company's type of production (constant) AND company's area of production (variable) AND extent of the danger for biodiversity perceived by the company (different sources, variable)
2	stakeholders ›care‹ for biodiversity AND stakeholders ›believe‹ that the company poses a danger to biodiversity AND the company has this information	›get along‹ with stakeholders is heuristics	convince stakeholders AND blame other companies	overall perception / awareness level of biodiversity among the stakeholders AND level of association of the company with biodiversity by stakeholders
3	there is a considerable public awareness for biodiversity (BDPA)	taking chances is style	taking chances	…

The triple focused on in this contribution.

diversity of resources has no (obvious) ›use value‹ but, at most, the resources themselves. For some economists this fact might raise the question whether companies facing triple 2 will act pro-biodiversely at all as they might not see an incentive to do so. However, interpreting the findings of Wagner (1999) and Meffert and Kirchgeorg (1998) I assume that the motivation to convince stakeholders (i) is based on the goal to *reduce the risk of negative economic impact* emerging from ecological risks or damage (ii) is mediated by the stakeholders' understanding of and opinion about biodiversity and that this motivation (iii) may well lead to pro-biodiverse action by virtue of various financial benefits.

It is easy to see that stakeholders play different roles for and are attached different importance by companies so that *triple two* describes a set of situations so heterogeneous that they can hardly be collected under one single label. Without reducing this heterogeneity and having a closer and more differentiated look at the social environment of the company, i.e., the company's stakeholders' (the public, NGOs, legislature etc.) knowledge, understanding and opinion of ›biodiversity‹ it is impossible to get hold of what ›convincing stakeholders‹ can possibly mean.

The theoretical framework to accomplish an analysis of the biodiversity related social environment is chosen to be the approach of Wagner (1999), dealing with *corporate environmental management*. Wagner (1999) argues that reducing the risk of economic negative impact resulting from self-induced environmental damage or loss is the goal of any corporate environmental management. This risk emerges from what Wagner (1999 : 361) classifies as (dimension 1) *potential* or *factual* and (dimension 2) *perceived* or *objectifiable* environmental damage or loss[10] which may be connected with the company in general, its area of production or other factors. These economical risks can be fuelled by a variety of either perceived or objectifiable impacts the company's production process has on the natural environment (e.g. combined production).

Against this theoretical background – which will be elaborated on throughout the following sections – I differentiate 3 subgroups of stakeholders with as different biodiversity related roles and importance for a company as possible: the legislature, the public and NGOs[11]. The selection of these three subgroups seems to be a ›good‹ simplification of the rather large array of stakeholders' subgroups stated above as they cover the prominent roles stakeholders can play in corporate environmental management: (i) the legislature and biodiversity related legislative measures (BDLMs). The legislature is a centralised institution with one final ›opinion‹ implemented as unambiguous BDLMs. BDLMs provide by all means the most important guidelines for corporate action. BDLMs will necessarily take the CBD's understanding of biodiversity as their reference point. A BDLM transforms any ecological risk to an objectifiable ecological risk and by this means simplifies its

[10] This categorisation is explained in Sect. 4. The original German terms are: potentielle / faktische / wahrgenommene / objektivierbare Umweltschäden.

[11] No internal stakeholders (see above) are considered here as their attitudes towards biodiversity can be seen as a mixture of the public opinion (as shareholders may well recruit from the public) and the one of an in profit-only interested homo economicus. They can easily be added to the model in a later stage.

quantitative evaluation and management. (ii) the public and biodiversity related public awareness (BDPA). The public is ›decentralised‹, the resulting BDPA is potentially varying from individual to individual, gradually changing over time. Any understanding of biodiversity, opinion towards biodiversity or perceived danger to biodiversity is potentially thinkable. Thus, with respect to the considerable *fuzziness*[12] and *openness*[13] of the public understanding, opinion and perceived dangers (and others) the characteristics of the BDPA is antipodal to BDLM. The BDPA determines the perceived environmental damage or loss associated with a company which is much harder to translate into a quantified negative economic impact. (iii) Biodiversity related NGOs. They have the role to promote biodiversity to either of these groups and may succeed in doing so in various degrees. In other words: they aim at lifting both the perceived and objectifiable environmental damage a company has to manage to push companies' effort to act pro-biodiversely. In addition, they are regarded trustworthy by the majority of the public when it comes to environmental friendly behaviour in general so that companies are forced to legitimate their actions by entering partnerships with NGOs.

The biodiversity-related ›attitudes‹ of NGOs, the public and the legislature are obviously interdependent. If the public opinion wasn't considered influential until approx. 30 years ago, it is attributed a continually growing power since then due to the growing influence and power of institutionalised stakeholders like NGOs over the last two decades. NGOs have the power to mobilise public and political pressure[14] and may provide the public and governments with information about biodiversity via their own communication channels, including public press or broadcast etc. The so pushed BDPA feeds back into the political decision-making system and this feedback mechanism has accelerated due to changed societal values[15].

Summing up, the interacting BDPA, BDLMs and biodiversity-related NGOs are regarded as a good first approximation of the relevant stakeholders of a company which is about to decide whether to take pro-biodiverse action or not. The subset of this understanding, opinion and desire to protect biodiversity relevant for a company situated in a certain social and legislative environment determines the outside pressure that the company has to manage. Explaining a company's pro-biodiverse action taking needs to consider a company as being situated in a dynamic and highly networked social environment.

2.3 How ›aware‹ are stakeholders of biodiversity?

To get hold of the risk driven pressure to engage in biodiversity conservation today and in the near future it is necessary to analyse the impact of biodiversity today

[12] The term ›fuzziness‹ refers to the feature that the elements of the set ›public understanding‹, i.e., the understanding of biodiversity of an individual, is not shared by everyone.

[13] The term ›openness‹ refers to the feature that the (fuzzy) set ›public understanding‹ may well change its elements over time.

[14] Cp. Meffert and Kirchgeorg (1998 : 94).

[15] For an elaboration on the change of societal values in regards to the natural environment cp. Meffert and Kirchgeorg (1998 : 104).

and in the near future for the legislature, the public and NGOs. After a very concise assessment of the current situation, a sketchy and necessarily incomplete estimation of possible future developments and uncertainty[16] attached to biodiversity is given.

The legislature – today

Existing biodiversity related policy measures are directed towards financial transfers from industrialised to biodiversity-rich developing countries[17]. In these countries ensuring the license to operate for companies may and sometimes already does require compliance with regulations of quasi-legislative status. However, national policy measures of industrial countries designed to take effect on the decision making of ›wrongdoers‹ on home soil can be characterised as nonexistent.

The public – today

In the public discussion biodiversity cannot be regarded as anything close to a major topic, too. In the competition of major world crises for public attention biodiversity is outperformed by, e.g., climate change and sustainability. Even in Germany, one of Europe's pioneering nations in terms of public awareness for ecological issues, BDPA is barely noticeable. Thus, biodiversity hasn't really walked out onto the public stage yet.

NGOs – working hard

However, much effort is undertaken by NGOs, the scientific community and some governmental organisations to get biodiversity both on the agenda of the legislature and into the minds of the public, and, therefore, the rather poor situation today may well change.

As an estimation of the future development of both the BDPA and BDLMs is key to determine the diffusion process in question, sketching the frame for the area of possible future development is attempted by an analysis of the CBD's formulation of both the concept[18] and goals[19] connected with biodiversity.

The goals of the CBD – conservation, sustainable use and equitable benefit sharing from access to genetic resources – are closely related to the concept of sustainability and can be partly interpreted as a transfer of the goals connected to sustainability to the area of biodiversity, an operationalisation of the idea of sustainability for biodiversity. With this reading, the CBD takes in the three-dimensional

[16] Uncertainty plays an important role in the management of environmental issues in a company due to its direct influence on the economic risk. An elaboration on biodiversity-related uncertainty can be found in section 4.

[17] Cp. Marggraf (in this volume) for an overview and assessment of existing international biodiversity policies.

[18] Cp. SCBD (2002 : article 2, paragraph 1).

[19] Cp. SCBD (2002 : article 1).

idea of sustainability[20] and extends it to all the subsystems of biodiversity, namely ecosystems, genes and species, innovatively emphasises the intra-generative responsibility as a part of the social dimension of sustainability and, in addition, demands the conservation of the variety of each of the biodiversity subsystems[21].

> »The Convention was – and still is – an ambitious undertaking.« SCBD (2001 : XI)

The legislature – tomorrow

For political decision-making these terms and goals have to be operationalised and adapted to national settings. Usually a minimal necessary requirement to succeed in doing so is the objectification and quantification of the value of biodiversity and its subsystems (ecosystems, genes and species) to be able to assign values to actions that change their state. Until now an agreement has not been reached on how to determine these values best. Various approaches compete with or complement each other[22]. After decades of research it has become obvious that there is no *one* way.

But as long as the various suggestions of how to evaluate biodiversity have not been combined in one way or the other to form a consistent biodiversity-valuation ›package‹, national governments will not be able to implement a mechanism to punish or reward biodiversity-related actions in a generally accepted way[23]. When the objectifiable impacts can be the target of legislative measures, the perceived non objectifiable impacts cannot. Unfortunately, this is the situation biodiversity conservation faces today and, as I argued above, will most probably face in the near future.

With this holistic and all-embracing initial definition of terms and goals, the CBD deprives biodiversity conservation of the opportunity to find a fast way into national legislation[24].

The public – tomorrow

Problematic these features may be for framing the handling of biodiversity via legislation, yet positive they are for an uptake of biodiversity in the public. In regards to its connection to sustainability, biodiversity can profit from the already existing public awareness for the concept of sustainability, the agenda 21, the world Summit

20 Sustainability regards social, ecologic and economic aspects as equally important.

21 The demand for the conservation of biodiversity in addition to its sustainable use explicitly underlines the CBD's assignment of an eigen-value to biodiversity, which, of course, is disputable.

22 Comparing Marggraf, Bräuer, Menzel and Fischer (all in this volume) gives an idea of the complexity and variety of approaches of how to evaluate biodiversity.

23 On an *individual* basis cases have been made against company action in regards to biodiversity damage in the US – cases which, however, cannot be generalised and therefore have no potential to be quantitatively taken as an example in any way suitable for this modelling.

24 For the situation in developing countries see section 2.3.

of Sustainable Development (WSSD), anti globalisation communities etc. Furthermore, one doesn't have to have any information about the CBD or its goals or its use of terms to seemingly understand what biodiversity means: ›bio‹ and ›diversity‹ both have an indisputably morally positive connotation. It is easy to find people ›on the street‹ with no or no proper knowledge of what biodiversity means in terms of the CBD who will nevertheless agree that biodiversity is to be protected and that it has an eigen-value for them. They don't have to bother about any unresolved questions of how to measure its value: they have an *opinion*[25]. As was argued above, leaving room for interpretation is problematic for both a scientific and a legislative approach to biodiversity. Leaving the same room for interpretation in the public discussion has potentially positive consequences: no one can be taught that he or she has ›got it wrong‹ whatever his or her understanding of biodiversity actually is. This option for broadening the interpretation of biodiversity is underlined by EARTHWATCH, IUCN and WBCSD (2002 : 8):

»In the end, biodiversity is everywhere and it is everyone's business.«

Thus, the questions where parts of nature happen to be regarded as biodiversity-rich and where biodiversity is endangered are decided in the minds of the people. Actually, individuals, groups and organisations already active in nature or environmental conservation do, may and will label their programmes as biodiversity conservation. In the public understanding pro-biodiverse activities do, may and will include conservation of not endangered domestic species, ecosystems that clearly wouldn't have been regarded very valuable in terms of the CBD or CBD experts[26].

With giving biodiversity a decisive broad meaning and an appealing morphology, the creators of the CBD deliberately or by accident gave up the central control over both the meaning of and the goals connected with biodiversity. This control is distributed to the hands and minds of everyone willing to make up his or her mind about biodiversity.

Summing up, biodiversity has a strong partner in sustainability, a powerful morphology with morally positive associations, and an conception open to give a variety of existing interests of a lot of people a new name. Hence, in the public arena biodiversity can play at its strengths.

3 Question II: Which companies have a motivation to act pro-biodiversely?

In this section the findings of the section 2 are used to clarify which companies (in terms of branches of industries) have or will have a motivation to act pro-biodiversely in principle, the set of which will be referred to as $C_{\mathtt{motivated}}$. It is

[25] Cp. Menzel and Fischer (both in this volume) for an application of the (direct) Contingent Valuation Method (CVM) to monetarise the public valuation of biodiversity.

[26] Cp. the contribution of Holl (in this volume) for an enlightening overview over the diversity of German home gardeners' understanding of ›biodiversity‹.

a subset of all industrial companies, referred to as C. In section 2 considerations were confined to the motivation associated with the goal of the reduction of negative economical impact from ecological risks (cp. triple 2 in Table 1) and it was found that this risk is transported by the social environment the company is situated in. To estimate the pressure a branch or a specific company has to manage it is to be clarified which industrial company's operational business poses a perceived or objectifiable risk to biodiversity from the perspective of either the legislature or the public (or the NGOs)[27]. I introduce the notion C_{legis} for the set of companies motivated by legislative and C_{public} for those motivated by public pressure. As I assumed the legislative and the public pressure to be a sufficient approximation for the direct[28] pressure to act pro-biodiversely a company has to manage, the motivated (and potentially active[29]) companies are the set union of C_{legis} and C_{public}: $C_{motivated} = C_{legis} \cup C_{public}$, thus. Trivially, the set of active companies, C_{active} is assumed to be a subset of $C_{motivated}$: $C_{active} \subset C_{motivated}$.

Let us first turn to the legislative perspective. It is straightforward that the companies that have a motivation to care for biodiversity due to today's and tomorrow's BDLMs based on access and benefit sharing, C_{legis}, are those operating internationally or globally in biodiversity rich regions: the branches of oil, gas, mining, timber, fisheries, seed breeding etc., i.e., the ones that are to be connected with biodiversity based on the CBD's (the ›official‹) understanding of the concept of biodiversity[30]. As there is no national legislation in industrial companies expectable,

---

[27] These representatives for the stakeholders (respectively the social environment) of a company were chosen in section 2.2 on page 90.

[28] NGOs are supposed to influence both the legislature and the public and thus to pressure companies indirectly.

[29] The amount of pressure resulting from different levels of motivation is assumed to correspond to the likelihood of pro-biodiverse action taking. *Remark*: If triple 1 (cp. Table 1) was the only one effectively driving pro-biodiverse actions of companies one would find only those companies investing in the conservation of biodiversity which rely on biodiversity in one kind or another as an input factor for their production process. As these branches only amount to a rather small part of all industrial branches it is a fortunate empirical circumstance – at least for those who ›believe‹ that biodiversity should be protected – that the majority of companies active in the conservation of biodiversity today recruit from those branches the core businesses of which are effectively or potentially *endangering* the diversity of ecosystems, genes and species; branches for which biodiversity is an *externality*, which massively extract natural resources (timber, fisheries) introduce invasive species (seed breeding) or that operate in biodiversity rich regions (oil, gas, mining). For these pioneering companies, stakeholders' pressure has become sufficient to drive pro-biodiverse behaviour. As not every company that may have this motivation actually has acted pro-biodiversely (or hasn't yet), there have to be certain internal parameters (e.g. the habit of risk-avoidance) which decide whether the pressure is sufficient for a individual company to take (considerable) action. These company internal decision mechanisms are not considered here. Therefore I keep the notion motivation and do not consider actions.

[30] For some examples cp. the remark in Footnote 29.

i.e., no qualitatively different BDLMs to the ones existing[31], I assume C_{legis} to be a stable and therefore approximately closed set of branches and companies. In other words the motivation to act pro-biodiversely because of BDLMs will not diffuse across the boundaries of the kinds of branches mentioned above.

Obviously, the situation is different for the set of companies that are potentially associated with biodiversity due to the *publicly perceived* damages of biodiversity they cause, C_{public}. When the elements and the boundaries of C_{legis} are rather clear cut, it is not that easy to determine them for C_{public} as this set certainly depends on the dynamics of the BDPA in regards to the *public understanding* of biodiversity and, of course, the overall awareness level. If the NGOs succeed in spreading the belief that ›biodiversity is everywhere‹[32] and in making biodiversity ›everyone's business‹[33] every company endangering natural resources one way or the other may become potentially associated with biodiversity. What follows is that in principle every and any company may encounter a situation in which it is motivated to take biodiversity-action in reaction to BDPA – the diffusion of public associations which link the company with a potential or factual biodiversity damage may, in principle, reach every company. However, this can only be regarded as an encouraging and valuable finding if it is clarified on which *paths* this diffusion might take place and if it will lead to a degree of pressure (motivation) sufficient to initiate companies' pro-biodiverse action. In other words, predicting any action on the basis of public pressure in industrialised countries will remain speculative or at least uncertain as long as the development of BDPA cannot be predicted or at least estimated.

To get hold of the likely paths the diffusion process may take it is essential to understand the creation process of the public association of a certain company with biodiversity, which determines C_{public} at time t. It is obviously the *public understanding* of the concept of biodiversity and not the intensity of BDPA that determines whether companies are or are not associated with biodiversity. Thus the understanding at time t determines C_{public} at time t. The development of the public understanding is unlikely to be utterly arbitrary, thus a path may be identified. I assume that this path is influenced by (1) the pro-biodiverse actions of pioneering companies which may initiate a diffusion along the lines of generalisation (2) the understanding communicated and promoted by NGOs which may guide the attention to certain focal points and (3) a rather fixed, predictable ranking of environmental activities ›naturally‹ associated with biodiversity.

I will not elaborate on (1) and (3) but I will try to explain what I mean by generalisation (2): consider company c_i of branch b_j acting pro-biodiversely and communicating what their activities in the public media in an industrialised country with a known awareness for ecological issues[34]. This causes at least two more

---

[31] For an explanation cp. Sect. 2.3 on page 93.

[32] Recall the quote given in the opening of section 1.

[33] See Footnote 32.

[34] It will be shown in section 4 that the pro-biodiverse activities of companies always include (in various degrees) communicating these activities to a broader community.

things to happen: (i) communicating pro-biodiverse activities deliberately or by accident increases BDPA in general and (ii) even if company c_k had not been associated with biodiversity at all before c_i acted pro-biodiversely, now (ii-a) c_k is publicly associated with biodiversity either because it shares the branch b_j with c_i (*branch related generalisation*) or just because it is an industrial company[35] (*sector related generalisation*) and (ii-b) c_k obviously has not done what c_i did, so c_k faces competitive disadvantage[36]. Obviously, to a certain extent companies have it in their own hands how much biodiversity will become a publicly known term and which companies will or won't be associated with the conservation of biodiversity.

The generalisation introduced in (ii-a) leads to what I refer to as a *diffusion of associations*. This diffusion is fuelled and given direction by public awareness initiatives (due to the public understanding of biodiversity) a kind of naturally perceived similarity of biodiversity to other environmental issues and, last but not least, pioneering actions of companies themselves. There is reason to believe that there will be a certain amount of diffusion in the future due to existing pioneering action and to the particularly high potential of biodiversity to be interpreted in many different and sometimes contradicting ways. Even if a certain path of diffusion was suggested, however, it is still uncertain how far-reaching this diffusion will be. In one extreme, public awareness will not be affected in any of the suggested ways, including that generalisation will not take place. Then, only individual companies that are operating in biodiversity rich regions will act pro-biodiversely. Then, $C_{public} = \emptyset$ and $C_{motivated} = C_{legis}$. In the other extreme association's diffusion is ›complete‹. Then, biodiversity will have become an umbrella term for a variety of environmental conservation issues and industrial companies in general are associated with biodiversity. Then $C_{public} = C$ and $C_{motivated} = C_{legis} \cup C_{public} = C$. Both situation are rather unlikely to fully describe future states of the problem space. The ›truth‹ will lie somewhere in the middle.

4 Question III: What are companies doing when they act pro-biodiversely?

Until now the term pro-biodiverse action was used as a possible but unspecified ›sequel‹ of a range of motivation (situation-goal-motivation triples) identified in Sect. 2. The question of this section is ›what is the range of possible pro-biodiverse actions of companies of $C_{motivated}$?‹. A special focus will be given to the potential of any such pro-biodiverse action to feed back into the social environment of the company and thus boosting the diffusion mechanism proposed in section 3.

A framework for identifying adequate strategies with respect to environmental issues is provided by Wagner (1999). Wagner (1999 : 361) suggests that the risks

[35] Companies are generally mistrusted and assigned a responsibility for nature conservation due to a slope of information from the company towards its stakeholders, in particular towards the public. Cp. Wagner (1999).

[36] The advantages of pioneering action discusses Meffert and Kirchgeorg (1998).

of negative economic impacts are managed differently whether they result from (dimension 1) *potential* or *factual* and (dimension 2) *perceived* or *objectifiable* environmental damage or loss caused by or attributed to a company.

Dimension 1 captures the states of a company of either having damaged biodiversity already or being susceptible to do so in the future, e.g., by virtue of its general operational business or concretely planned projects.

Strategies for (objectifiable and perceived) factual damage

Wagner's classification suggests that a factual damage is likely to entail institutionalising of transaction measures which usually include negotiations between those concerned by the damage and are characterised by a confined radius of effect[37]. The more objectifiable (the less perceived-only) a factual damage the more straightforward the compensation the more predictable the economic impact and the less risky the situation for the company.

Strategies for (objectifiable and perceived) potential damage

If the damage is classified potential, the selection of strategies heavily depends on the perceived/objectifiable status of the damage (Wagner 1999 : 65) and the two sets of appropriate strategies are rather different. If the damage is objectifiable a company may take various measures to deal with the resulting quantifiable economic risk. Appropriate strategies are discussed under the reference of avoidance, reduction, taking and shifting of economic risk (cp. Wagner 1999 : 365). These strategies are regarded as the traditional matter of interest of corporate environmental management. In general, these strategies mainly affect company-internal processes and decisions and do not necessarily attract public attention. This is different if it comes to potential perceived-only environmental damage or loss. In this case, appropriate strategies are directed to building up a company's reputation, which obviously embraces public awareness for the strategy to be effective.

Which strategies of the ones described by Wagner are appropriate for the issue of biodiversity? To get hold of the appropriateness of strategies in the context of biodiversity the preconditions for certain sets of strategies proposed by Wagner are now analysed against the background of the findings in section 2 and 3. To transfer Wagner's classification to the issue of biodiversity the discussion of the previous sections provides all that is needed to do so.

[37] Wagner (1997 : 61) considers these institutions to be ›rules (norms) for recurrent decision situations which pose as instruments for the solution of certain societal problems [...] These rules are to be reducing uncertainty [...] particularly by the provision of specific information about the decision situation and stable (reciprocal) expectations about the behaviour of other individuals and groups and thus about specific consequences of ones own decisions.‹. Original text in German, translation by me.

Objectifiable or perceived?

Judging whether biodiversity damage or loss is objectifiable equals judging whether there is a generally accepted way how to evaluate biodiversity. The question of evaluating biodiversity has been discussed in various sectors, e.g., science, business and politics. There are both intra-sectoral and inter-sectoral arguments about how to evaluate biodiversity and what this value shall be. Recalling our findings in section 2.3 on page 93 it is likely that a standardised evaluation of biodiversity will remain an unsolved problem. As a consequence, for biodiversity-issues the strategies mapped to objectifiable environmental damage or loss are not appropriate and companies focus on those linked to perceived-only biodiversity damage[38].

Factual or potential?

Based on the same argumentation judging whether biodiversity damage has already taken place or biodiversity is only potentially endangered is determined by the perception of the observer. As it is based on perception, the factual damage done to biodiversity is defined by those who *believe* that the damage has taken place. The occurrence of a biodiversity damage equals the occurrence of a perceived biodiversity damage. Thus, the occurrence of the biodiversity damage is dependent on the many facings and the dynamics of biodiversity-perception[39]. In general, the potential damage is a pre-stage of the factual damage (which does not have to occur, of course). Then, following the lines of the argumentation of the last paragraph, potential biodiversity damage has to be interpreted as the expectation that (i) more stakeholders become aware of biodiversity and that (ii) these stakeholders have a rather fuzzy understanding and that therefore (iii) the company is likely to be associated with damaging biodiversity (ex-post). In short, managing potential biodiversity damage is based on the anticipation of the diffusion mechanism.

Combining Wagner's proposed damage-strategy-linkage with my assumption about how biodiversity damages should be classified leads to conclusion that biodiversity management uses a mixture of strategies related to both perceived factual and perceived potential environmental damage and loss, i.e., it uses the strategies of institutionalising of multilateral transactional relations and signalling or other reputation related measures[40].

Of course, there is much more to say about how exactly companies do and will behave towards biodiversity, whether all the proposed strategies or only a subset of those are really appropriate and match empirical evidence etc – but at this point

[38] In general, the less objectifiable (the more perceived-only) a biodiversity-damage or loss appears for companies from $C_{motivated}$, the larger the fraction of communication of companies with the stakeholders, the public and NGOs will be and the likelier and stronger the association diffusion will take place.

[39] Please cp. the discussion in Sect. 3.

[40] Reality shows that both kinds of strategies have already been applied by companies.

enough information is gathered to decide over the initial question of this section, i.e., whether the applicable strategies are able and likely to feed back into the social environment. As all of the usual internal strategies could be ruled out and the remaining ones require the participation of the social environment in various scales the answer is quite straightforwardly ›yes, they do‹: the dynamics of BDPA and C_{public} are interconnected and fuel one another.

Having now verified that the mechanism defined in section 3 is likely to be fuelled by the pro-action taking of companies, it is ever more justified to assume that a diffusion of associations might take place in the way suggested. Then, both factual and potential damages will vary over time and so will the deployment of strategies.

5 A prototypical model specification: summary of answers

In the previous sections 2 to 4 I have described a prototypical model specification supposed to explain the dynamics of pro-biodiverse action taking of industrial companies on the basis of the concepts ›motivation‹, ›goal‹ and ›situation‹. In this section, the results are collected and combined. The following summary of findings is illustrated by figure 1.

The initial assumption of section 2 was that companies' behaviour towards biodiversity relevant for the problem ›how to achieve an explanatory adequate model of the aggregated actions of industrial companies in regards to biodiversity?‹ is determined by the dynamics of the company's social environment, namely its stakeholders. For reasons of simplicity three contrasting stakeholder subgroups were chosen to represent the social system a company is situated in: the legislature, the public and NGOs. The set of companies under legislative pressure was referred to as C_{legis}, the set of those under public pressure as C_{public}, the NGOs were assumed to have the potential to influence both the legislature and the public, thus influencing C_{legis} and C_{public} indirectly. These two sets are assumed to constitute the set of companies that potentially become pro-biodiversely active, so that $C_{motivated} = C_{legis} \cup C_{public}$, and $C_{active} \subset C_{motivated}$.

Activities of companies were assumed to reflect the state of this social system as this state corresponds to the pressure that companies have to manage and this pressure corresponds to the likelihood of action. The dynamics of $C_{motivated}$ was therefore assumed to be closely linked to the direction of the dynamics of the social system in terms of biodiversity, i.e., the dynamics of the biodiversity-related legislative measures (BDLMs) and the biodiversity-related public awareness (BDPA).

It could be shown that the concept of biodiversity, i.e., the CBD's definition of biodiversity, its use of terms and the goals attached to the use and non-use of biodiversity, hampers the operationalisation necessary for BDLMs, especially by a lack of a generally accepted valuation method for biodiversity. Therefore, it was assumed that BDLMs would not become either very definite or rigid or dynamic, in short, that they would not change qualitatively to what they are now: C_{legis} was therefore assumed to be closed and thus the motivation to act pro-biodiversely due

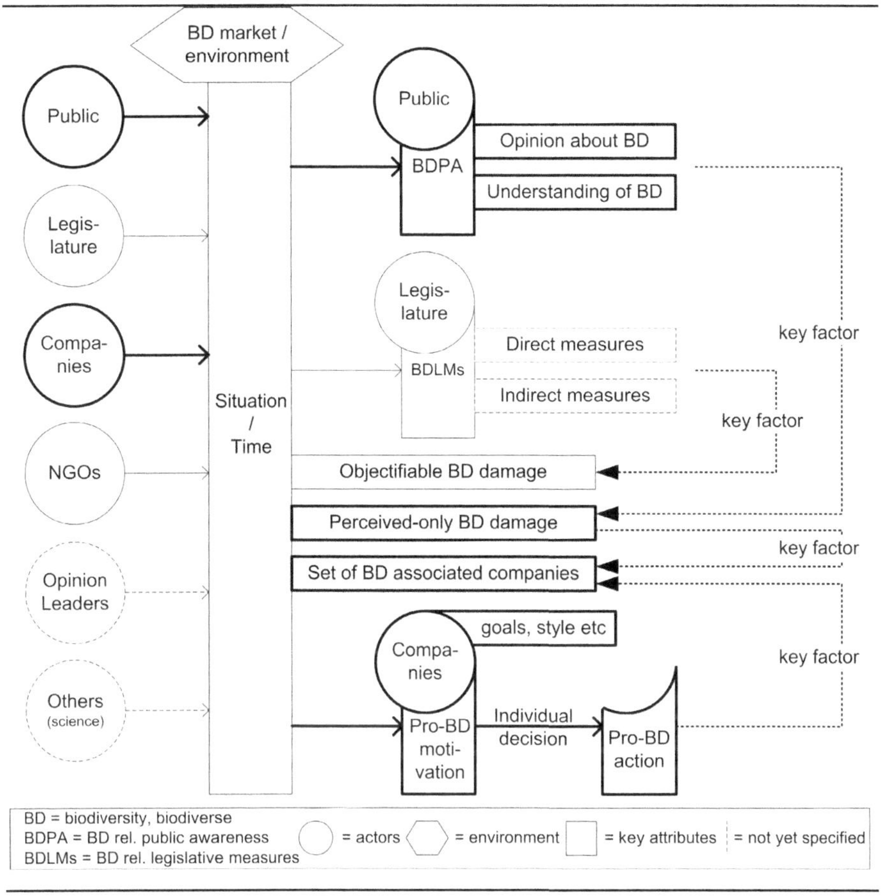

Fig. 1. Model summary: graphical representation

to legislative pressure was not susceptible to diffusion across xed (and rather narrow) boundaries. Thus, more dramatic dynamics of $C_{motivated}$ if there were any had to be determined by the dynamics of C_{public}. Fortunately in this sense, the same features of the concept of biodiversity hampering BDLM could be shown to facilitate the uptake of biodiversity in the public, BDPA. The resulting BDPA was shown to be likely to be fuzzy and it was concluded that the set of publicly perceived companies which damage biodiversity, C_{public}, will be so, too. A mechanism was suggested which both determines the development of BDPA and links it with C_{public}. This diffusion association mechanism is assumed to be threefold: (i) the pro-biodiverse actions of companies set initiation points for a diffusion along the lines of branch- or sector-wide generalisation (ii) NGOs are assumed to guide the public attention to certain focal points and are generally interested in a maximum overall diffusion and (iii) there is a rather fixed, predictable ranking of environmental activities ›naturally‹ associated with biodiversity.

An essential requirement for the assumed generalisation to take place – embracing or informing the companies' stakeholders – was provided by the findings of section 4: pro-biodiverse action of companies is necessarily based on a strategy mix of institutionalising transaction relations among stakeholders for certain projects of a company that are regarded as biodiversity damaging – this strategy is dominating for projects in biodiversity rich regions today – and signalling or other reputation related strategies for companies to cope with a perceived potential biodiversity damage. It follows that elements of $C_{motivated}$ at time t are dependent on the elements of C_{active} at time $t - 1$[41] due to (i) a feedback into the social environment of the actions at $t - 1$, (ii) a generalisation mechanism based on this action and therefore (iii) a changed BDPA in terms of companies associated with biodiversity[42]. This changes C_{public} and therefore $C_{motivated}$ at time t[43]. All (i), (ii) and (iii) are responsible for a *directional* diffusion of the issue of biodiversity in society and thus for the overall dynamics of the take-up of the concept of biodiversity and the dynamics of the actions related to biodiversity in the industrial sector.

6 Concluding remarks

The first step to answer the question whether pro-biodiverse actions of companies will remain an exception or whether they will successfully diffuse in the industrial sector has been taken. One of the early findings was that answers can only be obtained by focusing on the social system a company is typically situated in and to model the system's actors and rules. Relevant actors of this system were identified as the public, the legislature, NGOs and pioneering companies themselves. The prominent drivers of the system were assumed to be the associations made by the public which link companies with biodiversity damage and which follow certain generalisation rules making the diffusion directional. The model's specification was chiefly based on an analysis of the characteristics of the concept of biodiversity specified in the CBD in conjunction with corporate environmental management theory. If it was true that today's pioneering action is only the top of the iceberg and more and more companies have a motivation (an incentive) or even the duty to act pro-biodiversely, the likely paths this diffusion might take have been specified.

---

[41] Here I am assuming a discrete timescale only for reasons of presentation.

[42] Recall that the BDPA is assumed to be taking in both the public understanding of biodiversity and an opinion about biodiversity (reflecting the public valuation of biodiversity) and is therefore a key factor for the set of companies associated with biodiversity damage.

[43] Please note that all the sets are naturally heterogeneous as to the fact that, e.g., companies of C_{public} do share the qualitative feature of being under public pressure but the amount of pressure may well vary. So, even if the number of elements of C_{public} do not necessarily change by the suggested generalisation mechanism, the state of the elements may well do so.

7 Outlook

So far I have only started to represent the likely dynamics of the *positive* uptake of biodiversity in the industrial sector. Obviously there is more to consider than the solely enhancing drivers of the mechanism in question. The counter-forces to pro-biodiverse action taking will have to be dealt with and to do this a consideration of the ›competition of major world crises‹ is strongly required.

Finally, a computational representation of the model providing a testable and explorable positive theory of the problem space – the development of the role of biodiversity in the industrial sector – will be required. The use of multiagent based social simulation techniques will then provide the grounds on which the hopes of EARTHWATCH, IUCN and WBCSD (cp. the opening quotation) can be (re)built.

References

Earthwatch International (EARTHWATCH), World Conservation Union (IUCN), World Business Council for Sustainable Development (WBCSD) (eds) (2002) Business & Biodiversity : The Handbook for Corporate Action. Earthprint, `http://www.earthprint.com`

Meffert H, Kirchgeorg M (1998) Marktorientiertes Umweltmanagement : Konzeption – Strategie – Implementierung mit Praxisfällen, 3rd ed. Schäffer-Poeschel, Stuttgart

Secretariat of the Convention on Biological Diversity (SCBD) (ed) (2001) Global Biodiverity Outlook. SCBD, Montréal Québec

Secretariat of the Convention on Biological Diversity (SCBD) (ed) (2002) Convention on Biological Diversity : texts and annexes. SCBD, Montréal Québec

Wagner GR (1997) Betriebswirtschaftliche Umweltökonomie. Lucius & Lucius, Stuttgart

Wagner GR (1999) Umweltmanagement. In: Bitz M, Dellmann K, Domsch M (eds) Vahlens Kompendium der Betriebswirtschaftslehre, vol 2. Vahlen, München, chap E.3, pp 339–391

Part II

Local, regional and nationwide perspectives on the Convention on Biological Diversity: Examples from Germany

Problems and Prospects of the Conservation of Biodiversity in Germany

Matthias Schaefer

Ecology Group, Institute of Zoology, Anthropology and Developmental Biology, Berliner Str. 28, 37073 Göttingen, Germany, email to mschaef@gwdg.de

Summary. Germany is a small, densely populated country with necessity for agriculture, significant pressure on natural habitats and a high proportion of man-made open habitats. About 80.000 species of plants and animals exist in a diverse landscape with much anthropogenic disturbance of different kinds. Few endemic species occur; no hot spots of species diversity had developed. Thus the Central European region is a complex, in many parts steadily changing container filled with the biota. The population dynamics of the species is largely shaped by habitat fragmentation and by a dynamic subpopulation structure with distribution on habitat islands and by much influence of disturbance and stochasticity. I suggest a triple approach to the preservation of biodiversity – conservation of species and of habitats (with the problem of defining priorities and designing action plans) and a landscape/habitat/species or ›comprehensive‹ approach with the focus on the maintenance and enhancement of the diversity of the (natural and cultural) landscape.

1 Introduction

Conservation of biodiversity in Germany is rooted in the discipline conservation biology (Noss 1999), a flourishing branch of applied ecology, which addresses the problems of the many worldwide threats to the biota. The number of excellent treatises in this field is high and steadily increasing (Spellerberg et al. 1991; New 1995; Sutherland 1998; Primack 1998; Primack 2000; Aßmann and Härdtle 2002; Pullin 2002). Therefore it makes no sense to repeat general issues of nature conservation. Instead, I focus on the special situation in Central Europe and ask the following specific questions: Where is the place for biodiversity in a densely crowded country with only a small proportion of natural or near-natural habitats? Which are the threats to flora and fauna? How about the necessity of conserving the diversity of animals and plants? How can this be done? Are there specific Central European strategies?

›Biodiversity‹ is the variety of organisms considered at all levels, from genetic variants belonging to the same species through arrays of species, families, and still higher taxonomic levels; it includes the variety of ecosystems, which comprise both the communities of organisms within particular habitats and the physical conditions under which they live (Wilson 1992). ›Biodiversity‹ also covers the complex sets of structural and functional relationships among systems at the bioregional, landscape, ecosystem and habitat level (Heywood 1995). Often biodiversity is used in the sense of species richness.

›Conservation‹ is the preservation of natural (or cultural) systems with the focus on the whole (›nature conservation‹) or species (›species conservation‹). The term can be used in the narrower sense of preservation or maintenance of some or all the components of biodiversity, and in the broader sense to include also sustainable use of the components, or their recovery or restoration or both (Heywood 1995). ›Sustainable use‹ is the use of species or natural communities by man in such a way and at a rate that does not lead to the long-term decline of biological diversity, thereby maintaining its potential to meet the needs and aspirations of present and future generations (Heywood 1995; Sutherland and Reynolds 1998; Pullin 2002).

Which are the criteria for conserving animal and plant species, including inconspicuous lower plants or invertebrates? The four main broad (and overlapping) categories of values (Wilson 1992; New 1995) are (1) commodity (the role of a species in the market place), (2) amenity (improvement of the quality of peoples lives by species in a non-material way), (3) moral value (ethical implications of species conservation), (4) functional or ecological value (the role of species in ecological systems). We should not forget that for all species category (3) and for many species category (4) applies.

In the following short overview I will discuss conservation approaches in Central Europe from an ecological point of view and within the framework of present strategies in species and nature conservation. Emphasis will be more on animals. We will see that this region is a dynamic system strongly influenced by man. An important focus of nature conservation and management should be the diversity of the whole region, which is dominated by the cultural landscape with its negative and positive effects on the biota, intermingled with patches of more natural habitats.

2 Basic patterns: from natural to cultural habitats

The Central European landscape is densely populated by man with 225 inhabitants per km^2, and simultaneously it is a container filled with many species of plants and animals. It is a kind of template to which the biota had adapted itself in evolutionary and/or ecological time (Southwood 1988) and which is the environment for further adaptations (Stockwell et al. 2003). The diversity of natural and cultural habitats might mirror the diversity of the fauna and flora. Because of many colonization events out of the regional species pool, non-natural habitats must not be species-poor. Disturbance and other stress factors cause changes in population density and species diversity.

2.1 History

Four phases can be recognized in the history of the Central European landscape since the last ice age through the middle ages to modern time (Bundesamt für Naturschutz 1997; Mühlenberg and Slowik 1997; Plachter 1999a; Aßmann and Härdtle 2002; Pullin 2002).

Phase I. The landscape after the last ice age (14,000–12,000 BC) was dominated by forests; anthropogenic influences were restricted to hunting and to dispersal of seeds. Some species of the megafauna became extinct.

Phase II. About 5,000–4,000 BC, shifting cultivation as the dominant form of agriculture led to a reduction of the forest area to about 90 %.

Phase III. About 1,000 AD forest clearance was intensified. Thus, the pressure on forests became high (only 30 % of Central Europe were now covered with forests) and the importance of degraded habitats increased. On the other hand, the number of species continuously rose to a maximum for plants and animals between 1800 and 1850. Even many ›charismatic‹ species (such as the white stork) belong to this group of species inhabiting open, anthropogenic habitats.

Phase IV. At about 1850 human impacts on the landscape increased with intensive agriculture, modification and/or degradation of the more natural landscape (e.g., river regulation, reclamation of raised bogs), fragmentation of habitats, increase of industrialization.

Thus, the number of animal and plant species grew continuously by human action transforming forests into a more open landscape, with a culmination point of species diversity between 1800 and 1850. Afterwards as a consequence of the Industrial Revolution high anthropogenic pressure had increasingly negative effects on the diversity of the biota (Sukopp and Trepl 1987).

2.2 The cultural landscapes

At present, 54.7 % of the total area of Germany is agricultural land, 29.2 % are covered with forests, 11.7 % are taken up by settlements, industry, traffic, and 2.2 % are water bodies. Hence, Central Europe harbours a variety of man-made (›cultural‹) landscapes with specific – favourable or unfavourable – conditions for the biota, and the majority of habitats are artificial ecological systems with many kinds of stress for the biota and lack of coevolution between members of communities. Here main structuring forces for animal and plant communities are stochasticity and disturbance causing a pronounced spatial dynamics (cp. Sect. 3).

One of the values to be found in some traditional cultural landscapes is the presence of a sustainable pattern of land use. Many Europe-wide studies have confirmed the conservation and environmental values associated with such traditional landscapes (Phillips 1998). Three types of cultural landscapes can be found in Europe: designed landscapes (which are artificially created for aesthetic or related reasons), organically evolved, relict landscapes (where the process of landscape evolution came to a halt in the past) and continuing organic landscapes (where landscape evolution continues to this day) (Phillips 1998).

Table 1. A synopsis of species numbers of selected taxa in Central Europe (in most cases related to Germany). Some figures are rough estimates. Data from Bundesamt für Naturschutz (1996, 1998), Schaefer (1999, 2002), Aßmann and Härdtle (2002)

Taxon	Species number in Central Europe
Plants	
Macroalgae	>346
Fungi	>6,500
Lichens	1,691
Mosses (Bryophyta)	1,121
Higher plants (Pteridophyta, Spermatophyta)	>3,319
Animals	
Protozoa	3,500
Nematoda	1,100
Mollusca	440
Snails and slugs (Gastropoda)	350
Annelida	380
Enchytraeidae	100
Earthworms (Lumbricidae)	70
Arthropoda	34,000
Arachnida	4,000
Spiders (Araneida)	900
Harvestmen (Opilionida)	40
Mites (Acarina)	3,000
Gamasina	1,000
Oribatida	450
Crustacea	750
Isopoda	40
Millipedes and centipedes (Myriapoda)	210
Centipedes (Chilopoda)	50
Millipedes (Diplopoda)	130
Insects (Hexapoda)	29,000
Springtails (Collembola)	300
Hemiptera	2,300
Beetles (Coleoptera)	8,000
Carabidae	500
Staphylinidae	1,300
Wasps and bees (Hymenoptera)	10,000
Butterflies and moths (Lepidoptera)	3,000
Midges and flies (Diptera)	8,000
Vertebrates (Vertebrata)	648
Fishes (Pisces)	257
Amphibians (Amphibia)	21
Reptiles (Reptilia)	14
Birds (Aves)	256
Mammals (Mammalia)	100

For practical purposes the following categories should be distinguished: habitats far from nature, such as urban areas with ›green urban spaces‹ or roadside verges, intensely managed agro-ecosystems and the more traditional rural landscape of Central Europe (Mühlenberg and Slowik 1997; Plachter 1999b).

Thus nature conservation has to accept that besides natural or near-natural areas with only limited kinds of human impact many types of semi-natural and cultural landscape are likely to be of high conservation value (Rackham 1998). Conservationists are on weak ground when they criticize cultural landscapes for not being primaeval. Historical ecology helps to choose and designate conservable sites and identify appropriate conservation measures (Rackham 1998). We get to the point where we accept that the many cultural landscapes in Central Europe may help to conserve biodiversity.

2.3 General aspects of biodiversity in Germany

Central Europe is characterized by a high diversity of landscapes, habitats, sites with special structures and species. Riecken et al. (1994) are able to present a catalogue of 509 habitat types occurring in Germany.

As a very rough estimate, about 28,000–30,000 plant species and 45,000–50,000 animal species occur in Central Europe. For a selected set of taxa, species numbers are presented in Table 1.

The message taken from the synopsis of the German flora and fauna might be misleading. The list does not tell anything about population size and population structure. The patchy landscape of Central Europe abounds with small, often fragmented populations, which may suffer severe limitations (see below). Additionally, Germany is very poor in endemic species, and there are no hot spots of species diversity – a situation that is typical for many temperate regions and, by the way, may help in species conservation.

2.4 Biodiversity in natural and non-natural habitats

As a rule of thumb, a natural or near-natural terrestrial habitat in temperate latitudes contains 2,000 and more (probably rarely less) species of plants and animals. This has been documented for forests (Schaefer 1996, 1999), where up to 5–10 % of the species of the regional Central European pool occur, depending on the taxon in question (cp. Table 2). The species number of a community results from a balance between colonization and extinction. A source for colonization of a habitat is the regional species pool; colonization rates depend on abundance, geographic range, dispersal power and the degree of euryoecy of the species (Fig. 1). Species numbers in a habitat are determined by many abiotic and biotic factors; important are habitat size, favourableness of the habitat, spatiotemporal heterogeneity, productivity, diversity of food resources, interspecific competition, predation, disturbance and stochasticity and anthropogenic stress. The probability of extinction depends on abundance, genetic state of the population and stochastic influences.

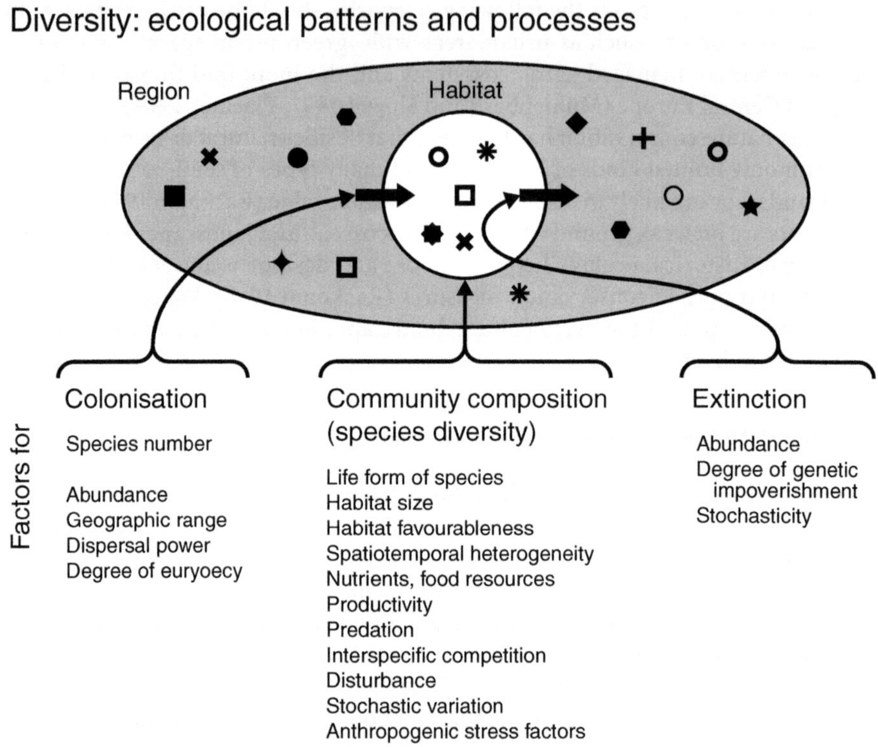

Fig. 1. Species diversity in a habitat as the result of colonization from the species pool, extinction and intra-habitat influences. Symbols = populations of different species

Interestingly, species diversity in non-natural man-made habitats of the open landscape, such as the agrarian landscape, roadside verges, urban green spaces or botanical gardens can be in the same order of magnitude as compared to more natural habitats. This is documented by the species numbers of selected taxa in agricultural land and in roadside verges (Table 2) (Schaefer 2003; Duelli and Obrist 2003).

Even very artificial habitats such as botanical gardens or the buildings in open-air museums may contain high numbers of species. Braun (1997) found 128 bee species (Hymenoptera: Apoidea) in the two botanical gardens of the city of Göttingen, that is 40 % of all species occurring in Lower Saxony; 23 species were specialists – oligolectic flower visitors. 187 species of wasps and bees (Hymenoptera Aculeata) - some of them rare specialists – inhabited an open air museum in the city of Detmold (Glöggler 1997), 38 % of all North Rhine Westphalian species.

2.5 Recent patterns of change in species diversity

The composition of the Central European fauna is continuously changing; populations may increase or decrease or become extinct. The causes are manifold: a dominant factor is anthropogenic disturbance from the local (destruction/ modification of habitats) to the regional scale (pollution, landscape changes); other influences are natural climatic changes or stochastic events. The general trend is a decline in species numbers in many taxa (Aßmann and Härdtle 2002); however, for lower plants and many invertebrates time-series analyses for the last decades are lacking, in contrast to many higher plants and vertebrates. One of the few noteworthy exceptions is the detailed documentation of population changes of 64 species of the Netherland carabid fauna by Den Boer (1990). As a consequence of habitat destruction and fragmentation, many carabid species with low dispersal power have become isolated in small remnants of their habitat and appear to be doomed to extinction. Biological invasions are another component of diversity dynamics. There is a reliable statistics about the neobiota for plants and animals; about 400 neophytes and 300 neozoa occur in Central Europe (Sukopp and Trepl 1987; Hartmann et al. 1994; Gebhardt et al. 1998; Kratochwil and Schwabe 2001; Kowarik 2003). For lack of space, I cannot go into more detail (cp. the reviews in Hawksworth 1974). Instead, I will shortly discuss some well-studied examples of diversity and population change for the German fauna.

Leafhoppers and planthoppers (Auchenorrhyncha)

At present, altogether 620 Auchenorrhyncha species are known from Germany, a reasonable estimate of the real species number should certainly exceed 650 and may even approach 700 (Nickel 2003). Five species are apparently irregular immigrants, three species are neozoa. Long-term population changes are difficult to substantiate. Population declines were documented for 35 species, mainly inhabitants of peatland, river banks, dry grassland and sparse vegetation on sandy soils (Nickel 2003). The Red Data Book for Auchenorrhyncha lists three species as extinct, 56 species as critically endangered and 72 species as endangered (according to IUCN categories) (Bundesamt für Naturschutz 1998).

Grasshoppers (Ensifera and Caelifera)

35 species of Ensifera and 44 species of Caelifera occur in Germany. Few species increase their range, many species go through population declines (Ingrisch and Köhler 1998). In Germany 249 cases of regional extinctions were documented (Köhler 1999). Habitat destruction and change were the main reason; however, about one third of the cases of extinction might have been caused by stochastic processes. Two species are extinct, 13 species are critically endangered, seven species are endangered. They are mainly inhabitants of river banks, sparsely vegetated dry habitats, alpine grasslands and wetlands (Bundesamt für Naturschutz 1998).

Table 2. Species numbers of some animal taxa in a forest, in the agrarian landscape and at roadside verges in Central Europe; most of the habitats were situated in southern Lower Saxony (from Schaefer 2003; see here for further details)

Taxon	Species number			
	Central Europe	Forest	Agrarian landscape	Roadside verges
Protozoa (only Testacea)	900	65	19–27	
Nematoda	1,100	90	30–90	
Gastropoda	350	30	low	
Enchytraeidae	100	36	17	
Lumbricidae	70	11	6–9	11
Araneida	900	102	169–194	196
Opilionida	40	8	15	17
Gamasina	1,000	80	28	
Oribatida	450	75	13	
Isopoda	40	6	1–13	
Chilopoda	50	10	low	
Diplopoda	130	6	10–19	
Collembola	300	48	40	
Carabidae	500	24	87–98	98
Staphylinidae	1,300	85	130–204	206
Diptera	8,000	299	224–440	

Butterflies and moths (Lepidoptera)

Many documented extinctions for regional lepidopteran faunas in Europe exist, as it is comparatively easy to monitor population changes for butterflies and moths (Heath 1974; Blab and Kudrna 1982; New 1997). For instance, in the county of Suffolk (UK), between about 1850 and the 1980s, 44 % of butterflies had become extinct (Thomas 1991). Habitat degradation and destruction is the single most important and widely recognized agent of butterfly extermination and decline (New 1997); categories of threat are in decreasing importance: intensification of agriculture, grassland abandonment, afforestations, floodplain destruction, civilization (housing, etc.), moorland drainage. Among the approximately 1450 German Macrolepidoptera species 34 are extinct, 99 are critically endangered and 161 species are endangered (Bundesamt für Naturschutz 1998).

Dragonflies and damselflies (Odonata)

80 odonate species occur in Germany, among them nine invasive species (Schorr 1990). Main threats to dragonflies and damselflies are the loss of water bodies,

modification of ditches and rivers, drainage, use of pesticides, water pollution and overstocking with fish (Buchwald 1992; New 1995). In Germany, threats to the odonate fauna are extraordinarily high because of its dependence on water bodies: two species became extinct, 12 species are critically endangered, 18 species are endangered (Bundesamt für Naturschutz 1998).

Birds (Aves)

Part of the Central European breeding bird fauna is highly threatened. 240 species are autochthonous, 27 species (among them 10 neozoa) appeared or reappeared between 1970 and 1994 (Bundesamt für Naturschutz 1998). The Red Data Book for Germany lists 256 species, 16 of them are extinct, 25 are critically endangered and 24 are endangered. Böhning-Gaese and Bauer (1996) analysed possible causes of changes in species abundance, range size, diversity, extinctions and colonisations for a bird community around Lake Constance in Southern Germany. For a time period from 1980–1981 to 1990–1992, changes in regional abundances of 151 coexisting bird species were influenced by breeding habitat and migratory status. Significant declines (or even extinctions) were found in populations of farmland species and long-distance migrants. Opdam and Wiens (2002) emphasize that changing land use and habitat fragmentation are the greatest threats to bird species in the temperate zone.

In summary, habitat destruction and habitat changes appear to be the dominant factors for faunal change. Hence, endeavours for conservation of biodiversity and possible management strategies should focus on the analysis of habitats and landscapes.

3 Problems: from the threats to the biota to the design of management strategies

Summarizing the basic patterns, Germany is a small country with a dense human population, necessity for agriculture, significant pressure on natural habitats and a high proportion of man-made habitats. On the other hand the diversity of the flora and fauna can be high, even in many anthropogenic habitats. However, there is a trend that species numbers and (in many cases) population sizes are steadily declining. It is necessary to categorize and evaluate the threats in order to find possible means of remediation.

In Central Europe three typical ecological situations for plant and animal populations exist: (1) occurrence of small populations in patches, liable to decline because of environmental stress factors or stochastic influences, with colonization and extinction events; (2) ›neutral‹ cases with no population trends; (3) development of large persistent populations patchily and temporally increasing because of their adaptedness to man-made cultural landscapes.

The job of the ecologically oriented nature conservationist is to draw the bow from basic ecological knowledge to preservation of species diversity. For conservation of biodiversity in Germany we should be preferably informed about the spatial dynamics of (mostly small) populations, anthropogenic stress factors, criteria for conservation and management measures for habitats. I will not discuss other important facets of the management of natural resources, such as legal activities (cp. Hellenbroich in this volume), economic aspects (cp. Bräuer in this volume), and the role of public policy and opinion (cp. Garrelts, Birner and Wittmer in this volume).

3.1 Population ecology: small populations and spatial dynamics

Small populations are assumed to be under the highest threat of imminent extinction and have been the focus of conservation scientists (Simberloff 1998b). Conservation biology is the ecology of small populations; hence it is no coincidence that many generalizations or principles guiding conservation rely on ideas offered by the theory of island biogeography (Noss 1999). The concept of minimum viable population size (MVP) implies that populations become extinct at some critical threshold, with demographic and environmental stochasticity as key forces.

Additionally, four main genetic processes threaten small populations disproportionately: inbreeding depression, genetic drift, mutational meltdown (the fixation of harmful alleles) and hybridization (outbreeding depression) (Simberloff 1998b; Aßmann and Härdtle 2002; Pullin 2002). However, contemporary evolution might compensate negative trends (Stockwell et al. 2003). Habitat degradation can lead to novel selection pressures, and certain species might be able to adapt to such a detrimental habitat change.

Population size is linked to habitat size. The effects of fragmentation upon the dynamics of populations and communities have thus become a major focus of conservation biology. Fragmentation has impacts above and beyond the simple loss of habitat (Hoopes and Harrison 1998). Within the smaller areas of habitats that remain, centres of populations are more isolated, leading to increased rates of local extinction and decreased rates of gene-flow and re-colonization. In turn, the local extinctions of species that function as pollinators, herbivores, predators, seed dispersers or decomposers may lead to secondary extinctions, community simplification and even changes at the ecosystem level. In recognition of such habitat geometry, three spatial ecological theories are relevant for conservation: metapopulation theory, source-sink dynamics and disturbance (Hoopes and Harrison 1998).

Metapopulation theory

examines the dynamics of sets of semi-independent populations connected by dispersal (Hanski 1999) and makes predictions for extinction and colonization in patches of a population. Local population extinctions appear to be a relatively common occurrence especially in small habitat patches. The probability of local extinction is related not only to the area of the local habitat patch and distance to the nearest source of colonists, but also to the distance of unoccupied suitable habitat patches (Mawdsley and Stork 1995). It leads to practical issues such as habitat connectivity, corridors and stepping-stones. However, it is extremely unlikely that metapopulation models can ever be the basis for defining safe strategies for habitat loss, given their requirements for large amounts of usually unobtainable data, and the inherent high degree of stochasticity in their outcomes (Hoopes and Harrison 1998). Additionally, the term metapopulation is too quickly applied to any case where populations or habitats are discontinuous (Reich and Grimm 1996).

Source-sink models

are applied for the dynamics of populations in habitat patches of different qualities: good habitats (sources) with net positive population growth and poor habitats (sinks) with net negative population growth. Local extinctions will be much more common in sink populations than source populations (Hoopes and Harrison 1998). The models also show that certain habitats may have a disproportionate importance that is not always evident from the species' abundance. A poor habitat can affect the viability of populations in adjacent good habitats, and small changes in the distribution of good and poor habitat may lead to relatively large changes in the species' viability. However, confusion arises because the »source-sink« label is too readily placed on any case in which population densities or habitat quality are spatially variable (Hoopes and Harrison 1998).

Disturbance dynamics

refers to the interplay between species diversity or abundance and the scale, rate and intensity of disturbance, which is defined as a rapid loss of a large fraction of the standing biomass of an area (Hoopes and Harrison 1998). »Intermediate disturbance« may lead to higher levels of diversity, because coexistence of species is promoted in a spatial mosaic of patches at different stages of succession. Disturbance dynamics is an important concept for natural (e.g., a river landscape) and anthropogenic habitats (e.g., dry grassland) because it sets the scene for management practices such as fire or grazing. Reserve size should be determined by the »minimum dynamic area«, i.e., the minimum area large enough to incorporate a »shifting mosaic steady state« of patches at different stages of recovery from disturbance (Hoopes and Harrison 1998).

Thus a scenario for a typical endangered population in Germany might be: it occurs in habitat patches with lower density, with local genetic changes, with immigration/emigration events (may be in a metapopulation situation) and local extinctions, with evolutionary changes as adaptation to a variable, stressful environment. There may be key habitats for subpopulations. The species might be dependent on early successional stages. However, it is doubtful if spatial theory as developed above can fulfil the needs of conservation of biodiversity in practice.

3.2 Anthropogenic stress factors on a macro- and mesoscale

Current human impacts are manifold. They can range from local to global effects (New 1995; Pullin 2002; Aßmann and Härdtle 2002). The most direct threat to species diversity comes from destruction of the habitat on which it depends. A further negative influence can be habitat fragmentation or monotonisation of the landscape. More indirect effects result from habitat change. A global phenomenon is climate change; regional stress factors are effects by immissions, above all eutrophication by nitrogen oxides and acidification by sulphur dioxide. More local influences are mechanical disturbance, environmental pollution and application of biocides. The introduction of exotic species can lead to changes in community composition and the extinction of native species. A very direct threat to populations comes from overexploitation or overcollecting. In summary, the combined effects of habitat loss and fragmentation, pollution, climate change and the introduction of exotic species represent an imminent threat to biological diversity in temperate latitudes.

A catalogue of possible stress factors might not be of much help for designing conservation measures in Germany. What counts are their relative importance and the mode of impact on populations. Patterns of response depend on the life-form of the species; it makes a difference if a member of the soil microfauna or a large carnivore is considered. A measure unit for the interaction system ›possible stress factor – target population‹ is the sensitiveness of the species. For animals, susceptibility to harm or extinction is determined by body size, temporal variability, dispersal ability, generation time, intrinsic rate of increase and the need for specific resources (Mawdsley and Stork 1995). Thus, there are sensitive species and less sensitive species; the first category is the supplier of bioindicators. Susceptibility to anthropogenic stress factors must not necessarily lead to increased mortality. Deterioration in a population can be quantified by fluctuating asymmetry (New 1995; Aßmann and Härdtle 2002) or the presence of heat-shock proteins (Sorensen et al. 2003). Once again it has to be emphasized that contemporary evolution may modify and modulate the ecological situation of a species, an important feature of the populations in the cultural landscape (Stockwell et al. 2003).

3.3 Criteria for conservation

Targets for conservation can be: species populations, plant communities, habitats (with the associated communities) and complexes of habitats (with the associated communities) (Kratochwil and Schwabe 2001). The standard procedure is to catalogue such biotic units and decide about possible strategies for conservation. For Germany, Red Data Books for plant and animal taxa (Bundesamt für Naturschutz 1996, 1998), for plant communities (only for parts of Germany) and for habitats (Riecken et al. 1994) have been be composed.

Conservation of species

There are major problems with targeting plant and animal species for conservation. One important criterion for conservation measures is the degree of vulnerability. Additional aspects are function in ecosystems, role as an umbrella population, charisma or aesthetic characteristics.

Two difficulties arise: (1) vulnerability and extinction assessment is not easy (cp. Sect. 4); (2) decisions have to be made which species should be conserved. Are there more precious species? Should we care about inconspicuous, ugly, detrimental species, too? Which is our attitude to undesirable pests? The question of ascribing a ›conservation value‹ to a species cannot be answered unambiguously. A more comprehensive concept is needed. This might be the habitat approach.

Conservation of plant communities, habitats and complexes of habitats

Many authors highlight the importance of the quality of the habitat in preserving populations of rare, endangered species; examples are butterflies (Thomas 1991), leafhoppers and planthoppers (Nickel 2003), grasshoppers (Ingrisch and Köhler 1998), carabid beetles (den Boer 1990). Thus, a possible conservation strategy might be centred on habitats and landscapes.

However, the questions of attributing a ›conservation value‹ to habitats are similar to those concerning species populations: Which sites should be conserved? Are there more precious sites? How about the values of anthropogenic sites? A criteria-based evaluation of sites includes: diversity, rarity, size, naturalness, representativeness and other relevant ecological factors (Bibby 1998; Kratochwil and Schwabe 2001; Aßmann and Härdtle 2002). Additionally, the ecological history of a habitat is important. For instance, a forest can be the result of different management histories; it can be a high forest, a coppiced wood with standards or a coppiced wood. Cultural criteria concern potential use for education, recreation, amenity or science; they are thus founded in the subjective evaluation by man and are not very satisfactory. Combining scores across different criteria is arbitrary and in many cases inappropriate. Important notions, such as representativeness have proved particularly intractable to measurement and to combination (Bibby 1998).

3.4 Necessity of managing habitats

In Central Europe human intervention has destroyed most of the pristine habitats (cp. Sect. 2). I discussed the key influence of disturbance (Hoopes and Harrison 1998) on biota and landscape in Germany in the context of population ecology and spatial dynamics. Part of the landscape in Germany was and is heavily influenced by man. Thus management must be an integral part of nature conservation and the conservation of biodiversity.

It is important to develop conservation measures on the basis of a restoration of original habitats and mimicking disturbance within the dualism natural disturbance – anthropogenic disturbance. Three objectives can be distinguished: (1) restoring original pristine habitats (such as *Sphagnum* bogs, coastal dunes, river banks); (2) allowing and/or imitating natural disturbance (mainly during succession); (3) allowing and/or imitating anthropogenic disturbance (such as coppicing, cutting of hedges).

A major reason for the need of intervention is that many species of conservation importance are dependent upon uncommon, early successional habitats which have to be maintained at an early ecological state by cutting, grazing, burning or disturbing (Morris 1991; Usher 1995; Sutherland 1998). Widespread cases of management concern the preservation of grassland and *Calluna* heath by cutting or grazing, coppicing of forests and hedges and the creation of mosaics of young and old forest stands. Management may try to mimic natural processes. However, in many instances it may result in a very different habitat, e.g., clearfelling imitating the effect of fire or windbreaks or cutting as a substitute for grazing.

The conservation strategies in the Lower Oder Valley National Park (*National Park Unteres Odertal*) comprise all three objectives and document the problems associated with habitat management (Rothenbücher, Bentlage and Just in this volume). Restitution of the annual flooding regime leads to the restoration of original wetland habitats and allows natural disturbance. When natural succession is allowed, agricultural land is superseded by shrubland and forested areas. Mowing of grassland vegetation prevents tree growth and helps to preserve anthropogenic habitats. There is the prospect for the whole park that a large number of species is preserved; however, the specialists preferring early successional phases might decrease in importance in many parts of the park.

4 Prospects: from the species to the landscape approach

Summarizing the discussion in the preceding sections, a complex set of factors is responsible for the patterns of species population and habitat change. In many instances, small size of populations, spatial dynamics and disturbance at early successional stages may be of paramount importance. For the conservationist, there are severe limits in coping with this ecological situation. A remedy should take into account ecological processes and be preferably based on ecological dynamics in the habitats.

4.1 Basis for conservation: the ecology of the species

An important step in species conservation is the knowledge about the species in question. If I take measures to protect a population against adverse, unfavourable, threatening influences, I should know much about the biology and ecology of the species. However, there are limits to understanding the position of species populations in the ecological web.

Much of the existing theoretical framework for investigating threats to species and probabilities of extinction is based on studies of large vertebrates and higher plants (Primack 1998, 2000; Pullin 2002), but the basic ecological characteristics of invertebrates and lower plants are very different from those of vertebrates and higher plant taxa (New 1995; Mawdsley and Stork 1995), with in many cases specific microhabitat requirements, peaks in mid-successional phases and more idiosyncratic responses to anthropogenic disturbances. Detailed ecological knowledge would be helpful or even essential for species conservation. For species depending on a specific biotic resource (e.g., herbivores on food plants) the precise knowledge of this interaction would support conservation measures. Rarity is a complex phenomenon; conservation of rare populations would be easier if the causes for rarity were clear. Species with a large home range, with complex life cycles and/or habitat requirements pose special difficulties, because they can only be preserved if their diverse interrelationship with the environment is understood. Examples for Germany are larger carnivores such as lynx, *Lynx lynx*, or otter, *Lutra lutra*, or larger birds such as capercaillie, *Tetrao urogallus*.

An illustrative example for a complex approach in species conservation is the pearl mussel *Margaritifera margaritifera*, which still occurs in some running waters, but is threatened with extinction, with much of the decline due to effects of eutrophication on young mussels, probably through increased sedimentation of material into the hyporheic zone (Bauer 1991; Silkenat et al. 1991; New 1995). Acidification is detrimental to the host of the parasitic glochidia, the brown trout *Salmo trutta*. Many Central European populations lack young mussels, which reach adulthood at the age of about 20 years. Management strategies include reduction of pollution (eutrophication and acidification), control of engineering of running waters, restocking rivers with brown trout and preventing the introduction of exotic salmonids.

In most instances the species conservationist has to get along without detailed ecological information about the species in question; a corollary of this is that predicting the extinction of specific populations is difficult (mainly among invertebrates or lower plants).

4.2 Programs and strategies for conservation

Instruments for the conservation in practice are available on several levels. Conservation strategies are developed from basic ecological data on species and habitats, if available. They have to orient themselves within the framework of local, regional and global legislation.

World-wide *legal instruments* include the activities of the International Union for the Conservation of Nature and Natural Resources (IUCN, founded 1948) publishing Red Data Books, the Convention on International Trade in Endangered Species of Wild Fauna and Flora (CITES, 1973) and the Environmental Summit in Rio de Janeiro (1992) with the Convention on Biodiversity. Regional important European treaties are two EU directives: the Wild Birds Directive (1979), which requires the establishment of special protection areas for birds and the Directive on the Conservation of Natural Habitats and of Wild Fauna and Flora (1992), which requires the establishment of special areas for conservation. Nature conservation in Germany has a long tradition dating back to the first half of the 19th century (Plachter 1999a, 1999b; Aßmann and Härdtle 2002). Important legislative outputs are the Federal Nature Conservation Law (*Bundesnaturschutzgesetz*) and the Federal Species Conservation Act (*Bundesartenschutzverordnung*).

It can be discussed if these legal rules are a sufficient basis for action in nature conservation, for appropriate *concepts and strategies* for the conservation of biodiversity. I recognize six (partly overlapping) approaches in the field of the preservation of species diversity, of biodiversity, three of them with the focus species populations, three of them with the focus habitat. The last procedure is a mixed habitat/species approach.

- Red Data Books and the protection of species
- Population viability analysis; the study of spatial dynamics of populations
- The study of focal (umbrella, flagship) species, in a functional context of keystone species
- Red Data Books and the protection, preservation of habitats
- Foundation of protected areas (from nature reserves to national parks)
- Maintenance and/or creation of a diverse landscape

4.3 The species approach: preservation of species; the problem of Red data books

Species belong to different categories. For instance, they can be sorted according to apparency (cryptic – non-cryptic), utility for man (useful – not useful) or harmfulness for man (detrimental – not detrimental). Conservation-focus species (›focal species‹) are umbrella species, flagship species, keystone species, threatened species (Heywood 1995; Simberloff 1998a).

Umbrella species are those whose occupancy area (plants) or home range (animals) are large enough and whose habitat requirements are wide enough that, if they are given sufficiently large area for protection, will bring other species under that protection. Flagship species are popular, charismatic species that serve as symbols or rallying points to stimulate conservation awareness and action. Keystone species are species whose impact on their community or ecosystem are larger and greater than would be expected from its relative abundance (Heywood 1995). Threatened species are species that are rare, often genetically impoverished, of low fecundity, dependent on patchy or unpredictable resources, extremely variable in population density, persecuted, or otherwise prone to extinction in human-dominated landscapes.

The fundamental problem is that the conservationist with the species approach has much to know about the species he wants to protect. As discussed above, many data are not easy to obtain, are not reliable; work is time-consuming and cannot be done for many species. Species vulnerability analyses are profitable in determining the most threatened habitats (in terms of species conservation) and the most serious threats to species, but uncovering the ecological attributes that make them prone to extinction is a much more difficult task. Much insect conservation has focused primarily on species, but despite considerable effort, very little is known of the vulnerability and conservation needs of most species (New 1999).

Population dynamics is determined by population size and population structure. In the field of species conservation the small population paradigm SPP and the declining population paradigm DPP prevail (Simberloff 1998b). Three of the most popular spatial ecological theories are metapopulation dynamics, source-sink dynamics and disturbance dynamics. Each of these theories has some important insights to offer for species conservation and management of populations (Hoopes and Harrison 1998) (cp. Sect. 3). However, it is difficult to apply them to specific real-world situations, such as a system of habitat patches with connecting corridors and stepping stones interspersed (Hoopes and Harrison 1998; Stockwell et al. 2003). Currently, not much is known about the functioning of corridors or habitat connectance. Summing up, it is very difficult to interfere in a complex ecological situation that is poorly understood. Additionally, many species appear to be stable and secure although they occur in low numbers and over very small ranges. Hence, rarity per se cannot be a criterion for conservation ranking (New 1999).

Thus two very practical difficulties arise: assigning priority to a species and conceiving species action plans. Priority is often defined on imminence of threats; other possible options are ›umbrella, flagship, keystone, apparent, useful species‹. Cultural values of species can also be important. However, there might be a difference between threat and priority (New 1999). If species action plans for vertebrates (and invertebrates or plants) can be designed, they are an important tool to guide immediate and long-term management, policy and decision-making for selected species. However, these types of species approach may fail, as it is the case for the capercaillie in the Harz National Park.

Widespread bases for species conservation are Red Data Books (RDBs). There are several criteria for compiling the lists, for example species range, degree of endemism, population density, population trend, rarity, habitat range, dispersal power, resilience to environmental changes (New 1995; Bundesamt für Naturschutz 1998; Kratochwil and Schwabe 2001). Red lists and protected species schedules for higher plants, mammals, birds, and other vertebrates have a high degree of reliability and consensus over priorities and generally represent far more than simple availability of names on a list. However, for the majority of taxa the number of shortcomings is high, and it is debatable, whether the Red Data Book approach indeed is optimal for most invertebrates or lower plants. I will mention only few of the reservations.

The data base can be insufficient. For Great Britain, Red Data Book status of insects is not clearly related to ecological factors; instead, it is mainly correlated with the two measures of distribution – the number of 10 km squares occupied and the number of different habitats occupied (Mawdsley and Stork 1995). The dependence upon distribution is expected because it is an important criterion used to assess RDB status. Thus, in essence this correlation is a vicious circle. Knowledge of population ecology of most of the listed species is scarce. This raises the question about the reliability of the entries.

A basic flaw in Red Data Books is their regional validity. There are Red Lists for the German States, for the whole of Germany, Europe, and the world. Many of the nominations for listing are widely distributed taxa whose ranges transcend political boundaries and which may be listed in only parts of their range, perhaps the geographical fringes (New 1999) or isolated populations (Gärdenfors 2001). Priority attention to saving isolated populations of species, which are secure elsewhere, may seem a poor use of resources. However, for setting conservation priorities the taxonomic, genetic or ecological uniqueness of the regional subpopulation has to be considered (Bourn and Thomas 2002).

Lists abound with species names, all of them candidates for conservation measures. However, available expertise and funds are grotesquely inadequate for the task of conserving all listed taxa (New 1999). The legal obligations to investigate the status of the listed species individually, to study them sufficiently, to design and put into practice optimal management and recovery plans for each, and to pursue those plans within a given period only rarely can be met under any currently feasible supply scenario.

The obvious thing to do is to use a short-cut (Simberloff 1998a). Because the threatened status of a species usually reflects changes in the habitat rather than direct human influences on the species itself, a habitat conservation approach by listing species of the habitat for priority can be adopted (New 1999). A further step and a true second option is the ›pure‹ habitat approach circumventing some of these difficulties, because knowledge of the species is not required.

4.4 The habitat approach: preservation of rare habitat, foundation of larger Protected Areas

The information basis is a catalogue of habitats; an important methodological tool is the grouping of plant communities. Riecken et al. (1994) list 509 types of ecosystems, 69 % of which are endangered.

For many *habitats* the conservation value is obvious, because they contain rare specialists. Examples are caves with troglobionts, running waters with a torrenticolous fauna, small water bodies with a distinct invertebrate fauna, forest with dead wood inhabited by saproxylic insects, dry grassland with xerophilous and thermophilous insects. However, in many cases the evaluation of a habitat in relation to its conservation value is difficult (cp. Sect. 3). A further problem are management strategies. For most management recommendation, it is difficult to find any objective evidence to justify them (Sutherland 1998).

The dispersion of the habitats and their connectance in the landscape can be a specific problem: Under habitat degradation and fragmentation, the restoration of population connectivity and gene flow might be a management option. However, if fragmented populations have diverged appreciably, efforts to initiate or restore gene flow could result in diminished adaptation and increased risk of extinction (Stockwell et al. 2003).

National Parks as *larger protected areas* might favour the biodiversity in habitat complexes. The non-existence of larger pristine or near-natural habitats, lack of space, high density of the human population in Germany are not a good starting point for the foundation of larger protected areas. Only few National Parks exist, for instance Harz, Unteres Odertal, Wattenmeer, Bayerischer Wald. Problems are: the inclusion of cultural habitats, diversity loss without management (e.g., by the increase of forest habitats), the role of agriculture, the danger of non-management (e.g., in the case of gradations of bark beetles). It is difficult to create and maintain core areas.

4.5 The comprehensive or landscape/habitat/species approach: preservation, restitution and/or creation of a diverse landscape

In practice, nature conservation still focuses on the protection of selected species and ecosystems. This goal should be pursued further. An *additional* safeguard for biodiversity is the preservation and the increase of the diversity of the whole landscape with natural, semi-natural, agricultural and artificial habitats interspersed. I call this the *landscape/habitat/species approach* or *comprehensive approach.*

The approach is simple. The basic strategy is the conservation of the diverse elements of the Central European landscape, thus preserving species, habitats, complexes of habitats and landscapes together with the associated processes. For that goal research is not a prerequisite. Helpful would be the creation of a network of habitats. Attention should be paid to rare, near-natural and high-quality cultural habitats interspersed in the landscape matrix. Additionally, many disturbed ecosystems are important to conservation (Phillips 1998; Pullin 2002). Restoration ecology should be included in biodiversity conservation (Young 2000). As even man-made cultural habitats may be a refuge for threatened species, an additional focus should be on the landscape pattern in agricultural land and in settled areas (Fry 1991). Obviously, a mixture of natural and disturbed/anthropogenic habitats favours the maintenance and development of biodiversity in cultural landscapes. According to studies of Duelli and Obrist (2003), the conservation of natural and semi-natural habitats, or the creation and maintenance of new semi-natural areas is the most prominent way to enhance or restore species richness in agricultural landscapes. 63 % of all animal species living in the agriculturally managed areas of the Limpach valley seem to depend on the presence of semi-natural habitats, which function as ›re-sources of biodiversity‹. The habitat blend supports mainly stenotopic species, dispersers, and ecotone species (Duelli and Obrist 2003) and encourages adaptations of species populations to the habitat conditions in the agrarian landscape.

A catalogue of landscape complexes might comprise: standing waters, running waters, coastal wetlands, inland wetlands, forests, alpine habitats, grassland, agricultural land, urban areas, green spaces in industrial regions or roadside verges. For each type of habitat complexes, it is not difficult to imagine specific conservation measures or management strategies to preserve or increase structural diversity on a mesoscale and a macroscale, with human interference ranging from no intervention to milder forms of management until drastic impacts. Examples for landscape elements potentially increasing biodiversity are mixtures of vegetation types, patches of vegetation of different age, mixtures of vegetation patches of different height and/or structure, ecotones, transition zones between different habitat/vegetation types, vegetation islands of different size, vegetation strips, mixtures of moist and dry areas, presence of water bodies (ditches, ponds), larger transition zones from wet to dry, piling up of dead organic material, dead wood, patches without vegetation, creation of rocky-type structures, heaps of stones, stone walls, creation and preservation of open sites, mixtures of different degrees and different types of disturbance (by cutting, grazing).

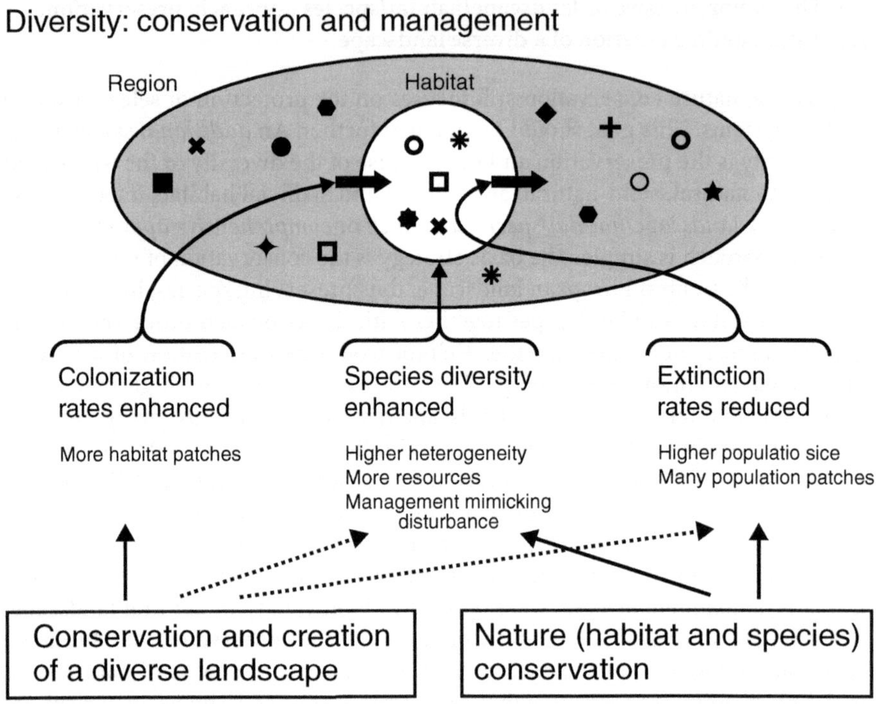

Fig. 2. The comprehensive approach in nature conservation: conservation and creation of a diverse landscape

The scale of the landscape is important for the maintenance of biodiversity, because a highly diverse landscape offers many chances for the persistence and the establishment of declining populations of threatened species and it offers habitat patches with possible colonists for habitats being in the focus of nature conservation (Fig. 2). A higher number of habitat and population patches enhances colonization rates and reduces extinction rates for species in certain habitats. Species diversity in these habitats is preserved and favoured by higher landscape heterogeneity and higher diversity of resources. Management mimicking disturbance should be included. Thus the wheel turns full circle: the basic ecological conditions for the development of diversity (cp. Fig. 1) can be translated into a landscape context of the preservation of diversity (cp. Fig. 2). A diverse landscape pattern increases the chance to preserve a number of threatened species without paying attention to them or even noticing them. The advantage is manifold: no expensive research is needed, no difficulties arise to decide for priority species or for priority habitats, no mistakes are made in managing threatened species populations.

5 Synthesis and conclusions

Nature conservation can only be a compromise between human demands and requirements, utilization of resources, unavoidable negative effects on the environment, (passive) non-utilization of parts of the landscape and active management of habitats and support for the biota. It may be presumed that the pressure on the biota can only partly be alleviated. There are limits for conservation of biodiversity.

The situation for conservation of biodiversity in Germany has positive and negative sides. About 80.000 species of plants and animals exist in a diverse landscape with much anthropogenic disturbance of different kinds. Few endemic species occur, no hot spots of species diversity had developed (a fact that helps in species conservation). The open landscape was mainly created by man and is characterized by a high diversity of a partly stenoecious flora and fauna. The Central European region is a complex, in many parts steadily changing container filled with the biota. The population dynamics of the species is largely shaped by habitat fragmentation and by a dynamic subpopulation structure with distribution on habitat islands and by much influence of stochasticity. It should be accepted that man is part of the system.

Thus conservation of biodiversity must happen in complex systems with in many instances complex ecological situations. The conservationist has to include the cultural landscapes, which can be species-rich and can harbour plant and animal populations of conservation value.

For the majority of species a precise population analysis is not possible because of the complex spatiotemporal dynamics in a fragmented landscape. In many instances no clear indication of vulnerability emerges. As a consequence, difficulties arise to identify criteria for conservation of species and/or habitats. I propose a solution: a holistic approach on several scales should be adopted and and many strategies should be pursued in order to maintain biodiversity. I suggest a triple jump of analysing and preserving species, habitats and landscapes.

The species approach within the framework of Red Data Books or focal species concept makes a contribution to conservation. However, it is not sufficient because it is often not possible to detect vulnerability and make priorities on the background of ecological data and theory. The habitat approach is important because habitats can be a container with a diversity of species. Preserving (and if necessary managing) habitats can mean preserving a number of species. However, it is difficult to assign conservation values and priority to specific types of habitats. For a *landscape/habitat/species approach* or *comprehensive approach* the diversity of the landscape (natural and cultural) is the focus of attention. Conservation of biodiversity should rely on the diversity of habitats, additionally to a distinct species or habitat approach. The man-made landscapes have to be included. In this context every type of non-natural habitats is important: urban areas – agricultural land – traditional cultural habitats – recreational areas – the part of the landscape devoted to traffic. The comprehensive approach, complementary to species and habitat conservation, helps to find comparatively simple solutions for the maintenance of biodiversity in Germany, without too much bureaucracy and planning.

Acknowledgements

I am grateful to Gerhard Wagenitz, who helped with species numbers of floral taxa. Judith Rothenbücher critically read the manuscript. Bernd Baumgart carefully draw the figures.

References

Aßmann T, Härdtle W (2002) Naturschutzbiologie. In: Härdtle W (ed) Studium der Umweltwissenschaften. Springer, Berlin, Heidelberg, pp 113–223

Bauer G (1991) Plasticity in life history traits of the freshwater pearl mussel – consequences for the danger of extinction and for conservation measures. In: Seitz A, Loeschcke V (eds) Species conservation: a population-biological approach. Birkhäuser Verlag, Basel, pp 103–120

Bibby CJ (1998) Selecting areas for conservation. In: Sutherland WJ (ed) Conservation science and action. Blackwell Science, Oxford, London, Edinburgh, pp 202–219

Blab J, Kudrna O (1982) Hilfsprogramm für Schmetterlinge. Naturschutz Aktuell 6. Kilda-Verlag, Greven

Böhning-Gaese K, Bauer HG (1996) Changes in species abundance, distribution and diversity in a Central European bird community. Conservation Biology 10:175–187

Bourn NAD, Thomas JA (2002) The challenge of conserving grassland insects at the margins of their range in Europe. Biological Conservation 104:285–292

Braun, C (1997) Die Wildbienen-Fauna (Hymenoptera: Apidae) zweier Botanischer Gärten. Untersuchungen zum Blütenbesuch an einheimischen und nicht einheimischen Pflanzen. Diploma thesis, Göttingen

Buchwald R (1992) Vegetation and dragonfly fauna – characteristics and examples of biocenological field studies. Vegetatio 101:99–107

Bundesamt für Naturschutz (1996) Rote Liste gefährdeter Pflanzen Deutschlands. Schriftenreihe für Vegetationskunde 28. Bundesamt für Naturschutz, Bonn-Bad Godesberg

Bundesamt für Naturschutz (1997) Erhaltung der biologischen Vielfalt. Wissenschaftliche Analyse deutscher Beiträge. Bundesamt für Naturschutz, Bonn-Bad Godesberg

Bundesamt für Naturschutz (1998) Rote Liste gefährdeter Tiere Deutschlands. Schriftenreihe für Landschaftspflege und Naturschutz 55. Bundesamt für Naturschutz, Bonn-Bad Godesberg

Den Boer PJ (1990) Density limits and survival of local populations in 64 carabid species with different powers of dispersal. Journal of Evolutionary Biology 3:19–48

Duelli P, Obrist MK (2003) Regional biodiversity in an agricultural landscape: the contribution of seminatural habitat islands. Basic and Applied Ecology 4:129–138

Fry GLA (1991) Conservation in agricultural ecosystems. In: Spellerberg IF, Goldsmith FB, Morris MG (eds) The scientific management of temperate communities for conservation. Symposia of the British Ecological Society 31. Blackwell Scientific Publications, Oxford, London, Edinburgh, pp 415–443

Gärdenfors U (2001) Classifying threatened species at national versus global levels. Trends in Ecology and Evolution 16:511–516

Gebhardt H, Kinzelbach R, Schmidt-Fischer S (1998) Gebietsfremde Tierarten. Ecomed, Landsberg

Glöggler, J (1997) Untersuchungen zur Hautflüglerfauna (Hymenoptera, Aculeata) eines Freilichtmuseums – Die Nutzung von Gebäudestrukturen als Nisthabitate. Diploma thesis, Göttingen

Hanski I (1999) Metapopulation ecology. Oxford University Press, Oxford

Hartmann E, Schuldes H, Kübler R, Konold W (1994) Neophyten. Ecomed, Augsburg

Hawksworth DL (1974) The changing flora and fauna of Britain. Academic Press, London, New York

Heath J (1974) A century of change in the Lepidoptera. In: Hawksworth DL (ed) The changing flora and fauna of Britain. Academic Press, London, New York, pp 275–292

Heywood, VH (1995) Global diversity assessment. Cambridge University Press, Cambridge

Hoopes MF, Harrison S (1998) Metapopulation, source-sink and disturbance dynamics. In: Sutherland WJ (ed) Conservation science and action. Blackwell Science, Oxford, London, Edinburgh, pp 135–151

Ingrisch S, Köhler G (1998) Die Heuschrecken Mitteleuropas. Die Neue Brehmbücherei 629. Westarp, Magdeburg

Köhler K (1999) Ökologische Grundlagen von Aussterbeprozessen – Fallstudien an Heuschrecken (Caelifera et Ensifera). Laurenti Verlag, Bochum

Kowarik I (2003) Biologische Invasionen: Neophyten und Neozoen in Mitteleuropa. Eugen Ulmer, Stuttgart

Kratochwil A, Schwabe A (2001) Ökologie der Lebensgemeinschaften – Biozönologie. UTB 8199. Eugen Ulmer, Stuttgart

Mawdsley NA, Stork NE (1995) Species extinctions in insects: ecological and biogeographical considerations. In: Harrington R, Stork NE (eds) Insects in a changing environment. 17th Symposium of the Royal Entomological Society of London. Academic Press, London, San Diego, pp 321–369

Morris MG (1991) The management of reserves and protected areas. In: Spellerberg IF, Goldsmith FB, Morris MG (eds) The scientific management of temperate communities for conservation. Symposia of the British Ecological Society 31. Blackwell Scientific Publications, Oxford, London, Edinburgh, pp 323–347

Mühlenberg M, Slowik J (1997) Kulturlandschaft als Lebensraum. Quelle & Meyer, Wiesbaden

New TR (1995) Introduction to invertebrate conservation biology. Oxford University Press, Oxford

New TR (1997) Butterfly conservation, 2nd ed. Melbourne, Oxford University Press, Oxford

New TR (1999) Limits to species focusing in insect conservation. Annals of the Entomological Society of America 92:853–860

Nickel H (2003) The leafhoppers and planthoppers of Germany (Hemiptera, Auchenorrhyncha): patterns and strategies in a highly diverse group of phytophagous insects. Pensoft, Sofia

Noss R (1999) Is there a special conservation biology? Ecography 22:113–122

Opdam P, Wiens JA (2002) Fragmentation, habitat loss and landscape management. In: Norris K, Pain DJ (eds) Conserving bird biodiversity. General principles and their applications. Conservation Biology 7. Cambridge University Press, Cambridge, pp 202–223

Phillips A (1998) The nature of cultural landscapes – a nature conservation perspective. Landscape Research 23:21–38

Plachter H (1999a) A central European contribution to a pan-European conservation strategy. La Cañada 10:11–13

Plachter H (1999b) The contributions of cultural landscapes to nature conservation. In: Hajos G (ed) Denkmal – Ensemble – Kulturlandschaft am Beispiel der Wachau. Berger, Wien, Horn, pp 93–115

Primack R (1998) Essentials of conservation biology, 2nd ed. Sinauer Associates, Freeman, Basingstoke

Primack RB (2000) A primer of conservation biology, 2nd ed. Sinauer Associates, Sunderland

Pullin AS (2002) Conservation Biology. Cambridge University Press, Cambridge, New York

Rackham O (1998) Implications of historical ecology for conservation. In: Sutherland WJ (ed) Conservation science and action. Blackwell Science, Oxford, London, Edinburgh, pp 152–175

Reich M, Grimm V (1996) Das Metapopulationskonzept in Ökologie und Naturschutz: eine kritische Bestandsaufnahme. Zeitschrift für Ökologie und Naturschutz 5:123–139

Riecken U, Ries U, Ssymank A (1994) Rote Liste der gefährdeten Biotoptypen der Bundesrepublik Deutschland. Schriftenreihe für Landschaftspflege und Naturschutz 41. Bundesamt für Naturschutz, Bonn-Bad Godesberg

Schaefer M (1996) Die Artenzahl von Waldinsekten: Muster und mögliche Ursachen der Diversität. Mitteilungen der Deutschen Gesellschaft für Allgemeine und Angewandte Entomologie 10:387–395

Schaefer M (1999) The diversity of the fauna of two beech forests: some thoughts about possible mechanisms causing the observed patterns. In: Kratochwil A (ed) Biodiversity in ecosystems: Principles and case studies of different complexity levels. Kluwer Academic Publishers, Dordrecht, pp 45–64

Schaefer M (2002) Brohmer – Fauna von Deutschland. 21. Aufl. Quelle & Meyer, Wiebelsheim

Schaefer M (2003) Die Bodenfauna in Agrarökosystemen: von der Diversität zum Nahrungsnetz. Nova Acta Leopoldina NF 87, Nr. 328:195–214

Schorr M (1990) Grundlagen zu einem Artenhilfsprogramm Libellen der Bundesrepublik Deutschland. Ursus Scientific Publishers, Bilthoven

Silkenat W, Silkenat M, Klupp R, Schmidt S, Wenz G, Eicke L, Bauer G (1991) Errichtung und Sicherung schutzwürdiger Teile von Natur und Landschaft mit gesamtstaatlich repräsentativer Bedeutung. Erfahrungen mit einem Projekt zur Rettung der Flußperlmuschel. Natur und Landschaft 66:63–67

Simberloff D (1998a) Flagships, umbrellas, and keystones – is single species management passe in the landscape era. Biological Conservation 83:247–257

Simberloff D (1998b) Small and declining populations. In: Sutherland WJ (ed) Conservation science and action. Blackwell Science, Oxford, London, Edinburgh, pp 115–134.

Sorensen JG, Kristensen TN, Loeschke V (2003) The evolutionary and ecological role of heat shock proteins. Ecology Letters 6:1025–1037

Southwood TRE (1988) Tactics, strategies and templets. Oikos 52:3–18

Spellerberg IF, Goldsmith FB, Morris MG (eds) (1991) The scientific management of temperate communities for conservation. Symposia of the British Ecological Society 31. Blackwell Scientific Publications, Oxford, London, Edinburgh

Stockwell CA, Hendry AP, Kinnison MT (2003) Contemporary evolution meets conservation biology. Trends in Ecology and Evolution 2:94–101

Sukopp H, Trepl L (1987) Extinction and naturalization of plant species as related to ecosystem structure and function. In: Schulze E-D, Zwölfer H (eds) Potentials and limitations of ecosystem analysis. Ecological Studies 61. Springer, Berlin, Heidelberg, New York, pp 245–276

Sutherland WJ (ed) (1998) Conservation science and action. Blackwell Science, Oxford:

Sutherland WJ, Reynolds JD (1998): Sustainable and unsustainable exploitation. In: Sutherland WJ (ed) Conservation science and action. Blackwell Science, Oxford, London, Edinburgh, pp 90–115

Thomas JA (1991) Rare species conservation: case studies of European butterflies. In: Spellerberg IF, Goldsmith FB, Morris MG (eds) The scientific management of temperate communities for conservation. Symposia of the British Ecological Society 31. Blackwell Scientific Publications, Oxford, London, Edinburgh, pp 149–197

Usher MB (1995) A world of change: land-use patterns and arthropod communities. In: Harrington R, Stork NE (eds) Insects in a changing environment. 17th Symposium of the Royal Entomological Society of London. Academic Press, London, San Diego, pp 371–397

Wilson EO (1992) The diversity of life. Belknap Press of Harvard University Press, Cambridge, Massachusetts

Young TP (2000) Restoration ecology and conservation biology. Biological Conservation 92:73–83

The Designation of National Parks in German Nature Conservation Law

Tobias Hellenbroich

Institute for Agricultural Law, Georg-August University of Göttingen, Platz der Göttinger Sieben 6, 37073 Göttingen, Germany, email to thellen@gmx.de

Summary. The declaration of rural areas which are valuable in an nature conservative context to national parks has been an important instrument for the protection of nature for a long time. Originated in the USA, the national park model has spread across the entire world. However, the same environmental circumstances as those existing in the USA do not predominate everywhere. Consequently, the US-American concept is not appropriate for many countries.

In the meantime fifteen national parks have also been established in Germany. In this context, the legal prerequisites for the designation of national parks in Germany are not always undisputed. Above all, the fact that Germany, as a densely settled country, no longer has the requisite extensive and pristine natural landscape causes problems. This contentious question culminated in the court case of the *Elbtalaue* National Park, which had already been established, with the result that the declaration of protection was rescinded. This problem is symptomatical for other national parks in Germany, e.g. the *Unteres Odertal* National Park, the German investigative area of the DFG Graduate Training Programme. As a consequence of this, the German federal legislators have eased the legal prerequisites for the establishment of national parks. To begin with, this contribution presents the currently valid preconditions for the designation of national parks in Germany.

Since the concept of national parks has meanwhile become established world-wide, standards for national parks have been developed on an international level. As an integral contribution within its sphere of competence, the IUCN's World Commission on Protected Areas promotes the establishment and the effective management of a world-wide representative network of protected areas. In the course of this task the IUCN worked out a definition and criteria for the national park protection category in good time and ratified a resolution on them at its 10th General Assembly in 1969. The Federal Republic of Germany, as an IUCN contract state, is bound by this resolution. As a consequence, the question as to whether the new statutory regulation still conforms to these international requirements will be positively answered.

1 Introduction

The declaration of rural areas, which are valuable in a nature conservation context to national parks, has been an important instrument for the protection of nature for a long time (Primack 1995 : 455). Originated in the USA, the national park model has spread across the entire world (Lang 1984 : 14ff). The concept of national parks has become more and more popular. Currently in Germany, bringing that instrument of nature conservation to areas such as former industrial areas is under discussion. In North Rhine-Westphalia, initial ideas have been developed to designate an Industrial Culture Ruhr Basin National Park (*Industriekultur Ruhrgebiet*) to protect and cautiously develop an industrial area that has existed since the 19th century (Oebbecke 2001 : 142ff).

Since the concept of national parks has been established, worldwide standards for national parks have been developed at an international level. As an integral contribution within its sphere of competence, the IUCN's World Commission on Protected Areas promotes the establishment and the effective management of a worldwide representative network of protected areas. The IUCN worked out a definition and criteria for the national park protection category and ratified a resolution at its 10th General Assembly in 1969. The Federal Republic of Germany, as an IUCN treaty state, should respect this resolution. Consequently, the question as to whether the new statutory regulation in the German nature conservation law still conforms to these international requirements also arises in this contribution. However, the same environmental circumstances as those existing in the USA do not predominate everywhere. Consequently, the US-American concept is not appropriate for many countries (for German cases Knake 2000 : 282; Zucchi 2001 : 376).

After much political discussion in Germany, efforts are currently being made to designate a few more national parks. The pleasing result of this political debate is the designation of the *Eifel* National Park and another national park that almost at once, fulfils the international criteria: This recently designated *Kellerwald-Edersee* National Park includes widespread beech groves (Zucchi 2002 : 376; Harthun 1999). Furthermore, other areas in Germany are being recommended national park status (FÖNAD 1997 : 285ff).

For this reason, the legal status for the designation of national parks in Germany and the actual international definition and claims for national parks should be presented and compared. If the results show that German law and the international claims are not similar, it will be necessary to discuss which conclusions can be drawn from the results for Germany.

2 The legal status in Germany

The designation of national parks in Germany is regulated in the legislation on nature conservation. In the federal structured Republic of Germany the legislative competence for the legislation on nature conservation is divided between the Federation itself (*Bund*) and the Federation States (*Länder*). The Federation has the power to issue framework legislation (Art. 75 (1) No. 3 Grundgesetz [GG]). The Federation States must enact a law that at least encompasses the general regulations, Art. 75 (1) No. 3 GG. As such, the Federation enacted the Nature Conservation Act (*Bundesnaturschutzgesetz* (BNatSchG)) on December 20th 1976[1]. Since that time there have been four amendments of that law, one small and three extensive ones. The latest amendment occurred in 2002 and it changed the rules on national parks. The legal status before 2002 should be illustrated with its problems in order to give the background for the prevalent rules.

2.1 Before the 2002 amendment

The pertinent rules

As shown, the Federation issued framework legislation on nature conservation in 1976. The *Reichsnaturschutzgesetz*[2], as its predecessor, did not include a protection category for national parks and thus there was a controversial discussion on whether a rule about the designation of national parks should be assimilated in the *Bundesnaturschutzgesetz* (Stich 1972 : 207).

Finally an appropriate rule, paragraph 14 BNatSchG, was admitted to the comprehensive body of regulations after Bavaria as a precursor brought the *Bayerischer Wald* National Park[3] into being (for this purpose Stock 2000 : 202). The wording of that rule was[4]:

> »**Paragraph 14 National parks**
>
> (1) National parks are areas designated on a legally binding basis as areas to be protected on a uniform basis that meet the following criteria:
>
> 1. The area concerned is an entity of major size with specifically characteristic features;
> 2. the criteria defined for nature conservation areas are meet in the greater part of the area,
> 3. the area concerned is in a status characterized by no or little human impact, and
> 4. conserve live and plant stocks, that are, where possible rich in endemic species.
>
> (2) The Federation States shall ensure that national parks receive the same level of protection as that afforded to nature conservation areas, taking into account the necessary exemptions required in view of the size of the area and its human stellements. As far as the protection purpose allows, national parks should be accessible to the public.«

So if more wilderness is desired by conservationists (e.g., Zucchi 2002), in general, the designation of national parks is the right instrument to achieve these desires (regarding the conflict between nature conservation and the use of national parks as recreational areas, see Arnberger and Brandenburg 2001).

---

[1] Bundesgesetzblatt I p 3574.

[2] Reichsnaturschutzgesetz (RNatSchG) from June 26th 1935 (Reichsgesetzblatt. I p 821).

[3] Statutory order from March 5th 1973, Gesetz- und Verordnungsblatt p 212.

[4] Translated by the author.

The criteria defined for nature conservation areas as cited in paragraph 14 (1) No. 2 are taken from paragraph 13 (1) BNatSchG. Nature conservation areas are areas designated on a legally binding basis as areas requiring special protection with regard to nature and landscape as a whole or with regard to components there of

1. in order to preserve biocoenoses or biotopes of certain species of wild fauna and flora
2. for scientific reasons for reasons relating to natural history or national heritage, or
3. because of their rarity, specifically characteristic features or outstanding beauty.

Beyond these material criteria as conditions for the designation of national parks, the Federation obligated the Federation States in paragraph 12 BNatSchG to rule formal and procedural Conditions for the designation of national parks. According to this, the declaration act of the national park had to include a description of the area, the protection purpose and the orders and prohibitions to be complied with to reach the protection purpose. Further, the declaration had to contain the respective management, development and restoration measures or the necessary empowerments. The Federation States were also obliged to set out rules of procedure in accordance with the rule of law (e.g. hearings, time limits, competences of government agencies etc.). Differing from the other categories of protected areas, the *Bundesnaturschutzgesetz* bound the Federation States in paragraph 12 (4) S. 2 BNatSchG to consult the Federal Ministry of Environment, Nature Conservation and Nuclear Safety and the Federal Ministry of Region Planning, Building and Urban Development.

The rules about the national parks – like the rules for the designation of protected areas in general – fulfil a double function (Gassner 1995 : 200). In Germany it is possible to incorporate privately owned land into national parks. So if it is possible to incorporate private pieces of land into national parks, the constitutional rights of the owners might be restricted. For example, estimated uses that the owner has planned might be forbidden in the future, e.g., constructing buildings. Even exercised uses might not be allowed after the designation, such as some kind of agricultural uses. That is why one function of the rule is to protect owners from the illegal restrictions of their property. The conditions, as stated in paragraph 14 BNatSchG, must at least be fulfilled to restrict the constitutional rights of the citizens. On the other hand, these rules also give directives, e.g., which actions and measures in the area concerned are required or forbidden. If paragraph 14 (2) BNatSchG obligates the Federation States to protect national parks to the same level as nature reserves, then the regulations of paragraph 12 (2) BNatSchG became part of the protection-concept of national parks. That means that in national parks any actions destroying, damaging or changing the area or its parts were forbidden, including actions causing lasting annoyances.

The Federation States enacted their own nature conservation acts to implement the general regulations of the Federation. Based on these acts, 13 other national parks were generated so that now 15 national parks exist in Germany. Because of the scope of the general regulations, the rules for the designation of national parks differ from State to State, initially in the legal way in which national parks are designated (Kolodziejcok 2000 : 252). This is the reason why in some states national parks are designated by law from parliament[5] and in other states by statutory order / administrative law from the government[6][7]. This makes a difference when somebody is taking legal protection against the designation of a park. If national parks are designated by parliamentary law, the only way to file a suit against it is to appeal on an institutional issue to the Constitutional Court. If the park is designated by statutory order from the government, in some Federation States it is possible to get a judicial review of the constitutionality of laws in front of a higher administrative court (paragraph 47 (1) No. 2 *Verwaltungsgerichtsordnung*[8] - VwGO).

The case of the Elbtalaue National Park

The Federation State Lower Saxony decided to designate national parks by statutory order from the government. In paragraph 25 (1) of its *Niedersächsisches Naturschutzgesetz*[9] (NNatG)[10] it ruled that areas which

1. are an entity of major size with specifically characteristic features,
2. meet the criteria defined for nature conservation areas in their greater part,
3. are in a status characterized by no or little human impact,
4. in particular, conserve live and plant stocks, that are, where possible rich in endemic species,

consistently should be protected and can be designated as a national park by statutory order from the supreme government agency for nature conservation (Ministry of Nature Conservation in Lower Saxony, paragraph 54 (2) 2nd sentence NNatG).

Thus, the nature conservation act of Lower Saxony is almost a copy of the conditions for the designation of national parks from the *Bundesnaturschutzgesetz*. More precisely, it stated that in Lower Saxony it is possible to make the designation by statutory order.

Based on this authority, the Ministry of Nature Conservation created the *Elbtalaue* National Park[11] which is nearly 82 km long and between 0.2 and 5.5 km wide

[5] Brandenburg, Hamburg, Schleswig-Holstein, Thuringia, Mecklenburg-Western Pomerania and after the *Elbtalaue* judgement also Lower Saxony.

[6] Bavaria (with agreement of parliament!), Hesse, Saxony, Saxony-Anhalt.

[7] No rules about the designation of national parks can be found in the nature conservation acts of Baden-Württemberg, Berlin, Bremen, Northrhine-Westfalia, Rhineland-Palatinate and Saarland.

[8] Regulations governing administrative courts.

[9] Nature conservation act of Lower Saxony.

[10] In the version from April 11th 1994 (Gesetz- und Verordnungsblatt p 155; revised Gesetz- und Verordnungsblatt p 267), last amendment of act from February 11th 1998 (Gesetz- und Verordnungsblatt p 86).

[11] Statutory order from March 6th 1998 (Gesetz und Verordnungsblatt 1998, p 178).

and about 10,900 ha. Among its other contents the statutory order restricted the exercised agricultural land use. Because of these restrictions, a private landowner took legal action against the park and got a judicial review of the constitutionality of laws in front of a higher administrative court of Lueneburg. In its judgement dated the February 22nd 1999, the court decided that the statutory order did not conform to higher-ranking rules and therefore it was declared null and void. The court gave in essence the following reasons for its decision:

The area of the national park designated by the Ministry did not fulfil the conditions required by paragraph 25 (1) No. 3 NNatG. In the opinion of the court the designated grounds were not characterised by no or little human impact. The court stated that 35.9 % of the designated area was used as grassland and 4.2 % as arable farm land. Even in the forests, (about 24.5 % of the national park), lots of anthropogenic forestal measures had occurred so it could not count as a non or near non anthropogenic affected area. In addition, there were lots of streets, a bridge, parking places, settlements, farmsteads and other buildings. Furthermore, the river *Elbe* is seen as a steady waterway constructed river and thus is in an anthropogenic affected condition.

The court determined whether it would be possible to look at these conditions in a broader way. But by all methods of interpreting legal norms, it came to the conclusion that a less strict position would be completely unfounded. The wording of the law requires that the areas have to be characterised by no or little human impact at the time of designation. On account of this, it is not possible to designate areas that are in an anthropogenic affected condition as a national park and develop them by re-creating habitats to a no or little anthropogenic affected status. Such a proceeding is impossible even if the areas include extraordinary large plots of various and near-natural biotopes and forms of life[12]. The systematic approach of the law does not state that any one of the conditions of paragraph 25 NNatG for the designation of national parks is more important than the others. A teleological point of view that is concerned about the rule resting without application is unfounded. On this suggestion, the court remarked firstly that the legislator had foreseen this problem when the law was enacted, and secondly, although they are strict conditions, the existence of the *Niedersächsisches Wattenmeer* National Park[13] is proof of application of that law – which reduces that law by an indeterminate legal conception to a one-application-law (Fisahn 1999 : 159). Finally, the court found no clues that the legislator wanted the law to be interpreted in an extensive way.

Lower Saxony tried to get the judgement of the higher administrative court of Lueneburg repealed and asked for an audit in front of the Federal Administrative Court as an appeal court. However, the Federal Administrative Court did not see any juridical importance in the case, and consequently an audit was inadmissible

[12] Against the opinion of the court, a lot of commentators hold this point of view, e.g. Gassner (1995 : 214), Gassner et. al. (1996 : paragraph 14 No. 8f), Blum et. al. (1996 : paragraph 25 No. 7); Fuchs (1999 : 449); Kolodziejcok (2000 : 252 ff).

[13] Lower Saxony Wadden Sea.

(paragraphs 132 and 133 VwGO)[14]. The Federal Administrative Court gave some reasons for its opinion of the lack of importance. It emphasised that neither the rule of paragraph 25 (1) No. 3 NNatG itself nor its interpretation by the higher administrative court of Lueneburg violate the reversible law. It explains that paragraph 14 (1) No. 3 BNatSchG is part of a fundamental principle right that leaves room for interpretation – whether by law or by judgement. The assessment of the higher administrative court of Lueneburg about the areas not being characterised by no or little human impact was correctly based on qualitative and quantitative criteria, so there was no room to criticise this point of view. The court also clarified that a designation of areas which are in an anthropogenic affected condition and developed them by recreation measures or development measures to a non or near non anthropogenic affected status would not conform with the fundamental principle right of paragraph 14 (1) No. 3 BNatSchG. The same can be said about areas including extraordinary large plots of various and near-natural biotopes and forms of life. Both conditions do not justify a designation of national parks.

Both judgements show that the rules of paragraph 14 (1) No. 3 BNatSchG and paragraph 25 (1) No. 3 NNatG have difficulties in their interpretation of two essential points: First of all it is not clear what kind of interpretation of nature is to be used as a scale. Secondly, it is ambiguous what time frame and what criteria define whether an area is characterised by no or little human impact – without considering its near natural state (Fisahn 1999 : 158f; Fuchs 1999 : 448).

The case of the Bayerischer Wald National Park

The case of the *Elbtalaue* National Park was the first to have a result of declaring the park null and void, but not the only one. Other courts had to decide about whether the area of a park was characterised by no or little human impact at the time of its designation, too. In September 1997 the first German *Bayerischer Wald* National Park was also brought to trial[15]. Although the Bavarian rule for the designation of national parks, Art. 8 Bayerisches Naturschutzgesetz (BayNatSchG)[16], differed from the wording of paragraph 14 (1) No. 3 BNatSchG in various points, the competent higher administrative court of Bavaria was bound to interpret the pertinent rule, Art. 8 (2) BayNatSchG, in consideration of the general federal framework law.

The higher administrative court of Bavaria had to answer the question of whether the forestry land use in the Bavarian forest (as an example of German forested low mountain ranges) gave these areas the characterisation of no or little human impact. Thus, this area might not have fulfilled the necessary conditions to be designated as a national park. On that topic the court stated[17]: »The use of forest areas by forestry alone as a matter of fact does not automatically exclude their

---

[14] Judgement from September 10th 1999–6 BN 1.99, published in Natur und Recht 2000, pp 43 ff.

[15] Higher administrative court of Bavaria, judgement from September 15th 1999 Az. 9 N 97.2686, published in: Bayerische Verwaltungsblätter (BayVBl.) 2000, pp 372ff.

[16] Bavarian Nature Conservation Act.

[17] Translated by the author.

nature related status. In fact, the character of a forest in which economic activity happens might be stamped by that kind of activity so that it is no longer possible to talk about a nature-related landscape. However, even with economic use of the forest a landscape might exist with nature-related communities of species. For example, if the appearance of the forest, its structure and the variety of its plant and animal species is not stamped by human economic activity, it can still have a nature-related appearance«[18]. However, later on in this statement, the court agrees with the higher administrative court of Lueneburg by finishing its exegesis on the problem of the near natural state of an area only indicating a characterisation of no or little human impact and that both of these aspects cannot be equated. This might bring us to the conclusion that the *Bayerischer Wald* National Park only survived this judgement because its appearance as a forest remains similar to the potential natural vegetation, forest.

Both examples, the *Elbtalaue* National Park and the *Bayerischer Wald* National Park are not alone with these problems. In particular, the *Unteres Odertal* National Park[19], the German investigative area of the German Research Foundation (DFG) Graduate Training Programme, must also be seen as a kind of ›development national park‹. Stock (2000) even calls it the epitome of a development-national-park in use (Stock 2000 : 204). The actual anthropogenic controlled water management[20] in the park would have been enough for the higher administrative court of Lueneburg to consider the area to be situated in an anthropogenic affected condition.

Finally, it should be noted that the problem of an almost totally cultivated landscape and the resulting conflict with the strict preconditions for the designation of national parks is not only discussed in judgements. In scientific articles there is an ongoing debate about whether the national park concept coming from the United States of America is appropriate for the German (and even the European) landscape (Lang 1984).

2.2 Since the 2002 amendment

The rescinding of the declaration of the *Elbtalaue* National Park was criticised by many conservationists (Kolodziejcok 2000; Stock 2000; Fuchs 1999; Wilkens 2001) and stimulated a discussion about national parks in Germany (Zucchi 2002 : 376ff; Stock 2000 with a lot of more references, Fisahn 1999; Fuchs 1999; Wilkens 2001).

In Lower Saxony, the legislators reacted by taking preventative measures to make sure that its two remaining national parks (*Niedersächsisches Wattenmeer* and *Harz*) could not be brought to trial at the administrative high court of Lueneburg (Knake 2000). For this reason, Lower Saxony re-designated both parks, this

[18] Higher administrative court of Bavaria, judgement from September 15th 1999 Az. 9 N 97.2686, published in: Bayerische Verwaltungsblätter (BayVBl.) 2000, pp 372ff (373); translated by the author.

[19] Gesetz zur Errichtung eines Nationalparks *Unteres Odertal* (NatPUOG) from June 27th 1995, Gesetz- und Verordnungsblatt I Brandenburg, p 114.

[20] Cp. Rothenbücher, Bentlage and Just in this volume.

time by law from parliament which is not controlled by an administrative high court[21].

As a consequence of the judgement on the *Elbtalaue* National Park, in the amendment of the *Bundesnaturschutzgesetz* in 2002, the German federal legislators have eased the legal prerequisites for the establishment of national parks on the disputed point of the anthropogenic influences (Gassner et. al. 2003 : before paragraph 22 No. 2). The wording of the rule now reads:

»**Paragraph 24 National parks**

(1) National parks are areas designated on a legally binding basis as areas to be protected on a uniform basis that meet the following criteria:
   1. The area concerned is an entity of major size with specifically characteristic features;
   2. the criteria defined for nature conservation areas are meet in the greater part of the area, and
   3. the greater part of the area concerned is in a status characterized by no or little human impact, or the area concerned is suitable for developing/being developed into a state which safeguards undisturbed ecosystemary interactions and their natural dynamic process to the extent possible.

(2) The aim of national parks is to safeguard, in the greater part of the area concerned, undisturbed ecosystemary interactions and their natural dynamic processes to the extent possible. Where and to the extent to which this is compatible with the protection purpose, national parks should also serve the purposes of scientific monitoring and surveillance, education in the field of natural history, biology and related subjects as well as enable the general public to experience nature.

(3) The Federation States shall ensure that national parks receive the same level of protection as that afforded to nature conservation areas, taking into account their particular protection purpose and allowing for exemptions required in view of the size of the area and its human settlements.«

As the new legislation shows, one very important point for the legislators was to differentiate between the preconditions for the designation of national parks on one hand and the protection purpose of the declaration on the other[22]. The existence of an area characterised by no or little human impact and the resulting natural dynamic processes are no longer imperative preconditions. Rather, it is the main aim of protection. This concept of development-protection-areas has not only been implemented in the protection category of national parks. It is also now found in other protection categories in the *Bundesnaturschutzgesetz*, for example the nature conservation areas. Due to the fact that in No. 2 of paragraph 24 (1) BNatSchG there is another precondition for the designation of national parks (the

[21] According to this, see above 2.1.2.

[22] Bundestags-Drucksachen. 14/6378, p 51.

criteria defined for nature conservation areas are to be met in the greater part of the area), the designation of national parks is thus promoted twice by the concept of development-protection-areas.

In connection with this, a paradigm shift has happened to the *Bundesnaturschutzgesetz* (Riecken 2002 : 397). Nature conservation is no longer ›supply-sided‹ but ›demand-oriented‹. Overall, this accommodates the realisation that the protection of biodiversity will not be possible in the actual protected reserves and existing areas that are characterised by no or little human impact. International law, like the convention on biological diversity and European community law like the habitats and birds directive require a coherent ecological network of special areas of conservation. This development was a necessary step in the right direction (Stock 2000 : 208f), although the eased preconditions caused juridical concern (Meßerschmidt 2001 : 245). Remembering that the preconditions in Germany fulfil a double function and that one of these functions is to protect people from illegal restrictions of their property, the conditions that guarantee the protection of property now seem to be at a very low level, although protection of private property is not completely withdrawn. The recourse on private grounds for nature conservation measures is not arbitrary because it has to redeem the conditions that have evolved in constitutional law. The constitution prescribes a weighting to the respective interests of the individual property on one hand and the public interest in nature conservation on the other hand. This substantiates the restrictions on individual property for the benefit of society. The weighting has to evaluate the situation of each property that is supposed to be incorporated into a protected area. The situation of land is – among other things – stamped by current and proposed uses. Additionally, land is marked by its qualification to evolve or be developed into a state which guarantees an undisturbed course of natural processes in its natural dynamic. However, in these cases the economic value and use are supposed to be profound. The agents for nature conservation will have to prove that the area should be a development-protection area, e.g., or a coherent network. If nature conservation outbalances the individual property, the owner as well as any established practice, can count on financial compensation. So the guarantee of individual property changes to a guarantee of its value. Thus, the juridical concerns about low level protection of individual property by the designation of development-national-parks are unfounded and can be rebutted.

Now we have answered the question of what preconditions make it possible to designate national parks. Finally, it should be deliberated whether there is an obligation to designate national parks. The wording of paragraph 22 (1) BNatSchG (»can be designated as a national park«[23]) does not obligate the Federation States to designate national parks (Bender et. al. 2000 : 203f; Kloepfer 1998 : No. 58). Because of European Community law in terms of the habitats and the birds directive, the Federal Republic of Germany (respective of the domestic authorities of the member states that are authorised to designate protection areas, meaning the Federation States) shall designate protection areas for certain habitats and species

[23] Translated by the author.

(Gellermann 2001 : 18). However, in the European directives it is not explicit which category of protection area should be chosen, thus the Federation States are not bound to designate national parks. The directives only commit the Member States to establish a high level protection standard. Because of this high and effective protection standard, and the vastness of national parks, reverting to the designation of national parks as European special protected areas is obvious. Anyway, nobody in Germany has a redress in law on the designation of national parks or that his land has or will become part of a national park – even if the land fulfil every precondition in order to be designated as a national park.

2.3 Juridical requirements for the configuration of national parks

Beyond the question, *if* a national park shall or could be designated, the question is *how* a national park should be configured in its Declaration Act. In this context, there is only room to analyse the requirements for the configuration of national parks set by the German nature conservation law. Another restriction must also be accepted. Because of the different legal forms, the Federation States can choose only the rules set in the *Bundesnaturschutzgesetz* for the designation of national parks as the federal nature conservation act will be examined. Because of the hierarchy of norms, the Federation States are bound to these general regulations regardless of which legal form they choose for the designation[24].

The requirements of paragraph 24 BNatSchG

First of all, the Federation States have to consider the rules set out in paragraph 24 BNatSchG. As they were illustrated in the previous section, here they will only be briefly recapped.

Once more the protection aim should be highlighted. This is to guarantee – as far as possible – an undisturbed course of natural processes in their natural dynamic. The main request is that the area of a national park shall evolve or be methodically developed back to its native state (Stock 2000 : 208). For this reason the declaration of national parks must state that the park should be returned to its native wilderness state. For each national park, the protection aim is to be put in individual concrete terms (e.g., forests, grassland, and estuary)[25]. To achieve these

[24] If the designation happens by statutory order from the government, the nature conservation acts of the Federation States are to be seen as laws which stand higher in the hierarchy of law and therefore should be strictly adhered to. In the case of designation by law of parliament, the nature conservation acts are on the same level in the hierarchy of norms and only can be seen as a kind of voluntary self obligaton because it is enacted by the same legislator.

[25] The protection aim of the *Unteres Odertal* National Park is for example, according to paragraph 3 sentence 1 NatPUOG, to protect, maintain, conserve and develop in their natural functions the Lower Oder Valley with its Central Europe special flood-plain landscape, its variety of animal and plant species, its numerous wetland habitats, meadows/grass-land and flood-plain forests and the forests on the slopes accompanying the river's floodplain in connection with other forests and the arid habitats.

protection aims will be the most important reference value (Gassner 1995 : 194). As far as the protection aim allows, national parks should also serve as a scientific environment for observation, natural history education and nature experiences for the public. In addition to this, the declaration acts have to forbid all actions that could cause any destruction, damage or diversification to the park or its components or any lasting disturbances. The legislators have not ignored the fact that the vast dimensions and settlement areas have to be taken into account when specifying necessary requirements and prohibitions, and it is likely that some exemptions will be needed (Kloepfer 1998 : 733).

The declaration act of the *Unteres Odertal* National Park gives some examples of exemptions, e.g., the public streets and ways and waterways in the park are opened for traffic according to the conditions laid down (paragraph 8 (1) No. 4 NatPUOG) and the operation, maintenance and repair of the existing water supply installations in some parts of the park (paragraph 8 (1) No. 8 NatPUOG). Important exemptions are made for hunting and fishing in the national park area (paragraphs 11 and 12 NatPUOG), although these activities have to respect the protection aim of the national park.

Other requirements of the Bundesnaturschutzgesetz

The designation of national parks as a planning act

This section will discuss what further requirements for the designation of national parks the Federal Nature Conservation Act contains other than paragraph 24 BNatSchG.

For this purpose, it is necessary to know that the configuration of protection areas such as national park is a kind of planning act – at least concerning the realignment of boundaries and the professional fixing of priorities of the protection contents (Bender et al. 2000 : 203; Kloepfer 1998 : 729). This is why the legislator is allowed some latitude and has some scope in terms of development. This scope can be illustrated by the example of the *Bayerischer Wald* National Park. There was a big dispute about the infestation of the forest by bark beetles. The legislator had restricted all pest management measures in the statutory order – they were only allowed in the outlying areas of the refuge. An owner of forest grounds brought the case to trial because he was of the opinion that the restriction did not conform to the protection aim to conserve the forest and that measures should have been taken against the bark beetles[26]. In its judgement, the higher administrative court of Bavaria tolerated the restriction set out in the statutory order and declared that the method of achieving the protection aim (in this case by allowing nature to help

[26] Presented in the judgement from the Higher administrative court of Bavaria, judgement from September 15th 1999 Az. 9 N 97.2686, published in: Bayerische Verwaltungsblätter (BayVBl.) 2000, pp 372ff (374f).

itself – a kind of ›proceeding protection‹) should be left to the legislator[27]. It agreed with the legislator that nature has the potential to help itself.

Because of the principle of the separation of powers, only some aspects can be fully controlled by the court, namely if the components incorporated into the park, fulfil the preconditions of a protected area which means they are worth being protected and their protection is necessary (Bender et. al. 2000 : 203). Thus, the legislators are given the discretion to realign the boundaries and set the priorities of the protection requirements professionally (Gassner 1995 : 199). Here, courts are restricted in their investigation of the declarations of national parks. They only can check whether the legislator has abused his discretions. It is the competence of the legislator that decides the expanse of a national park, its protection aims and the way those aims are to be reached.

The legislator regularly commissions experts to analyse the status of the area, propose conservation protection measures and, if necessary development measures. Based on this expert advice, and considering other juridical requirements, a draft bill or statutory order can be worked out.

Similar to other planning acts or decisions of discretion, a weighting is necessary to designate a national park (Kloepfer 1998 : 731f; means the weighting with the repectable interests of individual property). So, all relevant aspects must be investigated, evaluated by their objective importance and finally be brought to a fair balance. In the case of the designation of a national park, the Federal Nature Conservation Act requirements for each nature conservation measure in general must be accomplished. According to paragraph 2 (1) BNatSchG, in individual cases the aims of nature conservation and landscape conservation are to be realised as far as necessary, possible and appropriate by applying a weighting to the aims of paragraph 1 BNatSchG and to other requirements on nature and landscape. The criteria to find the objective weight of an aspect are to be taken from the law itself and further, from the level of achievement of the objectives and the circumstances of each case (Gassner 1995 : 205).

Imperative requirements and prohibitions

The legislator is not entirely free in the configuration of national park declarations. In the following section, the juridical requirements of the law itself as one of the three criteria to find the objective weight of an aspect will be considered in detail. In doing so, a differentiation can be made between imperative requirements and prohibitions that do not have any disposition in the weighting and other valuations that are found in law, e.g., that some subjects of protection are emphasised in law.

The imperative requirements that cannot be compromised in the weighting process are: 1. the rules on interventions in nature and landscape (paragraphs 18ff BNatSchG), 2. the rules that implemented the European habitats and birds directive into German law (paragraphs 32ff and 39ff BNatSchG), 3. the rules for the legal protection of natural areas (paragraph 30f BNatSchG).

[27] Higher administrative court of Bavaria, judgement from September 15th 1999 Az. 9 N 97.2686, published in: Bayerische Verwaltungsblätter (BayVBl.) 2000, pp 372ff (374f).

The rules on interventions in nature and landscape are an instrument of the German nature conservation law to keep the status quo of nature and landscape. The main content of these rules is, that in certain conditions interventions are forbidden and that in other cases the intervening party shall be obligated to primarily endeavour to offset any unavoidably impairment through measures of natur conservation and landscape management (compensatory measures), or to offset them in some other way (substitute remediation). Both, compensatory measures and substitute remediation have to take the functions of the damaged ecosystem into account. The relevant interventions are defined by the law as changes to the shape or appearance or utilization of land or changes to the groundwater table with its close correlations to inhabited soilcompartments that may significantly impair the ecosystem, or the natural scenery. Even measures taken in the name of nature conservation might cause an impact which can be illustrated with the following example[28].

The building of an industrial area in Hamburg-Finkenwerder to assemble the Airbus A3XX caused the destruction of the *Mühlenberger Loch*, a freshwater tideland which was very important from a nature conservation perspective. As compensation for the destruction there was a plan to build a freshwater tideland in another place. However, the land selected already had a high conservation value (international bird area) and therefore was unsuitable for the planned development measures. Thus, the case was brought to trial by NGOs. The administrative court of Schleswig supported the thesis that this kind of nature conservation measure should be treated as an intervention in nature and landscape. The court pointed out that according to law, it is less important that the new land status might be favoured by authorities, but rather that the planned compensation measure will change the actual status and function of the selected land (according to Schrader and Hellenbroich 2002 : 215ff).

The benchmark to analyse whether a measure is to be treated as an intervention in nature and landscape is the efficiency of its ecological functions and the temporary connection between the actual intervention and the ensuing compensatory measures.

Considering these criteria, the planned compensation would probably have generated a new freshwater tideland. However, because the selected land had a high conservation status, introducing the planned compensatory measures would have caused a failure of functions of the ecosystem even though the result would have been a highly valued freshwater tideland. The higher administrative court of Schleswig agreed with that opinion. So, if it is possible that well meaning but wrong measures in the name of nature conservation can be a disadvantage, it cannot be excluded that such disadvantages might also be the result of the wrong set of protection aims or maintenance and development schemes in national parks.

---

[28] It is the case of a decision from the administrative court of Schleswig from October 10th 2001–12 B 10/01, published in: Zeitschrift für Umweltrecht 2002, pp 213ff The decision was affirmed by the higher administrative court of Schleswig, judgement from February 12th 2002–4 M 93/01, published in Natur und Recht 2002, pp 695ff.

One of the first results of the changing of land uses in a national park's development scheme might violate the rules on interventions in nature and landscape and therefore be illegal.

Stricter than the rules on interventions in nature and landscape in general, are the rules of paragraph 30 BNatSchG that obligate the Federation States to forbid the destruction or significant or lasting adverse impact on certain habitats. Among the habitats protected by these rules are many types that are the result of anthropogenic land use and that need special maintenance measures (Riecken 2002 : 398).

The juridical obligation of a national park concept based on native wilderness state runs the risk of colliding with the prohibition to destroy or damage such kinds of protected habitats. For these habitats, special regulations must be implemented in the declaration of national parks. Otherwise it must be checked whether the designation of the national park fulfils the criteria that might lead to an exemption of the protection of these habitats. Such exemptions can be made if adverse impacts on the biotopes can be offset or if the measures concerned are necessary for reasons of overriding public interests.

In almost the same way there might be conflicts with those acts which implemented the European Community Law of the habitats and birds directive into German law, paragraphs 32ff BNatSchG (Gellermann 2001). The European Community obligates its Member States to designate protection areas for certain habitats and species. The habitats that are to be protected are listed in Annex I of the habitats directive. Among this list are a lot of types that result from anthropogenic land use, even though the directive calls them natural habitat types. These habitat types rank among the site's conservation objectives (Gellermann 2001 : 70f) of which the Member States shall take appropriate steps to avoid deterioration. Naturally, nature conservation measures also have to conform to these rules.

The rules on the protection and management of wild fauna and flora species include imperative requirements and prohibitions. For example, paragraph 42 (1) Nr. 1 BNatSchG prohibites to damage or destroy nestings or breeding sites, other living quarters or inhabited sites or any other places of refuge of ›special protected species‹[29]. As far as habitats of species of naturally occurring birds in the wild state in European territory are concerned, in terms of the birds' directive or habitats of species of community interest in the habitats directive, the rule accomplishes the implementation of community law. An example of a violation of these rules might be the aim of developing an alluvial forest in the habitats of species of the open landscape, e.g., birds breeding in meadows.

It might be disputed whether directed measures of native state conservation as a kind of omission can infringe upon one of these prohibitions. In favour of this, it could be said that this would be a kind of planned development based on a decision to designate a national park when the legislator was aware of the negative consequences for protected habitats or species. On the other hand, it should be noted

[29] Which species are specially protected is defined in paragraph 10 (2) No. 10 BNatSchG. In Germany, the status of protection is differentiated between three categories: wildlife species in general, special protected species and strictly protected species.

that the wording of the rule about destruction or damage can be interpreted such that a direct action is necessary to violate the rules. Also, it is a matter of fact that in Germany there are no general rules in the nature conservation acts that obligate anybody to take maintenance measures of habitats. Both arguments sustain a point of view about the designation of a national park that native state development, as an act of planning, does not directly violate the present rules. Enacting a declaration is not a destruction or damage of a habitat. Although it is imaginable that the planning act might require an illegal act and therefore might be illegal itself.

Regardless of these reflections, such a designation might be illegal because it would not respect the assessment of the legislator who set the level of achievement of the objectives – which is one of the criteria for objective weighting. The weighting of how species and habitats are to be protected and how the protection can be guaranteed in a national park, have to follow the weightings that the legislator has implemented into superior law such as the Federal Nature Conservation Act. For example, the legislator has highlighted that some species and habitats are to be specially protected. The declaration for the designation of a national park as an implementation of nature conservation in general, must not go against this legislative weighting. In the selection of species for which protection measures will be taken in a national park, the weighting of the legislator has to be reflected. The protection of ›strictly protected species‹ regularly has higher priority than the protection of special protected species which in turn generally has a higher priority than the protection of wildlife species. For priority natural habitat types in terms of the habitats directive, stricter protection activities regularly take place rather than for natural habitat types of community interest or which are ›only‹ on a red list. As a matter of course, this will just be a general orientation and there might be deviations in individual cases. There is room to bear in mind other criteria such as the responsibility of Germany for the protection of a species in an international context (according to, e.g., Boye and Bauer 2000 : 71ff).

To sum up, it can be said that lots of juridical requirements of the nature conservation law are to be respected when designating national parks. This is not only in regard to the question of whether a national park should be designated at all. There are also a lot of rules for the configuration of a national park which must be adhered to, especially for boundaries of the area, fixing the protection aim and establishing a maintenance and development scheme. Some of the juridical requirements to be considered are aimed at the protection of habitats that are anthropogenically stamped biotopes as a result of regular and extensive human land use. Thus, the concept of wilderness which is to be implemented in national parks might be contrary to these regulations.

Finally, for the sake of completeness it should be remarked upon that there are still a lot of requirements, apart from those established by the nature conservation law, that have to be adhered when designating national parks. For example, the principle of equal treatment, the necessity of clarity and definitiveness etc. which cannot be presented and analysed here (Gassner 1995 : 203ff; Bender et. al. 2000 : 208).

3 Conformity of German law and international requirements

3.1 International requirements

Since the concept of national parks has become established worldwide, it is not surprising that standards for national parks have recently been developed at an international level. Above all, the endeavours of the IUCN are to be accentuated. On one hand, in 1969 a definition of the abstract term »national park« was compiled. By that definition, a national park was a relatively large area:

1. where one or several ecosystems are not materially altered by human exploitation and occupation, where plant and animal species, geomorphological sites and habitats are of special scientific, educative and recreative interest or which contains a natural landscape of great beauty, and
2. where the highest competent authority of the country has taken steps to prevent or eliminate as soon as possible exploitation or occupation of the whole area and to enforce effectively the respect of ecological, geomorphological or aesthetic features which have led to its establishment; and
3. where visitors are allowed to enter, under special considerations, for inspirational, educative, cultural and recreation purposes.

On the other hand, criteria were worked out for areas that Member States cannot designate as a national park. In addition to the definition, the United Nations publish a list of National Parks and Protected Areas which is updated at intervals. From a juridical point of view, it is important that these criteria and definitions are not hard and fast international law (Lang 1984 : 14). Rather, they are guiding principles, directed to the IUCN treaty states to which the Federal Republic of Germany belongs. As an IUCN state, the Federal Republic of Germany should familiarise itself with these criteria and definitions which are international requirements.

Soon after the agreement of the definition, it became clear that a lot of States did not have such natural areas without material alterations by human exploitation and occupation. That is why the criteria and definition were revised several times since they were first drawn up.

The list of national parks currently in force (World Conservation Monitoring Centre and IUCN World Commission for protected areas 2003 : 1) is based on IUCN Guidelines for Protected Area Management Categories (IUCN 1994). It had ten different categories of protection areas and included the criteria for the classification of each category.

Because of the peculiarities of Europe, (relatively small continent, high allotment of historically developed cultural landscape, merely existing unaltered natural areas, dense settlement etc.), with the assistance of UNEP World Conservation Monitoring Centre (WCMC), the IUCN's World Commission on Protected Areas (WCPA) and the EUROPARC Federation, special Interpretations and Applications of the IUCN Management Categories for Protected Areas in Europe were written (EUROPARC and IUCN 2000). These guidelines present six categories of protection areas which differentiate the main protection aims and anthropogenic influence. The protection aim is the main indicator which determines which category an area belongs to (EUROPARC and IUCN 2000).

The actual definition of the term »national park« is:

> »**National Park: protected area managed mainly for ecosystem protection and recreation**
> Natural area of land and/or sea, designated to
> (a) protect the ecological integrity of one or more ecosystems for present and future generations,
> (b) exclude exploitation or occupation inimical to the purposes of designation of the area and
> (c) provide a foundation for spiritual, scientific, educational, recreational and visitor opportunities, all of which must be environmentally and culturally compatible.«
> (EUROPARC and IUCN 2000 : 24)

The term »natural« included in the definition is to be understood as follows:

> »Ecosystems, where since the industrial revolution (1750) human impact (a) has been no greater than that of any other species, (b) has not affected the ecosystem's structure. Climate change is excluded from this definition.« (EUROPARC and IUCN 2000 : 14)

The IUCN has also recognised that the states of Europe have problems fulfilling the criteria for designating protection areas of the categories I (Strict Nature Reserve / Wilderness Area), II (National Park) and VI (Managed Resource Protected Area) because of the widespread anthropogenic influence. As a consequence, the criteria for Europe were adapted. Thus, in Europe for the IUCN, it is satisfactory that exercised land uses will be dropped immediately or at least as soon as possible (EUROPARC and IUCN 2000 : 16).

Anyhow, the IUCN learned that although the definition and criteria had been adapted and eased since its initial establishment, a couple of areas which do not even fulfil the eased criteria have still been designated as national parks. To counteract the resulting confusions, the IUCN established a kind of certification system. If an enquiry is made, the IUCN send an affirmation that an area fulfils the criteria to be designated as a national park (EUROPARC and IUCN 2000 : 13).

3.2 Comparison with German law

Concerning the accordance of German law with the international criteria, henceforth the following two results can be presented:

- If the international criteria require the designation of a natural area of land and/or sea, they are high requirements that will predominate only in a few areas of Germany. Referring to the time of the industrial revolution, the international requirements are not as strict as those supported by the higher administrative Court of Lueneburg, which in its judgement on the *Elbtalaue* National Park even factored in the influences from the middle ages.

- The initial strictly established criteria for the designation of national parks in Germany have been eased in German nature conservation law. Also, the IUCN has eased the definition of national park criteria in respect to the specific landscape in Europe. Under the precondition that exercised land uses will be dropped immediately or at least as soon as possible, it has enabled the designation of national parks including areas in which human land uses occur or have occurred. This is in order that the international criteria will accept ›development-national-parks‹ as well.

If even the IUCN, with good reasons eases the criteria of the requirement of a non or near non anthropogenic affected condition of the area, the question must be raised whether it makes sense to keep this criteria? The question should be seen in the context that the biodiversity existing in Central Europe today owes more to anthropogenic extensive land uses such as grazing management, and less to natural dynamic processes. In answering the question, the IUCN reverted to the creation of a new category of protection area. In doing so the ›Managed Resource Protected Area‹ was generated; an area containing predominantly unmodified natural systems, managed to ensure long term protection and maintenance of biological diversity, while at the same time providing a sustainable flow of natural products and services to meet community needs (EUROPARC and IUCN 2000 : 32).

Both concepts of nature conservation – the protection by native state and protection using maintenance and development measures – benefit different species and habitats. That is why both concepts are needed to guarantee a fully successful protection of biological diversity. The protection of areas where anthropogenic influences predominate might be designated as biosphere reserves in Germany. A criticism of the designation of the criteria of the character of no or little human impact of the area in general, has to be refuted. Both of these concepts are valid and necessary.

4 Conclusion

In conclusion it can be said that the German nature conservation law for the designation of national parks, although easing the preconditions by differentiating between requirements and protection aims, still conforms to the international IUCN criteria. The reason for this conformity is that the IUCN eased the international criteria for the designations in respect of the special status of Europe, and tolerates ›development national parks‹ as well. Thus, the designation of the areas recommended to be declared as national parks should be brought forward.

Altogether the capacities for national parks in Germany seem to be restricted. Last but not least, this depends on the fact that biological diversity is also a result of extensive land use and that its subsequent habitats have their value and are also protected by law. It would be a pity if these valuable habitats and species were disadvantaged by the wrong set of nature conservation measures and that wilderness was developed in the wrong places.

References

Arnberger A, Brandenburg C (2001) Der Nationalpark als Wohnumfeld und Naherholungsgebiet. Naturschutz und Landschaftsplanung, 33:157–161

Bender B, Sparwasser R, Engel R (2000) Umweltrecht, 4th ed. Müller, Heidelberg

Blum P, Agena CA, Franke J (1996) Niedersächsisches Naturschutzgesetz, Kommentar, Loseblattsammlung, Kommunal- und Schulverlag, Wiesbaden

Boye P, Bauer HG (2000) Vorschlag zur Prioritätenfindung im Artenschutz mittels Roter Listen sowie unter arealkundlichen und rechtlichen Aspekten am Beispiel der Brutvögel und Säugetiere Deutschlands. In: Binot-Hafke, Gruttke, Ludwig, Riecken (eds) Bundesweite Rote Listen - Bilanzen, Konsequenzen, Perspektiven. Bonn-Bad Godesberg, pp 71–88

EUROPARC and IUCN (2000) Richtlinien für Managementkategorien von Schutzgebieten - Interpretation und Anwendung der Management Kategorien in Europa, 2nd ed, Grafenau

Fisahn A (1999) Annotation to higher administrative court of Lueneburg, judgement from February 22nd 1999. Zeitschrift für Umweltrecht 10:158–159

Fuchs W (1999) Naturschutz im Abseits. Natur und Recht 21:446–450

Föderation der Natur- und Nationalparke Europas, Sektion Deutschland e.V. (FÖNAD) (1997) Studie über bestehende und potentielle Nationalparks in Deutschland. Landwirtschaftsverlag, Münster

Gassner E (1995) Das Recht der Landschaft. Neumann, Radebeul

Gassner E, Bendomir-Kahlo G, Schmidt-Räntsch A, Schmidt-Räntsch J (2003) Bundesnaturschutzgesetz, Kommentar, 2nd ed. Beck, München

Gassner E, Bendomir-Kahlo G, Schmidt-Räntsch A, Schmidt-Räntsch J (1996) Bundesnaturschutzgesetz, Kommentar, 1st ed. Beck, München

Gellermann M (2001) Natura 2000, 2nd ed. Blackwell, Berlin

Harthun M (1999) Aktuelle Entwicklungen zum geplanten Nationalpark Kellerwald nach dem Regierungswechsel in Hessen: Nationales Naturerbe oder Wirtschaftsforst? Jahrbuch Naturschutz in Hessen 4:217–224

IUCN (1996) Guidelines for Protected Area Management Categories. Gland and Cambridge

Kloepfer M (1998) Umweltrecht, 2nd ed. Beck, München

Knake M (2000) Ein Großschutzgebiet wird demontiert. Naturschutz und Landschaftsplanung 32:281–282

Kolodziejcok K (2000) Nationalparke am Wendepunkt? Natur und Recht 22:251–254

Lang A (1984) Welchen Anforderungen muß ein Nationalpark in der Bundesrepublik Deutschland genügen? Natur und Recht 6:14–18

Meßerschmidt K (2001) Wiedervorlage oder Innovation? Zum Entwurf einer Gesamtnovellierung des Bundesnaturschutzgesetzes vom 2. Februar 2001. Zeitschrift für Umweltrecht 12:241–246

Oebbecke J (2001) »Nationalpark« Industriekultur Ruhrgebiet? Natur und Recht 23:142–146

Primack RB (1995) Naturschutzbiologie. Spektrum, Heidelberg, Berlin, Oxford

Riecken U (2002) Novellierung des Bundesnaturschutzgesetzes: Gesetzlich geschützte Biotope nach § 30. Natur und Landschaft 77:397–406

Schrader C, Hellenbroich T (2002) Annotation to administrative court of Schleswig, judgement from October 10th 2001–12 B 10/01. Zeitschrift für Umweltrecht 13:215–217

Stich R (1972) Notwendigkeit und Inhalt eines modernen Naturschutz- und Landschaftspflegerechts: Deutsches Verwaltungsblatt 87:201–211

Stock M (2000) Nationalparke in Deutschland: Den Entwicklungsgedanken gesetzlich absichern und konkretisieren! Zeitschrift für Umweltrecht 11:198–210

Wilkens H (2001) Gibt es weiterhin Nationalparke in Deutschland? Naturschutz und Landschaftsplanung 33:358–360

World Conservation Monitoring Centre and IUCN World Commission for Protected Areas (2003) 2003 United Nations List of Protected Areas. Thanet, Gland and Cambridge

Zucchi H (2002) Wildnis als Kulturaufgabe – ein Diskussionsbeitrag. Natur und Landschaft 77:373–378

Conservation management of target species or conservation of processes – Winners and losers of two different conservation strategies

Judith Rothenbücher[1], Kai Bentlage[2], and Peter Just[3]

[1] Institute of Zoology, Ecology Group, Georg-August University of Göttingen, Berliner Str. 28, 37073 Göttingen, Germany
[2] Centre for Nature Conservation, Georg-August University of Göttingen, Von-Siebold-Str. 2, 37075 Göttingen, Germany
[3] Institute of Geography, Landscape Ecology Group, Georg-August University of Göttingen, Goldschmidtstr. 5, 37077 Göttingen, Germany

Summary. Studies on insect and amphibian communities as well as on selected meadow bird species were carried out in the floodplain of the Lower Oder Valley National Park. The aim of these three studies was to analyse the effects of the present land use and management of the flooding regime on the selected species. The floodplain mainly consists of grassland habitats. Parts of the grassland are extensively used for farming whereas other parts were taken out of use in 1995. Currently, dykes bordering the river Oder prevent natural inundation of the floodplain during the summer months. During winter, floodgates, that are integrated into the dykes, stay open. Thus, the typical winter flooding can inundate the floodplain. The national park is still in the developmental stage. In plans for the future development of the national park, two main conservation strategies are discussed. On one hand conservation management is proposed with the aim to preserve and generate suitable habitats for the selected target species, corncrake (*Crex crex*). On the other hand, process conservation is suggested intending to create habitats suitable for the typical wetland fauna and flora by leaving some areas to natural processes and succession. On the basis of the three case studies the effects of the two conservation strategies on the selected taxa are discussed. On one hand Auchenorrhyncha and Anura will not necessarily benefit from conservation effort carried out for the benefit of the target species corncrake. These results call the selection of the corncrake as umbrella species into question. On the other hand, conservation of processes negatively affects corncrake populations while the anuran community is neither positively nor negatively affected and insect communities benefit from the conservation strategy.

1 Introduction

In 1993 Germany ratified the Convention on Biological Diversity (CBD) and thus committed to translating the content of the convention into national legislation. This includes, among other things, the protection of biological diversity within protected areas in Germany, i.e., the conservation of ecosystems and natural habitats as well as the maintenance and recovery of viable populations of species in their natural surroundings[4]. About ten years after the world summit of Rio de Janeiro in 1992, it is interesting to investigate whether conservation of biological diversity is successful within protected areas in Germany. In the context of the graduate training programme *Conservation and Valuation of Biodiversity*, three case studies were carried out in the Lower Oder Valley National Park. The study carried out by Peter Just analysed the habitat parameters that are crucial for the successful establishment of breeding corncrakes in the national park. In the second study Kai Bentlage investigated the impact of flooding on the amphibian community, whereas in the third study carried out by Judith Rothenbücher the impact of land use and flooding on the diversity of insects was the centre of interest. In the national park conservation management for target species and conservation of processes are realised side by side which provides the opportunity to discuss and compare the effects of the two conservation strategies on the selected taxa within the context of the protection of biological diversity.

2 Two conservation strategies

Conservation management for selected target species and conservation of processes are two strategies in nature conservation that have been largely discussed (Jedicke 1995, 1998; Mühlenberg and Hovestadt 1992; Vogel et al. 1996; Scherzinger 2002; Schneider 1998).

It is the aim of the target-species approach to preserve and – if necessary – manage the habitat of the selected target species in such a way that it can maintain at least a minimum viable population (MVP) of the species. The target-species approach is not restricted to just one selected habitat. By choosing several target species as representatives for different sites in a habitat mosaic, a larger area can be managed. For the successful use of this approach in nature conservation it is crucial to select the appropriate target species. Vogel et al. (1996) name four criteria for the selection of target species: degree of vulnerability, chance of survival, ›windfall gain‹ (*Mitnahmeeffekt*) and popularity. Taking the degree of vulnerability into account, emphasis should be laid on species that have their centre of distribution in central Europe and that are threatened on a nation-wide, regional or local level. Furthermore, the particular species should have a reasonable chance of survival in the selected area. To reach a good windfall gain preferably umbrella species, key-stone species or species with a large home range should be chosen as it is assumed that quite a number of other species will profit from conservation efforts (Mühlenberg and Hovestadt 1992; Vogel et al. 1996). Beside these biological criteria, popularity is important too, although it is more a political criterion. Public support for a conservation programme is gained more easily if charismatic species are chosen as target species. An advantage of this approach is that by choosing a target species the conservation aim is more clearly defined. This easily allows

[4] Convention on Biological Diversity, June 12nd 1992, Bundesgesetzblatt II 1993, p 1742.

evaluation of the conservation efforts and optimisation of the conservation strategy if necessary. Disadvantages of this strategy are that it can be time consuming and expensive. One crucial point of this approach is the assessment of the target species' chance of survival in the selected area. The suitable scientific approach is a population viability analysis (PVA). For this analysis, information on long term dynamics and habitat factors are needed. To gain a meaningful PVA, data of at least three years field work should be analysed (Vogel et al. 1996). For some species permanent habitat management, e.g., regular cutting might be necessary to maintain suitable habitats which might include long term costs. Furthermore, there is a lack of investigation on the postulated windfall gain.

It is the aim of the ›conservation of processes‹ strategy that the functions and processes in communities and ecosystems are preserved under preferably natural conditions for a long period of time (Schaefer 2003). Thus, a habitat is chosen to become a protected area and from this moment on the area is left for succession. Anthropogenic impacts are restricted to a single measure with the aim of initiating a favoured stage of succession (*Initialmaßnahmen*), e.g., planting willows to initialise the development of an alluvial forest. Whereas most authors restrict the concept of process conservation to the conservation of anthropogenically uncontrolled dynamics, Jedicke (1998) includes man made dynamics, e.g., land use in the concept. The former is known as a segregative, the latter as an integrative strategy (Jedicke 1998). In this contribution process conservation alludes solely to the segregative concept. Jedicke (1998) names two prerequisites for a successful integration of the strategy into nature conservation programmes. Firstly, the protected area must comprise an area of sufficient size. To assess the appropriate size population dynamics must be taken into account which might include among others, the application of the concept of MVP and the theory of island biogeography. The ideal case would be that the area should be able to host all seral stages side by side if possible. Secondly, any anthropogenic impact must be forbidden in the area which includes accepting unwelcome and unexpected developments, e.g., a large outbreak of pest species as happened in the Bavarian Forest National Park (Pongratz 1995). The advantages and risk of this conservation strategy is that it is comparably cheap to realise. Thus, it must be carefully assessed whether the approach is appropriate for the protected area in question depending mainly on the size of the area. The assessment of the appropriate size is the sore point of the strategy. In contrast to the target-species approach no ›target‹ is defined which can be used to assess the needs to host a MVP successfully. Furthermore, evaluation of the conservation aim is more difficult than in the target-species approach. Nevertheless, the strategy of conserving processes seems at present the only approach that includes the preservation of habitat dynamics into nature conservation hence being a chance to preserve naturalness in a largely man-made landscape.

In a nutshell, the target species approach aims at preserving a habitat for a selected species whereas process conservation aims at preserving a habitat and its dynamics.

3 Study area

For the case studies the Lower Oder Valley National Park was chosen as study area (Fig. 1). The river Oder is a typical lowland river in central Europe with a three to four km wide floodplain adjacent to the river. This grassland dominated floodplain is the dominant habitat of the national park.

Until the middle of the 19th century, the present national park area was more or less untouched by man. Afterwards the river Oder became canalised and an artificial waterway was built running parallel to the river Oder. River dykes were built around the area between the two waterways and the former floodplain was drained. Since 1931, flooding has been regulated in the newly formed polder areas. In the so called ›dry polder‹ the dykes prevent inundation throughout the whole year. In the ›wet polders‹ flood gates that are integrated into the dykes are opened between November 15th and April 1st each year so the typical winter flooding can inundate the polder area. After closing the flood gates to prevent inundation during spring and summer the water is pumped out to make land use possible (Fig. 2) (Dohle et al. 1999). Within 80 years the natural floodplain had been transformed into pasture land. During the time of the GDR on the one hand land use was intensified (Schalitz and Petrich 1999) on the other hand first protected areas were designated within the context of the Ramsar Convention of Wetlands[5]. After the political change in 1989 many areas valuable for nature conservation were declared as protected areas - among these was the Lower Oder Valley (cp. Garrelts et al. in this volume). In 1995 the area was designated as a national park by the parliament of Brandenburg. At present the national park is still in the developmental stage as the flooding regime is regulated and a large proportion of the grassland is used for fodder production and grazing. Of these two, fodder production by mowing is presently the major land use in the national park. Most meadows are cut twice a year, i.e., at the beginning of June and September. Only about 10 % of the national park area is currently declared as core zone. Here the flooding regime is still regulated, but the sites have been abandoned since 1995 and left for succession (Fig. 3) (Jehle and Pankoke 1999).

The aims and purpose of the Lower Oder Valley National Park are outlined in the national park law that was passed on June 22nd 1995.

The purpose of the national park is:

- to protect, to manage, to preserve and to develop the natural functions of the Lower Oder Valley which includes the floodplain, flora and fauna, wetlands, meadows, flood-plain forest and dry grassland
- to protect and to develop natural processes on a large area which are undisturbed by anthropogenic impacts
- to preserve and regenerate the natural flooding regime
- to allow environmental friendly recreation
- to develop tourism
- to conduct environmental education[6]

[5] Ramsar Convention on Wetlands, February 2nd 1971, `http://www.ramsar.org/key_conv_e.htm`

[6] Lower Oder Valley National Park Designation Act (*Nationalparkgesetz Unteres Odertal* – NatPUOG) June 27nd 1995, GVBl I, p 114.

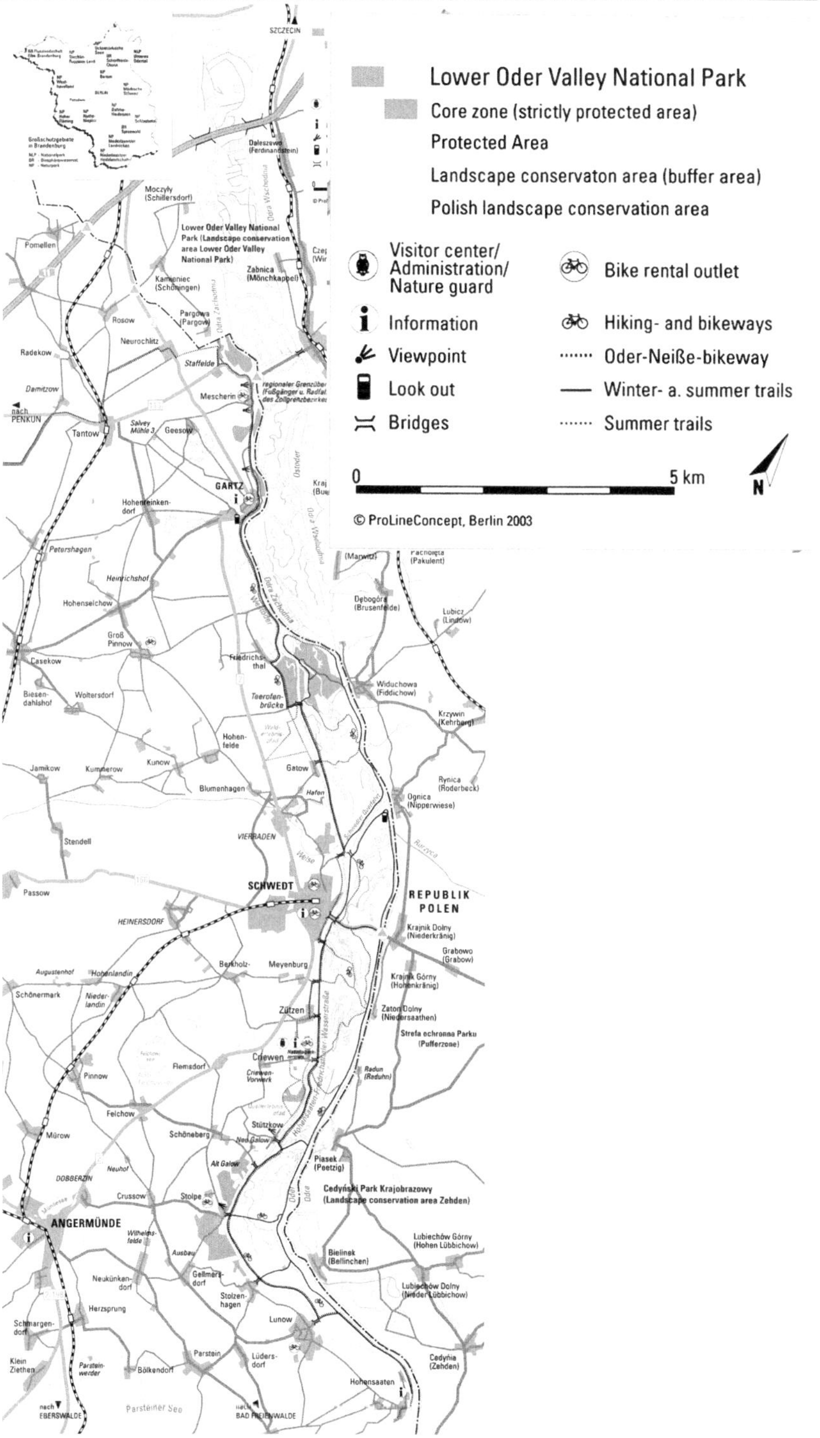

Fig. 1. Map of the Lower Oder Valley National Park (© Pro Line Concept 1997)

Fig. 2. ›Wet polder‹ (left) and ›dry polder‹ (right) separated by a dyke. The picture at the top was taken in March, the one at the bottom in July. (Photos: H. Lüssow, J. Rothenbücher)

Fig. 3. Core zone near Teerofenbrücke in May (Photo: P. Just)

Additionally, the Lower Oder Valley National Park area has been federally funded as a project area in the context of a programme for riparian nature conservation (*Gewässerrandstreifenprojekt*) since 1992. In the long run it is planned to restore the floodplain to semi-natural conditions. To accomplish this aim the flooding regime will be altered to a preferably large area as close to natural flooding dynamics as possible, i.e., longer and more frequent flooding events. Concrete measures have been worked out in detail by an environmental planning agency (IUS Weisser & Ness 1998). Furthermore, it is planned among other things, that by 2010 50 % of the national park area will be declared as core zone and left for succession. In other parts of the park breeding and foraging habitats for endangered species (e.g., the corncrake) will be optimised by conservation management. Thus, current plans for national park development try to realise the two conservation strategies side by side.

4 Three case studies from the Lower Oder Valley National Park, Germany

4.1 Impact of mowing on the suitability of grassland as habitat of corncrakes (*Crex crex*) in the Lower Oder Valley National Park

Introduction

The corncrake (*Crex crex* L.) is a bird species, a rail (Gruiformes: Rallidae), that is considered under threat of global extinction (Groombridge 1994). The birds return in late April and early May from the south and southeast of Africa and start breeding on the ground of marshland, hay and silage meadows and other tall vegetation in central Europe in the middle of May (Glutz von Blotzheim et al. 1973). The world-wide population has been estimated as 1.5 million individuals in 1997 (Green et al. 1997). Losses of nests, chicks and adults occur during the mowing of meadows. Former studies indicate that the mechanisation of agriculture, mowing earlier in the year and loss of habitats with tall vegetation are causes of the decline (Green 1996). With the process of mechanisation in agriculture, especially in western Europe, the corncrake population has been decreasing for more than a century (Cadbury 1980). The bird has its centre of distribution in Europe. In Germany the population consists of less than 3000 individuals (Green et al. 1997) having its greatest population with 100-300 individuals in the Lower Oder Valley National Park. As the species is categorised as critically endangered in the Red List of Germany (Witt et al. 1998) as well as in the Red List of Brandenburg (Dürr et al. 1997), Germany has a certain responsibility for the conservation of this species. Hence, it is necessary to investigate corncrake's habitat requirements to enable successful conservation programs. The bird species comply in a special way with the criterions for target species, proposed by Mühlenberg and Hovestadt (1992) and Vogel et al. (1996):

- the species is threatened with global, nationwide and local extinction (Schäffer and Münch 1993; Fischer et al. 1999).
- the threat of extinction is founded in changing landscapes, not in direct persecution by humans (Green and Stowe 1993).
- the species has its main breeding area in Europe (Glutz von Blotzheim et al. 1973).
- corncrakes are habitat specialists, who are bound by special environmental conditions (Schaefer 2002, personal notification).
- corncrakes have a great home range (Stowe and Hudson 1991; Schäffer and Münch 1993; Helmecke 2000), so they may be considered as an umbrella species.

The hypothesis of this study is that the non use of grassland, as proposed for wide areas in the national park, will have negative impact on the suitability of grassland as corncrakes' habitat. The impact of land use changes in floodplain grassland on the suitability of the corncrakes' habitat was investigated. Additionally, two further questions have to be answered:

- What are the main factors for the colonisation of meadows by corncrakes in May?
- What are the main factors for the establishment of corncrakes in June?

Methods

The places of singing corncrakes in the national park have been counted and marked in maps since 1994 by regional ornithologists. In May 2002, the structure of vegetation was determined in 58 plots (1 m^2) in 36 territories of corncrakes, and in 45 plots (1 m^2) in the dominant grassland associations of the national park, five in each vegetation unit. In the middle of June, the same variables were estimated in 58 plots in 53 territories of corncrakes and 50 plots in 10 plant associations. The 15 variables were:

- height of vegetation
- total cover
- height and cover of the upper, middle and lower strata of grassland
- height and cover of dead phytomass
- proportion of grass and herbs in the upper, middle and lower strata
- penetrability
- land use

Height of vegetation was defined as the height of a line on a table (1x1.5 m^2), upstanding in the grassland, which was more than 1/3 covered by plants, seen from a distance of 5 m and a height of 1.75 m. Cover was estimated as a percentage. The height of strata was defined as the centre of strata in cm above ground.

Penetrability was acquired by the resistance the vegetation offered to an effortless swinging foot in a 6-ary ordinary scale, by the author. Land use has been mapped since the strictly protected areas in the national park were founded in 1996, in three categories: meadow, fallow and willow.

The data were analysed by stepwise multiple logistic regression. Logistic regression methods are useful in the analysis of relationships between a response variable and one or more explanatory variables and allows the use of metric and categorical scaled variables (Hosmer and Lemeshow 2000; Kleyer et al. 1999/2000). Binary logistic regression predicts the probability of occurrence of a dichotomy response variable as a function of one or more independent variables. Thus, it can be used to predict the probability of occurrence of a species as a function of environmental variables. The probability of occurrence of a species can be interpreted as the suitability of defined areas as a habitat (Schröder 2000). The comparison of regression coefficient of significant standardised variables gives insight into the relative importance of environmental variables as habitat parameters (Peeters and Gardeniers 1998). The backstep variation of logistic regression takes stepwise variables out, as long as there is no more significant debasement of the model (Schröder 2000). In statistical modelling of the species-habitat relationship, the binary response variable (1 or 0) is based on presence and absence observations of species in the field. In this study absence was defined as plots where there was no singing male within a range of 200 m. Presence and absence can be defined as a species-specific parameter, so that additionally the abundance of small species can be defined as a dependent variable (Kleyer et al. 1999/2000).

Results

Figure 4 shows the development of the number of corncrakes within the polder including the highest density of corncrakes, in comparison to the number of individuals within the core zone. The core zone, untouched since 1995, has a range of 200 ha within the polder with a range of 1800 ha in total. It is shown that the number of corncrakes in the core zone has decreased since 1997. Since 2000, there were scarcely any, and in 2002 there were no more corncrakes, whereas the number of birds in polder 10 increased after a long period of depression in the last three years from 12 to 30 individuals. In sum, there is great variability in the number of individuals of corncrakes in the national park.

What are the main factors for the colonisation of grassland in the study area by corncrakes?

The significant ($p = 0.05$) of the 15 standardised variables acquired in the middle of May, analysed by binary logistic regression: *cover of upper strata* and *height of upper strata*. These two variables were analysed by backstep multiple logistic regression (Table 1).

Table 1 shows that the variable *height of the upper strata* affects the strongest impact on the distribution of corncrakes in the middle of May. Additionally, the variable *cover of the upper strata* significantly separates between the presence and absence. A regression model with these two variables predicts 86.1 % of presence and absence. The regression model reaches a Nagelkerkes R^2, of 0.643 and a –2 Log Likelihood of 78.448.

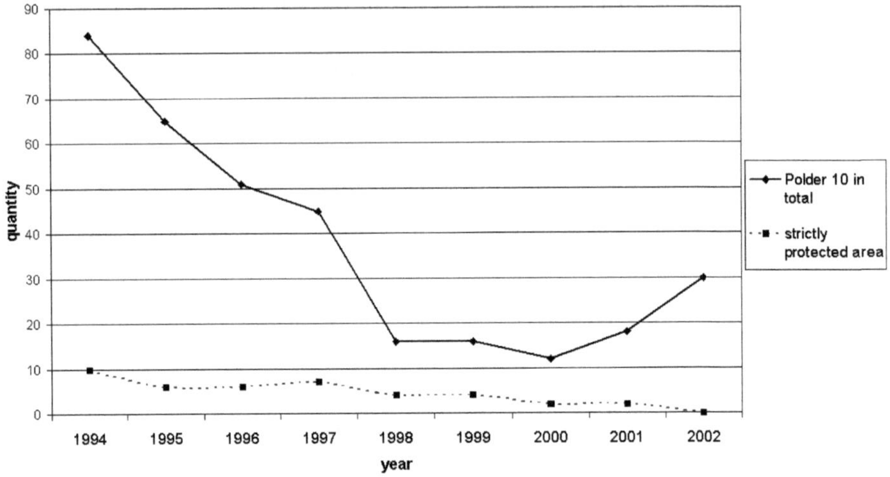

Fig. 4. Number of singing males in polder 10 and in the core zone

Table 1. Variables in equation after backstep logistic multiple regression including the standardised significant variables acquired in mid May (p = 0.10)

Variables	Regression coefficient (ß)	Standard deviation	Significance
Height of upper strata	3.729	0.793	0.000
Cover of upper strata	1.150	0.342	0.001
Constant	0.177	0.304	0.560

These two variables, i.e. height and cover of the upper strata of grassland, were the most impacting variables in the structure of vegetation for the establishment of singing males in May, when they try to attract females with their loud and buzzing shouting at night.

In the middle of June the same statistical method containing the same variables, leads to very different results: Nine variables significantly (p = 0.05) separate between the presence and absence of corncrakes in the study area:

- height of vegetation
- total cover
- cover of the lower strata
- height of dead phytomass
- cover of dead phytomass
- proportion of herbs in total

- proportion of herbs in the middle strata
- proportion of herbs in the lower strata
- land use

The backstep multiple logistic regression leads to a more complex regression model (Table 2). Three variables: *cover of the lower strata*, *proportion of herbs in total* and *land use* were let out, in three steps.

Table 2. Variables in equation after backstep logistic multiple regression including variables acquired in the middle of June (p = 0.10)

Variables	Regression coefficient (ß)	Standard deviation	Significance
Total cover	0.689	0.333	0.039
Height of dead phytomass	0.654	0.437	0.134
Cover of dead phytomass	-1.557	0.461	0.001
Proportion of herbs in the middle strata	−0.789	0.291	0.007
Proportion of herbs in the lower strata	1.060	0.335	0.002
Height of vegetation	−0.528	0.277	0.056
Constant	0.168	0.254	0.507

In the middle of June the variable *cover of dead phytomass* (ß = −1.557) has the strongest impact on the distribution of corncrakes in June. Additionally, the variables *proportion of herbs in the lower strata* (ß = 1.060), *proportion of herbs in the middle strata* (ß = −0.789), *total cover* (ß = 0.689), *height of dead phytomass* (ß = 0.654) and *height of vegetation* (ß = −0.528) affects the distribution of corncrakes in the Lower Oder Valley in June. The regression model with these six variables predicts 69.9 % of the corncrakes and Nagelkerkes R^2 amounts to 0.473, the −2 log Likelihood a value of 98.676.

Discussion

In May the height and the cover of the upper strata were the significant variables separating between the presence and absence of corncrakes. Another backstep logistic regression analysis, only using metric scaled independent variables (not the categorial scaled variable *land use*) confirms this result. The significant variables with the highest regression coefficient were *height of the upper strata* (ß = 4.743), *cover of the upper strata* (ß = 1.873), *total cover* (ß = 1.729) and *height of the lower strata* (ß = 1.019). So the height and the cover of grassland were the main variables in the structure of vegetation for the habitat selection of corncrakes in the study area, after the return of the birds. The higher the meadows grow in the national park and the higher the cover of the meadows, the higher the probability of occurrence of corncrakes in May. The results of this study comply with the results of Schäffer (1997), who studied habitat selection of corncrakes in the Biebrza National Park in Eastern Poland. Schäffer (1997) found that in May the height and the cover of grassland vegetation are the determining factors for singing corncrakes. He detected a minimum height of grassland of 20 cm and a total cover of more than 50 % to be suitable as habitat (Schäffer 1997). »An important feature of selected habitat, is that they are tall enough to provide cover« (Tyler 1996). In the northern part of the breeding range in western Scotland, corncrakes select stands of tall herbs like *Urtica dioica* or marsh vegetation *Iris pseudacorus, Phragmites australis* and *Phalaris arundinacea* as habitat in May and early June, because hay and silage meadows provide too little cover. Only when the height of grass exceeds about 20 cm in mid June, the meadows are used by corncrakes (Cadbury 1980).

Most parts of grassland within the national park exceed much more than 20 cm in height in mid May, so in particular, the dominant plant association *Phalaridetum arundinaceae* and *Caricetum gracilis* are suitable for the corncrakes. In mid May there are no indications in regression analysis that fallows in the core zone are less suitable for singing males than used meadows.

In mid June, there is the highest correlation between the variable *cover of dead phytomass* and the distribution of corncrakes (ß = –1.557). The lower the cover of dead phytomass, the higher the probability of the occurrence of corncrakes. The variable *total cover* (ß = 0.689) indicates that certainly the grassland has to offer a high cover in June. For the other highly significant variables in the regression model: *proportion of herbs in the lower strata* (ß = 1.060) and *proportion of herbs in the middle strata* (ß = –0.789) there are no indications in the literature on the ecology of corncrakes. Perhaps herbs in the lower strata offer less resistance to the birds than closely spaced blades of grass. Green et al. (1997) found that in most selected habitats of corncrakes, the annual production of plants is removed by mowing, grazing or winter floods. »Stands of dead grasses from previous years tend to be avoided« (Green et al. 1997). Additionally, meadows where the grass is pushed to the ground by wind and rain are avoided because they are too difficult to penetrate (Schäffer 1997). In the Biebrza National Park, Schäffer (1997) observed no corncrakes in grassland where the meadows were too dense to be penetrated by an effortless swinging foot and where the resistance of grassland near the ground was higher than 20 Newton. The findings of Tyler (1996) confirm these results. »Another important feature of selected habitats is, that they are not so dense that they are difficult for the birds to walk through« (Tyler 1996). Schäffer (1997) presumes that closely spaced grasses in fertilised meadows may also be too dense to be penetrated by corncrakes.

With the growths of meadows in June, they become more dense and less penetrable. The cover of dead phytomass, as an indicator for fallow grassland, is the most correlated variable on the distribution of corncrakes in mid June. The higher the cover of dead phytomass, the lower the suitability of grassland as corncrakes' habitat. So, there is a great impact of mowing on the suitability of grassland of corncrakes in the Lower Oder Valley. The results shown in Fig. 4 presume that only after non use for more than four years does the cover of dead phytomass finally become too dense to be penetrated by corncrakes at all, and the fallows become totally unsuitable to be colonised by the species.

In a nutshell, in the study area we find a preference of used meadows adverse to fallows in June. Thus, nature conservation management on corncrakes as a target species has to include the mowing of the meadows. They have to be mown to inhibit the accumulation of dead phytomass (Green and Stowe 1993) and the date of mowing has to be considered. Corncrakes start to breed in mid May after the last egg is laid. The breeding takes 15–18 days. It takes until the end of July, until the chicks are able to fly, when the meadows are mown by fast and heavy machines. There will be almost no reproduction of the species if the meadows are mown earlier than the end of July. In Britain and Ireland the species was absent if the average date of mowing of hay meadows was earlier than late July (Green et al. 1997). In summary, in the parts of the floodplain where process conservation in form of fallows takes place, in a few years time there will be no suitable habitats for corncrakes.

One aim of the national park's development plan is that by 2010 50 % of the national park's area will be declared as core zone. The core zone will be predominantly situated in the northern parts of the national park. Unfortunately, most corncrakes occur at present in the area of the future core zone. It is assumed that by creating new appropriate habitats by the help of management, the national park's corncrake population will accept and colonise the new suitable habitats. However, this raises the following questions: In the last few years, in these parts of the national park there were scarcely any corncrakes. Which parts of these areas are the most suitable and how are these areas to be managed to offer corncrakes optimal habitats and provide for the reproduction of the species? These questions will be answered in future by the development, application and validation of spatial explicit statistical habitat models (Kleyer et al. 1999/2000; Schröder 2000) including further parameters like, i.e., plant associations, soil moisture or the exact point in time when the meadows are used.

4.2 Impact of flooding on the amphibian community

Introduction

The Lower Oder Valley National Park is one of the last remaining floodplains in central Europe. With its numerous collections of different waterbodies in form of old river arms, ditches, ponds, pools and puddles, it represents suitable habitats for many amphibian species present in Germany. The different requirements of the fourteen species of Anura resident in Germany, are given in the different natural landscape structures in the more than 10,000 ha large area. As already mentioned in the description of the study area, there are two different polder systems in the national park which are different with regard to the regulation of the flooding taking place. One is exposed to flooding until the middle of April, the other is excluded from flooding the whole year. The aim of the work was to investigate the occurrence of amphibians and their dominance structures in the polder area of the national park. A special interest applied here for the comparative view between waterbodies in the ›dry‹ and the ›wet polder‹. The work was mainly based on the following hypothesis: The different management measures in the polder system, which lead to different flooding conditions, affect the community of the amphibians which are highly dependent on water regimes. This takes place in a specific way corresponding to the preferences of each amphibian species. In order to be able to represent the differences between the ›dry‹ and the ›wet polder‹ on the basis of the occurrence of amphibians, a defined number of waterbodies was examined in two areas of the national park and the frequency of amphibians occurring there, was determined. On this basis, statements about the effects of flooding, concerning amphibians could be made. Future planning, which concern the flooding and the water management in the polder system, can be judged on the basis of the investigations in an amphibian-specific manner.

Material and Methods

Study area

The area of investigation, with a total of 24 examined waterbodies, is limited to two ranges of the national park. The investigation area *A*, the so-called ›dry polder‹ (Lunow-Stolper Polder), lies in the southern part of the Lower Oder Valley. The investigation area *B*, the so-called ›wet polder‹ (Criewener Polder – Polder A), is subjected to an artificial flooding regime. Gates, which are opened between the middle of November and the middle of April, lead to a direct contact with the river Oder and the high level water in the flooded polder. When closed in April, the polder area becomes drained.

Selection of the waterbodies

Twelve waterbodies in the ›wet‹ and twelve waterbodies in the ›dry polder‹ were selected. In each case, five were permanent waterbodies, four were temporary waterbodies and three were ditch-sections of 100 m length. The selection of waterbodies took place on the basis of coincidental criteria and a comparability of the study areas in ›wet‹ and ›dry polder‹ was given.

Times of data acquisition

Each of the 24 waterbodies was examined nine times in total between May and September 2002. The investigation of all waterbodies took place in regular time blocks fourteen days apart. Each one was examined at different times during the day and night.

Mapping of adults and juveniles

To take stock of the amphibians existing in the study area, each waterbody was orbited for 30 minutes by foot and all adult and juvenile amphibians were caught and determined. Searching took place water-near, within a radius of 25 meters distance to the edge of the water. Moreover the calling activities of adult male frogs in the water were examined. On the basis of the calling males, the place of residence of the frogs in the water was determined and registered. Both male and female frogs were registered and visually estimated. The registered individual numbers referred therefore not to the calling males, but to the visually detectable frogs. The occurrence of juvenile froglets was taken up and rated as reproduction proof at the specific waterbody. Only freshly metamorphosed froglets were regarded and defined as juveniles.

Mapping of larvae

The occurrence of larvae (tadpoles) of different amphibian species and their frequencies were determined on the basis of dipnetting. Four catches were performed at four different places in each waterbody. Generally, the selection of the catching-places took place on the basis of five selected vegetation structure zones. The catching took place both at the edge and within the waterbody and also near the surface and in up to 1 m depth. The determination of the amphibian larvae was made directly after the catch in the field. The determination of the larvae of *Rana temporaria* L. and *R. Arvalis* Nilss. was made on the basis of the different morphology of their mouthparts. A distinction of the larvae of *R. Ridibunda* Pall. and *R.* kl. *esculenta* L. was not possible in the field, so all larvae of these species were taken up as larvae of the water frog complex.

Water frog complex

The so-called water frog complex consists of three species of central European green frogs. Berger (1967) discovered that *Rana* kl. *esculenta* is a hybrid form, which appears after the crossing of two other species of water frogs, *R. Ridibunda* and *R. Lessonae* Cam. (Eikhorst and Rahmel 1986). Nearly all water frog populations occurring in Germany consist of mixed populations of *R.* kl. *esculenta* and *R. Ridibunda* or of *R.* kl. *esculenta* and *R. Lessonae*. An occurrence of all three species of water frogs together is rare (Günther 1990; Plötner 2001). Due to the morphological similarity of water frogs, which can be differentiated only with difficulty (Eikhorst 1984), they were not separately regarded, but designated as the water frog complex (RID/ESC). The composition of the water frog complex in different parts of the national park, was investigated in a separate study (Ogielska personal notification): a large sample of frogs was caught over a two night period and identified.

Results

In the study area, the polder system of the Lower Oder Valley National Park, a total of eight species of amphibians could be found as adults, juveniles or tadpoles. Three of these species were found in the area of the ›wet polder‹ and seven were found in the area of the ›dry polder‹. From the adult frogs found in the ›dry polder‹, only individuals of *R. Ridibunda* (lake frog), *R.* kl. *esculenta* (edible frog), *R. Arvalis* (moorfrog) and *Bufo bufo* (L.) (common toad) could be found in larger individual amounts. *Bombina bombina* (L.) (fire-bellied toad) and *Hyla aborea* (L.) (European tree frog) could only be found as isolated adults or, in case of the *R. Temporaria* (common frog) only as tadpoles. Juvenile froglets of the following species could be proven: the common toad, the moorfrog, the common frog and the frogs of the water frog complex (Table 3).

Table 3. The occurrence of amphibian species in the polder system of the Lower Oder Valley National Park and the frequency of adult and (in brackets) larval individuals; x: occurrence of juveniles. The numbers shown refer to the total number of individuals found during the eight or nine (larvae) investigations

Species	Wet polder	Dry polder
R. ridibunda and R. kl. esculenta	2663 (970) x	1917 (304) x
R. arvalis	0 (0)	209 (403) x
R. temporaria	0 (0)	0 (47) x
B. bufo	0 (0)	27 (167) x
B. bombina	0 (0)	3 (0)
H. aborea	0 (0)	1 (0)
P. fuscus	1 (0)	0 (0)
Number of species	3	7

The frogs of the water frog complex, both adults and tadpoles, were the only ones in the ›wet polder‹ to be found in larger amounts. Together with one caught individual of *Pelobates fuscus* (Laur.) (common spadefoot) only these three species could be proven. Compared with the result of the caught adults and tadpoles, only juveniles of the water frog complex could be found.

The dominance conditions of the seized adult frogs in the ›wet‹ and ›dry polder‹ are shown in Fig. 5 and 6. In the ›wet polder‹, the frogs of the water frog complex (RID/ESC) are the almost exclusively occurring species. With six species proven in the ›dry polder‹, it exhibits a larger diversity of species than the ›wet-polder‹. The species of the water frog complex and *R. Arvalis* are called ›dominant species‹ and thus ›main species‹, while the other occurring species are classified as ›recedent‹ and ›sporadically‹ species (Engelmann 1978).

In the case of the ›wet polder‹, the occurrence of tadpoles is limited to the larvae of the water frog complex (RID/ESC). Other species cannot be proven in the waterbodies (Fig. 7 and 8). In the ›dry polder‹, tadpoles of the species RID/ESC, *R. Arvalis*, *R. Temporaria* and *B. Bufo* can be registered. *R. Arvalis* is the ›dominating‹ species, followed by the frogs of the water frog complex which are called ›eudominant‹ and the species *B. Bufo* and *R. Temporaria* which are classified as dominant and subdominant (Engelmann 1978).

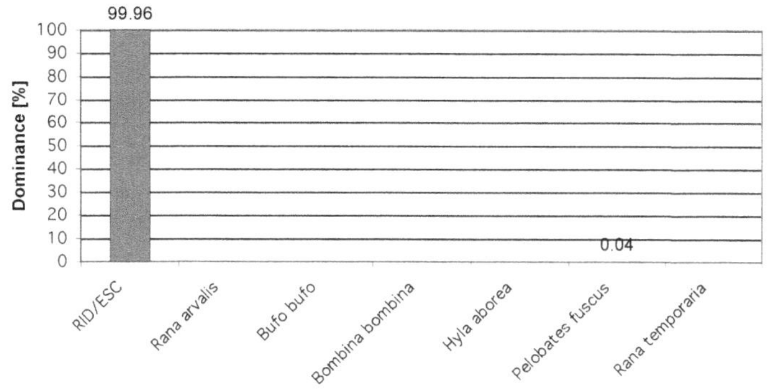

Fig. 5. Dominance structure of adult amphibians found in the investigated waterbodies of the ›wet polder‹ in the Lower Oder Valley National Park

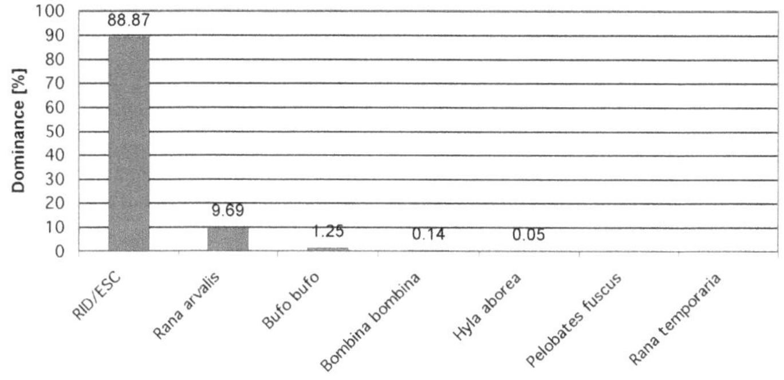

Fig. 6. Dominance structure of adult amphibians found in the investigated waterbodies of the ›dry polder‹ in the Lower Oder Valley National Park

The water frog complex population in the Lower Oder Valley consists of the lake frog *R. Ridibunda* and the edible frog *R.* kl. *esculenta*. With a portion of 70 % in the ›dry polder‹ and 90 % in the ›wet polder‹, *R. Ridibunda* is the dominant species in each area (Ogielska personal notification) (Fig. 9 and 10).

Discussion

With eight anuran species occurring in the polder system of the Lower Oder Valley, the national park represents a suitable habitat and reproduction area for many species of amphibians occurring in Germany (Blab 1986). Following further studies about this area, it can be assumed that there are still four further species of

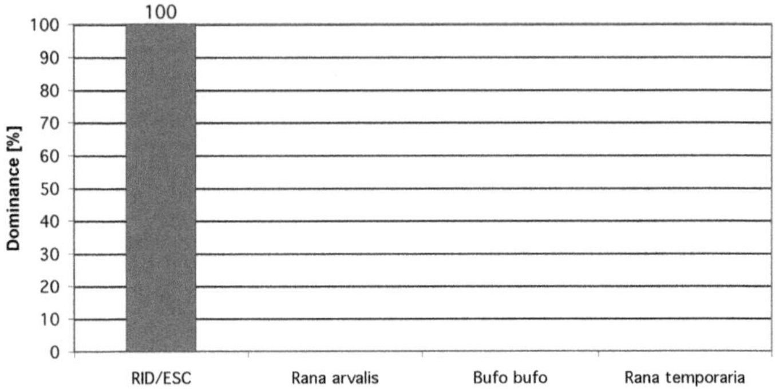

Fig. 7. Dominance structure of amphibian larvae in the waterbodies of the ›wet polder‹ in the Lower Oder Valley National Park

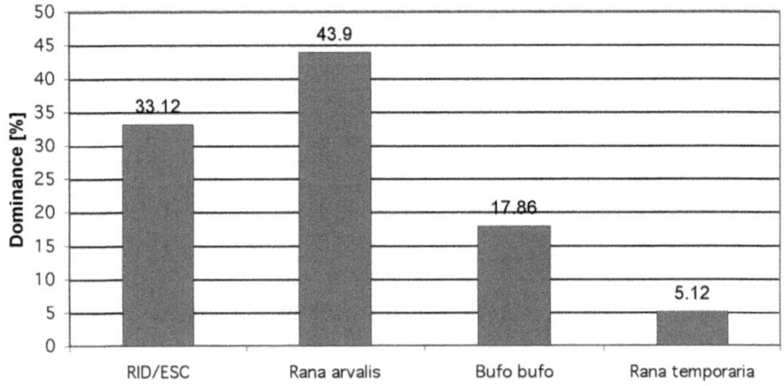

Fig. 8. Dominance structure of amphibian larvae in the waterbodies of the ›dry polder‹ in the Lower Oder Valley National Park

amphibians within parts of the national park and its direct environment. The following species could be found: *B. Viridis* Laur., *R. Lessonae*, *Triturus vulgaris* (L.) and *T. Cristatus* (Laur.) (Mädlow and Adam 1998; Wilke 1995). This diversity of amphibian species in a closely limited area is rarely observed in Germany (Schmidtler and Gruber 1980).

Looking at the distribution of the amphibian species in the different areas of the national park, it becomes clear that the majority of the proven amphibians occur only within the smaller part of the ›dry polder‹. Within the area of the national park which is affected by flooding (and constitutes its largest part), only the frogs of the water frog complex seem to appear.

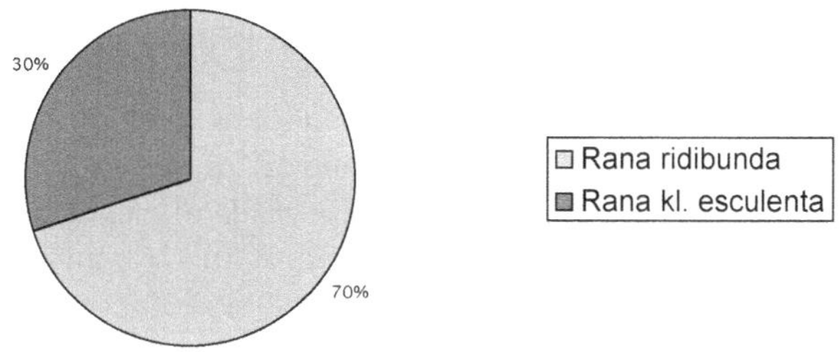

Fig. 9. Composition of the water frog complex of the ›dry polder‹ in the Lower Oder Valley National Park (n=20)

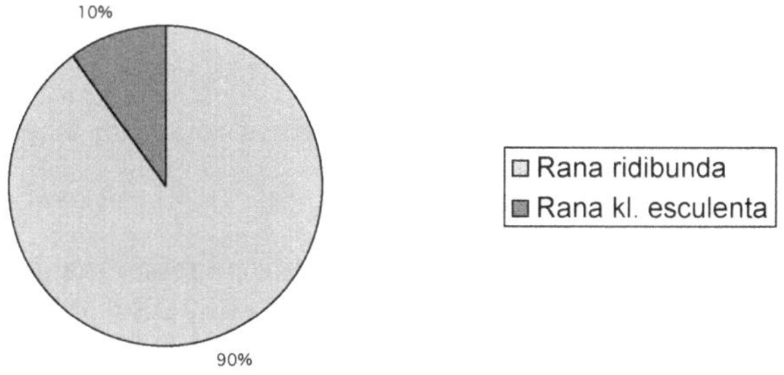

Fig. 10. Composition of the water frog complex of the ›wet polder‹ in the Lower Oder Valley National Park (n=39)

After the flooding of the ›wet polder‹, during which it appears in a sea-similar condition lasting until the middle of April, it is not possible for other species of amphibians (except the water frog complex) to find suitable places for hibernation in winter and waterbodies for reproduction in spring. While the frogs of the water frog complex have the possibility to hibernate under water (Günther 1996a, 1996b), the other occurring species of amphibians search for accommodation ashore (Nöllert and Nöllert 1992). These are underground, under stones or under leaves and wood structures (Blab 1986). Only the common frog is known to be able to hibernate under water as well (Schlüpmann and Günther 1996). The second reason for the absence of amphibian species in the ›wet polder‹ is the absence of suitable waterbodies for breeding. They are firstly formed after the drainage of the flooded polder at the beginning of May. At this time, most of the amphibians have already begun to migrate to their breeding places. *R. Arvalis*, *R. Temporaria*, *B. Bufo* have already passed the breeding period (Blab 1986). Thus, a possible hibernation in the dyke of the ›wet polder‹ would not make the use of waterbodies as breeding places possible because the right time conditions are not met. For later breeding species like *B. Bombina* and *P. Fuscus*, a migration to the waterbodies would still lie in the phase of water running off the polder, which would lead to suboptimal conditions (Grosse 1994, Schneeweiß 1996). Water frogs start their mating period in May and mating can take place until July. In contrast to other amphibian species, they do not have the problem of a long distance migration and thus can find suitable waterbodies for breeding in the ›wet polder‹ in their specified time period (Blab 1986; Günther 1996a, 1996b).

In the ›dry polder‹ there are a lot of different types of waterbodies with a different structure and vegetation. In principle, the whole area offers the possibility for all proven amphibians to use it as a suitable habitat and for reproduction (Lindeiner 1994). The human influences, in form of the artificial lowering of the ground-water level and agricultural use, make it harder for some of the proven amphibians to develop larger populations. This concerns, above all, the two species *B. Bombina* and *H. Aborea* which are both threatened by extinction (Borgula 1995; Schneeweiß 1996).

4.3 Impact of land use and flooding on the diversity of insects

Introduction

As outlined above, land use and flooding are currently the two major factors affecting species living in the floodplain of the Lower Oder Valley National Park. Both factors are planned to be altered in the future. At the moment there are already areas in the national park differing in land use as well as in flooding duration and frequency. Comparing the communities occurring in these sites gives us the opportunity to analyse the impact of land use and flooding on the species assemblages. On this basis, it is possible to assess the impact of the plans for future national park development on the diversity of insects.

On a global scale, insects comprise more than 50 % of all described living species (Wilson 1992). Thus, when talking about diversity, this group should not be neglected. However, dealing with such a highly diverse group also leads to restrictions as one single investigator is not able to study the whole group. Thus, plant and leafhoppers (Hemipetra: Auchenorrhyncha) were chosen as a group typical for grassland habitats. Plant and leafhoppers occur in high numbers of species and individuals in wet grassland (Achtziger and Nickel 1997). Additionally, they can easily be sampled and information on the ecology of most species is available in the literature which can be included for further analyses.

Materials and methods

Samples were taken throughout the vegetation period of the years 2001 and 2002 by a motor-driven suction apparatus on study sites differing in land use as well as in flooding period and frequency (Table 4). While sites situated on the river bank are subjected to winter and summer flooding, sites in the dry polder are protected against inundation throughout the whole year. The grassland of the wet polder is flooded only during the winter months. Here, it is differentiated between sites subjected to long winter flooding that are located in ditches and sites subjected to short winter flooding that are situated on more elevated locations in the wet polder.

Table 4. Study design. For each pair of influencing factors the number of study sites investigated is listed

	Location	River bank	Wet polder		Dry polder
	Time of flood	Summer & win.	Winter (long)	Winter (short)	No flood
Year	Flooding impact	High	Medium	Low	No
2001	Mown grassland	1	5	2	2
	Fallow	2	2	5	1
2002	Fallow	3	3	3	3

The investigation on the effects of land use was carried out in the first year of the study. It concentrated on the impact of mowing which is currently the major land use in the national park. It was presumed that flooding duration and frequency affect the communities as well, thus *flooding* was included as a second factor. In 2002 the investigations were focused on the impact of flooding. On each study site the suction apparatus (STIHL SH 85, diameter of the suction tube: 14 cm) was placed onto the ground ten times at random for approximately ten seconds to gather a mixed sample of each study site. The samples were put in a cold box while in the field and afterwards stored in a deep-freeze compartment for several days. Invertebrate animals were sorted and stored in 70 % ethanol. Adult Auchenorrhyncha were identified to species level. The nomenclature for all Auchenorrhyncha except Cicadellidae follows Holzinger et al. (2003) and for Cicadellidae Nickel and Remane (2002). Species that were difficult to identify were checked by H. Nickel (Göttingen). For the analysis, all samples taken throughout each year were summed up for each study site.

Results

Between June and September 2001, 55 species of plant and leafhoppers with a total of 1,564 individuals were collected. To find out whether the impact of mowing or the impact of flooding has a stronger effect on the communities, a canonical correspondence analysis (CCA) was carried out including mowing and flooding as environmental factors (Fig. 11).

Each dot in the diagram represents the community of a study site. Study sites with similar species composition are plotted next to each other whereas those dif-

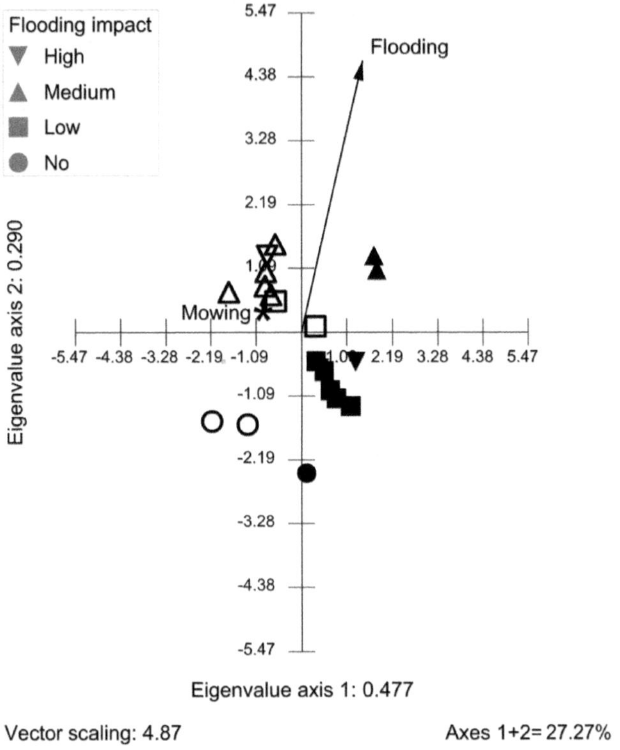

Fig. 11. Ordination diagram of a Canonical Correspondence Analysis (CCA). The CCA is based on abundance data of the plant and leafhopper species caught with more than 2 individuals in the study area and the environmental factors *mowing* and *flooding*. Open markers represent mown plots, solid markers fallows. The arrows indicate the environmental variables

fering in species composition are plotted apart. Both influencing factors, i.e., flooding and mowing are plotted as vectors. The dots are grouped together firstly in respect to land use and secondly in respect to flooding influence. Thus, mowing has a stronger impact on the arthropod communities than flooding. The communities of mown study sites show just a few differences due to flooding influence. Only study sites with no flooding impact differ notably from all other plots. In fallows we find a different pattern. Here the study sites are arranged along the flooding gradient. Surprisingly, those sites that are affected by short winter flooding (low flooding impact) and those affected by summer and winter flooding (highest flooding impact) are clustered together, whereas sites subjected to long winter flooding (medium flooding impact) are situated at the top end of the gradient.

Furthermore, it is interesting to analyse in what respect the species assemblages differ in regard to different land use and flooding influences. For the analysis, numbers of species, numbers of individuals, diversity as well as the proportion of species specialised to certain moisture conditions and food plants are taken into account.

As land use seems to have a stronger impact on the communities than flooding, the influence of mowing on the species assemblages is analysed first. The highest number of species (20) was found on fallows with no flooding impact whereas the lowest number of species (10) was collected on mown plots with high flooding influence. No significant differences between the mean numbers of species on fallows and on mown grassland were identified.

Comparing the numbers of individuals caught on mown sites with those on fallows revealed again no significant difference. The lowest (67 individuals per m^2) as well as the highest density (268 individuals per m^2) were both found on fallows. The species diversity (Shannon-Wiener Index) combined with the evenness gives additional information about the communities. The diversity index combines the number of species with the relative abundance of the species. The lowest diversity (0.647) was found on fallows with medium inundation influence. The highest diversity (2.562) occurred on fallows with no flooding impact. Again, no significant differences in diversity could be identified between mown plots and fallows, but the range between the values of fallows is much higher than of mown study sites, i.e., 1.915 and 0.957 respectively.

For a further analysis species were classified according to their habitat preferences (Table 5). Auchenorrhyncha species were assigned to the classes by H. Nickel, Göttingen (personal notification). The classification takes the adaptation to a certain degree of moisture into account, thus the proportion of hygrophilous species reflects the degree of specialisation of the communities to floodplain habitats.

Table 5. Definition of habitat preferences for Auchenorrhyncha of open landscapes

Classification	Definition	Abbreviation
Hygrophilous	Species occurring preferably in moist to wet habitats	h
Euryhygric	Species occurring in nearly all habitats relatively independent of the degree of moisture	e
Xerophilous	Species occurring preferably in dry habitats	x

Comparing the proportion of hygrophilous species occurring on fallows to those found on mown grassland shows that there are significantly less hygrophilous species on mown grassland than on fallows (Fig. 12). Study sites subjected to land use were dominated by euryhygric species whereas the proportion of xerophilous species was very low in both.

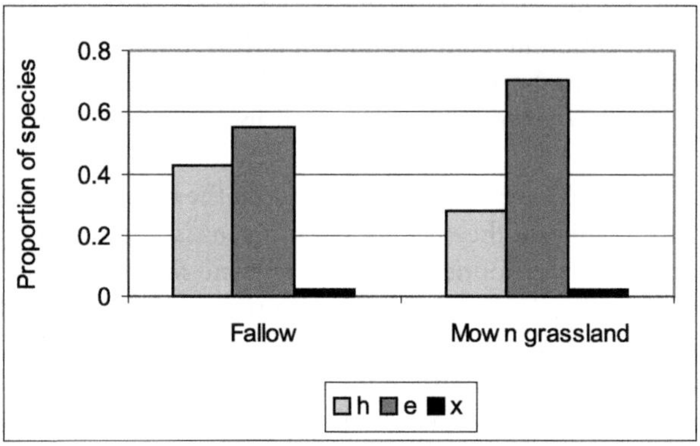

Fig. 12. The proportions of hygrophilous (h), euryhygric (e) and xerophilous (x) species for fallows and mown grassland. For all three categories a Mann-Whitney U-Test was carried out separately. Hygrophilous species: $p = 0.04$, euryhygric species: $p = 0.007$, xerophilous species: not significant (n.s.)

Besides habitat preferences, the degree of diet breadth is a second measure giving information on the specialisation of species assemblages (Table 6). For all species caught in the study area, it is possible to find detailed information on the food plant specialisation (Nickel and Remane 2002).

Table 6. Definition of diet breadth from Schaefer (2003)

Classification	Definition	Abbreviation
Monophagous	Species feed on plants of one host genus	m
Oligophagous	Species feed on plants of one host family	o
Polyphagous	Species feed on plants of several families	p

For the proportion of individuals of monophagous species no significant differences between fallows and mown grassland were identified. However, the proportion of oligophagous species as well as of polyphagous species differed significantly in respect to land use (Fig. 13). Whereas fallows were dominated by oligophagous species, mown study sites host a higher number of polyphagous individuals than fallows.

In 2002 the investigations were focused on the impact of flooding, hence only fallows that were subjected to different flooding influence were chosen as study sites. Samples were taken on 12 fallows differing in inundation influence. Between

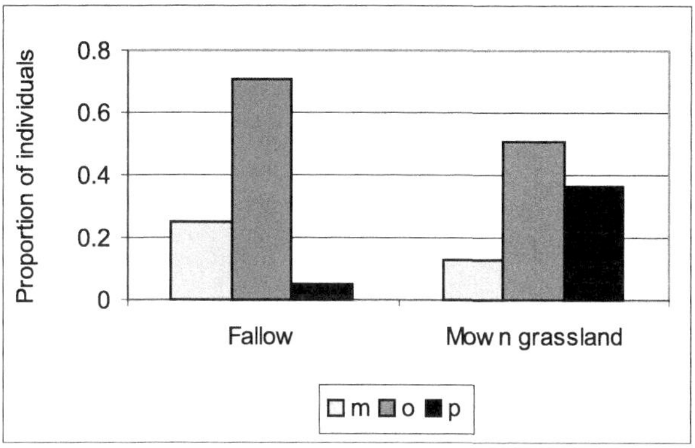

Fig. 13. The proportion of individuals of monophagous (m), oligophagous (o) and polyphagous (p) species for fallows and mown grassland. For all three categories a Mann-Whitney U-Test was carried out separately. Monophagous species: n.s., oligophagous species: $p = 0.03$, polyphagous species: $p < 0.001$

May and September, 55 species with a total of 1,339 individuals were sampled. Ten of these species had not been found in the previous year. The highest number of species was found on sites with no flooding influence. A significantly lower number of species was caught on sites with medium flooding influence (Fig. 14).

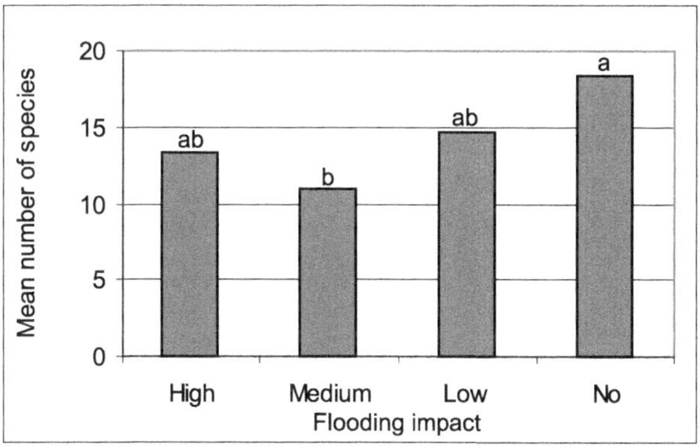

Fig. 14. The influence of flooding on the mean number of species. Analysis of variance (ANOVA): $F = 4.03$, $p = 0.05$. The letters in the diagram indicate the Tukey groupings. Bars with different letters are significantly different

The mean number of individuals per m^2 did not differ significantly in respect to inundation impact. As in the samples taken in 2001, we find the lowest diversity value (Shannon-Wiener) on fallows with medium flooding influence. A significantly higher diversity value was found on sites with no flooding influence (Fig. 15). The diversity values on fallows with high and low inundation impact lie in-between.

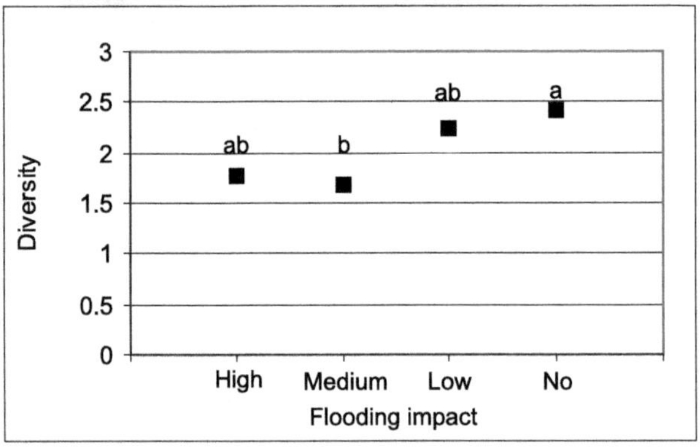

Fig. 15. The influence of flooding on the mean value of the Shannon-Wiener diversity index. Analysis of variance (ANOVA): $F = 5.26$, $p = 0.02$. The letters in the diagram indicate the Tukey groupings. Dots with different letters are significantly different

The comparison of the proportions of hygrophilous species between study sites subjected to different flooding influences shows that on fallows with medium inundation impact, significantly more individuals of hygrophilous species were collected than on all other study sites (Fig. 16). Euryhygric species on the other hand dominate on sites with high flooding impact.

In Fig. 17, the proportion of individuals of monophagous, oligophagous and polyphagous species is plotted per flooding influence. Analysis of variance revealed that there were significantly more individuals of monophagous species on fallows with medium flooding influence than on all other study sites. The highest proportions of oligophagous species were caught on sites with low and sites with no flooding impact, respectively. The proportion of oligophagous individuals caught on the river bank lie in-between. For polyphagous species, no differences were identified.

It is often argued that in particular specialists are endangered as specialisation makes species more vulnerable due to, e.g., habitat loss than generalists (Mühlenberg and Slowik 1997). Achtziger and Nickel (1997) divided the plant and leafhoppers occurring in wet meadows into specialists and generalists. The classification takes into account habitat preferences, food plant specialisation, mobility and the number of generations per year. Specialists are restricted to a narrow degree of

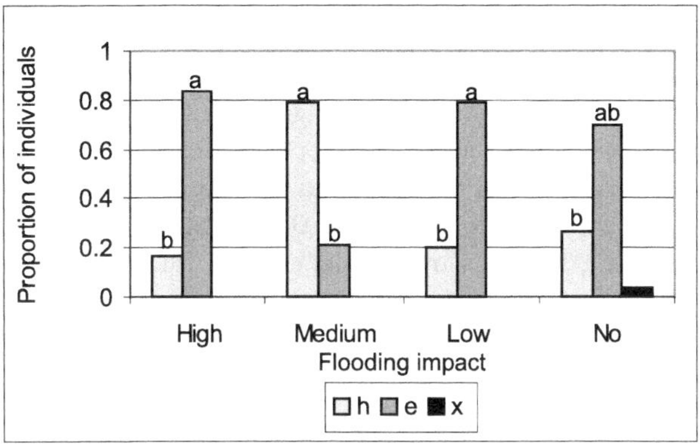

Fig. 16. The influence of flooding on the proportions of individuals of hygrophilous (h), euryhygric (e) and xerophilous (x) species. For all three categories an analysis of variance (ANOVA) was carried out separately. To gain homogenous variances data was transformed by arcsin (x0.5). Hygrophilous species: F = 6.40, p = 0.02, euryhygric species: F = 6.05, p = 0.02., xerophilous species: n.s.. The letters in the diagram indicate the Tukey groupings. Bars with different letters are significantly different

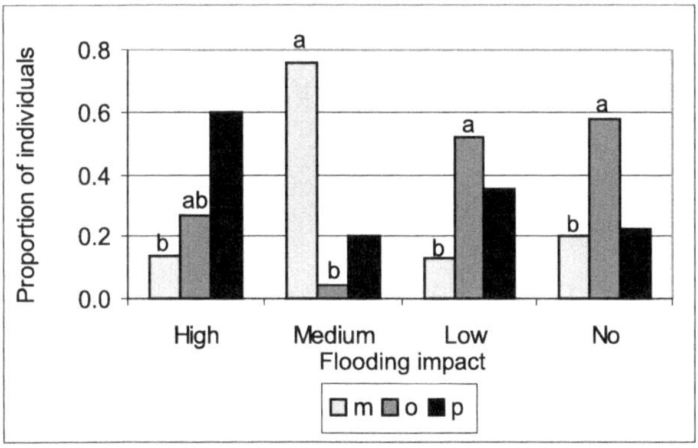

Fig. 17. The influence of flooding on the proportions of individuals of monophagous (m), oligophagous (o) and polyphagous (p) species. For all three categories an analysis of variance (ANOVA) was carried out separately. Monophagous species: F = 7.75, p = 0.01, oligophagous species: F = 5.74, p = 0.02., polyphagous species: n.s. The letters in the diagram indicate the Tukey groupings. Bars with different letters are significantly different

moisture conditions and to a narrow range of food plants. Furthermore, they are often brachypterous, hence not capable of flight and pass mostly through just one generation per year. Generalists on the other hand are mostly euryhygric, polyphagous, macropterous and pass through more than one generation per year. Comparing the proportion of endangered species in specialists with their proportion in generalists, reveals that about 30 % of the specialists found in the study area are listed as »near threatened« in the Red Data Book (Nickel and Remane 2002), whereas none of the generalists are threatened (Fig. 18).

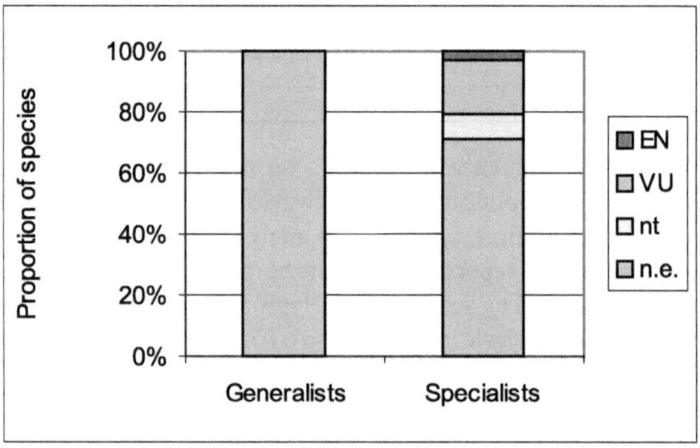

Fig. 18. The percentage of species listed in the Red Data Book for generalist and specialist plant and leafhopper species caught in the study area. EN: endangered; VU: vulnerable; nt: near threatened; n.e.: not endangered

Discussion

It was obvious from ordination that mowing appeared to be the most important variable affecting the distribution of species assemblages. At present a large proportion of the floodplain grassland of the Lower Oder Valley National Park is cut twice a year for fodder production. Future plans for national park development include reduction of land use, i.e., a large proportion of the currently mown grassland will be abandoned and left for succession (IUS Weisser & Ness 1998). Presently, we find sites in the national park that were already set aside some time ago. Comparing species assemblages of mown study sites with those of fallows gives the opportunity to assess the changes in Auchenorrhyncha communities when regular mowing will be stopped. Furthermore these findings can be evaluated from a nature conservation point of view.

In this investigation no significant differences in numbers of species and individuals, as well as in diversity, were found between fallows and mown grassland. Taking the habitat preferences of the plant and leafhoppers into account revealed that significantly more hygrophilous species were found on fallows than on mown grassland, whereas a higher proportion of euryhygric species were caught on mown sites. Furthermore, a higher proportion of individuals of oligophagous species was caught on fallows than on mown grassland while the proportion of individuals of polyphagous species was higher on mown plots.

The conclusions drawn from these findings depend on the conservation objectives. A national park is a protected area sensu CBD thus, one aim is the conservation of biological diversity. This might be understood as preserving high numbers of species and high diversity. Against this background it seems to make no difference whether the grassland is cut twice a year or left for succession as no differences in numbers of species and diversity between fallows and mown study sites were found. Many studies revealed that with decreasing land use intensity the numbers of species and individuals increase (Morris and Lakhani 1979; Klieber et al. 1995; Nickel and Achtziger 1999), but decrease again with longer fallow period (Nickel and Achtziger 1999). Thus, for increasing the numbers of species in the area one might think about reducing mowing, e.g., to one cut a year or every two years but abandoning grassland would be out of the question.

In the declaration act of the Lower Oder Valley National Park[7], the aims and purpose of the national park are outlined. Among other things, it is stated that the purpose of the national park is to protect and to develop natural processes on a large area which is undisturbed by anthropogenic impacts. This leads to a different conservation aim, i.e., a development towards more naturalness which includes abandoning the grassland and leaving it for succession. The proportion of hygrophilous species can be used as a measure of the degree of specialisation of the communities to floodplain habitats. It is interesting to compare the proportion of hygrophilous species on mown study sites to that of fallows. A significantly higher proportion of hygrophilous species was caught on fallows than on mown study sites whereas significantly more euryhygric species were caught on mown grassland. Fallows seem to host the species assemblages more typical for floodplain grassland than mown plots. Hence, giving up mowing activity will favour the typical Auchenorrhyncha floodplain fauna.

Additionally, a higher proportion of oligophagous species was found on fallows than on mown study sites while on mown grassland a higher proportion of polyphagous species was caught than on fallows. Thus fallows host more species not only specialised to the typical wetland conditions but moreover to a narrow food plant range. On mown plots, the species that are most successful are those that can populate a wide range of habitats both in terms of moisture conditions as well as diet width.

Apart from the reduction of land use, the plan for future national park development includes a change in flooding regime towards more natural flooding conditions, i.e., longer and more frequent flooding events. To assess the impact of these

[7] Lower Oder Valley National Park Designation Act (*Nationalparkgesetz Unteres Odertal* - NatPUOG) June 27th 1995, GVBl I, p 114.

changes on the plant and leafhopper communities, fallows that were subjected to different flooding influence were chosen as study sites. The river bank seems at first sight to be the habitat in the national park closest to natural conditions. But the areas between the dyke and the river Oder are very narrow, thus the water level rises and falls rapidly and the grassland and the Auchenorrhyncha species are subjected to very variable water levels and high currents. The former natural floodplain was much wider thus inundation was a slower process with slower currents. The long lasting floods occur in the Lower Oder Valley typically during winter. Summer flooding is a more rare event and high water levels that are comparable to winter flooding occur on average only once in 10 years (Vössing 1998). Thus, the sites with long lasting and regularly occurring winter floods, i.e., with medium flooding impact, seem to be closest to natural conditions.

Analysis of the data revealed that fallows without any flooding influence host species rich and diverse communities whereas on sites with medium flooding impact the lowest number of species and the lowest diversity value were found. Additionally, flooding duration and frequency affects the occurrence of hygrophilous species. On fallows with medium flooding influence the proportion of individuals of hygrophilous species was 80 % and differed significantly from all other study sites. Less than 20 % of the individuals caught on the river bank were hygrophilous species, the habitat was dominated by euryhygric species. Like the fallows on the river bank, sites with low flooding impact were dominated by euryhygric species. Comparing the species in regard to their food plant specialisation gives further information on the effect of flooding on the Auchenorrhyncha communities. A significantly higher proportion of individuals of monophagous species were found on fallows with medium inundation impact than on all other study sites. Oligophagous species occurred in high numbers of individuals on sites with no and low flooding impact, respectively, whereas on fallows with medium flooding influence the lowest proportion was found.

The analysis revealed that fallows host more specialised communities in terms of moisture and food plant specialisation than mown plots. Within fallows, ordination shows that the species assemblages differ in regard to flooding influence. Interestingly, communities of sites with high and low flooding influence are quite similar and are grouped in the middle of the flooding gradient, while sites with medium inundation influence are situated at the top end of the gradient. Sites subjected to high and low flooding impact have in common that the flooding influence is not very predictable. Sites on the river bank are subjected to regular winter flooding while summer floods are less predictable events. Sites subjected to low flooding influence are situated on more elevated locations in the wet polder, hence they might not be inundated every winter. On the other hand, sites subjected to medium flooding influence are inundated regularly every winter. Thus, it seems that regularity and predictability of flooding are important factors influencing species assemblages. Furthermore, most specialised species, i.e., monophagous and hygrophilous species were found on sites with medium flooding influence. Plans including the reduction of land use and a change towards more natural flooding conditions seem to favour specialised plant and leafhoppers but reduce species diversity. Furthermore, as all endangered species found in the national park are specialists, the plans for future national park development can be approved as positive from a nature conservation point of view.

5 Winners and losers of the two conservation strategies

Combining the results of the three case studies makes it possible to assess on one hand the effect of conservation management for the target species corncrake on the diversity of plant and leafhoppers. On the other hand, the consequences of process conservation on the corncrake population as well as on the Anura and Auchenorrhyncha communities can be assessed.

Currently, conservation management for the corncrake follows a very pragmatic approach. No management plan officially regulating the mowing activity in the national park is in use. Nesting sites of the corncrake are mapped every year by voluntary local ornithologists. Based on this survey, instructions for mowing are given to the farmers because the present mowing practice, i.e. first cut at the beginning of June, would lead to the destruction of eggs or the death of freshly hatched chicks. Thus, small areas around the identified nesting sites are left uncut. In the long run it is intended to alter the mowing practice. The first cut is planned to be not before August 15th and larger patches of grassland will be left uncut each year within the mown meadows (IUS Weisser & Ness 1998).

The comparison of the vegetation structure in meadows and fallows revealed that *cover of dead phytomass* was the most correlating variable for the differentiation between suitable and unsuitable nesting sites for corncrakes. By mowing, phytomass is taken out of the meadows, thereby reducing its accumulation. Thus, mowing seems to be an appropriate management practice to preserve suitable habitats for corncrakes. Green et al. (1997) states that the amount of dead phytomass in grassland is also reduced by winter floods. These findings could not be confirmed by this study as both fallows and mown grassland are regularly subjected to long winter floods, but clearly differed in suitability for corncrakes.

The study on Anura did not take the factor mowing into account but Blab (1996) and Beebee (1996) state that reducing land use has a positive effect on amphibian communities.

The study on plant and leafhoppers showed that the current mowing practice does not affect numbers of species and individuals but significantly reduces the number of specialists. As many specialists are listed in the Red Data Book (Nickel and Remane 2002), mowing has a negative effect on the Auchenorrhyncha species assemblage from the point of view of nature conservation. One might argue that the planned change in mowing practice might have a positive effect on Auchenorrhyncha because Achtziger et al. (1999) showed that with decreasing land use intensity the number of species, individuals and specialists increased. However, not only the number of mowing events but also the timing plays an important role. It seems that mowing in autumn or early in spring leads to species rich and specialised Auchenorrhyncha communities (Achtziger et al. 1999; Morris and Lakhani 1979). Mowing early in spring is not debatable when talking about conservation management for the benefit of the corncrake as the birds need sufficient cover when establishing their mating territories. If both corncrake and plant and leafhoppers should profit from the conservation management, it seems that mowing should take place in autumn. But current data is not sufficient to answer this question satisfactorily. An experimental approach would be necessary to investigate the impact of mowing at different times in the year on the diversity of plant and leafhoppers as well as on other insect groups. In addition, a socio-economic study would be necessary to assess the costs for compensating the farmers for the loss in fodder quality due to later mowing activity, i.e., for conservation management. It seems that by changing the mowing practice, plant and leafhoppers will not necessarily be the ›losers‹. It is more a question of whether the farmers or the plant and leafhoppers will be the ›winners‹.

Against these results, one can call the selection of the corncrake as target-species into question. Corncrakes, having a large home range, are considered to be an umbrella species and thus it is supposed that a large number of smaller species will profit from the conservation efforts made for the corncrake. However, the results presented above do not support this assumption. At least Auchenorrhyncha and Anura, being typical species of wet grassland habitats, do not profit from the conservation management. On the contrary, conservation measures even negatively affect the selected taxa.

Besides conservation management for target species conservation of processes will be realised in the national park. It is planned that until 2010 50 % of the national park's area will be declared as core zone[8]. These areas will be set aside and left for succession. Furthermore, process conservation includes that flooding will be altered towards a more natural flooding regime, i.e., longer and more frequent flooding events. The artificial lowering of the ground water level is planned to be stopped as soon as the area of the respective polder is completely set aside.

For the corncrake, the results show that regular mowing events are essential to maintain suitable breeding habitats. Thus setting aside grassland will surely make grassland unsuitable to be used by corncrakes' population after a few years. As the case study of the amphibian community presented above focused on the impact of the long winter flooding on the Anura, it is not possible to assess the impact of reduced mowing activity on the amphibians. However, the reduction of land use might even have a positive effect on the amphibian community (Blab 1986; Beebee 1996). For Auchenorrhyncha, it was previously stated that stopping mowing activity will lead to a community consisting of only a few, but specialised species.

At present it is not clear how long the floodplain would be inundated without regulations as the flooding regime is still managed and ideas of how to realise the change in flooding regime are still under discussion. Independent of how the changes will be realised they will lead to longer flooding time in spring.

[8] Lower Oder Valley National Park Designation Act (*Nationalparkgesetz Unteres Odertal* - NatPUOG) June 27th 1995, GVBl I, p 114.

The study on Anurans revealed that the current flooding regime already negatively affects amphibian diversity. In the ›wet polder‹ two species of the water frog complex were found whereas in the ›dry polder‹ seven species were caught. But even longer flooding periods might not negatively affect the water frog complex. Although their mating time starts in May, it can continue until July. Furthermore, in contrast to the frogs occurring in the ›dry polder‹, species of the water frog complex are able to hibernate under water which makes hibernation in the ›wet polder‹ possible (Günther 1996a, 1996b). Thus, they do not have the problem of long distance migration and can find suitable waterbodies for breeding in the ›wet polder‹ during their mating period (Blab 1986; Günther 1996a, 1996b). A reduction of the artificial lowering of the ground water level throughout the ›dry polder‹ might have a positive effect on the amphibians. Besides others species, the fire-bellied toad and the moor frog which are bound to high levels of ground water, will find more waterbodies suitable for breeding and thus have the possibility to spread (Günther and Nabrowsky 1996; Schneeweiß 1996). Unfortunately the ›dry polder‹ is the area where conservation management for corncrakes is planned, thus mowing must still be possible and this includes maintaining the regulation of the ground water level.

For Auchenorrhyncha, the study revealed that the species assemblages on fallows differ in regard to flooding period and frequency. Significant differences in specialisation of the species to moisture conditions and diet width were identified. Communities on sites that are subjected to long winter floods are dominated by specialised plant and leafhoppers, while sites that are subjected to short winter floods or to summer and winter floods host more generalists. Ideally the change in flooding regime will lead to a patchy environment that is subjected to dynamic changes due to flooding influence. High summer floods are a more rare event in the study area. High water levels similar to those of the winter floods occur on average only once in ten years (Vössing 1998). Changes towards longer flooding in combination with higher ground water levels throughout the whole year will therefore increase the number of patches in the wet polder, similar to the areas subjected to long winter flooding. Additionally, areas subjected to winter and summer floods that are comparable to the study sites on the river bank will increase as the more regularly occurring smaller summer floods will no longer be restricted to the river bank. The more elevated areas are comparable to those study sites subjected to low flooding impact. The changes will favour on one hand specialised plant and leafhoppers, but will host on the other hand a number of generalists that can cope with the more unpredictable flooding conditions. Thus, besides the preservation of specialists and endangered species, process conservation might lead to a high β-diversity.

To summarise, conservation management for the target species corncrake will certainly have a positive effect on the corncrake population. Its effect on the anuran community is not assessable by this study because the impact of mowing on amphibians was not investigated. For the plant and leafhoppers, as an example of typical grassland insects, benefit and loss seems to be dependent on the date of mowing which is in turn, dependent on the money that can be spent for conservation management. However, current plans concerning the conservation management for the target species corncrake seem to have negative effects on Auchenorrhyncha. The studies reveal that the corncrake seems an unsuitable umbrella species for wet grassland habitats. Process conservation, on the other hand will clearly have negative effects on the corncrake population. The anuran community in the ›wet polder‹ seems to be neither positively nor negatively affected by a change towards more natural flooding conditions. In the ›dry polder‹ a change towards a higher ground water level might be beneficial for many anurans. The changes in the Auchenorrhyncha community due to an altered flooding regime seems to be positive for both preservation of specialists and diversity.

With these results, it seems that the idea of realising the two conservation strategies side by side can be a successful approach to combine conservation of rare endangered species as well as of biological diversity. But for the Lower Oder Valley National Park as well as for other protected areas following a similar approach, some open questions still remain. Is the protected area big enough to host (in just a part of its area at least), a MVP of the selected target species? Is the remaining area sufficient to host all seral stages created by natural dynamics side by side? Are the revitalised dynamics sufficient to create a patchy environment comparable to natural conditions? How can, e.g., farmers currently using large parts of the floodplain be integrated into the conservation plans? The answers to these questions are out of the scope of the studies presented above and thus further investigations are necessary.

6 Acknowledgements

We would like to thank G. Gerold and M. Schaefer for reading and commenting on the manuscript and the administration of the Lower Oder Valley National Park for permission to work on their sites and their support. J. Sadlik and the Ornithological Working Group Uckermark (OAG) provided invaluable ornithological data. We appreciated advice on various aspects of our work from G. Gerold, M. Mühlenberg and M. Schaefer. Assistance in the field was provided by D. Kowalczyk.

References

Achtziger R, Nickel H (1997) Zikaden als Bioindikatoren für naturschutzfachliche Erfolgskontrollen im Feuchtgrünland. Beiträge zur Zikadenkunde 1 : 3–16

Achtziger R, Nickel H, Schreiber R (1999) Auswirkungen von Extensivierungsmaßnahmen auf Zikaden, Wanzen, Heuschrecken und Tagfalter im Feuchtgrünland. Schriftenreihe des Bayrischen Landesamtes für Umweltschutz 150 : 109–131

Beebee TJC (1996) Ecology and Conservation of Amphibians. Chapman & Hall, London

Berger L (1967) Embryonal and larval development of F1 Generation of green frogs different combinations. Acta Zool. Cracov. 12 (7) : 123–160

Blab J (1986) Biologie, Ökologie und Schutz von Amphibien. Schriftenreihe für Landschaftspflege und Naturschutz 18. KILDA Verlag, Bonn – Bad Godesberg

Borgula A (1995) Langjährige Bestandsentwicklung einer kleinen Metapopulation des Laubfrosches (*Hyla aborea* L.) in der Schweiz. In: Geiger A (ed) Der Laubfrosch (*Hyla aborea*) – Ökologie und Artenschutz. DGHT, Bonn, 8–26

Cadbury CJ (1980) The status and habitats of the corncrake in Britain. Bird Study 27 : 203–218

Dohle W, Weigmann G, Schröder T (1999) Das Untere Odertal – Charakterisierung des Untersuchungsgebietes. In: Dohle W, Bornkamm R, Weigmann G (eds) Limnologie Aktuell. E. Schweizerbart'sche Verlagsbuchhandlung, Stuttgart, 1–13

Dürr T, Mädlow W, Ryslavy T, Sohns G (1997) Rote Liste und Liste der Brutvögel des Landes Brandenburg. In: Landesumweltamt Brandenburg (ed) Naturschutz und Landespflege in Brandenburg 6 (2) (Beilage)

Eikhorst R (1984) Untersuchungen zur Verwandtschaft der Grünfrösche. Dissertation, Universität Bremen

Eikhorst R, Rahmel UAP (1986) Zur Verbreitung des Kleinen Teichfrosches *Rana lessonae* und des Seefrosches *Rana ridibunda* in Niedersachsen. Salamandra 22 (1) : 79–92

Engelmann HD (1978) Zur Dominanzklassifizierung von Bodenarthropoden. Pedobiologia 18 : 378–380

Fischer S, Dittberner W, Helmecke A, Sadlik A, Bellebaum J (1999, unpublished) Bewertung des Beitrages nationaler und internationaler Naturschutzvorhaben in Deutschland zur Erhaltung stark gefährdeter Vogelarten auf landwirtschaftlich extensiv genutzten Flächen. Projektbericht im Auftrag des Bundesamtes für Naturschutz (BfN)

Glutz von Blotzheim U, Bauer KM, Bezzel E (1973) Handbuch der Vögel Mitteleuropas 5, Galliformes und Gruiformes. Akademischer Verlag, Frankfurt am Main

Green R (1996) Factors affecting the population density of the corncrake in Britain and Ireland. Journal of Applied Ecology 33 : 237–248

Green R, Stowe TJ (1993) The decline of the corncrake in Britain and Ireland in relation to habitat change. Journal of Applied Ecology 30, 689–695

Green R, Rocamora G, Schäffer N (1997) Populations, ecology and threats to the Corncrake in Europe. Vogelwelt 118 : 117–134

Groombridge B (1994) IUCN Red List of Threatened Animal. World Conservation Monitoring Centre, Cambridge

Grosse WR (1994) Der Laubfrosch. Die Neue Brehmbücherei Nr. 615. Westarp Wissenschaften, Magdeburg

Günther R (1990) Die Wasserfrösche Europas. Ziemsen Verlag, Wittenberg Lutherstadt

Günther R (1996a) Teichfrosch – *Rana* kl. *esculenta* LINNAEUS, 1758. In: Günther R (ed) Die Amphibien und Reptilien Deutschlands. Gustav Fischer Verlag, Jena, 455–474

Günther R (1996b) Seefrosch – *Rana ridibunda* PALLAS, 1771. In: Günther R (ed) Die Amphibien und Reptilien Deutschlands. Gustav Fischer Verlag, Jena, 490–507

Günther R, Nabrowsky H (1996) Moorfrosch – *Rana arvalis* NILSSON, 1842. In: Günther R (ed) Die Amphibien und Reptilien Deutschlands. Gustav Fischer Verlag, Jena, 364–388

Helmecke A (2000) (unpublished) Raum- und Habitatnutzung des Wachtelkönigs (*Crex crex, L.*) im Unteren Odertal. Diplomarbeit Institut für Biologie der Humboldt Universität Berlin

Holzinger WE, Kammerlander I, Nickel H (2003) The Auchenorrhyncha of Central Europe – Die Zikaden Mitteleuropas. Volume 1: Fulgoromorpha, Cicadomorpha excl. Cicadellidae. Brill Publishers, Leiden

Hosmer DW, Lemeshow S (2000) Applied logistic regression. Wiley, New York

IUS Weisser & Ness (1998, unpublished) Pflege- und Entwicklungsplan Unteres Odertal, Ziele und Maßnahmen im Kerngebiet. IUS Weisser & Ness, Heidelberg, Potsdam, Kandel

Jedicke E (1995) Ressourcenschutz und Prozeßschutz – Diskussion notwendiger Ansätze zu einem ganzheitlichen Naturschutz. Naturschutz und Landschaftsplanung 27 : 125–133

Jedicke E (1998) Raum-Zeit-Dynamik in Ökosystemen und Landschaften. Kenntnisstand der Landschaftsökologie und Formulierung einer Prozeßschutzdefinition. Naturschutz und Landschaftsplanung 30 : 229–236

Jehle P, Pankoke, K (1999) Die Pflanzengesellschaften ausgewählter Totalreservatsflächen im Nationalpark Unteres Odertal. In: Dohle W, Bornkamm R, Weigmann G (eds) Limnologie Aktuell. E. Schweizerbart'sche Verlagsbuchhandlung, Stuttgart, 123–146

Kleyer M, Kratz R, Lutze G, Schröder B (1999/2000) Habitatmodelle für Tierarten: Entwicklung, Methoden und Perspektiven für die Anwendung. Zeitschrift für Ökologie und Naturschutz 8 : 177–194

Klieber A, Schröder U, Irmler U (1995) Der Einfluß der Mahd auf die Arthropoden des Feuchtgrünlandes. Zeitschrift für Ökologie und Naturschutz 4 : 227–237

Lindeiner von A (1994) Die Amphibienlaichgewässer des Naturparks Schönbuch. Zeitschrift für Feldherpetologie 1 : 135–152

Mädlow W, Adam S (1998, unpublished) Pflege- und Entwicklungsplan Unteres Odertal. Erfassungsbericht Amphibien. Weisser & Ness, Heidelberg

Morris M, Lakhani K (1979) Responses of grassland invertebrates to management by cutting. I. Species diversity of Hemiptera. Journal of Applied Ecology 16 : 77–98

Mühlenberg M, Hovestadt T (1992) Das Zielartenkonzept. NNA-Berichte 5 : 36–41

Mühlenberg M, Slowik J (1997) Kulturlandschaft als Lebensraum. Quelle & Meyer Verlag, Wiesbaden

Nickel H, Achtziger R (1999) Wiesen bewohnende Zikaden (Auchenorrhyncha) im Gradienten von Nutzungsintensität und Feuchte. Beiträge zur Zikadenkunde 3 : 65–75

Nickel H, Remane R (2002) Artenliste der Zikaden Deutschlands, mit Angabe von Nährpflanzen, Nahrungsbreite, Lebenszyklus, Areal und Gefährdung (Hemiptera, Fulgoromorpha et Cicadomorpha). Beiträge zur Zikadenkunde 5 : 27–64

Nöllert A, Nöllert C (1992) Die Amphibien Europas: Bestimmung, Gefährdung, Schutz. Franckh-Kosmos Verlag, Stuttgart

Peeters E, Gardeniers J (1998) Logistic regression as a tool for defining habitat requirements of two common gammarids. Freshwater Biology 39 : 605–615

Plötner J (2001) Struktur und Dynamik einer Seefrosch/Teichfrosch-Männchen-Population (*Rana ridibunda, Rana esculenta*) in der Oderaue bei Frankfurt/Oder. Zeitschrift für Feldherpetologie 8 : 253–264

Pongratz E (1995) Eine Jahrhundert-Tragödie? Die Erweiterung des Nationalparks Bayerischer Wald. Nationalpark 3/95 : 6–9

Schaefer M (2003) Wörterbuch der Ökologie. Spektrum Akademischer Verlag GmbH, Heidelberg Berlin

Schäffer N (1997) Habitatwahl und Partnerschaftssystem von Tüpfelralle (*Porzana porzan*) und Wachtelkönig (*Crex crex*). Dissertation, Universität Würzburg

Schäffer N, Münch S (1993) Untersuchungen zur Habitatwahl und Brutbiologe des Wachtelkönigs Crex crex im Murnauer Moos/Oberbayern. Vogelwelt 114 : 55–72

Schalitz G, Petrich G (1999) Möglichkeiten der Renaturierung intensivierter Grünlandflächen an der unteren Oder unter den Bedingungen verlängerter Überflutung. 9 : 442

Scherzinger W (2002) Stehen Biotop- und Artenschutz in Widerspruch zur Idee des Prozess-Schutzes? Artenschutzreport 12 : 1–5

Schlüpmann M, Günther R (1996) Grasfrosch – *Rana temporaria* LINNAEUS, 1758. In: Günther R (eds) Die Amphibien und Reptilien Deutschlands. Gustav Fischer Verlag, Jena, 412–453

Schmidtler JF, Gruber U (1980) Die Lurchfauna Münchens. Schriftenreihe Naturschutz und Landschaftspflege 12 : 105–139

Schneeweiß N (1996) Zur Verbreitung und Bestandentwicklung der Rotbauchunke in Brandenburg. In: Krone A, Kühnel KD (eds) RANA Sonderheft 1: Die Rotbauchunke (*Bombina bombina*) Ökologie und Bestandssituation. Natur & Text, Rangsdorf, 87–103

Schneider R (1998) Das Zielartenkonzept aus verhaltensbiologischer Sicht. Artenschutzreport 8 : 26–30

Schröder B (2000) Zwischen Naturschutz und Theoretischer Ökologie: Modelle zur Habitateignung und räumlichen Populationsdynamik für Heuschrecken im Niedermoor. Selbstverlag, Dissertation, Universität Braunschweig

Stowe TJ, Hudson VA (1991) Radio telemetry studies of Corncrakes in Great Britain. Vogelwelt 112 : 10–16

Tyler GA (1996) The ecology of the Corncrake, with special reference of the effect of mowing on breeding production. PhD thesis, University College Cork

Vogel K, Vogel B, Rothaupt G, Gottschalk E (1996) Einsatz von Zielarten im Naturschutz – Auswahl der Arten, Methode von Populationsgefährdungsanalyse und Schnellprognose, Umsetzung in der Praxis. Naturschutz und Landschaftsplanung 28 : 179–184

Vössing A (1998) Der Internationalpark Unteres Odertal Ein Werk und Wanderbuch. Stapp Verlag, Berlin

Wilke HJ (1995) Frösche, Kröten, Unken, Molche. Schwedter Jahresblätter 19 : 19–29

Wilson E (1992) Der gegenwärtige Stand der biologischen Vielfalt. Spektrum Akademischer Verlag, Heidelberg, Berlin, New York

Witt K, Bauer HG, Berthold P, Boye P, Hüppop O, Knief W (1998) Rote Liste der Brutvögel (Aves). In: Bundesamt für Naturschutz (ed) Rote Liste gefährdeter Tiere Deutschlands. Schriftenreihe für Landschaftspflege und Naturschutz, 55 : 40–47

Valuation of ecosystem services provided by biodiversity conservation: an integrated hydrological and economic model to value the enhanced nitrogen retention in renaturated streams

Ingo Bräuer

Department of Economics, Sociology and Law, UFZ Centre For Environmental Research, Permoserstr. 15, 04318 Leipzig, Germany, email to `ingo.braeuer@ufz.de`

Summary. The importance of ecosystem functions for humankind is well known. But only few attempts have been undertaken to estimate the economic value of these ecosystem services. In particular, indirect methods are rarely used, even though they are most suitable for the task. This discrepancy is because quantitative knowledge of changes in ecosystem functions is scarce. This paper presents a user-friendly procedure to quantify the increased N-retention in a renaturated river using easily available data. In a case study of the renaturated River Jossa (Germany) the benefits of increased nitrogen retention caused by beaver reintroduction are determined by using the replacement cost method. The quantification of chemical processes is discussed in detail, as well as the problems of defining an adequate reference scenario for the substitute costs. Results show that economic benefits from the evaluated ecosystem service (€ 12,000/annum) equal 12 % of the total costs of the corresponding conservation scheme.

Key words: Biodiversity conservation programmes, cost-benefit-analysis, replacement cost method, ecosystem services, nutrient retention

1 Introduction

Nature conservation has to cope with two problems which are coming to a head. On one hand there is an increasing pressure on remaining natural habitats from alternative land use (Mitch and Gosselink 2000; Turner et al. 2000). On the other hand, budgets for conservation agencies to cope with the accelerating problems are decreasing (Stratmann 2002).

In this situation, efficiency and acceptance of conservation policy are crucial. Efficiency, as tested by cost-benefit-analysis, is to make sure that money is spent according to public preferences and that no money is wasted (Zander 2000; Brouwer and Slagen 1998; Marggraf and Streb 1997). Acceptance will help to ensure further public spending in the field of nature conservation. One way to promote acceptance is to visualise all the benefits of conservation programmes some of which have been neglected so far. The feedback to economic studies underlines this assumption (e.g., Pimentel et al. 1997; Costanza et al. 1997). Thus, acceptance and efficiency can be promoted by cost-benefit-analysis.

To undertake an environmentally sensitive cost-benefit-analysis, all relevant benefits of nature conservation have to be taken into account and hence estimated. This study focuses on the evaluation of indirect use values (according to Turner et al. 1993 : 112). The procedure will be demonstrated in a case study of a beaver reintroduction programme in the centre west of Germany. One of the assumed benefits of the programme is the enhanced nitrogen retention in rivers, which have been altered by beaver activity. To monetise the changes in that ecosystem service, the Replacement Cost Method (RCM) is used.

So far, economic evaluations have concentrated on either direct use values or non-use values. Particularly in the field of nature conservation, where the focus has been set on species and landscape evaluation, the valuation of indirect use values derived from ecosystem services have been of less interest. Often it has been claimed that these values have already been considered by the use of the Contingent Valuation Method (CVM) (Gren 1995; Costanza et al. 1997). In some studies this might have been the case, depending on the design of the CVM. But, due to the complexity of ecosystems it is questionable if the CVM is the appropriate method for a sound evaluation of particular ecosystem functions.

The Replacement Cost Method allows us to focus on single ecosystem functions. This makes the importance of single ecosystem services more obvious. A further advantage of this valuation procedure is the use of technical substitutes and their market prices. Together with the easily understandable valuation procedure this will ensure an improved public acceptance of the results, which cannot be taken for granted. The political input of many Contingent Valuation Studies suffers from low acceptance. The main reason for this is the welfare-economic concept of the method and the measurement of consumer's surplus which is difficult to communicate. Additionally, values estimated by the Replacement Cost Method will always mark the lower boundary of the true value, because the consumer's surplus is not implied.

The value of the ecosystem function »nutrient retention«

Since the nutrient import in aquatic ecosystems has increased considerably during the last few decades, the self purification potential of river systems becomes more and more important to guarantee ecosystem integrity (Mitch and Gosselink 2000). Furthermore, the nitrate content in stream or ground water often exceeds critical values of drinking water legislation. Nitrogen removal becomes an increasingly important factor with regard to the cost of drinking water purification (Grünebaum 1993). Thus, it is not astonishing that in studies which deal with ecosystem services, the self purification potential is often evaluated (Byström 2000, 1998; Dehnhardt 2002; Gren et al. 1995; Folke 1991).

Various studies have shown that the self purification potential of a river depends on its structure and its surrounding buffer strips. Restoration of the old stream structures can help to improve the nitrogen removal in these systems. The reasons for this are manifold (Vought et al. 1994). One important factor is the stream velocity. The higher the residence time of the water in the river system, the higher the denitrification (Behrendt and Opitz 2000).

Another factor is the area of flooded river banks. Flooded soils have a denitrification potential which is up to 100 times higher than under dry conditions (Gäth et al. 1999). Thus, flooding causes an increase in denitrification by enhancing the microbiological activity in the soils.

2 Investigated conservation programme and study site

The study area is the Spessart Mountains in Hesse, Germany, where a beaver reintroduction programme was launched in 1987/88. In the context of this scheme, 18 beavers were released in the local rivers and buffer strips were purchased. The population numbers are now around 200 individuals and dam building occur in 16 home ranges (Loos pers. com.).

The significant impact on structure and hydrology of highland brooks by beavers is well known (Woo and Waddington 1991; Smith et al. 1991) and similar results were reported in the Spessart study (Harthun 2000).

3 Materials and Methods

A higher self purification potential of beaver modified rivers is one of the assumed benefits of the investigated beaver conservation programme. To assess the economic benefits of this improved ecosystem function, the Replacement Cost Method (RCM) is used. This technique looks at the costs of replacing a damaged asset, e.g., water quality standard to recover its original state (Pearce and Moran 1994).

The RCM requires a three step procedure:

1. Quantification of the retention effects (*estimation of the ecosystem function*)
2. Definition of the reference scenario (*substitute and its marginal costs*)
3. Economic valuation (*estimation of the ecosystem service*)

These three steps will be described in detail.

As mentioned above, nitrogen retention occurs in the river as well as in the flooded areas of the floodplains. Beavers affect both areas. Their dam building activities lead to an increase of flooded area along the river sides and to a decrease of flow velocity in the river. Both effects have to be estimated separately (Fig. 1). Direct measurements of nitrogen retention are very complicated to conduct (Peterson et al. 2001). In this study, parameters are measured which are known to influence the denitrification (flooded area, running velocity) instead. In a second step, effects of these changes in the river system on the denitrification are estimated. For the floodplain and the river two different approaches are used (Fig. 1).

As for all economic evaluations, only gradual changes in the evaluated goods are relevant and suitable for valuation. In this particular case only changes caused by the conservation programme are of interest. Relevant parameters are the additional flooded area in the floodplain and the decreased running velocity in the river. To estimate the latter, changes in the overall area of the river system are assessed by analysing aerial views of the study side from the year 2000 by means of geographic information system (ARC-VIEW).

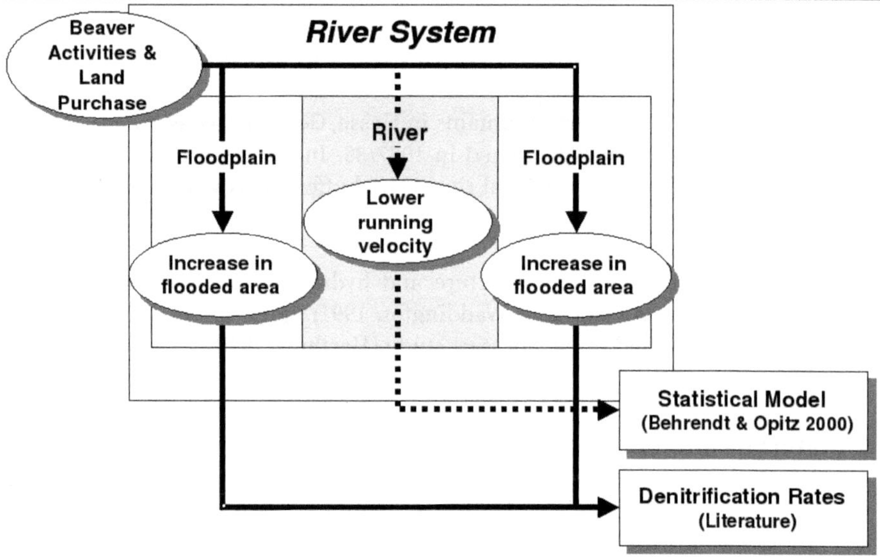

Fig. 1. The hydrological model: Influence of the beaver reintroduction scheme on the river system and the different quantification procedures of nitrogen retention. (Effects are caused by impounding by beavers and additional buffer strips)

To quantify the influence of the altered flow velocity on the nitrogen retention within the river, a statistical model from Behrendt and Opitz (2000) is used. For this model the N-retention is derived by comparing the net transport of the river with the theoretically expected loads from the nitrogen emissions inventory. The nitrogen retention (R_N) is linked to the Emissions (E_N) and the load weighted retention potential (R_L) by Equation 1.

$$R_N = E_N - \frac{E_N}{1 + R_L} \qquad (1)$$

A regression analysis of 100 river basins in Europe shows that the load weighted retention potential (R_L) depends on the hydraulic load of the river. For river basins with a size of less than 1,000 m^2, Behrendt and Opitz (2000) derived the following equation (Eq. 2)

$$R_L = a * x^b \qquad \text{The specific coefficients are } a = 3.3;\ b = -0.65. \qquad (2)$$

This model explains 44 % of the variance in the observed nitrogen retention.

To estimate the denitrification in the flooded areas, denitrification rates from literature are used. The recorded rates vary between 30 and 1200 kgN/(ha*a) (e.g., Byström 1998; Dörge 1994; Gale et al.1993). For the following calculations a mean denitrification rate of 300 kgN/(ha*a) is taken as a basis.

To deal with the high variances of both the statistical model and the nitrification rates for soils, two different variants are calculated (Table 1). In the minimum variant the lower bond of the 90 % prediction intervals is used. In the maximal variant the upper bond of the 90 % prediction intervals is used. A minimal denitrification rate of 100 kgN/(ha*a) is assumed, for the maximal variant a rate of 500 kgN/(ha*a). Both numbers are common appraisal values for either conservative estimations (Gren et al. 1995; Kronvang et al. 1999) or average nitrification rates (Gren et al. 1995).

Table 1. Assumptions for the calculation of the minimal and the maximal variant

Area	Procedure	V_{Min}	V_{Max}
river	statistical model	lower bound 90 % interval	upper bound 90 % interval
floodplain	denitrification rates	100 kgN/ha*a	500 kgN/ha*a

The fieldwork is done at one river only, the River Jossa. This is due to the availability of site-specific data, but the river is a very suitable reference. The four beaver territories along its course differ significantly from each other. They represent the whole spectrum of dam building consequences. For the whole study area the results of the Jossa are projected. For this procedure, the ratio of territories with dam building activities in the Jossa compared to the entire investigated area is taken into account.

To assess the replacement cost value of the estimated nitrogen retention the costs necessary to provide an equivalent service have to be determined. Here, costs reported in the literature come into operation.

4 Results

According to the three step procedure of the Replacement Cost Method, the contribution is divided into three parts: the quantification of the ecosystem function, the definition of the technical substitute and its marginal costs, and finally, the valuation step.

4.1 Quantification of the ecosystem function

The four beaver territories along the river Jossa have significant influence on the morphology of the river. The dam building causes 11 pools and ponds and 2 secondary ponds (Table 2). These ponds increase the total river surface by 17 %. The measure relevant for denitrification, the hydraulic load, diminishes by 15 %. In addition, 15,500 m^2 of floodplain are flooded.

Table 2. Morphological changes in the investigated river system

		Jossa (original)	Jossa (Beaver)	Changes	
				abs.	rel.
Runoff	[m^3/s]	1,58	1,58	–	–
River area	[m^2]	122.268	142.816	20.548	+17 %
Hydraulic load	[m/a]	408	349	59	–15 %
Flooded area	[m^2]	– [a]	15.505[a]	15.505	

[a] To estimate the benefits of the reintroduction programme only the additional flooded area is of interest.

These changes in river structure result in significant changes in self purification (Table 3). The increased hydraulic load leads to an additional nitrogen retention of 700 kgN/a in the river according to the model from Behrendt and Opitz (2000). Taking the 90 % confidence interval into account, values range from 520–900 kgN/a. Depending on the assumed denitrification rates, the additional retention in the floodplain amounts to 160–780 KgN/a. The mean is 470 KgN/a. The total additional retention in the Jossa and the surrounding floodplains total up to 1,200 kgN/ year.

To scale these results up for the investigated area, the ratio of territories with dam building activities are taken into account. In the investigated area there are a total of 16 beaver territories with dam building. Assuming that the four investigated territories are representative, the overall effect adds up to ca. 4700 kgN/a in the whole project area (Table 3).

4.2 Definition of the substitute and its marginal costs

The valuation of the assessed ecosystem enables the identification of the potential technical substitute and ascertainment of the costs to provide an equivalent service. The following selection is in accordance with the criteria of Bockstael et al. (1998): (a) the substitute must be able to cope with the main problem and (b) cope in the most cost efficient way. Theoretically, there are three potential substitutes to improve the water quality: drinking water purification, sewage plants and political programmes to avoid nitrogen emissions.

Table 3. N-Retention in the river and the floodplain of the river Jossa and the whole investigated area. Results of minimum and maximum setting and the mean

		Mean	$V_{Min}{}^{a}$	$V_{Max}{}^{a}$	+/-
River	[kg N/a]	700	520	900	26,8 %
Floodplain	[kg N/a]	470	160	780	66,6 %
Jossa	[kg N/a]	1,200	670	1,670	42,7 %
Spessart	[kg N/a]	4,700	2,690	6,680	43 %
Project duration[b]	[kg N]	115,633	66,354	164,991	

[a] Variants according to the assumption see Table 1.
[b] Population development is considered during the project duration (1987–2012).

The substitute most commonly used to value self purification potential of rivers is sewage plants (Byström 2000; Gren 1995; Dehnhardt 2002). Here, the functional process and the final effect are very similar: microbiological activity improves the water quality of the river. However, emissions from agriculture are the main source of nitrogen input in aquatic ecosystems (Vought et al. 1994). These non-point loads can only be eliminated by drinking water purification or political programmes and not by sewage plants. Drinking water treatment offers a higher quality standard than necessary in rivers, thus it is not cost efficient. The most cost effective alternative to reduce non-point agricultural nutrient loads in the river is their avoidance by political strategies. Most common are agricultural schemes to compensate farmers for special regulations for the application of fertilisers. This compensation refers either to the technique or the amounts of application.

The choice of an adequate substitute does not define the replacement costs automatically. There is a big variety of measures and marginal costs within the agricultural schemes to reduce nitrogen emissions (for a review see Bräuer 2002 : 214 and Hennies 1996 unpublished).

Depending on the production system and the intended reduction level, the replacement costs vary between € 1–23 /kgN (Hennies 1996). For the calculations in the presented study, average costs of € 2.56 /kgN are assumed to guarantee a conservative calculation (the original value for the calculations was 5 DM). To make this number comparable: for nitrogen removal in sewage plants marginal costs of € 5–8 /kgN are reported in Germany (Grünebaum 1993).

When using replacement costs in a cost-benefit analysis (CBA) it has to be kept in mind that distorted markets exist. Thus, either shadow prices have to be used or – if the CBA includes changes in prices and income only – the economic costs of the respective agricultural schemes are the right figure. In the latter case, the excess-burden of the expenditures have to be added. This would increase the benefits of the ecosystem service. In these calculations excess-burden is not taken into account for two reasons: (i) values in literature vary between 7 and 28 % (Musgrave et al. 1993 : 113) and (ii) to ensure conservative calculations.

4.3 Valuation of the ecosystem service

If marginal costs of € 2.56 /kgN are assumed, the economic value of the mean estimated additional nitrogen retention in the Jossa has a value of € 3,000 (Table 4). The replacement value of the ecosystem service in the whole investigated area accounts for € 12,000 in 2000 (Table 4). According to the different assumptions in the estimation of the chemical processes (Table 1 and Table 3), the deviation from the mean value is plus minus 43 %.

For a better judgement of the results it is necessary to compare the calculated benefits with the costs of the conservation programme. Therefore, the retention processes of the year 2000 have to be extrapolated for the whole project duration. The project duration is assessed to last 25 years (1987–2012). A definition of a project duration is necessary to account for the standing expenses (e.g., land purchase at the beginning of the project) with the variable costs. The total costs of the reintroduction scheme amount to € 1.89 million (Bräuer 2002).

Table 4. Economic value of measured nitrogen retention

		Econ. Value	Deviation[b]
Jossa (2000)	[€]	3,000	1,700–4,300
Investigated area (2000)	[€]	12,000	6,900–17,100
Project duration[a]	[€]	250,300	143,700–356,900
Portion of project costs	[%]	12	8–17

[a] Values for the whole project duration (1987–2012). Population development is considered.

[b] Deviation according to the results of V_{Min} und V_{Max} in Table 3.

The increased nitrogen retention in rivers is directly correlated to the beaver activities. To extrapolate the results of the year 2000 to the project duration, they have to be adjusted with the size of the beaver population. For this adjustment the population size has been computed using a linear regression. The model describes a highly significant correlation between population size and time after the reintroduction ($y = 12,45 * x - 24770$; $r^2 = 0,97$; $p < 0,001$; $F = 297$; $n = 10$). According to this model, the beaver population will expand until ca. 320 individuals in the year 2012. This population size and hence the statistical model is realistic because there are enough habitats in the area.

The corrected retention potential in the investigated area is estimated at 115,600 kgN (± 49,300 kgN) for the period between 1987–2012. The economic value of this service corresponds to € 250,300 (± € 106,600). The economic benefits of the ecosystem service nitrogen retention amount to 12 % of the total investment costs. (Sewage plants as reference scenario (see Sect. 4.2) would result in a value of € 35,980 /annum. This corresponds to 40 % of the project costs.)

5 Discussion

Results show that one ecosystem function alone makes up a significant part of the total costs of the investigated conservation programme, even though the calculations have been very conservative (choice of the replacement costs, waiving of excess-burden etc.) and consumer surplus has not been considered. The results are surprising because improving the self purification potential has not been the main objective of the beaver conservation programme. However, the results underline the importance of a systematic consideration of all consequences of nature conservation programmes. So far this is an exception. If at all, economic evaluations of species conservation programmes apply the Contingent Valuation Method only where ecosystem functions are not taken into consideration explicitly.

Unfortunately, the results show a high variance. The presented results have to be discussed in two ways. On one hand there are problems with the quantification of chemical processes. Are the methods applied suitable? On the other hand, the valuation step has to be discussed as well. Here it has to be clarified whether the used technical substitute is appropriate.

The hydrological processes have been estimated using a statistical model and general denitrification rates from literature. Both values are only rough estimates because local characteristics which are relevant for the denitrification are not taken into account. To narrow the results down, conservative assumptions have been made to guard against overestimation.

An even more important factor for the final results is the definition of the technical substitute. Theoretically, there are two other plausible technical substitutes next to the chosen substitute *agricultural policies*: *drinking water purification* and *sewage plants*. The marginal costs per unit abated nitrogen are much higher in these cases than for the chosen substitute. These substitutes would lead to results three to twenty times as high. The applied scenario *agricultural policies* is the most cost-efficient alternative, but it offers less services than the ecosystem, because atmospheric emissions are not included. This results in an underestimation of the ›true value‹. Once again, the chosen valuation procedure prevents an overestimation by all means. Thus, the estimates denote the lower bound and can be used in practice as a minimum value.

Conservative calculations are essential to ensure credibility of the valuation procedure, but they open the door for a massive and dangerous underestimation of ecosystem services. In this context it is essential to point out exactly what the valuation subject was. For example, it might be tempting to equal the evaluated nitrogen retention with the value of the renaturated river. That's not the case at all. There are several ecosystem services which have not been considered so far, like an increased flood control or higher groundwater recharge. Even the ecosystem service of self purification is only evaluated from one point of view: the nitrogen retention. Yet other aspects of self purification in the river such as phosphate retention, have been neglected.

If these limitations are considered and the results are interpreted with caution, economic valuation can offer big advantages. Results of economic evaluations do not only help to decide whether a given programme has been efficient or not, but also whether enough money has been spent for a special task (Brouwer and Slagen 1998; Zander 2000). The knowledge also helps to design new programmes and offers the possibility of new approaches in landscape design, respectively. Examples for the latter are studies by Byström (2000) or Gren et al. (1997), which have shown that restoration or building of new flood plains are cost-efficient alternatives to wastewater treatment plants to meet the goal »reduction of nitrogen loads to the Baltic Sea«.

6 Conclusions

Benefits derived from ecosystem services, in this case the enhanced nitrogen retention potential of a renaturated river, can play an important role in the judgement of conservation programmes. The estimated benefits of a single ecosystem service, which has not been in the centre of the species conservation programme, already amount a considerable part of the total costs of that programme. This observation underscores the importance of an evaluation of all relevant consequences of conservation programmes. A systematic procedure, which uses an adequate evaluation procedure for each particular category of consequences, is essential for this task. To gather all use and non-use values of a biodiversity conservation programme a combination of Contingent Valuation and Replacement Cost Method seems to be essential.

The RCM is particularly suitable for the evaluation of ecosystem services because it offers a straightforward and easily understandable procedure. Nevertheless, the results are only rough estimates and should be interpreted with caution. To ensure that no overestimation takes place the calculations are based on conservative assumptions (at the level of the chemical processes as well as at the level of the choice of the technical substitute).

The model from Behrendt and Opitz (2000) offers the possibility to calculate denitrification rates in a relatively easy way. This will help to reduce the costs of estimating the benefits of an ecosystem function, a figure very complicated to measure directly. Cost and time savings of the presented procedure will help to establish an economic evaluation of nature capital in the political decision making process.

Acknowledgements

Additional financial support from the scholarship programme of the German Federal Environmental Foundation (DBU) is gratefully acknowledged.

References

Behrendt H, Opitz D (2000) Retention of nutrients in river systems: dependence on specific runoff and hydraulic load. Hydrobiologia 410 (1) : 111–122

Bockstael NE, Freeman III AM, Kopp RJ, Portney PR, Smith VK (1998) On Valuing Nature. Unpublished manuscript, (Raymond J Kopp Working Papers), Resources for the Future

Bräuer I (2002) Artenschutz aus volkswirtschaftlicher Sicht. Die Nutzen-Kosten-Analyse als Entscheidungshilfe. Hochschulschriften, vol 76, Metropolis, Marburg

Brouwer R, Slangen LHG (1998) Contingent Valuation of public Benefits of agricultural Wildlife Management: The case of Dutch peat meadow land. European Review of Agricultural Economics 25 : 53–72

Byström O (1998) The nitrogen abatement cost in wetlands. Ecological Economics 26 : 321–331

Byström O (2000) The Replacement Value of Wetlands in Sweden. Environmental and Resource Economics 16 : 347–362

Costanza R, d'Arge R, deGroot R, Farber S, Grasso M, Hannon B, Limburg K, Naeem S, O'Neill RV, Paruelo J, Raskin RG, Sutton P, van den Belt, M (1997) The value of the world's ecosystem services and natural capital. Nature 387 : 253–260

Dehnhardt A (2002) Der ökonomische Wert der Elbauen als Nährstoffsenke: Die indirekte Bewertung ökologischer Leistungen In: Dehnhardt A, Meyerhoff J (eds) Nachhaltige Entwickung der Stromlandschaft Elbe. Nutzen und Kosten der Wiedergewinnung und Renaturierung von Überschwemmungsauen. Vauk, Kiel, pp 185–218

Dörge J (1994) Modelling nitrogen transformation in freshwater wetlands. Estimating nitrogen retention and removal in natural wetlands in relation to their hydrology and nutrient loadings. Ecological Modelling 75/76 : 409–20

Folke C (1991) The Social Value of Wetland Life-Support. In: Folke C, Kaberger T (eds) Linking the natural Environment and the economy: Essays from the Eco-Eco Group, Kluwer, pp 141–171

Gäth S, Antony F (1999) Bewertung der standortlichen Denitrifikationsleistungen und N-Vorratsdüngung von Böden und Bodennutzungssystemen. Mitteilungen der deutschen bodenkundlichen Gesellschaft 91 : 1213–13

Gale PM, Dévai I, Reddy KR, Graetz DA (1993) Denitrification potential of soils from constructed and natural wetlands. Ecological Engineering 2 : 119–30

Gren IM (1995) The value of investing in wetlands for nitrogen abatement. European Review of Agricultural Economics 22 : 157–172

Gren IM, Groth KH, Sylvén M (1995) Economic Values of the Danube Floodplains. Journal of Environmental Management 45 : 333–345

Gren IM, Söderqvist T, Wulff F (1997) Nutrient Reductions to the Baltic Sea: Ecology, Costs and Benefits. Journal of Environmental Management 51 : 123–143

Grünebaum T (1993) Stoffbezogene Kosten der kommunalen Abwasserreinigung. In: Neue Ansätze im integrierten Umweltschutz – Eine Herausforderung für

Staat, Wirtschaft und Bürger. Schriftenreihe Gewässerschutz - Wasser- Abwasser. Aachen, vol 139 (23) : 1–15

Harthun M (2000) Einflüsse der Stauaktivität des Bibers (*Castor fiber albicus*) auf physikalische und chemische Parameter von Mittelgebirgs-Bächen (Hessen, Deutschland). Limnologica 30 : 21–35

Hennies H (1996) Ökonomische Beurteilung von baulichen Auflagen für Güllebehälter. Diplomarbeit, Institut für Agrarökonomie, Universität Göttingen (unpubl.)

Kronvang B, Hoffmann CC, Svendsen LM, Jensen JP, Dörge J (1999) Retention of nutrients in river basins. Aquatic Ecology 33 : 29–40

Marggraf R, Streb S (1997) Ökonomische Bewertung der natürlichen Umwelt: Theorie, politische Bedeutung, ethische Diskussion. Spektrum, Heidelberg and others

Mitsch WJ, Gosselink JG (2000) Wetlands, 3rd. Ed.; Wiley, NY and others

Musgrave RA, Musgrave PB, Kullmer L (1993) Die öffentlichen Finanzen in Theorie und Praxis, vol 2. UTB für Wissenschaft, Tübingen

Pearce D, Moran D (1994) The Economic Value of Biodiversity. Earthscan, London

Peterson BJ et al. (2001) Control of Nitrogen Export from Watersheds by Headwater Streams. Science 292 : 86–90

Pimentel D, Wilson C, McCullum C, Huang R, Dwen P, Flack J, Tran Q, Saltman T, Cliff B (1997) Economic and Environmental Benefits of Biodiversity. BioSience 74 (11) : 747–757

Smith ME, Driscoll CT, Wyskowski BJ, Brooks CM, Cosentini CC (1991) Modification of stream structure and function by beaver (*Castor Canadensis*) in the Adirondack Mountains, New York. Can. J. Zool. 69 : 55–60

Stratmann U (2002) Aufgabenspezifische Erfassung der Naturschutzausgaben von Bund und Ländern (1985-2001) Methode, Analyse und Ergebnisse - Gutachten im Auftrag des Bundesamtes für Naturschutz (unpubl.)

Tuner RK, Pearce DW (1993) Sustainable economic development: economic and ethical principles. In: Barbier EB (ed) Economics and Ecology. Chapman & Hall, London, pp 177–94

Turner RK, van den Bergh JCJM, Söderqvist T, Barendregt A, van der Straaten J, Maltby E, van Ierland E (2000) Ecological-economic analysis of wetlands: scientific integration for management policy. Ecological Economics 35 (1) : 7–23

Vought LBM, Dahl J, Pedersen CL, Lacoursiére JO (1994) Nutrient Retention in Riparian Ecotones. Ambio 23 (6) : 342–48

Woo MK, Waddington JM (1990) Effects of Beaver Dams on Subarctic Wetland Hydrology. Arctic 43 (3) : 223–230

Zander K (2001) Der Einfluss von Information auf die Zahlungsbereitschaft. In: Elsasser P, Meyerhoff J (eds) Ökonomische Bewertung von Umweltgütern: Methodenfragen zur Kontingenten Bewertung und praktische Erfahrungen im deutschsprachigen Raum. Metropolis, Marburg, pp 141–160

Zander K (2003) Ökonomische Bewertung des Streuobstbaus aus einzelbetrieblicher und gesellschaftlicher Sicht. Schriften zur Umweltökonomik, Vauk, Kiel

Towards sustainable land use: Public demand for plant diversity in agricultural landscapes of central Germany

Anke Fischer

Department of Agricultural Economics, Georg-August University of Göttingen, Platz der Göttinger Sieben 5, 37073 Göttingen, Germany, email to afische6@gwdg.de

Summary. In the stress field between World Trade Organization (WTO) policies to reduce agricultural overproduction and the public desire to preserve agricultural landscapes, secure the livelihood of inhabitants of rural regions in Germany and at the same time, to foster biodiversity and conservation issues in agriculture, agri-environmental payment schemes seem to be a promising perspective for agriculture in the decades to come. How can these schemes be designed and implemented in a way that ensures sustainability? This contribution resumes the results of a case study on a rural region of central Germany with regard to public demand for environmental services of agriculture. The population's willingness to pay for a specific ecological good was elicited by means of a contingent valuation survey. At the same time, the motivation of the local population to participate in this survey as an instrument of public decision making was evaluated. The contribution concludes discussing the suitability of the contingent valuation method and complementary techniques as an approach to sustainable policy making that takes into consideration ecological, economic and social criteria alike.

Key words: public demand, payments for environmental services, agriculture, contingent valuation method, sustainability

1 Introduction

In the stress field between land use and conservation, payments for environmental services may be an appropriate instrument to encourage sustainable land use (see Convention of Biological Diversity (CBD) Art. 10; 11). Incentive measures in the sense of Art. 11 (CBD) have to meet three types of criteria. The concept of sustainability requires the consideration of all three (i) social, (ii) economic and (iii) ecological aspects of actions (cp., e.g., Jörissen et al. 1999; Huber 1995).

In actual German policy practice several types of incentive measures are currently applied. These include instruments based on police law, maintenance contracts specifically aimed at nature conservation, and market solutions such as payment schemes (for a discussion see, e.g., Randall 1972; Nellinger 1996; Heißenhuber and Lippert 2000). The concept of payments for environmental services (see also Máñez Costa and Zeller in this volume) has – despite its drawbacks – several advantages that make it a particularly suitable incentive measure. If payment schemes could be designed carefully and were compatible with ecological, economic and social aspects of sustainability, they could be quite a powerful instrument used to promote sustainability. Somewhat simplified, this goal requires (i) choosing environmental services based on ecological criteria, (ii) an economic mechanism which ensures efficient pricing of these services and (iii) a public frame which ensures transparency and acceptance of these measures.

The *ecological* soundness of incentive measures requires a provision of knowledge on ecosystem functions, ecological links and conservation biology in general. As a matter of fact, this knowledge often includes normative aspects that have to be made transparent. In most cases, ecological knowledge has to be provided by experts in the field.

Unlike contracts with farmers on specific aspects of habitat management, agri-environmental payment schemes are based on the creation of markets for environmental goods. As a consequence, farmers are encouraged to act as entrepreneurs, and the provision of these goods is designed to follow the principles of supply and demand (see, e.g., Hespelt 2002; Fischer, Hespelt and Marggraf 2003). Thus, environmental goods can be provided in an *economically* efficient way.

Social aspects with regard to incentive measures in contrast, are often judged best by the affected population, that is, by the stakeholders. These stakeholders may include the general public as well as particular groups such as the farmers, landowners or local conservationists. Relevant social issues address, for example, distributional impacts of incentives, the acceptance of public decision making processes and the transparency of these processes. The issues are likely to depend on the opportunity for stakeholders to participate in the decision making.

How can these three criteria be accounted for when designing incentive measures such as payment schemes? The following case study illustrates an attempt to consider ecological, economic and social aspects at the same time. The remainder of this contribution is structured as follows. Sect. 2 introduces an agri-environmental payment scheme designed by a research group from Göttingen University, together with a local advisory board in the project region. Sects. 3, 4 and 5 present the methodology and results of a contingent valuation study which might contribute to a more sustainable design of the payment scheme. In Sect. 6, the suitability of this approach for public decision making with regard to sustainability criteria is discussed.

2 Payments for environmental services of agriculture: a pilot scheme in central Germany

In recent trends in both scientific and political approaches to agriculture and rural development, *multifunctionality* of agricultural activities has increasingly come to the fore. The agricultural sector is no longer considered to be only providing food and renewable energy sources. ›Side-effects‹ such as impacts on the natural environment and shaping and maintaining a scenery are also recognised as services. The latter – in contrast to food and energy sources that are regular market goods – traditionally are public goods and thus can be interpreted as external effects of agricultural activities (see Máñez Costa and Zeller in this volume). For the time being, farmers do not get sufficiently remunerated for the provision of these services. As a consequence, supply of environmental services, apart from being partly regulated through governmental measures such as laws and by-laws and maintenance contracts, depends entirely on chance and goodwill of farmers. In addition, agricultural subsidies in many cases seem to give incentives in an opposite direction, fostering, e.g., overproduction rather than the provision of environmental services. How can the supply of joint products, e.g., water quality, scenic landscapes and biodiversity in agriculture be encouraged?

Apparently, providers of environmental services have to be remunerated as directly as possible to ensure clear incentives for the production of well defined amenities (e.g., Randall 2002). Direct compensation would be easier if supply of environmental services was decoupled from the production of market commodities. But in most cases, environmental amenities are external benefits that at present do not have a market price. Consequently, the internalisation of external effects – in other words, the creation of markets for these services – could lead to a provision of these amenities which is both more efficient and easier to control.

The decoupling of agricultural food production and the provision of non-market goods contributes, on one hand, to a more sustainable land use (see Sect. 1). In addition, it can also be seen in a more concrete policy context. In Germany as well as in other central European countries, the disaffirmation of ›amber box‹ subsidies for farmers by WTO agricultural agreements and the consequential tendencies in EU agricultural policy have led to the development of ›green-box‹ payment schemes.

Contiguous to these political and scientific processes, a particular payment scheme designed by researchers from the University of Göttingen, is currently being established in the rural county of Northeim in central Germany. The core of this programme is the creation of a market for ecological goods (Bertke, Isselstein and Gerowitt 2002; Gerowitt and Marggraf 2001; Hespelt 2002)[1]. These goods were chosen as indicators for bundles of environmental services and in a pragmatic way, serve as their respective proxies (see also Murtough, Aretino and Matysek 2002). Bertke, Isselstein and Gerowitt (2002) define three groups of ecological commodities. The first group contains goods in arable land and mainly focuses on weed species diversity. This is similar to the second group of grassland goods which are based on several degrees of plant diversity. The third group comprises so-called non-productive elements in agricultural landscapes such as ponds, fal-

[1] A main feature of the scheme is the principle of remunerating actual *outcomes*. In contrast, many comparable programmes chose an obviously less efficient, *action*-oriented approach, thus discouraging entrepreneurial activities of farmers. This particular aspect, albeit important in the policy context, will be neglected here.

low strips, hedges and tree rows. For each type of goods two different qualitative classes are stipulated: (i) a basic class with quality specifications yet considerably higher than the current common standard for agriculture (›good practice‹, ›*gute fachliche Praxis*‹), and (ii) a high-quality class which requires particular attributes, e.g., the presence of endangered or scarce plant species. As a matter of course, remunerations for high-quality goods are planned to be higher than those for basic class goods.

In the next step, markets for these goods are created which like any market, consist of a demand and a supply side. Supply is provided by the local farmers who in an open competitive bidding process (Hespelt 2002) offer to produce as much as they like of the commodity in question. Opportunity cost calculations for the production of the defined commodities are made accessible in an electronic database by the research team and farmers have the opportunity to predicate their bids on these data. This means that farmers take part in the scheme by their own decision as entrepreneurs in agriculture.

The local population has to be seen as the beneficiary of these environmental goods and as such, is represented by a Local Advisory Board on the project. This group, consisting of representatives of the political parties in the council, environmental NGOs, the district administration and farmers, decides which and how much of the commodities to demand and which offers to accept. On the whole, trading ecological goods in this scheme gets quite close to comparable processes involving invitations to bid with a public demand side (Hespelt 2002).

However, the demand for ecological goods could be specified even more precisely if it was based on an assessment of benefits deriving from these goods. After all, it is still the population on the demand side of this market. An appropriate instrument to measure individual benefit by eliciting willingness to pay (WTP) for the good in question might be the contingent valuation method (CVM), a widely applied and recognised technique to evaluate the benefits of non market, environmental goods (for an overview, see, e.g., Bateman et al. 2002, see also Menzel in this volume). Based on the concept of consumer sovereignty, the application of the CVM implies that the respondent is considered a consumer of an ecological commodity, and that he or she is sovereign in his or her decision. The respondent is assumed to act as a *homo oeconomicus*, i.e., rationally. Therefore, aggregated stated WTP values can be interpreted as an expression of the population's demand for the ecological amenities in question. In addition, supplementary questions on the respondents' attitudes and opinions with regard to environmental goods may provide valuable qualitative information on the local population's view on the goods and the payment scheme proposed.

An application of this technique in the project context described above has several advantages: Both university researchers and the Local Advisory Board get information about the demand for the valuated ecological commodity and estimates about the local population's preferences. On an aggregated level, the utility deriving from the provision of a particular quantity of the good can be calculated. Moreover, results with regard to the respondents' attitudes and an analysis of decision processes illustrate the relevance of the topic for the local populace. On this basis the Advisory Board are then able to:

- modify the definition of the good corresponding to the preferences of the population, if necessary and ecologically sound
- align the actual demand for the good with the potential utility deriving from it, in other words, orientate the demand according to a Pareto-criterion of efficiency on the basis of a cost-benefit-analysis
- ensure acceptance of public spending for environmental goods of agriculture through participation of the population, public information on the payments and transparency of decisions.

Following this approach, payments for ecological commodities might fulfil both the WTO requirements – as compensations are decoupled from current production levels – and the main ideas of sustainability: the local population itself decides on an informed basis about the allocation of their community's budget and therefore is less likely to reject the payments later. Besides, compensation payments can be adjusted to regional conditions, decisions are transparent and can be adapted gradually if circumstances change.

It has to be noted though, that from a conservationist perspective the application of the CVM bears several risks. The quantification of the demand for ecological goods and thus the economically efficient level of the goods' provision depends on the population's preferences, i.e., on consumer sovereignty. Consequently, these choices might in principle interfere with nature conservation issues or, in addition, fairness aspects as perceived by minorities.

3 Study objectives

The objectives of the study were thus twofold and included;

- an assessment of the demand for a particular ecological good in monetary and, where possible, in verbal terms
- an analysis of the suitability of the contingent valuation method to provide an instrument for sustainable public decision making.

4 Methods

In order to reach these goals, a two-part study was designed, with a first module consisting of a face-to-face contingent valuation (CV) survey with an open-ended WTP question. The commodity to be evaluated was the planting and the cultivation of 100 km of farm hedgerows in the Northeim district, with a single payment for the next 10 years as a payment vehicle. Hedges were chosen as the good to be valued as they were defined by the research team as one of the landscape features to be remunerated (cp., e.g., Bertke, Isselstein and Gerowitt 2002). At the same time, hedges seemed to be familiar enough to be evaluated by the general public (see, e.g., Mitchell and Carson 1989 : 296). The main idea of the payment scheme was outlined in a scenario and two photographs of two different qualitative specifications of hedgerows were shown to the respondents (see Fig. 1). Hedgerow type 1 was meant to denote the payment scheme's ›basic quality‹ type of hedges (see Sect. 2) and was depicted as a uniformly grown hedge. Type 2 denoted ›high quality‹ hedges with a higher diversity in shrub and tree species.

Fig. 1. Hedgerows type 1 (left) and type 2 (right)

Additionally, participants were encouraged to ask for as much additional information from the interviewer as they liked. Interviewers gave standardised answers to these questions, which nevertheless – as turned out when pilot-testing the interview procedure – did not affect the normal interview atmosphere. Finally, participants were asked to state their general willingness to contribute to this programme, and then, to express their WTP for 100 km for each of the two hedgerow types in their county. The second module consisted of a mixed-technique decision behaviour study. In order to measure the time participants spent on their decisions and to quantify any additional information needed, interviews were recorded in an mp3-format. The recorder started automatically as soon as the WTP question was posed and an offer to provide more information was made. The recording was stopped after respondents had stated their bids, or if they were not willing to contribute, after they had given their reasons for not being willing to contribute to the programme. In addition, after completing the CV, respondents answered a questionnaire which asked for their involvement with the topic, knowledge, their attitude towards financing the programme, the perceived complexity of the valuation task, socioeconomic and other variables. Additionally, interviewers had to observe respondents and to rate their involvement with the topic and their pressure of time in the interview situation on a -3 to $+3$-scale. Study design for both parts was tested in a pilot-test with a sample of $n = 41$ and was improved correspondingly.

5 Results

In the following section survey results are summarised. The first paragraph deals with WTP values stated by the respondents; the second describes the decision behaviour of participants in terms of communication parameters.

5.1 Stated willingness to pay

The study was conducted with a representative sample concerning gender, age, income, and municipalities of residence to avoid biases in any of these characteristics, with a sample of n = 299 drawn from a population of 123,131 persons of full age. Respondents had to be residents of the project region in order to participate in the study as WTP was asked exclusively for hedges in this county. The lack of relevance of the CV subject for inhabitants of other regions obviously would have biased the results in a strong way.

Overall, 183 participants (61.2 %) of the sample expressed a positive WTP. Three of them, however, did not see themselves in a position to state actual WTP amounts. 116 respondents (38.8 %) refused contributions from their own household budgets. Just 14 persons in this group justified their decision claiming that there was no need for additional hedgerows. However, most of the reasons given for rejecting payments made reference to personal budget restrictions or to the payment vehicle:

- 25 persons stated that they could not afford contributions
- 28 opted for voluntary, uncompensated plantations of hedges by the farmers
- 22 individuals declared that the public should not be charged for ecological services, as that is what existing taxes and subsidies are for.

Protest zero bids are payment rejections that are neither a result of free-riding behaviour nor of an actual zero WTP for additional farm hedges (Bateman et al. 2002). Rather, they express objections to some feature of the survey or the valuation design. Thus, the zero bid does not capture the actual value the individual assigns to the good. In this survey, the categories *hedges should be planted voluntarily*, *the public should not be charged* and *not sure if money will actually be spent on hedges* are interpreted as protest responses. Overall, 20.4 % of the sample gave protest bids. In other words, 52.6 % of the rejections referred to protest reasons. WTP bids are shown in Table 1. After removing one single case with outlier bids of € 10,000 from the sample, mean and median WTP values for both types of hedgerows were calculated, including persons who rejected payments (replacing their bids by 0; n = 116) as well as persons who felt unable to state monetary values. These latter bids (n = 3) were replaced by the mean.

Eighty individuals (44.4 %) valued both types of hedges equally, stating the same WTP for each type. Twenty respondents gave higher bids for 100 km of type 1-hedges. Bid differences amounted between € 1 and 125. Correspondingly, 80 participants valued type 2-hedges higher in monetary terms. Bid differences

Table 1. Stated willingness to pay for two different types of hedgerows (n = 298)

	Min	Max	Mean	Standard deviation	Median
WTP hedgerow type 1 [€]	0	1,000	35.72	100.97	5
WTP hedgerow type 2 [€]	0	2,000	58.35	188.48	10

amounted between 20 cents and € 2,000. The average difference between bids for Type 1 and Type 2 hedges in the whole sample equalled € 36.78.

Six individuals (2 % of the total sample) did not act consistently and stated the same WTP for both types while at the same time expressing verbally their preference for one particular type of hedge. In two cases this was due to competing preference aspects (»I think that one's looking better. But this one's more natural«). Four individuals did not express their aesthetic preferences in monetary terms (»I would pay € 10 for either hedge. But I like that one better«).

5.2 Information processing

As a means of quantifying the qualitative information captured on the mp3-recordings, the following were extracted as numerical variables:

(i) number of queries for additional information after a respondent had heard the valuation scenario. Apart from the total *number of queries*, two more variables were derived from that measure. The *number of ›new‹ questions* included all queries for information not provided in the valuation scenario. The *number of ›redundant‹ questions* in contrast, included all queries which asked for information already given in the scenario, for example, »Was that a single payment?« or »How many km of hedges are they going to plant?«.
(ii) the decision time, ascertained automatically.

160 (53.5 %) participants asked for additional information. The decision time ranged from 20 seconds to almost 17 minutes. Other basic parameters are shown in Table 2. From a total of 352 queries, 206 (58.5 %) were questions which required information not provided in the scenario (›new‹ information). Correspondingly, 146 enquiries (41.5 %) were recall questions which demanded information already given in the scenario.

Table 2. Decision behaviour parameters (n=299)

	Min	Max	Mean	Standard deviation
Number of questions (total)	0	10	1.18	1.53
Number of queries for information already given in the scenario (›redundant questions‹)	0	4	0.49	0.80
Number of queries for information not provided in the scenario (›new questions‹)	0	7	0.69	1.096
Decision time (min)	0.33	16.48	2.18	1.80

5.3 Interpretation

WTP and decision parameters indicate that overall;

- respondents were willing to participate. Unfortunately, no figures on non-response could be obtained. However, only three of 299 participants mentioned that they considered the hedge topic irrelevant. This observation suggests an unexpectedly high interest of the local population in this subject.
- respondents were sufficiently motivated to spend more than two minutes on average (see Table 2) on their personal decision on whether and how much to contribute to the scheme. The high percentage of questions that asked for new information (59 %) underscores the notion of a rather high motivation of the general public to participate in these decision processes.
- respondents processed the information on different qualities of hedges provided by the photographs, since 55.5 % of the individual bids distinguished between the two types of hedge (see Sect. 5.1). This finding implies that at least this portion of participants did perceive a difference between hedge types. The remainder of the sample also might have perceived a qualitative difference, but did not value these qualities differently in monetary terms.

Additional measures of prior knowledge and emotional involvement (see Sect. 4) support the conclusion that the population in general is both able and motivated to take part in public decision making with regard to ecological goods.

Still, is the contingent valuation method an adequate technique to involve the general public in these decision processes? In comparison to the CVM, focus group discussions as carried out in the design phase of the survey may allow deeper insights into the population's perceptions of agricultural and landscape issues. Qualitative information is regularly not captured in CV surveys. The survey presented here, in contrast, suggests a way to collect more information on the opinions of the stakeholders than a standard CV does. Participants were given the opportunity to ask for additional information on the proposed scheme. Audio recordings of the subsequent dialogues proved to be useful to get a grip on the respondents' doubts, attitudes and their particular interests. The following section discusses the CVM's potential in this regard.

6 Discussion: participation of the local populace in agri-environmental decision making

As outlined above, the presented study is part of an interdisciplinary research project which addresses the design and the implementation of an agri-environmental payment scheme in the Northeim district in central Germany. For feasibility, the demand side – the local population – is represented by a Local Advisory Board (Hespelt 2002). However, it is the demand of the population that constitutes the actual demand for these goods in the project district. Thus, an elicitation of this demand, i.e., of the expected utility or the preferences for these goods, respectively would be the first best approach to determine an *economically efficient* level of the provision of these goods.

From the background of the sustainability concept (see, e.g., Barbier 1987; Jörissen et al. 1999), however, public decisions should not only meet economic, but also ecological and social[2] criteria (see Fig. 2).

In other words, public decision making with respect to agri-environmental issues should take into consideration:

(i) estimates based on ecological findings with regard to, e.g., ecosystem functions, safe minimum standards, threshold levels etc.
(ii) the perceived fairness of payment schemes, aiming at a transparent and socially accepted[3] way of designing and implementing these schemes, including the issue of resource distribution
(iii) the economic efficiency of resource allocation.

Derived from these aspects, targets to be met by these decision processes are:

1. they should be ecologically sensible and ensure a sustainable use of nature and natural resources
2. they should be as representative as possible (Bateman et al. 2002 : 40) and ensure the equality of votes (see below); at the same time however, the interests of minorities are to be considered. Decision processes should be comprehensible and traceable by the public
3. they should overcome market failures.

Fig. 2 lists several techniques which seem to be appropriate to meet these criteria[4]. Since contingent valuation was originally developed to determine the utility

[2] As a matter of course, in a representative democracy most decisions are made by the elected representatives. In the case of the Northeim project this might be the government of Lower Saxony or the county council. However, especially in environmental issues the scope for decision making is wide and the preferences, i.e., the mandate of the population often unknown. In this case, participatory approaches are particularly relevant and might foster the population's backup with regard to the project being undertaken (Hellenbroich and Stratmann forth.).

[3] In a pragmatical sense, ›socially accepted‹ means that a decision is at least tolerated by the population.

[4] For applications of budget games see, e.g., Ziehlberg (1999) and Menzel and Reinhardt (2002); for an application of a combination of CVM and participatory rural approach

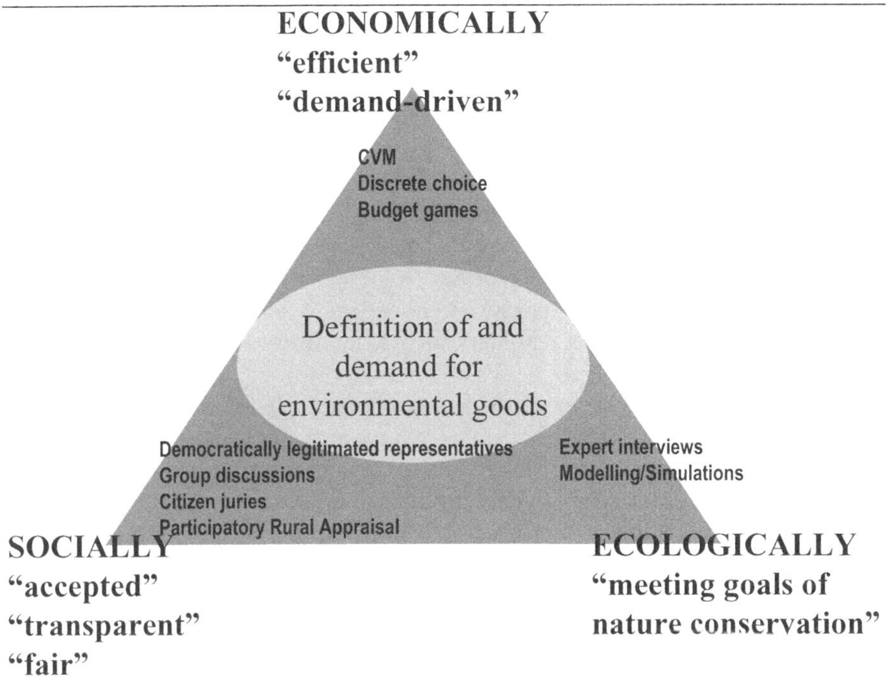

Fig. 2. Examples of techniques for public decision making using the background of the sustainability concept (source: own illustration modified from Huber 1995 : 43; see also Máñez and Renner in this volume)

derived from the provision of a particular good, it matches the economic criterion best and naturally falls short to some extent with respect to the demand of the social and ecological criteria. In a CV survey for instance, ecological features of the good are given, ideally based on a thorough investigation of ecological findings with regard to the good and in collaboration with the respective experts. However, during the survey the definition of the good is not subject to the respondents' preferences anymore. It could be however, if in the previous phase of the survey design the population took part in the development of the valuation scenario and the process of defining the good. In part, this was actually carried out in this survey. Based on the preliminary findings of landscape ecologists such as Bertke et al. (2002), definitions of farm hedges and their functions, merits and demerits were discussed in focus-groups. An ideal way to bring together experts in ecology and the population

see e. g., Putri (2002). Juries as an aid to decision making are described, e.g., by Brown, Peterson and Tonn (1995).

in order to meet ecological and social goals at one time, might consist of setting up discussion groups including participants from all relevant parties[5].

With respect to the social implications of agri-environmental issues, this is equally valid. Social aspects such as the perceived property rights and fairness are often discussed in focus-groups during the design phase of the CV survey. In fact, they contributed substantially to the definition of the valuation scenario in this survey. However, there was still a relatively high percentage of protest zero bids in this survey (see Sect. 5.1) which could have been reduced by a higher degree of participation in the valuation process. As indicated in Fig. 2, citizen juries and similar techniques might be appropriate remedies to these shortcomings. In particular, representatives of stakeholders such as landowners and landusers should be involved in these activities.

Still, these methods do not provide any data suitable for cost-benefit-analysis as their outcomes do not necessarily have a monetary format. The same applies for expert interviews which focus on the ecological aspects and in addition, often neglect social aspects as perceived by the local populace.

If the ecological and social dimensions had a greater weight within the CVM, this technique might be a suitable instrument for public decision making on the background of the sustainability concept. What does the survey presented here reveal with respect to the suitability of the CVM in an agri-environmental context?

Results show that the chosen sample of the population in the project region, a rural county in central Germany, did account a positive value to an additional 100 km of hedgerows in their county. Concerning information processing, the demand for additional information and the distinction the majority (i.e., 100 out of 180 respondents who were willing to contribute financially) made by valuating the two types of hedges differently, suggest that information offered in the scenario is received and handled adequately. In the case of a regional good evaluated by the local population, decision behaviour of the vast majority of respondents seems to be adequate to the valuation task. Thus, the CVM appears to be a valid and feasible approach to public decision making on payments for ecological goods.

However, certain limitations of this method also have to be taken into consideration. In most cases, for practical reasons only a sample of the population can be asked to valuate the commodity. To start with, the public opinion expressed by means of the survey cannot be considered a *democratic* vote, since not every citizen in the population was given the chance to state his or her valuation.

---

[5] Apart from experts in ecology and a cross section of the population, discussion groups might include experts in other fields such as agriculture and administration and stakeholders such as landowners. In this sense, the Local Advisory Board is a panel of experts – not necessarily in ecology, but in other fields of regional relevance. The Board consists of several democratically legitimated members of the county council, thus, the population is represented on the panel. However, the weight of both efficiency and social aspects in decision making could possibly be enhanced if (i) the portion of the population participating in the process increased and (ii) discussion groups were institutionalised and the population took part in the design process of a CV survey.

Moreover, these valuation processes are based on limited knowledge on the commodity, even though respondents have access to further information. At the same time, the survey respondent having listened to the valuation scenario text, probably possesses more knowledge on the environmental good than an average citizen who does not participate in the survey. If preferences are affected by information and are constructed in the very moment of the interview, the preferences of the survey participants are likely to differ from the preferences of the remainder of the population who were not provided the scenario information (e.g., Carson 1998; Chilton and Hutchinson 1999). However, this argument does not hold because in a proper referendum every citizen would get the same amount of information. Thus, preferences of survey participants might still differ from those held by the population which are based on less information. They can be considered closer to the ›true‹ preferences when compared to a referendum setting where citizens get a reasonable amount of information through the referendum scenario.

A more substantial objection however, refers to the lack of democracy and equality of votes in stated preference techniques. Since these methods are market-based, the amount of money an individual bids is what counts. One single high bid from one individual could thus outweigh the lower bids of several other individuals (Mitchell and Carson 1989 : 50). In democracy, each individual is supposed to have one vote, regardless of his or her fortune, income, or wealth. Obviously, public decision making based on stated preference techniques is not in line with democratic principles given through a state's constitution (for an analysis see, e.g., Hellenbroich and Stratmann forth.; Hespelt and Krebs forth.)[6]. This could be the case however, if WTP bids were corrected for income levels, i.e., if respondents were asked to either accept or reject the contribution of a fixed proportion of their income (Cansier 1993 : 81; Boadway and Bruce 1984 : 292).

Spash (1997 : 403) mentions an entirely different reason why stated preference techniques such as the CVM; »...in terms of policy, environmental management on the basis of totalling economic values is liable to be undemocratic because of the systematic exclusion of a section of the general public«. In Spash's opinion, this is because the majority of citizens hold concepts and ideas of ›value‹ which differ substantially from those of the economists who design the survey. These values however, are not elicited by economic surveys, since they are not a field of the investigation. Survey participants either do not manage or just are not willing to translate their values to economic terminology. This is especially evident with regard to biocentric or intrinsic (Marggraf and Streb 1997) values of nature which are considered incommensurable with monetary values by the respective individuals (see also Clark, Burgess and Harrison 2000). If individuals express protest bids or even refuse to participate because they feel misunderstood, the stated preference approach indeed excludes particular parts of the public. As already mentioned

[6] The rule ›one person-one vote‹ is not necessarily valid in group-discussion based valuation methods either, since individuals differ with respect to their competences and skills in discussing. Here, it might not be the »big wallet, but the big mouth« (Caspian Richards/Clive Spash pers. comm. Aberdeen, April 28th 2003) that counts.

above, this shortcoming could probably be mitigated by an enhanced participation of the population during the design process of the survey and the setting up of the valuation scenario.

From a philosophical point of view, it could be argued that »it is wrong to reach collective decisions by adding up self-interested preferences« (Bateman et al. 2002 : 19). Public decisions should be made in settings that allow the consideration of public interests, e.g., in citizen juries, and not based on the principle of consumer sovereignty. However, it is still not clear what an interface between transactions of public and private goods should be like. Bateman et al. (2002 : 20) avoid this issue and emphasise »that CBA is not a substitute for the political process; it merely provides information to the actors in this process«. The WTP values elicited in this survey are actually used exactly in this way. The Local Advisory Board of the Northeim project has been informed about the population's WTP for farm hedges and is now free to take these values into consideration when allocating the money, e.g., on the different types of hedges.

Despite these objections, CV surveys can be considered a valuable contribution to regionalised, participatory decision making in a sustainability context. However, they have to be complemented by other techniques (see also Carson 1998; Enneking 1999). As Bateman et al. (2002 : 80) observe, »To avoid paternalism . . . , secure a citizen-based CBA and some form of expert assessment together«. In the Northeim survey, as well as in the studies conducted by Clark et al. (2000), respondents appreciated the opportunity to participate in the public decision making process and were willing to »contribute meaningfully to decisions concerning their own local environment« (Clark et al. 2000 : 60). The following issues still have to be investigated in more detail:

1. if the CVM and its results, based on *individual* valuation processes, or rather elicitation methods which involve *collective* valuation processes, are more appropriate (Clark, Burgess and Harrison 2000) for ›sustainable governance‹,
2. how all three sustainability dimensions (see Fig. 2) can be represented at the same time in the decision process.

7 Summary

In this contribution, the suitability of stated preference techniques as an approach to public decision making which considers all three sustainability dimensions was discussed. First, an innovative agri-environmental payment scheme, developed in a particular project region in central Germany was introduced. As payment schemes are flexible with respect to their ecological objectives and at the same time, take advantage of market principles, they might provide appropriate incentives to foster conservation issues even in high-intensity agricultural land use.

However, the design of the payment scheme requires decision procedures that take into account ecological, social and economic criteria alike. The contingent valuation method is suggested as an instrument to help design these schemes, i.e., as a means to elicit preferences with respect to ecological goods and take into consideration social aspects such as the perceived fairness of project proposals. As an illustration, results of a CV survey in the project region are presented. These results indicate that the local population is (i) willing to contribute financially, (ii) willing to participate in the decision process and (iii) able to express their preferences and opinions on the proposed ecological goods. However, a standard CV survey still lacks an adequate linkage between ecological, economic and social aspects, even though in recent approaches both ecological and social issues tend to gain importance, particularly by combining quantitative and qualitative techniques and including, e.g., group discussions to the conventional standardised interview techniques. The last section discussed future potential and limitations of these approaches with regard to sustainable public decision making.

References

Barbier E (1987) The concept of sustainable development. Environmental conservation 14 : 101–110

Bateman I et al. (2002) Economic Valuation with Stated Preference Techniques: A Manual. Edward Elgar, Cheltenham

Bertke E, Isselstein J, Gerowitt B (2002) Ökologische Güter der pflanzlichen Biodiversität in einem Konzept zur ergebnisorientierten Honorierung ökologischer Leistungen der Landwirtschaft. In: Bundesamt für Naturschutz (ed) Treffpunkt Biologische Vielfalt II. Landwirtschaftsverlag, Bonn

Boadway R, Bruce N (1984) Welfare Economics. Blackwell, Oxford

Brown TC, Peterson G, Tonn B (1995) The values jury to aid natural resource decisions. Land Economics 71 : 250–260

Cansier D (1996) Umweltökonomie. Lucius & Lucius, Stuttgart

Carson RT (1998) Valuation of Tropical Rainforests: Philosophical and Practical Issues in the Use of Contingent Valuation. Ecological Economics 24 : 15–29

Chilton SM, Hutchinson WG (1999) Exploring Divergence Between Respondent and Researcher Definitions of the Good in Contingent Valuation Studies. Journal of Agricultural Economics 50 : 1–16

Clark J, Burgess J, Harrison C (2000) »I struggled with this money business«: respondents' perspectives on contingent valuation. Ecological Economics 33 : 45–62

Enneking U (1999) Ökonomische Verfahren im Naturschutz. Peter Lang, Frankfurt a.M.

Fischer A, Hespelt SK, Marggraf R (2003) Ermittlung der Nachfrage nach ökologischen Gütern der Landwirtschaft – das Northeim Projekt. Agrarwirtschaft 52 : 390–399

Gerowitt B, Marggraf R (2001) Naturschutz als Produktionszweig für die (ökologische) Landwirtschaft – Neue Wege durch die ergebnisorientierte Honorierung. In: Bundesamt für Naturschutz (ed) Naturschutz und Ökolandbau – Handlungsbedarf für die AGENDA 2000 und gemeinsame Vorbereitung auf die AGENDA 2007. BfN-Skripten, Bonn-Bad Godesberg

Heißenhuber A, Lippert C (2000) Multifunktionalität und Wettbewerbsverzerrungen. Agrarwirtschaft 49 : 249–252

Hellenbroich T, Stratmann U (forth.) Zahlungsbereitschaftsanalysen als Entscheidungshilfe für die Verwaltung? In: Marggraf R, Bräuer I, Fischer A, Menzel S, Stratmann U, Suhr A (eds) Ökonomische Bewertung bei umweltrelevanten Entscheidungen – Einsatzmöglichkeiten von Zahlungsbereitschaftsanalysen in Politik und Verwaltung. Metropolis, Marburg

Hespelt SK (2002) Marktwirtschaftliche Einbindung eines ergebnisorientierten Honorierungssystems für ökologische Leistungen der Landwirtschaft. In: Bundesamt für Naturschutz (ed) Treffpunkt Biologische Vielfalt II. Landwirtschaftsverlag, Bonn

Hespelt SK, Krebs M (forth.) Perspektiven von Zahlungsbereitschaftsanalysen in partizipativen Verfahren und als Medium der Politikberatung. In: Marggraf R, Bräuer I, Fischer A, Menzel S, Stratmann U, Suhr A (eds) Ökonomische Bewertung bei umweltrelevanten Entscheidungen – Einsatzmöglichkeiten von Zahlungsbereitschaftsanalysen in Politik und Verwaltung. Metropolis, Marburg

Huber J (1995) Nachhaltige Entwicklung: Strategien für eine ökologische und soziale Erdpolitik. edition sigma, Berlin

Jörissen J, Kopfmüller J, Brandl V, Paetau M (1999) Ein integratives Konzept nachhaltiger Entwicklung. Forschungszentrum Karlsruhe, Karlsruhe

Marggraf R, Streb S (1997) Ökonomische Bewertung der natürlichen Umwelt. Spektrum Akademischer Verlag, Heidelberg

Menzel S, Reinhardt FJ (2002) Can economic valuation of biodiversity contribute to the protection of biodiversity? – Examples of valuation by German citizens. In: Gesellschaft für Ökologie (ed) Verhandlungen der Gesellschaft für Ökologie Band 32. Die Werkstatt, Göttingen

Mitchell R, Carson R (1989) Using Surveys to Value Public Goods: The Contingent Valuation Method. Resources for the Future, Washington DC

Murtough G, Aretino B, Matysek A (2002) Creating Markets for Ecosystem Services. `http://www.pc.gov.au/research/staffres/cmfes/cmfes.pdf`

Nellinger L (1996) Zur Weiterentwicklung der EU-Agrarreform unter einkommens-, markt- und umweltpolitischen Gesichtspunkten. Agrarwirtschaft 45 : 295–302

Putri EIK (2002) Integration von Kontingenter Bewertungsmethode und partizipativen Ansätzen am Beispiel des Gunung Gede Pangrango Nationalparks in Indonesien. Cuvillier, Göttingen

Randall A (1972) Market Solutions to Externality Problems: Theory and Practice. American Journal of Agricultural Economics 54 (2) : 175–183

Randall A (2002) Valuing the outputs of multifunctional agriculture. European Review of Agricultural Economics 29 : 289–307

Spash CL (1997) Ethics and Environmental Attitudes with Implications for Economic Valuation. Journal of Environmental Management 50 : 403–416

von Ziehlberg R (1999) Präferenzen für Naturschutz in der Agrarlandschaft. Heydorns, Uetersen

Narrating diversity: Plants, personal knowledge and life stories in German home gardens

Anne Holl

Institute for Rural Development, Waldweg 26, 37073 Göttingen, Germany, email to aholl@gwdg.de

Summary. In the vast field of biodiversity research social sciences still have a difficult standing. We find widespread misconceptions of the role sociology and cultural studies should play when studying potentials for nature conservation. Often the »social view« is regarded as something subordinate to hard scientific facts. Even researchers exploring local, everyday knowledge on biological diversity or human interaction with nature tend to operate with concepts and methods not adequate to their research subjects. Frequently the research into folk knowledge equals a collection or stocktaking of those kinds of knowledges that might be useful for science. The employment of a scientific conception of knowledge, and also of biodiversity becomes problematic. For the idea of biodiversity developed in a rather detached small sphere of society and hardly has relevance in everyday life. Thus, when studying people's knowledge on »biodiversity«, such research methods should be employed which allow people to develop their personal view on the subjects in question so that informants can give their own personal meanings to the world. For my own research on home gardeners' knowledge I opted for episodic interviews with a strong narrative and biographical component, supported by participant observation and group discussions. The gardeners' stories make clear that their nature perception and management is inseparably linked to their individual biography. Their knowledge on nature and »biodiversity« is woven into their total, constantly changing knowledge on the world – it forms part of their identity. Not only nature is perceived as diverse but also knowledge itself. The gardeners reflect that their knowledge depends on time and space, and that in other situations than their own other ways of thinking and acting might be appropriate.

Key words: knowledge, science, sociology, research methodology, home garden

1 Introduction

»There is no use in studying folk knowledge on biodiversity in Germany because diversity is so low.«

»You could ask what gardeners know on particular taxa and whether they appreciate biological diversity. Compare the answers to the real diversity in people's garden. Identify the gaps between knowledge and action.«

»What definition of biodiversity do you apply in your research?« – »I have none. The definition is what I want to discover. Or put it this way: In my project biodiversity is what it is to the people studied.« – »But that is no science! You have to have a definition. A scientist must know what she is talking about!«

These statements taken from personal discussions with other (mainly natural) scientists illustrate only some of the fallacies – circulating in academy itself – about what role social sciences and social phenomena can or should play in biodiversity research. Moreover, the views expressed reveal an underlying claim of science to hegemony in knowledge production.

In my own qualitative research I study gardeners' everyday knowledge on nature and its diversity. Moving between two worlds, science and everyday life, I soon became aware of striking different concepts of knowledge and knowing, or »styles of thinking« as Ludwik Fleck (1935) named them. This fact alone is not so surprising but what follows, from being woven into a particular »think tank«, earns a closer look if we want different social groups to communicate. Sect. 2 of this contribution explores how the concepts of »knowledge« and »biodiversity« are used in science and are imposed on both social research and social life. We shall see that these concepts and the research methodology derived are often inappropriate to the subjects studied. Thus, the challenge social studies on biodiversity face lies in the design of alternative research strategies that account for the character of knowledge in special and society in general. Sect. 3 therefore will review literature on what knowledge is or might be. In my own study the narrative format of gardeners' knowledge on nature and the world became apparent. My semi-structured interviews more and more developed into episodic ones telling stories of people and plants. This development, and thoughts on research methodology, is sketched in the fourth section, followed by statements drawn from gardeners' interviews. From the statements we can conclude that science could even learn from everyday knowledge and its producers.

2 Misreading the social in biodiversity research

Despite decades or even centuries of anthropological and social research into human attachment to nature science follows its own pathway in the pursuit of knowledge as if these social studies and their findings had never existed. Especially in environmental and conservation studies scientists accumulate heaps of the most detailed knowledge on organisms and natural processes, improve their models on the world and seem to hope that their authority alone will produce insight in the non-scientific world, i.e., will stimulate change in human action towards the natural world. Sociology, anthropology, science studies and their double message seem to be overheard in the natural sciences. The first part of the message could be described as: All human attachment to nature is culturally defined, as we can share our perceptions and sensations only by the use of a common language, systems of meanings and symbols. And exactly the participation in systems of meaning is what creates a culture. This observation constitutes the second half of the message: ›cultural glasses‹ are not only imposed on nature but also on the means by which we perceive it (and other things), i.e., on knowledge processes.

Few scientists today would dissent that the meaning of ›water‹ or ›tree‹ differs from culture to culture. It is also widely accepted among scientists that scientific knowledge applied in development projects has often produced negative socio-economic and ecological consequences in the past. That, in certain circumstances, one should better rely on local people's knowledge – again a product of culture and a particular natural environment – to solve a problem than to import pat solutions. In the field of biodiversity research the idea has inspired a multitude of studies into ›traditional‹ or ›indigenous‹ knowledge in non-western countries, studies which often include cultural and cosmological dimensions. Such investigations find support in both politics – the Convention on Biological Diversity (CBD) is the most prominent ally – and science.[1] As we shall see, even some of these sociological and anthropological studies lack a sensibility for the social world, and hence, can only give a distorted view on it.

Turning to Western knowledge on biodiversity – of which science forms a part – people's everyday experience and its cultural embeddedness seem no longer of relevance. The view that society and culture are unrelated to biodiversity issues is hardly ever made explicit but can be read out of research programmes, funding priorities, scientific conferences or students' text books on the topic: The chapters on social dimensions of biological diversity remain blank, for reasons one can only speculate on.

Most scientists, however, reject such a radical view, and admit that sociological studies are either a *must* or *useful* for biodiversity research. Advocates of the first opinion have recognised that their most detailed scientific knowledge on nature and its management must be put into practice. Social studies should provide insight on how this task can be best achieved. Indeed, the last ten years saw a lively research into biodiversity issues in the ›hard‹ social sciences. Especially topics like the equitable sharing of benefits from the use of biological resources, the settlement of conflicts between various stakeholders, questions of intellectual property rights or the adjustment of national regulations to the requirements of the CBD are covered in disciplines like economics, political sciences and law. Most of these studies are based on biodiversity knowledge originating from biology and take over its definitions or value judgements, e.g., how (from the biological point of view) an endangered species can be best protected. The conservation of biodiversity is reduced to a technocratic problem.

Another branch of social research turns to the social or human dimensions of nature conservation on the micro level. Since the 1980s environmental sociologists and psychologists have been developing models on the links between environmen-

[1] For completeness, a third party should be added, the life science industry, which even sends out its own researchers ›bio prospecting‹. These economic interests in biodiversity research are excluded in this contribution for they would open another vast dimension of the problem.

tal knowledge, awareness, values and action. We find observable discrepancies between what people can tell about nature and environment and how they act, so sociology and cognitive sciences should explore the reasons for this incongruence. Among the models and explanations for environmental unfriendly human action there are two major trends, which often intermingle. One tendency stresses ›outer‹ reasons, i.e., the setting or social circumstances which do not favour non-polluting decisions (e.g., Wüstner and Stengel 1998). This party calls for low-cost situations for individual environment protection and social sanctions of dropouts.

In the view of the other party, mainly cognitive psychologists, conservation of nature and the environment originates in the individual. People's environmental knowledge, awareness and values are too weak for effective nature protection. Although this model has hardly been applied to biodiversity issues but to topics as climate change, pollution or waste management, its ideas have penetrated the whole discourse on sustainability and nature protection in Germany. In earlier times a rather simple model, it has been developed into a complex web of relations between different types of knowledges, attitudes, perception, problem awareness (a product of the former three), emotions, demographic factors and situational factors. (Gerster-Bentaya 1999)

Statistically measured, the influence of knowledge on ecological action is rather small. Kaiser and Fuhrer (2000 : 53f) argue for an underestimation of knowledge in quantitative studies. One reason, in their opinion, lies in the missing differentiation of distinct knowledge types. For environmental friendly action the individual needs knowledge on ecological processes, on options how to act in a certain situation, on effectiveness of action and social knowledge on motivations and intentions of other individuals. For Kaiser and Fuhrer it is not the amount of knowledge that determines ecological action but the convergence of the different knowledges towards a common goal. Whether one trusts empirically tested (quantitative) models and follows the authors' computational view of human mind or nor, even their theoretical considerations on knowledge types leave some questions open. In their research design, only knowledges that are closely connected to the problem studied: environmental protection, are of relevance. What about all other spheres of life? Can knowledges on them be excluded? Is it possible to sort knowledge into these distinct categories? Is it only this bundle of ›environmental knowledge‹ that influences ›environmental action‹? Kaiser and Fuhrer at least admit the delicacy of the term ›environmental‹ or ›ecological‹ action because it determines the objective of acting from outside. From a psychological viewpoint one could only speak of an environmental friendly act if the individual fulfils its own intention.

Advocates of nature conservation still choose public education of environmental knowledge and values as their main strategy. In 2002, ten years after the Rio Earth Summit and the Convention on Biological Diversity, the German Federal Ministry of Environment launched a big ›Biodiversity Campaign‹ (*Biodivkampagne*)[2]. Under its label research institutions, NGOs, other action groups, companies and private persons should create public awareness for the necessity of biodi-

[2] www.biologischevielfalt.de.

versity conservation. Again, the coalition of science and politics hopes for a change of action via the information channel in a simple stimulus-response model. You tell the people why biodiversity is important for life, how it can be protected in daily business and the loss of genes and species will slow down. The biodiversity scouts, however, do not take into account knowledge processes, i.e., how given information is transformed into personal and social knowledge. In addition, the knowledge transferred – most obviously the Whys and Hows of conservation – contains value judgements that are not necessarily shared by recipients. Both ethics and sociological research assert that an individual will most likely adopt a value and consider it in action if the individual participated in the value creation.[3]

So, environmental education has developed a second strategy that allows for participation in the knowledge building process: the direct experience of nature through all human senses. People, especially children, discover the secrets of nature by themselves, get touched by its beauty, realise the living character of nature. Bögeholz (1999), again in a quantitative study, found the significance of direct contact to nature in everyday action more than six times higher than the significance of environmental knowledge. She pleads for a combination of long-term direct experience of nature and theoretical ecological education. Condensing all positions which see biodiversity knowledge as a must in society, for them knowledge has the function to lead human action into a desirable direction. What biodiversity knowledge is, is externally defined, by science.

Let us now turn to the third set of positions, which acknowledge the *usefulness* of studying society for biodiversity research. Driven by the insight that scientific knowledge could not be applied successfully in all contexts in the past one should value folk's traditional knowledge on the things for better problem solving (e.g., DeWalt 1994). In the case of biodiversity the motivation is slightly altered. People's knowledge is not studied for their own sake but to enrich science and to assist its project of complete understanding of biodiversity. The vocabulary of research titles already hints at science's intentions: »Use of indigenous knowledge for rapidly assessing trends in biodiversity« (Hellier et al. 1999) or »Using indigenous knowledge to improve agriculture and natural resource management« (DeWalt 1994). In worst cases the life science industry expropriates indigenous people of their knowledge and claims patents for ›mental property‹. Public science is not much better: in general ›knowledge diggers‹ will publish ›their‹ findings but indigenous people are hardly ever asked whether they want to share their knowledge with the world or not. Most probably people will not even recognise their former knowledge for it is ripped out of context, transformed by science and used for completely different purposes. Even if leaving these crucial ethical considerations behind, many studies into ›local‹ or ›traditional‹ knowledge on biological diversity must subject to critique from a sociology of knowledge standpoint. For it is again science that decides what kind of knowledge is useful or of value and should be extracted, and not the knowing people. »What kind of knowledge« here refers to both content and form

[3] See, e.g., Molitor (1989). Another option for value or norm enforcement would be social sanctions in case of non-conformist behaviour.

of knowledge. Researchers set the agenda and ask for widely unknown species or indigenous classification of plants or animals – subjects that fit their research interests. Furthermore the structure of folk knowledge studied should be compatible with scientific categories. Such knowledge research strategies are extremely popular for the so-called ›ethno‹ sciences (e.g., ethnobotany, -medicine, -ecology)[4] and in applied environmental disciplines including geography and agriculture. On the other hand, lots of social anthropological studies have been conducted that apply a very broad definition of knowledge and also describe the meanings of knowledge in certain contexts, societies and cultures (e.g., Berkes 1999; Warren et al. 1995).

From this overview on relations between biodiversity research, social studies, society and especially society's knowledge we can extract a three-point summary:

1. In biodiversity research social phenomena do not go beyond a subordinate status, so social studies have a hard time being heard in the coalition of natural sciences and politics. Social studies are insofar subordinate as they are seen as mere service disciplines which moderate between science and society. They should on one side create acceptance for science's ideas in public and on the other side inform science on useful knowledge circulating in society.
2. In biodiversity research ideas, theories and methods of social studies are not perceived in their complete range. Only convenient, familiar ideas and methods are accepted, i.e., the complementary but not competing thoughts to natural sciences, e.g., knowledges that easily fit. Among social research methods especially the exact, comparable techniques to natural science methods are accepted. Simple models explaining cause and effect are imposed on the social world.
3. The core of the problem, in our case: knowledge on biodiversity and its application in practice, lies in the concept of knowledge as defined by science. Or its not-definition because technical and natural science disciplines hardly ever reflect their own knowledge production. The tested and proofed counts towards scientific facts that are claimed to be universal, value-free and independent of any personal or social interference. From science's standpoint thoughts and beliefs circulating in society acquire only the status of knowledge if they hold firm scientific criteria. Knowledge is seen as a definable entity that is related causally to action.

So, after all this criticism and the deconstructive task, the central question is: What is knowledge, in science and society, in general and on biodiversity in special? Only then we can continue with assessing people's knowledge on nature.

[4] The term ›ethno‹ science could be somewhat misleading because not knowing persons but academics from outside define it as science and structure the knowledge for a greater audience. So if indigenous people hold a certain knowledge on plants and their identification, ethnobotany is the scientific account of this ethno knowledge (see, e.g., Martin 1995), and it might nourish scientific botany. The greatest project of this kind is the »Global Taxonomy Initiative« under the CBD which aims at compiling knowledge on species and varieties from all over the world.

3 Knowledge, action and the world ›out there‹

The question what knowledge is has driven mankind for millennia and yet we have not found an easy definition up to present. In epistemology, the question for knowledge is intrinsically woven into the question for truth. Here knowledge is what can be called true from an ›objective‹ viewpoint. The distinction between ›objectivity‹ and ›subjectivity‹ finds its roots with the great ancient Greek philosophers, especially Aristotle. He distinguished at least three different kinds of knowledge, which are needed in three separate activities: *Episteme* is the universal knowledge on the order of the world, the eternal truths, which are accessed via ›motionless‹ observation and immersion. *Episteme* is needed when doing *theoria* and equals science. For *poesis*, the fabrication of things, we need another type of knowledge: *techne*, skill. The third kind of knowledge Aristotle calls *phronesis* is a prudent understanding of what has to be done in different situations, including moral standards. It is the practical knowledge for human action (*praxis*) and learnt by experience.[5] The Aristotelian distinction between episteme and phronesis remains alive in the claims of natural sciences and positivist epistemology. Social sciences, however, have at least theoretically broken with positivism and study scientific knowledge production in a social context. No knowledge can only be seen theoretically, there is always a practice of knowing.

What might on one hand serve as a reference for the divide between objective science and subjective human action on the other hand helps us to question the drawn line: Aristotle tells us that there is not only one knowledge but different modes to know the world and all have a right to exist. And for the relationship between theoretical and practical knowledge he comments: »Some people who do not possess *episteme* are more skilful in action than others with their knowledge. These are especially those people with a lot of experience.« (Aristotle, Nicomachean Ethics, par. 1141b, cited in: Reichert 2000)

The short trip into philosophy has raised some topics for our search for a suitable knowledge definition. If there are different kinds of knowledges, not necessarily the Aristotelian ones, how are they structured and related to each other? Where do we locate knowledge: on paper, in persons or in a community? How is knowledge produced and transferred? Does knowledge change its quality in time, e.g., through personal experience? What are the knowledges good for? Let us leave the ancient thoughts behind and look at contemporary sociology, how it views knowledge, both everyday ›lay‹ and scientific knowledge.

When reviewing the literature on ›local knowledge‹ (or ›folk‹ or ›traditional knowledge‹) again we face a multitude of implicit and explicit explanations of what this knowledge is. As a starting point for our exploration I chose a definition by the social anthropologist Uta Schultze. She describes local knowledge as »knowledge, skills and views of life which arose in a particular natural environment and a particular cultural context, and which are subject to change.« (Schultze 1998 : 3)

[5] Aristotle distinguishes two other types of knowledge, *nous* (reason) and *sophia* (wisdom), which support *episteme* in the production of theory.

Although she concentrates on locality for characterising the knowledge she emphasises its personal character. In some cultures, she explains, there does not even exist a noun *knowledge* but only the active verb *to know*. In these cultures knowledge can never be seen separate from a knowing person, knowledge is no object one can ›possess‹ (p 7). It is not enough, however, to locate knowledge in a single person but in a community: it is both constructed in human interaction (pp 4ff) as well as socially transferred and distributed (pp 19f). The Convention on Biological Diversity and some ›technical‹ publications on local knowledge ignore its social distribution and conceptualise it as a collective good of a community that can be accessed by all members, regardless of gender, age, status etc. But exactly these social divides also describe the borderlines of knowledge. »Different types of local and non-local people hold many divergent, sometimes conflicting, interests and goals, as well as differential access to vital resources. Knowledge, which is diffuse and fragmentary, emerges as a product of the discontinuous and inequitable interactions between these competing actors. Through their respective ›discursive‹ networks, different kinds of information and processes are communicated and legitimated« (Scoones and Thompson 1994 : 21). Scoones and Thompson see the reason for the fragmentation of local knowledge right in its production process: In the interactions, in which knowledge is produced actors with different resources and interests meet. The knowledge negotiated expresses power relations between the actors involved. In this context, Long and Long (1992) even speak of »battlefields of knowledge« where actors struggle »against each others' different social and cognitive worlds« (Long and Villareal 1994 : 44). On these battlefields or »interfaces« between knowledge senders and recipients, if even distinguishable, the knowledge changes: »Although knowledge creation/dissemination is in essence an interpretative and cognitive process entailing the bridging of the gap between a familiar world and a less familiar (or even alien) set of meanings, knowledge is built upon the accumulated social experience, commitments and culturally-acquired dispositions of the actors involved.« (Long and Villareal 1994 : 42) Because of these dynamics the authors do not want to speak of (closed) knowledge systems but of networks whose participants and communication channels again are subject to change. The networks must not be woven too tight so that innovations can penetrate, and »gate keepers« to other epistemic communities facilitate (or impede) the information flow (Long and Villareal 1994 : 46f).

According to Peter Antweiler (1995) local knowledge can furthermore be characterised by its »cultural location«, i.e., knowledge gets its meaning only within a culture. Following Clifford Geertz (1983) culture can be defined as a web of meanings through which people interpret their experience and which guides their action. So one could say knowledge *is* culture, for the socially negotiated meanings constitute culture. »(...) All knowledges, whatever else they do, operate as systems of meaning; (...) they provide categories and conceptions that enable their users to understand their worlds *as something*.« (Percy 1958, cited in: McCarthy 1996 : 108f) Like knowledge culture cannot be seen as a uniform collective good. »Culture is diverse, many-layered, and multicoded.« (McCarthy 1996 : 25) From the cultural location of knowledge follow its qualities of being situated and context sensitive:

knowledge is created in particular cultural, social, economic, ecological etc. contexts. The meaning of certain knowledge is comprehensible only within these contexts, as with its application, i.e., action, that makes sense only in a particular constellation (Schultze 1998 : 10f). This knowledge also includes beliefs, values, myths, traditions and so on, anchored in the respective culture.

Scoones and Thompson (1994 : 24) sum up: »The process of knowing should be seen as engaged, value-bound and context determined, rather than detached, value free and independent of context.« When the authors characterise knowledge, or better: knowing, as a process, they refer to its change through time, both within a person and a community. By new experience (experiments, observation, communication) the already existing knowledge is constantly tested, enlarged, corrected, judged – it is in permanent motion. For example peasants' knowledge on cropping is a constant adaptive performance to a changing natural and social environment (Drinkwater 1994 : 33). Scoones and Thompson (1994 : 20) add that peasants' knowledge production and application take place in »real time« – in contrast to science's laboratories. On the fields knowing and acting, theory and practice coincide.

For the transfer of local knowledge it follows that it is learnt mainly in practice. We are talking about knowledge that knowing persons often neither are aware of nor can explain to others, because it is so natural (Schultze 1998 : 9; Drinkwater 1994 : 40). This knowledge is expressed in action, routines, traditions and habits. You learn it in everyday life through tales and stories of the others or when participating in an activity. You learn it while acting, while solving a problem and not in a theoretical instruction. (Schultze 1998 : 13) Here we find another reason for the uneven distribution of local knowledge within a community: its oral tradition and its »learning by doing« character restrict the transfer of knowledge to the participating audience.

A branch of cognitive psychology also postulates the coincidence of knowing and acting, or at least its close relationship. Gerstenmaier and Mandl (2000) want to clarify the link between knowledge and action with the help of an intentional theory on human agency and criticise the overweight of functional theories within psychology. The functional theories in most cases see knowledge as a precondition for action and ask on the syntactical level, how knowledge explains, controls, governs (p 301) human action. But in the authors' opinion the analysis of the knowledge-action-link must start at the knowledge construction by the acting subject, and concentrate on the semantic dimension of knowledge, the meaning (p 290). Drawing on pragmatism and social constructivism, in their concept knowledge is created in the context of action and gets meaning by its use. They see knowledge acquisition and application as situated learning, so that context gets more weight and no longer counts as disturbing factor. Gerstenmaier and Mandl agree with Jerome Bruner's position for whom meaning forms the central concept of psychology. For our knowledge, experience and action are formed by intentional states, and these intentional states get a shape only by participating in cultural symbolic systems (language, discourse, narratives). We will return to Bruner's thoughts on knowledge and the narrative as central tool for the construction of meaning when discussing research methodology below.

Referring to Michael Polanyi Bengt Molander calls knowledge, which is learned and created in context, through personal experience and practical training, »tacit knowledge«. »The core of such knowledge does not consist of verbal or mathematical formulations; it consists of abilities to *make judgements* and to *do things* in practice, skilfully and with insight. The knowledge is *in* the judging and the doing.« (Molander 1992 : 11) Polanyi (1985) himself uses the term »tacit knowing« for emphasising the active, personal process involved. For him, knowing is a process that takes place between two different levels: from the level of detail recognition we shift our attention to a higher, more complex level where the overall meaning can be grasped (pp 16ff). So integrating pieces of information into the big whole creates knowledge. But the information pieces change their meaning in the knowledge process as we concentrate on the overall picture, so that the initial experience of details remains in the dark, it becomes »tacit« knowledge hard to verbalise. This way of knowing depends on our body (pp 22ff) for we experience it through all our senses and, only the body can express this knowledge. The learning person must observe, imitate, feel the meaning of the knowledge performance, whether theoretical or practical knowledge.

Molander concludes that all knowledge has tacit aspects. If we divide knowledge on a particular subject into three types, propositional knowledge (articulated statements and theories), practical knowledge (skills) and knowledge of familiarity (with practices and phenomena), then practice and familiarity fall into the tacit category and become precondition for knowing. »There is a possibility of knowledge only if one *understands* the concepts used and the contexts in which the sentences are normally used, and that is not the same as having the ability to repeat the sentences parrot-fashion. (...) The tacit – our understandings and our actions – gives meaning to the words we use.« (Molander 1992 : 14) On the other hand, no knowledge is exclusively tacit, and no learning will do without words. A description will help in most cases, whether identifying a face, learning to cycle or a profession. At the same time there will always be a difference between cycling and explaining someone how to cycle.

In the relation between knowing and acting, Molander points at another problem without being able to give a satisfying solution: »What is knowledge in action? Confidence and/or critical consciousness?« (p 23) Polanyi had already characterised personal knowledge as acritical, i.e., neither critical nor uncritical but it evades any doubts or critical questions (Polanyi 1978, cited in: Molander 1992 : 24). »When we learn a skill by training, at the same time we learn confidence, we learn to trust in what we do. Skilful acting and confident acting are inseparable in practice.« The more routine we get in applying knowledge, the more we become experts, the lesser we ask whether our knowledge is appropriate in a particular situation. The culture or community in which we grow up exerts a similar force on our knowledge. It expects us to take over ideas and beliefs without prior critical questioning (p 27). Society determines what counts towards knowledge, and each of us can participate in knowledge and develop it further.

Here Molander's ideas reminds us of Ludwik Fleck's (1935) theory on »styles« and »collectives« of thinking – that might be institutionalised research groups, academic disciplines, professions, sub-cultures, neighbourhoods, families or any other kind of groups. The doctor and sociologist Fleck postulated that there is no possibility to produce knowledge without particular assumptions on the subject studied. These assumptions, according to Fleck, do in no case exist a priori but are the historical and sociological product of an active collective thinking – there are no such things as unconditional seeing and observing! In his major publication Fleck describes the development of the syphilis concept in medicine from the 15th century to the present. Throughout history the hegemonic influence of scientific disciplines changed: in the beginning the illness was explained by astrology, later by religion and chemistry until the scientific community convened on a rational, serological view. The development to the present syphilis definition did not happen linearly but equals a zigzag line, constantly fed by new ideas. As Fleck observes, some of these ideas persisted for an extremely long time, although some scientist had already falsified them. He sees the cause in the inertia of once founded thinking collectives, which reinforce themselves by their research. According to Fleck, real progress in knowledge, or scientific revolutions, happen only by hazard (e.g., an experiment in the laboratory ›fails‹) or when persons participating in more than one think tank introduce new thoughts in a collective (pp 40ff). At the same time, knowledge, seen as a collective process, offers the possibility for discovering/creating more/other insights than the sum of all individuals involved, for the thinker can build on the knowledge of others. The individual will hardly ever notice in this process how the collective style exerts power and pressure on his/her own thinking (pp 54ff). For everyday life Jerome Bruner (1990 : 11ff) postulates a similar relation between individual and social thought. When we are born into a culture its symbolic systems for the construction of meaning are already there. We learn them so that we can participate in culture and knowledge. Even if we dissent certain positions we can communicate our otherness only by referring to shared meanings. Thomas Kuhn's book »The Structure of Scientific Revolutions« (1962) was inspired by many of Fleck's ideas. Both Kuhn and Fleck have demonstrated for the natural sciences that all knowledge and insight is constructed by man and a product of context, culture and history.

Scientific biodiversity knowledge must face the same critique. The definition of what biological diversity is is not only determined by what is ›out there‹ in nature but also by the observing, thinking and concluding scientists. Researchers involved in this project cannot deny personal values and interests and their research takes place in a certain political environment, might they be aware of it or not. Possibly the most prominent account of the science-politics-interference in the construction of the biodiversity problem is David Takacs book on »The idea of biodiversity« (1996).

Giving an overview of the complete scientific biodiversity knowledge accumulated during the last decades would lead us too far for our question. A range of issues are already covered in other contributions in this volume. So I shall concentrate on the essential points, which facilitate the comparison to lay knowledge on diversity. To begin with, there is the definition of biodiversity as given in the CBD and widely accepted among scientists, as the variability of living organisms on the level of genes, species and ecosystems. One could say diversity is the quality of being different. Biological diversity can be measured on different levels, e.g., as relative or absolute abundance. Scientists have not yet agreed whether higher biodiversity is in general better than lower, as some ecosystems poor in species prove extremely stable. There is also disagreement on the ways the global biodiversity can be best protected, for conflicts already exist on the biological level. To protect one species might extinguish another, so many conservationists prefer ecosystem conservation to species protection. The different strategies in biodiversity management also root in differing motivations for conservation. Ethics argue for an intrinsic value of nature, we should respect living creatures independent of its usefulness in society and culture. Natural sciences point out the functions biodiversity fulfils for human life: different ecosystems compensate various environmental problems of the planet, species counteract each other before they become a plague for mankind and genetic resources for food, medicine, fuel etc. might free the earth of hunger, diseases or energy scarcity. The strongest argument for biodiversity conservation is found in uncertainty. Presently, science has only little knowledge on the estimated 10-20 million species of the world and their genes and how they all interact in the landscape. So, as long we do not know about the human problems the future will bring and about the function of a biological entity we should better conserve it for we might need it.

In contrast to philosophy it makes no difference from a sociology perspective whether this knowledge is ›true‹ or not, in the sense whether it describes nature, genes, species, ecosystems, their functions and interactions in an essentially ›real‹ correct way, as they exist independently from human discovery. For the sociologist »the world exists only in so far as it is represented to us.« (Durkheim 1895, cited in: McCarthy 1996 : 2) Here the question is whether knowledge counts as real within a society or group, and through which social processes it achieves the status of ›real‹ knowledge. The term ›reality‹ by itself is used in a different way in social sciences: it always means a social reality that members of society both face and create. To be precise, one must even speak of realities because what is ›real‹ might be different for each of us. The »unified mental world« has disappeared (McCarthy 1996 : 19). And, as Berger and Luckmann have pointed out, realities and knowledges are reciprocally related and socially generated. Summing up these points, from a sociology of knowledge perspective »knowledge refers to any and every set of ideas accepted by one or another social group or society of people, ideas pertaining to what is accepted as real.« (McCarthy 1996 : 2) So when studying people's knowledge for the social researcher ›real‹ is what it is to the people studied.

4 Getting close to people's knowledge

In my own research I wanted to study people's everyday knowledge on nature in general and on ›biodiversity‹ in special. There was the underlying assumption that the people studied were not acquainted with the scientific definition of biological diversity, its meanings and implications. The research was about ›endogenous‹ conceptions of diversity and natural processes comparable to the scientific concept. How do people in daily practice make sense of their experience with nature? Empirical research was restricted to only one sphere of daily life: home gardening. Two reasons for this choice were more quantitative in character. Gardens are reported to give home to high biological diversity compared to other land uses (Howard-Borjas 2001 : 14f). Especially for tropical home gardens, the multitude of used plant species and varieties has been described throughout the last decades (see, e.g., Watson and Eyzaguirre 2002; Vogl 1998; Landauer and Brazil 1990). For Central Europe the publications are fewer but two recent studies from Tyrol farm gardens indicate high diversity levels (Vogl-Lukasser 1999; Heistinger 2001). In the case of home gardens it becomes clear that man can be actively involved in creating biological diversity. A plot is not by hazard rich in species and varieties but because the gardener sowed, planted or did not weed. Science would speak of agrobiodiversity found behind the garden fence. For some theoretical considerations I do not want to draw a line between biological diversity found in ›wilderness‹ and in agroecosystems. The question would be where to draw the line. The untouched, unspoiled (by mankind) nature and the totally controlled monoculture field form the two opposite poles of a continuum. In practice, nature and culture both influence particular landscapes to differing degrees. The home garden itself underlines its positioning between nature and culture. Throughout history the garden has served as passageway from wilderness to the cultivated fields and back to wilderness. All crops we know were first settled in the kitchen garden, tested and improved before they were transplanted to the field (Inhetveen et al. 2002). Today we can observe a movement in the opposite direction: Former crops, e.g., cabbage, beans, certain cereals, or flax redraw from the field into the garden (Vogl-Lukasser 1999). Some former garden plants, e.g., *Aristolochia clematis*, nowadays have run wild (Inhetveen 2000). Then again people plant species they found in wilderness in their garden, or leave wild flowers or herbs in the patch.

A second quantitative reason for choosing the home gardens as a research field lies in their wide spread throughout German society: 58 percent of all households own or rent a garden (Noelle-Neumann and Köcher 2002). One could conclude that it forms the interface where most Germans actively interact with nature. This gives us a qualitative reason for studying gardening. In the garden people experience nature ›first hand‹, gardeners can directly observe and draw their conclusions, as they can influence plants and natural processes. Thus, the garden allows for studying ›knowledge in action‹.

Social research pleads for methods in data collection appropriate to the subject studied. The particular qualities of the subject imply particular methodological considerations. As we have seen above, the knowledge we want to explore is not necessarily to be put in words. The knowing persons often even cannot consciously reflect their thought and action. Furthermore, we are looking at ›situated knowledge‹, highly sensitive to context, culturally determined and socially construed. There is no sense in simply asking, »What do you know on biodiversity/on particular species?« In my opinion, there is no other choice than opting for qualitative research strategies if we want to get close to meanings people attach to the world. Qualitative research is based on two characteristic principles, openness and communication between researcher and the researched. Openness means that the subject studied is not or only partly structured by the researcher prior to data collection but in the research process by the researched. The key to grasping people's structuring and meaning is communication. This requires the researcher to return to data collection over and over again until she or he understands meanings given by the counterpart. This can be a simple question after the main part of an interview, in the style of »You mentioned X. Could you explain that to me in more detail?« The understanding of research as a circular process has also been successfully extended to the complete research design by the fathers of ›grounded theory‹, Barney Glaser and Anselm Strauss. In their view sociology's main task lies in the generation of theory and not in – mainly quantitative – testing of old theories because neither verification nor falsification leads social sciences any further. Glaser and Strauss' central idea is to ›ground‹ a new theory in data, i.e., to develop theory out of observations, interviews, statements, surveys etc. When starting research, e.g., when analysing the first interview, the developing theory is rather fragmentary and preliminary. However, when it is contrasted to another case, it is tested, extended, corrected, condensed. This process goes on until the theory is ›saturated‹, when the inclusion of another case will no longer produce new aspects of the theory. Sampling is also based on theoretical considerations, i.e., the sample is not fixed before the beginning of data collection but based on the developing theory. What case(s) might be of interest next, considering certain aspects of my theory? Shall I look for maximal or minimal contrast? This strategy requires a circular research design. Data collection, analysis and interpretation of data and theoretical sampling alternate constantly and flow into each other.

The impatient – natural or quantitative social – scientist at latest at this point will exclaim: »This no science! Such a research is from beginning to end driven by the researcher's subjective decisions! Where are the objectivity and the true knowledge?« What other perspectives on science want to eliminate, the subjectivity and personality of the researcher, Glaser and Strauss see as an asset: For generating good, innovative theory the researcher should draw on his/her experience and knowledge of context. Whenever a related idea – from an earlier study, a theory, or private background – comes to the researcher's mind he or she should write it down and include it in the analysis (e.g., Strauss 1998 : 36f). The criterion for the ›truth‹ or suitability of a theory is whether it can be proved in social reality or not. Nevertheless, qualitative researchers subject their research to the criteria of reliability and validity (e.g., Kirk and Miller 1986; Flick 1995 : 239ff). The ways of achieving these criteria, however, differ obviously from those in quantitative research. One strategy for reaching reliability is the triangulation of different data types, e.g., comparing findings from observation, interviews and survey data. In qualitative research reliability is mainly a ›procedural‹ one, guaranteed via standardised research procedures, for example by accurate, standardised documentation of data, testing of once found analytical categories on other texts or interviews, and constant reflection on how particular data were produced. A good qualitative study contains a documentation of research processes and the researcher's reflections so that the reader can judge himself where accounts of the subject end and the researcher's interpretation begins. The reflection on data collection situations and its documentation also serves as a source of validity. Which distortions were operating in an interview situation and how can I account for these distortions in the interpretation of data? Furthermore, qualitative research draws on validation via communication. Before a theory is published and presented to a critical audience the people under study validate the researcher's interpretation.

The decision for ›grounded theory‹ does not automatically imply certain methods for data collection. Here, the above-cited appropriateness should be the criterion. For my own research I chose participant observation in gardens and interviews with gardeners, gardener groups and ›experts‹. Participant observation – which often developed into observing participation – for studying practical and tacit aspects of knowledge. I sometimes had the chance to experience the ›knowing how‹ in practice, getting taught by a skilful gardener. But as it was tacit knowledge ›of the body‹ I am not better off than the gardeners, I am unable to give a detailed description. So when working in the garden together or when I was just observing I could only ask critical questions and discover ›blind spots‹ in gardener's knowledge. I can only tell there is a veil at a certain point but not what is behind.

Group interviews were conducted twice, one with a mother and son who share a garden, the other with three neighbour gardeners of pensioner age. The group situation aimed at participating in situ in the social construction of knowledge.

The main data source however, were individual interviews with home gardeners, for they offer the possibility to study both personal and social knowledge processes. Among the range of qualitative interview techniques so-called »episodic« interviews (Flick 1996) seemed to be the most suitable to generate data on knowledge in context. According to Flick – who draws on a distinction made by Tulving (1972) – our knowledge on the world is organised in two different formats: narratives that recall concrete situations and personal experience we made, and a more abstract semantic knowledge, which covers the routines and generalisations of these episodes. Both modes of knowing interact, episodes condense in rules and the rules determine the narrated episode. Moreover, episodic and semantic knowledges cannot be strictly separated; they are two poles of a continuum (Flick 1996 : 147ff). In an episodic interview the researcher can tap on both ways of knowing. The interviewee is asked for both episodes and general meanings on different level of abstraction: concrete situations, »repisodes« (situations that reoccur regularly), examples of a generalisation, subjective definitions or theoretical argumentations (Flick 1996 : 155).

Whereas in Flick's own research into everyday knowledge on technology he covers all of these data types in balance, my personal interviews soon attained a narrative overweight. After I had analysed the first two ›pre-test‹ interviews I became aware that on the basis of abstract accounts I could not conclude whether information had achieved the status of knowledge or not, I simply lost context. On the other hand, when asking people to narrate garden situations they experienced, they also produced stories rich in generalisations. I found confirmation in the writings of Jerome Bruner for whom the narrative format is the organising principle of all our knowledge. Narratives include both, the normal and the special. Culture imposes the ›normal‹ on our lives and has created a ›folk psychology‹, »a set of more or less connected, more or less normative descriptions about how human beings ›tick‹, what our own and other minds are like, what one can expect situated action to be like (...)« (Bruner 1990 : 35). Bruner compares our life to a theatre play: When we enter human life »others on stage already have a sense of what the play is about, enough of a sense to make negotiations with a newcomer possible.« (Bruner 1990 : 34) Only when constituent beliefs of this folk social science are violated narratives are constructed. These narratives explain why in a certain situation one did not act according to the cultural norm. Narratives forge links between the exceptional and the ordinary, they can deal with both. The reasons we give for deviant thought or action are by no means logical but normative interpretations we offer to the world.

Bruner freed me from my worries of how to assess the social side of knowledge within a certain community or locality. The social or cultural is already included in personal stories. With his view on knowledge as cultural codified meanings he also freed me from my worries how knowledge is linked to action. Bruner (1990 : 16) criticises verificationist tendencies in social sciences to allege a discrepancy between what people say and what they actually do. »Saying and doing represent a functionally inseparable unit in a cultural oriented psychology« (p 19) because »the meaning placed on most acts by the participants in any everyday encounter depends upon what they say to one another in advance, concurrently or after they have acted« (p 18). For me as a researcher this means to interpret gardeners' actions in the light of their narrative explanation.

The principal challenge of my interviews was in how to incite gardeners to tell stories about an idea probably unknown to them, biodiversity. Direct questions like »Do you appreciate species richness in your garden?« or »How do you conserve your varieties?« were no solution as these ideas and components of biodiversity originate from science, too. Again, it would have been a scientist who had decided on the importance of topics. On the contrary, the topics mentioned in my interviews were intended to be chosen and structured by the gardeners to the largest extent possible. So my impulse questions were rather broad in scope, all experience connected to gardening was of relevance. For example, I asked for the history of the garden as well as for the personal development as gardener, for the characteristics of a beautiful garden, for relationships to plants or how people deal with problems. Questions were not restricted to gardening but also raised other, probably relevant topics: experiences with nature, environmental problems perceived and judgements on policies. Only ex post the interview passages on nature, its diversity and management were analysed within the context of the overall story.

Interviews were conducted with home gardeners from two different regions in Germany. As a home garden I defined a plot cultivated mainly for private purposes, independent of location (at the house, in a colony, on a field) and patterns of use (fruit and vegetables for home consumption, lawn, trees and flowers for recreational purposes etc.). The study started in the very East at the Polish border, the so-called *Oderbruch* region, about 60 km northwest of Berlin. This area looks back on a rich professional and private gardening tradition. For the last 200 years it has been producing a large share of vegetables for the Berlin market. From a cultural viewpoint the Oderbruch is highly interesting for a study into local knowledge. Until the middle of the 18th century the land was inundated most of the year by the Oder River and therefore sparsely populated. When Friedrich II ascended to the Prussian throne in 1740 he ordered the region to be drained and settled families from all over Europe. The settlers originated from Poland, Saxony, Sweden, France, Switzerland, Palatinate and other southern German states, and their all brought their particular gardening tradition with them. Up to present you find a diversity of garden styles in the 40 different villages that Friedrich II founded. In this research region my study focused on traditional, socially distributed knowledge on gardening. For a number of reasons – some personal, others originating from the field and the research design itself – such a study proofed difficult to conduct.

Therefore a new research region in the former West Germany was chosen. *Rheinhessen* is located at the river Rhine and close to the city of Mainz, belonging to the Land of Rhineland-Palatinate. Still an agricultural landscape producing mainly wine, fruit and crops, for the last 30 years the region experienced urbanisation due to its proximity to the Frankfurt economic area. Today a mixture of commuting workers, Turkish migrants and farm families inhabit the villages and small towns of Rheinhessen. Together with the research region the central research question shifted from the analysis of local knowledge to the personal level of knowledge processes concerning the garden, nature and biodiversity. The results presented below draw on interviews with both Oderbruch and Rheinhessen gardeners.

Before getting to the garden stories I want to make two final remarks on the spot light character of my study, in the sense that it will and cannot describe a complete knowledge ›package‹. One reason lies in the character of knowledge itself, the other in the character of qualitative social research. As sketched in Sect. 3, knowledge is no closed system, but a highly dynamic web of meanings that constantly changes on the personal and social level. So I can only give a punctual, momentary version of meanings gardeners attributed to their experience. The processes, however, in which knowledge is attained, negotiated, transferred and transformed, will be quite similar throughout time. Hence, this study primarily aims at analysing the processes of knowing: How do gardeners know what they know on nature, diversity and their management?

Qualitative research does not aspire generalisability or representativity in the sense of quantitative studies. Qualitative social research, especially in the tradition of ›grounded theory‹, wants to discover the range of possibilities and condensate them into conceptual categories. These concepts form the core of emerging theories. Or, in accordance with Glaser and Strauss (1998 : 47), in contrast to theory-testing methods which aim at identifying regularities and little variation, when generating theory the most different categories on many conceptual levels are desirable. The world is complex and cannot be reduced to common places or linear relationships. In research practice the sociologist must learn how few types of social behaviour can be adequately grasped in already existing theoretical concepts. So, it should be sociology's first task to develop new theories and concepts for largely unexplored spheres of social life.

The theory I am trying to build and ground in the data claims to be one of medium range. This means, it should not be wantonly transferred to other social spheres than the home garden. When gardeners have developed a certain view of nature and a way to act towards living organisms, before and beyond the fence, their attitudes do not necessarily apply in the case of other social groups. Speaking with Glaser and Strauss (1998 : 42ff), my expected result will be a »material theory« developed for a particular empirical field. For a »formal theory« applying to recurring concepts and patterns in society, the gardeners' cases had to be contrasted with cases from totally different material fields, e.g., non-gardeners, farmers, industrial workers, leisure hunters, city people, scientists and so on. Therefore I hope my study to be a first step into the qualitative study of personal knowledge on nature.

5 Stories of difference

> »For sure I read or heard about it. Biological diversity, yes.« – »What do you associate with the term ›biological diversity‹?« – »Well, you know, a community of life: animal, plant, human being. That is what I guess! I do not know what it means!«

> »Biological diversity? That is something like a mixed forest, I would say. That is my first thought, a mixed forest. There you can find everything, deciduous trees and conifers. That is how I understand biological diversity. On the other side it also means bringing plants to the region which do not belong here.«

> »When I hear the term ›biological diversity‹ I think of the monocultures here in our region. But for economic reasons you cannot do anything about it ... « – »Do you think there is biological diversity in your garden?« – »Definitely not, I only grow crops. I want crops only but no flowers.«

> »At least I heard of biological diversity, in principle.« – »What is biodiversity? Or what do you imagine it to be?« – »Well, actually, well, ehm, how I actually want my garden to be, I would say. That is what I think it is. Everything grown all over the place. I do not know whether I got it right or not. Tell me what it is.«

> »To be honest, I have got no idea what biological diversity is.«

In most cases my last interview question on the meaning of ›biodiversity‹ produced unsureness in the gardeners. They seemed to feel uncomfortable about the change of roles taking place. In the hours before they had been the gardening experts, their experience had been in the centre of communication. Then suddenly I confronted them with the language of science and now it was me who was seen as the expert. Communication began to flag.

I had the concept of biodiversity expected to be hardly known in everyday life but its zero familiarity surprised me. Independent of sex, age, education level and garden type, none of my eleven interviewees could tell me in the year 2003 what exactly science and politics understand biological diversity to be. Gardeners were evidently well informed on other globally discussed environmental risks like the thinning of the ozone layer, water scarcity or pollution. Eleven years after Rio and one year after the huge German biodiversity campaign, science and politics had not succeeded in communicating the worldwide loss of species, genes and ecosystems. The unawareness of the politically defined problem, however, does not imply that diversity, as a quality of nature is unknown to the gardeners. Most of them appreciate luxuriance and actively enhance diversity among the plants.

»For me the incarnation of a beautiful garden is individuality. The diversity, one loves it this way, another one loves it that way. The diversity, at every time another blossom, different colours, lots of colours. Not uniform.«

»I grow a lot of dill. The whole village gets dill at my place, for the cucumber pickles. For I do not sow the seeds – as other do – in a row but just throw them into the garden. And then the dill grows wherever it wants to grow. I leave the dill there, even if it grows among the flowers. I don't care. Sometimes it looks rather desolate and chaotic but this is an example where I prefer wildness.«

»I always mix vegetables and flowers. My son, however, wants the strawberries to stand in rows. I cannot do it, in my garden the strawberries grow in the flower patch. Besides, I do not master all the slips, they spread wherever they want.«

»Every year I try new vegetable varieties. Some years ago, I got me ten different kinds of tomatoes. One variety was called Janosch, like the guy with the tiger-duck. I liked the name and the photograph. These were black tomatoes. But I realised that Janosch are not my taste. I am a person who tries lots of things.«

At the same time, gardeners commit themselves to the conservation of plants.

»In the garden of my dreams mainly those plant species grow which are nearly extinguished here in the Rhineland-Palatinate. Not necessarily plants that come from China. There would be for example, a chestnut tree. In the pond I would grow various smaller grasses typical for the region. Furthermore iris, you can still find it here in the wilderness. And lots of lilacs.«

The main motivation for conservation of certain plants has its roots back in the past and is connected to dear memories of people and situations:

»We have got this cherry tree my grandfather planted. A ›glass cherry‹, that is how we call it, white, bright, with a red cheek. This year and last year the starlings came and took them all before ripeness. We did not harvest a single cherry. Should we cut the tree down now? I still see my three children standing round the tree, how happy they were. We took the ladder and picked some cherries for them. Now the idea of cutting the tree hurts. We simply cannot do it.«

Gardeners' knowledge on nature and diversity is, in most cases, woven into their knowledge of the world and their selves, as their experience of nature comes in remembering their own biography. Images of nature are a historical product, influenced by ancestors, episodes and sensual experience. The reasons for choosing plants and a particular level of diversity root in personal history. In general, gardeners do not refer to healthy ecosystems or intrinsic values of plants or biodiversity but to motivations that originate in their daily life. Among the reasons for certain attitudes and certain ways in the handling of nature two bigger sets of determinants can be distinguished. One set covers traditions taken over from parents, grand parents or other ›teachers‹. Whether in the grandfather's old roses, the walnut tree planted by the father who left the family or the ivy transplanted from the mother's grave into the front garden: in the plants beloved people live on.

> »Larkspur is one of my favourites. I had it for all my life. Once my husband said to me, he was just about to deliver the milk, he said to me: ›Today your special friend gave me a love letter for you.‹ This guy had written on a scrap of paper: ›Please come to such and such a place, the larkspur is blooming.‹ So I grabbed the kid, put it on the bike and cycled to the blooming larkspur. And my friend was overjoyed when someone had a look at his flowers.«

Preferences are not only conditioned by personal tradition but also by the times and places of childhood. The interviewees from the former East of Germany for example, showed striking differences in their choice of tomato varieties, depending on the gardeners' age. The pensioners who had grown up in the pre-GDR-era favoured old German breeds, e.g., *Lukullus*. Gardeners in their 30s to 50s, however, still cultivated GDR breeds, especially *Harzfeuer*. The gardeners from the former West were rather indifferent about the varieties, they cultivated what the market offered and liked to experiment.

One 34-year-old woman who grew up in Romania misses the mulberry trees: »The next tree I plant will be a mulberry but I do not know where to get it. We had plenty of these trees down there in Romania. There were long tree-lined roads between the villages. Old trees, huge trees, and they carried plenty of fruit. I always ate enormous quantities. Furthermore we made a lot of liquor.«

The second set of factors, which influence gardeners' nature concept, its management, and the choice of plants includes all kind of sensual, first hand experience. In the most functional cases vegetables and fruit a chosen according to the family's taste and culinary traditions:

> »We still have these old apple trees we planted right after the war, six different kinds. Today there is not much to harvest in autumn but each apple is a pleasure. These new varieties you can buy in the supermarket, they look fantastic I have to admit, but I do not like their taste. There are people however, who prefer the dull taste. I have heard that these apples are not good for making juice. So they want to propagate the old sourish varieties.«

> »I have to add that my wife is from Thailand. She tried to grow a lot of vegetables from over there but most things failed, for example these giant beans. At least we

have pepper, chilli, eggplant and various herbs. My wife she puts peppermint leaves in all dishes, can you imagine?«

Another way of choosing both vegetables and flowers are visual impressions. One 18-year-old male gardening enthusiast grows red Brussels sprouts, white carrots, yellow beetroot, lychee tomatoes and lots of other specialities. He is a convinced conservationist and stresses that these are all old forgotten varieties he gets via an alternative seed exchange service. Another male gardener decides upon the maxim ›big is beautiful‹:

> »I have got giant kohlrabi. Other people do not want it because they think the smaller kohlrabi tastes better. But I do not care, I put it in the soup and use more spices. Two years ago I had one of eight kilograms. Kohlrabi can become even bigger. I am dreaming of breeding a super kohlrabi! I take delight in such things.«

Gardeners also admit beauty to weeds and sometimes leave them among the cultivated plants. One woman says: »I do not weed before I can identify the plant. I wait. I have to see the blossom. If I like the blossom I let it stay.«

The same woman describes gardening as a feast for all senses: »I think the garden calms me. Just watching the plants, I like to do it while watering. I take a close look whether a new blossom has opened in the garden, how things grow. I love to be outside in the garden. Only for feeling the wind at my nose. Seeing how the wind moves the plants. Yes, and how the plants smell. I even love to touch them.«

Out of sensual experience some gardeners have developed quite personal relationships to their plants: »See, these are all my kids, I speak to all of them. I set my heart on each plant or bush or tree. Each of them is part of the miracle of nature. The older you grow, the more you feel it«, explains an 80-year-old women who still keeps a huge garden. She is not alone with her feelings, a women gardener, nearly 50 years younger, gives a similar account:

> »It is really hard when you see a plant dying. For example the *Clematis alpina*, it was so tragic, because ... It is a plant of the alpine vegetation. It had blossomed in fabulous blue, such a wonderful soft blue, and there were blossoms all over. Then the plant budded and I was looking forward to the next year, to the blossom experience. It did not come, it had just died! This was really hard but even a plant once reaches its final age. I feel sad when I look at the poor new plant with which we replaced the old one. I still have the image of this unbelievable blue in my head.«

All interviewees except two have learned gardening in their childhood from their parents or grandparents. Most gardeners explain their activities today with their socialisation: »For me it is natural that I keep a garden. I do not know it any other way.« Or: »I could not live without it.« In most cases, the parents served as models but sometimes the gardeners have left their ancestors' path and developed their own gardening style. One gardener in his late 60s recalls the horrors of childhood: »We always had to help our parents but they did not wake my passion for gardening. They treated us in a nasty way. When we acted clumsy, my father was quick to laugh at us. Although he did not understand very much of gardening. We were also lacking the theoretical instruction. My father just ordered: ›Weed here! Water there!‹« Only at the age of fifty did this man start digging his plot again. The woman from Romania tells another story:

> »Of course I helped with the garden but I never had interest in remembering what you do at what time. I still do not know when you sow what kind of vegetables. As a kid I realised that a certain order was observed, but that was beyond me! And I always had the impression my mother did not even think how she had to do something. When she came home after work, she changed clothes, grabbed for example the seeds in spring and just did it off-hand. And here in my garden I think for hours: ›Does this plant need sun? Can I plant it here? How big will it grow?‹ Such things I must look up in a book. I forget most things, I have to admit. So I have to read over and over again. Fortunately the seed bags nowadays carry instructions. But my mother, she had to collect her seeds. In her case she had to know all the things. It would have also been a shame if she had not known.«

Long-term experience seems to be an asset for successful gardening for one develops a feeling for what to do at what moment without thinking twice:

> »How do you know that the balance of plant and pest tips over?« – »Oh, that is a question of experience. Let us suppose, there is a rare annual plant and all blossoms are full of beetles. From experience I know that the following week they will have eaten all the blossoms and, hence, the plant will produce no seeds anymore. In this case I would use a pesticide. For other plants, for the lotus for example, I use my fingers and remove the lice from the buds. How do I make out what is to do? It is kind of having an eye for something, an overall look. And I know how fast the parasites multiply, and according to the plant's stage I react or wait.«

Though gardeners appreciate to learn from other experienced colleagues – »Copy what the older people do.« – »I often ask the neighbour couple. They are both professional gardeners and I really can trust them.« – they are aware of the importance of own trials.

> »How gardening is learnt best?« – »I think by doing. By trying. Of course you need some knowledge in advance of the plants and the techniques. But in the end gardening depends on the locality. On the soil and the light. Probably it also depends on what one expects from the garden. There are some things I do not have to know in detail. Furthermore I do not want pretentious plants.«

> »When I was a kid I had my own ten square metres. Quite often things failed. For example, my parsley never did germinate. And then my mother told me you have to keep the seeds humid all the time. I tried again and it worked. The failures incited me to learn more about the plants, what they need at all stages of growth. I remember well all the catastrophes, my mother scolded a lot. She would have liked to work my garden herself.«

A recurring topic in the gardeners' narratives is ›difference‹, as they name it. The feature could as well be called diversity, and the gardeners apply it to nature but also to the human world: preferences, knowledge, and modes of action.

»I tried to cultivate this flower over and over again, and each time I was unfortunate. I always bought it in the garden store but I think the plant does not prosper in every soil.«

»It is rather different over the years, depending on weather we can start gardening only in April but sometimes in March already.«

»In general, men are not passionate about flowers, but then there are men who love flowers. You cannot tell, it is quite diverse.«

»Once my neighbour came and brought me flowers that had developed poorly in her garden. And with me they prospered wonderfully.«

»One of my neighbours, the one with the perfect lawn, has a completely different attitude to gardening. The appearance of the garden is extremely important to him. For him it is a tragedy if the lawn does not look green. Once he said the lawn was brownish when it seemed green as always to me. He works the garden in a different way; he uses a lot of fertiliser. Now and then I ask him about a problem but what he answers I rarely put into practice. We simply do not share a common goal.«

Gardeners' statements cited in this section are by far not complete; pages of transcripts could be added. Nor do I claim that these declarations stand for a coherent all-embracing theory on how home gardeners know nature and manage ›biodiversity‹. Theory is still in the making and more interviews will be conducted until theoretical saturation is achieved. Nevertheless, these statements alone challenge the missionary character of both biodiversity research and policy. On their own gardeners have discovered the principle of diversity. Often they actively enhance diversity for a number of reasons irrelevant to science and reaching beyond utilitarism. Knowledge on nature and plants comes along in another shape than in science, in stories and actions.

6 Towards a new culture of knowledge

At least among home gardeners, natural scientists, politicians and conservationists could find numerous allies for biodiversity conservation. Probably the number of everyday ›biodiversity‹ experts – they will name it in some other way – in German society will even exceed the 58 percent of households that dig a garden. Out of this dialogue between science and the public, the exchange of knowledges, new strategies for biodiversity conservation could be formulated. Strategies that are democratically legitimated and rooted in people's life world. A real dialogue however, has not started yet for the word dialogue demands equal partners communicating. Who, in the past exerted more power in biodiversity issues – science or the public, is hard to decide. On one hand natural scientists have been extremely successful in influencing environmental and especially biodiversity policy for the last decades (Haila 1999; Peuhkuri and Jokinen 1999). Conservation interests, justifications and management programmes developed by the scientific community are fixed, to a certain degree, on the formal level, i.e., in both national and international law and policy. On the other hand, the public masses seem unimpressed by biodiversity loss scenarios that thinkers and governors have sketched, might possibly be due to insufficient media coverage or difficult immediate perceptivity of the problem. Public life goes on as if a biodiversity problem has never been formulated.

Science, however, can not expect the public to be interested in a topic hardly perceived as problem in everyday life. So it should be science looking for dialogue and communication but its traditional authoritative self-image impedes courage. For centuries science's monopoly in knowledge production was not questioned. In today's democratic societies, however, the knowing elites have broadened: experts and ›protoexperts‹ have spread in all spheres of life and due to higher education standards scientific findings face a critical, thinking public (Felt and Nowotny 1993; Nowotny 1993). Nowadays science has to produce »socially robust knowledge« (Nowotny) that stands firm to both cognitive testing and social acceptance. Science must admit the public as an equal partner in knowledge creation. Social sciences could play a mediating role but the success of the project will at first depend on the will of the partners involved. In the United Kingdom, biodiversity scientists, amateur naturalists and policy makers are already experimenting with new knowledge coalitions. The British Economic and Social Research Council funds a three-year project called »Amateurs as experts: harnessing new knowledge networks for biodiversity«, located at Lancaster University and the Natural History Museum London. Sociologists and anthropologists monitor knowledge processes inside science and at the science-public interface.

A starting point for the communication of science and ›lay‹ knowledges could be the reflection on how knowledge is produced and represented in respective sphere. Up to the present, natural sciences have delegated this task to the science studies but have hardly ever exercised it within their disciplines. Gardeners, as amateur natural scientists, already practise reflection on their knowledge. They have accepted that their knowledge depends on time and space, and that in situations other than their own, other ways of thinking and acting might be appropriate. By doing so, they have found pragmatic solutions. Reflection on the ways of knowledge creation does not automatically lead to total deconstruction or relativism. It offers the potential however, to become aware of underlying values and personal decisions in our knowledge and helps to overcome claims for pat solutions. If science is earnest in its will to conserve biological diversity it should not blindly reject suggestions originating from alternative knowledge networks.

Gardeners' knowledge on nature could serve as a prototype for science in a second sense: the knowledge consciously includes both, natural *and* social aspects. What gardeners know and do is not only determined by observing nature but also by unique personal trials, socially determined judgements and cultural imprints. A ›great divide‹ of the natural and social does not exist. As Bruno Latour (1998) noted, modern science has enforced this division into nature and culture since its naissance in the 17th century. Currently, natural and social sciences claim their own laws independent of each other. Nevertheless, ›independently‹ produced scientific facts from the laboratory always generate consequences perceptible in society. Who would deny that the first nuclear fission had an impact on the political world order? Or that its trial was driven by prior (political) interests? Whether we are talking about the ozone layer, waste management or the loss of biodiversity – they all show natural, technical and social dimensions that often are tightly woven into each other. Latour warns against ignoring these sociotechnical webs for this very ignorance has led mankind to produce ›monsters‹, technologies that cannot be controlled. To impede the expansion of risks, society – including science – must acknowledge the existence of ›hybrid‹ spaces where the social cannot be separated from nature. Latour (1998 : 59) invites us to use the pre-modern societies that anthropology studies, as an example. These societies did not bring about monsters for they consider gods and culture when interacting with nature. If nature and culture/society are inseparably linked then natural and social sciences would do better when working together in the hybrid space of biodiversity conservation.

All quotations taken from German references and gardeners' interviews were translated by the author.

References

Antweiler P (1995) Lokales Wissen: Grundlagen, Probleme, Bibliographie. In: Honerla S, Schröder P (eds) Lokales Wissen und Entwicklung. Verlag für Entwicklungspolitik, Saarbrücken, pp 19–52

Aristotle, Nicomachean Ethics

Berger PL, Luckmann T (1999) Die gesellschaftliche Konstruktion der Wirklichkeit, 16th edn [The social construction of reality, 1966]. Fischer, Frankfurt/Main

Berkes F (1999) Sacred ecology: traditional ecological knowledge and resource management. Taylor and Francis, Philadelphia

Bögeholz S (1999) Qualitäten primärer Naturerfahrung und ihr Zusammenhang mit Umweltwissen und Umwelthandeln. Leske + Budrich, Opladen

Bruner J (1990) Acts of meaning. Harvard University Press, Cambridge, London

DeWalt BR (1994) Using indigenous knowledge to improve agriculture and natural resource management. Human Organization 53 : 123–130

Drinkwater M (1994) Knowledge, consciousness and prejudice: adaptive agricultural research in Zambia. In: Scoones I, Thompson J (eds) Beyond farmer first: rural people's knowledge, agricultural research and extension practice. IT Publications, London, pp 32–41

Durkheim E (1895) Les règles de la méthode sociologique. Alcan, Paris

Felt U, Nowotny H (1993) Science meets the public: a new look at an old problem. Public Understanding of Science 2 : 285-290

Fleck L (1999) Entstehung und Entwicklung einer wissenschaftlichen Tatsache: Einführung in die Lehre vom Denkstil und Denkkollektiv, 4th edn [1935]. Suhrkamp, Frankfurt/Main

Flick U (1995) Qualitative Forschung: Theorie, Methoden, Anwendung in Psychologie und Sozialwissenschaften. Rowohlt, Reinbek

Flick U (1996) Psychologie des technisierten Alltags: Soziale Konstruktion und Repräsentation technischen Wandels. Westdeutscher Verlag, Opladen

Geertz C (1983) Dichte Beschreibung: Beiträge zum Verstehen kultureller Systeme [The interpretation of cultures: selected essays, 1973]. Suhrkamp, Frankfurt/Main

Gerstenmaier J, Mandl H (2000) Wissensanwendung im Handlungskontext: Die Bedeutung intentionaler und funktionaler Perspektiven für den Zusammenhang von Wissen und Handeln. In: Gerstenmaier J, Mandl H (eds) Die Kluft zwischen Wissen und Handeln. Hogrefe, Göttingen, pp 289–321

Gerster-Bentaya M (1999) Biotop oder Psychotop? Untersuchungen zum Konzept des Naturgartens und zu seiner Akzeptanz im Stadtbereich Stuttgart. Margraf, Weikersheim

Glaser BG, Strauss AL (1998) Grounded theory: Strategien qualitativer Forschung [The discovery of grounded theory, 1967]. Hans Huber, Bern

Haila Y (1999) Biodiversity and the divide between culture and nature. Biodiversity and Conservation 8 : 165–181

Heistinger A (2001) Die Saat der Bäuerinnen: Saatkunst und Kulturpflanzen in Südtirol. Loewenzahn, Innsbruck Bozen

Hellier A, Newton AC, Ochoa Gaona S (1999) Use of indigenous knowledge for rapidly assessing trends in biodiversity: a case study from Chiapas, Mexico. Biodiversity and Conservation 8 : 869–889

Howard-Borjas P (2001) Women in the plant world: the significance of women and gender bias for botany and for biodiversity conservation. Inaugural Address, Wageningen University

Inhetveen H (2000) Wurzbüschel – ein Dokument traditionellen Kräuterwissens von Landfrauen. In: Holl A, Meyer-Renschhausen E (eds) Die Wiederkehr der Gärten. Studienverlag, Innsbruck, pp 196–216

Inhetveen H, Schmitt M, Spieker I (2002) Die Gärten der Frauen. Biodiversität aus Sicht der ruralen Geschlechterforschung. Georgia Augusta 1/2002 : 72–78

Kaiser FG, Fuhrer U (2000) Wissen für ökologisches Handeln. In: Gerstenmaier J, Mandl H (eds) Die Kluft zwischen Wissen und Handeln. Hogrefe, Göttingen, pp 51–71

Kirk JL, Miller M (1986) Reliability and validity in qualitative research. Sage, Beverly Hills

Kuhn TS (1976) Die Struktur wissenschaftlicher Revolutionen [The structure of scientific revolutions, 1962]. Suhrkamp, Frankfurt/Main

Landauer K, Brazil M (eds) (1990) Tropical home gardens. United Nations University Press, Tokyo

Latour B (1998) Wir sind nie modern gewesen: Versuch einer symmetrischen Anthropologie. Fischer, Frankfurt/Main

Long N, Long A (eds) (1992) Battlefields of knowledge: the interlocking of theory and practice in social research and development. Routledge, London, New York

Long N, Villareal M (1994) The interweaving of knowledge and power in development interfaces. In: Scoones I, Thompson J (eds) Beyond farmer first: rural people's knowledge, agricultural research and extension practice. IT Publications, London, pp 41–52
Martin G (1995) Ethnobotany: a methods manual. Chapman and Hall, London
McCarthy ED (1996) Knowledge as culture: the new sociology of knowledge. Routledge, London, New York
Molander B (1992) Tacit knowledge and silenced knowledge: fundamental problems and controversies. In: Göranzon B, Florin M (eds) Skill and education: reflection and experience. Springer, London, pp 9–31
Molitor B (1989) Wirtschaftsethik. Vahlen, München
Noelle-Neumann E, Köcher R (eds) (2002) Allensbacher Jahrbuch der Demoskopie 1998–2002, vol 11. K.G. Saur, München
Nowotny H (1993) Socially distributed knowledge: five spaces for science to meet the public. Public Understanding of Science 2 : 307–319
Percy W (1958) Symbol, consciousness and intersubjectivity. Journal of Philosophy 5 : 631–41
Peuhkuri T, Jokinen P (1999) The role of knowledge and spatial contexts in biodiversity policies: a sociological perspective. Biodiversity and Conservation 8 : 133–147
Polanyi M (1978) Personal knowledge: towards a post-critical philosophy. Routledge and Kegan Paul, London
Polanyi M (1985) Implizites Wissen [The tacit dimension, 1966]. Suhrkamp, Frankfurt/Main
Reichert D (2000) Die schwarze Katze in der Kohlenkiste: Suche nach einer Wissenskonzeption in der Literatur. In: Reichert D, Fry P, Held C, Steinemann U (eds) Wissenschaft als Erfahrungswissen. Deutscher Universitäts-Verlag, Wiesbaden, pp 141–186
Schultze U (1998) Lokales Wissen und Entwicklungszusammenarbeit – Eine Einführung. In: Pasquale S, Schröder P, Schultze U (eds) Lokales Wissen für nachhaltige Entwicklung: Ein Praxisführer. Verlag für Entwicklungspolitik, Saarbrücken, pp 1–56
Scoones I, Thompson J (1994) Knowledge, power and agriculture – towards a theoretical understanding. In: Scoones I, Thompson J (eds) Beyond farmer first: rural people's knowledge, agricultural research and extension practice. IT Publications, London, pp 16–32
Strauss AL (1998) Grundlagen qualitativer Sozialforschung, 2nd edn [Qualitative analysis for social scientists, 1987]. Fink, München,
Takacs D (1996) The idea of biodiversity: philosophies of paradise. Johns Hopkins University Press, Baltimore
Tulving E (1972) Episodic and semantic memory. In: Tulving E, Donaldson W (eds) Organization of memory. Academic Press, New York, pp 381–403
Vogl B (1998) Hausgärten der Mayas: Zwischen Tradition und Moderne. Brandes and Apsel, Frankfurt/Main

Vogl-Lukasser BN (1999) Studien zur funktionalen Bedeutung bäuerlicher Hausgärten in Osttirol basierend auf Artenzusammensetzung und ethnobotanischen Analysen. Final Report BMWV and BMLF, Wien

Warren DM, Slikkerveer LJ, Brokensha D (eds) (1995) The cultural dimension of development: indigenous knowledge systems. IT Publications, London

Watson JW, Eyzaguirre PB (eds) (2002) Home gardens and in situ conservation of plant genetic resources in farming systems. IPGRI, Rome

Wüstner K, Stengel M (1998) Wissen, Wollen und Handeln: Einführung in die ökologische Psychologie. In: Günther A, Haubl R, Meyer P, Stengel M, Wüstner K (eds) Sozialwissenschaftliche Ökologie: eine Einführung. Springer, Berlin Heidelberg, pp 219–280

Aspects of bird valuation in Brandenburg-Prussia: Towards the significance of socio-economic conditions for biodiversity perception between the 16th and 20th century

Johannes Klose

Environmental History Unit, Department of Zoology and Anthropology, Georg-August University of Göttingen, Bürgerstraße 50, 37073 Göttingen, Germany, email to jklose@gwdg.de

Summary. This contribution aims to give insight into the perception of biodiversity in the past by looking at a historical case study. The study aims to illustrate how people in Brandenburg-Prussia perceived and valued birds between the sixteenth century and around 1930. It also aims to identify the driving forces of bird perception to find out about factors determining the perception of biodiversity. The historical information and evidence discussed in the case study comes mainly from research conducted on archival sources held in the former Prussian State Archive in Berlin. The archival sources reveal five aspects of bird perception: (1) Birds have been valued as nourishment or delicacy, as hunting objects and trading goods; (2) Grain-eating and crop-devastating birds have been feared as vermin. Birds of prey have been feared as a danger to game and livestock; (3) Insect-eating birds have been valued for being beneficial to agriculture; (4) Birds have been subject to anthropocentric and non-anthropocentric ethical considerations; and finally, (5) birds have been valued for their aesthetics. All five aspects will be illustrated by giving some historical examples in the three main sections. In the final section, hypotheses on the significance of the five aspects of bird perception will be discussed. The findings show that the perception of birds is strongly dependent on the level of nutrition of the majority of the population. When there was a shortage of grain and other agricultural produce in the eighteenth and the first half of the nineteenth century, most bird species were predominantly valued as nourishment whereas grain-eating birds were feared as vermin in agriculture. It was only once nutrition of the people improved considerably in mid-nineteenth century, that other aspects of bird perception came to the fore. Aims to preserve insect-eating birds as beneficial animals were expressed and gradually realised. Apart from that, humane-ethical and later, bird-centred ethical considerations fostered bird conservation aims. Aesthetic arguments generally played a secondary role and became significant only in the twentieth century. The food shortage during World War I caused a set-back in conservation thinking and brought basic utilitarian aspects of bird perception to the fore again. Otherwise, the diversity of bird species was generally not an aspect worth discussing. Rather, people were aware of the changes in the abundance of song-birds in general and of certain bird species in particular. From my point of view, the perception of biodiversity follows the same driving forces that are relevant for the perception of birds.

Key words: bird valuation, biodiversity perception, Brandenburg-Prussia, fare birds, vermin, beneficial animals, song-birds

1 Introduction

Nowadays, the public attributes natural diversity a high value, worth conserving for its own sake. Non-governmental organisations (NGOs) and the Convention on Biological Diversity (CBD) agreed upon by the governments of numerous nations in 1992 challenge governments and the public to take action on biodiversity conservation. Analysing people's perception of natural diversity in the past and asking for basic principles which determined people's attitudes and actions towards nature in former periods of time, and which may determine their respective values and behaviour in the future, is therefore not only interesting from a historical perspective, but also useful for making conservation policies sustainable in the long-run. This is particularly true as the conservation of biodiversity implies considerable conflicts between ecological, economic and social goals.

This contribution aims to illustrate how people in Brandenburg-Prussia perceived and valued birds between the sixteenth century and around 1930 and, at the same time, wants to show how the socio-economic significance of birds varied over time. The example of birds serves as a case study to figure out the prerequisites and factors determining humans' perception of biodiversity in the last four hundred years. However, due to the character of the archival sources analysed in this research it is necessary to operationalise the term ›biodiversity‹. Therefore, I allude to the diversity of animal and plant species. This is because petitions, claims, complaints and value judgements with regard to certain animal and plant species are the sort of information we can expect in the historical tradition. For obvious reasons, there is no information on the genetic diversity of species manifested in the historical record before the twentieth century. The diversity of ecosystems was generally not discussed either[1]. Overall, the archival stocks contain a huge amount of material about those animal and plant species which had become relevant to people for basically two reasons. Firstly, the mentioned species were relevant as a resource (e.g. for hunting, nutrition and trade) or for their utility. Obviously, resources are subject to arguments about allocation and entitlement. Thus, written documents which express economic interests with regard to animal and plant species are likely to be found in the archival material. Considerations about the utility of certain species may be just another facet of economic interest – I am thinking of beneficial animals – but may also express aesthetic or ethical needs e.g., with regard to song-birds. Secondly, the mentioned species were harmful in some respect and thus challenged people to take action to prevent detriment to their livestock, fields and belongings. This resulted in complaints about ›vermin‹ and ›predators‹, in proposals on how to defeat them and in respective legal action taken by the government. Again, this brought about considerable archival records.

[1] However, the historical documents give some hints of the landscape character of selected areas. In the case of the Oder river marshlands (*Oderbruch*), which were subject to drainage, melioration and agricultural intensification in the eighteenth century, there are detailed planning maps and landscape descriptions which allow some conclusions about the historical ecosystems (cp. Herrmann 1997).

From the considerable amount of records dealing with animals and plants, I selected birds as they are well-documented in the archival material and clearly show a transformation of values and nature perception, due to socio-economic development. The records I investigated are filed in the former Prussian State Archive (i.e., *Geheimes Staatsarchiv Preußischer Kulturbesitz Berlin-Dahlem* = GStA PK), and stem from the late sixteenth century until about 1930. They represent juridical and administrative files and reflect the interests of the people. However, in the early times it is the interests of predominantly the nobility and the government officials. The core reference area of my research is the *Mark Brandenburg* (in short: Brandenburg), the centre and heart of the Brandenburg-Prussian state[2].

This contribution is structured in five main sections. Sect. 2 discusses the research approach, historical background and hypotheses. Sects. 3, 4 and 5 present some historical evidence on the valuation of birds in Brandenburg-Prussia over the course of the centuries. Whereas Sect. 3 deals with birds as game, trading goods, food and delicacy, Sect. 4 discusses the significance of birds as vermin. Sect. 5 illustrates the development of bird conservation for functional, ethical and aesthetic reasons. Sect. 6 will summarise the general findings, discuss the hypotheses posed, and draw some conclusions.

2 Research approach, historical background and hypotheses

In my research approach, I assume that the basic principle of bird perception and the fundamental trend of time-dependent changes in perception depend on the predominating needs of individuals and society related to or affected by the respective bird species. This may be the need of secured nourishment, the need to safeguard against any sort of harm and threat, aesthetic needs and ethical needs. In turn, the predominance of certain needs depends on the socio-economic prerequisites. The needs relevant to the valuation of birds are reflected in the written documents held in the archive. The most significant aspects of bird perception which I recognised in the archival records can be grouped into five groups:

1. Birds have been valued as nourishment or delicacy, as hunting object and trading goods.
2. Grain-eating and crop-devastating birds have been feared as vermin. Birds of prey have been feared as a danger to game and livestock.
3. Insect-eating birds have been valued for being beneficial to agriculture.
4. Birds have been subject to anthropocentric and non-anthropocentric ethical considerations.

[2] When territories came under the administration of the Brandenburg electors and later Prussian kings, the structure of administration changed and hence, there are records stemming from other parts of the kingdom which needed to be considered as well. For example, there are requests, reports and initiatives from the Rhine Land which had an impact on the conservation of birds in the whole Prussian state, including Brandenburg.

5. Birds have been valued for their aesthetics. This applies to small song-bird species in particular.

I expect that the significance of each single aspect varied over time, due to the nutrition of the people and the general circumstances in agriculture. Therefore, it is necessary to outline the Brandenburg-Prussian history of agriculture, population growth and nourishment first. For this purpose, I will summarise the historical development in three time sections: (1) Population growth, dearth and famines (ca. 1700–1850), (2) Agricultural reforms and the improvement of nutrition (ca. 1840–1870), (3) The shortage of nourishments during World War I (1914–1918). After each of the three sections, hypotheses about the significance of birds for nourishment, as vermin and beneficial animal, respectively, and for ethical consideration and aesthetics will be posed.

(1) Population growth, dearth and famines (ca. 1700-1850)

During the eighteenth century, the population grew faster than in previous centuries. In Brandenburg, population density increased from 630 people per km^2 in 1701 to 1,100 people per km^2 in 1740, mainly due to birth surplus rather than to immigration. Population growth led to the resettlement of afore abandoned villages since the end of the seventeenth century, and to the foundation of new settlements in the course of the eighteenth century, particularly since mid-eighteenth century. Such inner colonisation was supplemented by land reclamation. Between 1718 and 1725, the marshlands of the *Havelländisches Luch* and the *Rhinluch* west of Berlin were partly drained and settled. The drainage and melioration of the Oder river marshlands, the *Oderbruch*, since 1747 represented the largest colonisation project in Brandenburg. During the reign of Friedrich II, about 100 settlements were founded all over Brandenburg. Almost half of the settlers were from ›abroad‹, that means from outside Prussia. Immigration into Brandenburg became a decisive factor of population growth only after 1740. Colonisation not only led to population growth but accounted to significant changes of the social structure. In the colonisation areas in the *Neumark*, that is the Eastern parts of Brandenburg, proper farm holdings and small peasant holdings almost balanced each other, whereas in the colonisation areas west of the Oder river, 78 % of the plots were small peasant holdings. In addition, the rise of population was particularly strong among the lower classes in the countryside that had small holdings or did not even possess any farm land. In the parts of Brandenburg west of the Oder river, both aspects accounted to an increase of households with no farm land or small holdings from 47.4 % in 1748 to 55.8 % in 1798. The increase of the lower classes in absolute numbers as well as in relative portion of the total population resulted in growing social disparities. Apart from that, the overall population of the *Kurmark*, that is the Western parts of Brandenburg, increased from 283,566 people in 1701 to 834,080 in 1801 and thus, almost tripled within one hundred years. Despite the fact that agricultural production was increased by land reclamation and the intensification of land use, agriculture was not capable of providing sufficient food for the quickly growing population. As a consequence, prices of grain, legumes, meat, eggs, venison and other produce rose. The dearness affected the growing number of poor people most severely and benefited only a small number of affluent farmers and squires, especially in the fertile regions of the *Uckermark* in northern Brandenburg and the *Oderbruch* after melioration (Materna and Ribbe 1995). Overall, the rising prices for agricultural produce resulted in the deterioration of nutrition. Nourishments which were voluminous, rich in carbohydrates and, despite dearness stayed comparatively cheap in price, such as grain, legumes and later the potato, superseded foodstuffs which were rich in proteins, concentrated and rich in flavour but relatively expensive, such as meat, butter, cheese, eggs and venison. Meat consumption which amounted to more than 100 kg per capita and year in late medieval Germany decreased to less than 20 kg in early nineteenth century Prussia, Saxony and presumably the whole of Germany. Similarly, the consumption of eggs, butter, poultry, venison and wine was much lower in late eighteenth and early nineteenth century than in the late Middle Ages. In the early nineteenth century, when people talked about food scarcity they were referring to the lack of grain, legumes, vegetables. In the following decades, this included the potato, the cheapest of all nourishments. Food scarcity and dearness resulted in an increase of dearth among the masses, such as small peasants and urban craftsmen, particularly however, among the landless and old people, goliards and beggars. In addition to the secular trend of nourishment deterioration, famines were a constant threat people had to face. Famines were mainly caused by crop failures, leading to a significant increase in the price of produce. Landless people who were dependent on buying their foodstuffs in the market suffered the most. Food shortage resulted in typical hunger diseases like dysentery and caused higher mortality. Pre-industrial society was generally not able to overcome food shortage in one region by importing produce from other regions with surplus yields as transportation was slow and costly. Apart from that, feudal interests were often an obstacle to provide the lower classes with food in times of crisis. It seems evident that people tried to utilise all possible sources of comestibles. In the 1816/17 famine in *Württemberg*, south-west Germany, people used wild herbs, roots, vegetables and the meat of horses, dogs and cats to fill their stomachs (Medick 1985 : 101). Most probably, this was also true for the famines in Brandenburg in 1771/72 and 1846/47. It is therefore very likely that fowling was an important and widespread means to improve the insufficient and monotonous menu by an additional source of meat. Similarly, it is very probable that people strengthened their aims to protect their crops from vermin and intensified pest control.

Hypothesis I.1 *I assume that in times of hunger and economic misery, birds are predominantly given material value and mainly valued as food or source of income. Therefore, I presume that the significance of small birds as food and trading goods increased during the eighteenth century, was highest in the last quarter of the eighteenth century and the first quarter of the nineteenth century, decreased afterwards and became insignificant in the second half of the nineteenth century.*

Hypothesis II.1 *I expect that whenever grain and agricultural produce is scarce people strengthen their commitment in fighting against vermin and pests in agriculture. This is because food rivals cannot be tolerated by people who live in constant fear of starvation. I expect that the fight against birds as vermin in agriculture increased during the eighteenth century, was most intense in the last quarter of the eighteenth century and the first quarter of the nineteenth century, decreased afterwards and became insignificant in the second half of the nineteenth century.*

(2) Agricultural reforms and the improvement of nutrition (ca. 1840–1870)

In the 1840s to 1860s, the nutrition of most people in Central Europe improved significantly. In Prussia, this was mainly due to the agricultural reforms which were initiated by the ›land regulation decree‹ from 1811 and the declarations from 1816 and 1821. According to the decree and the declarations, farmers were allowed to acquit themselves of feudal subservience and become free and independent agrarian entrepreneurs. However, this required capital and annuity payments as well as assigning land to the former seignior, often leading to the impoverishment of farmers and to the development of large-scale farm enterprises, respectively. Between 1840 and 1850, most of the farmers became independent by ransoming themselves from socage constitution (Wehler 1996 : 409–428). Apart from, but associated with the social changes in the countryside, the Prussian agricultural reforms comprised mainly four aspects:

(a) The restructuring of the leas:

Due to the declaration of 1821, the three-field-rotation farming system was abolished. The three large fields, which used to consist of land parcels owned by various owners and farmed with the same crop at the same time, were separated and often the land was reallocated. Small and narrow field parcels were joined together to establish quadrate plots instead; these plots were then farmed by farmers individually. Before the realignment, a farm used to own 60 to 100 scattered field parcels but the number of parcels was reduced by about 74 % between 1850 and 1860. For example, a farmer who used to till 96 parcels of land had to encounter only 24 parcels after reallocation. Hence, the farmer saved a lot of time and energy when accessing the fields. This was also made possible by laying out new field paths (Materna and Ribbe 1995 : 417).

(b) The cultivation of badlands:

Directly and indirectly, the agricultural reforms brought about the cultivation of wastelands, leading to an increase in crop land of about 40 to 45 % until 1860/1870. According to some estimates, the agricultural crop land area in Brandenburg even doubled in the respective period of time (Materna and Ribbe 1995). There are a few reasons for the substantial increase of the cultivated land. The common lands formerly mainly used as pastures and for timber-harvesting were separated, transformed into private property and henceforth farmed intensively. This was promoted by the fact that farmers had assigned land to the former seigniors in order to ransom themselves and thus, required new land to cultivate. Additionally, the vast population growth, especially among small peasants and landless people, increased the imperative to cultivate badlands.

(c) The introduction of new crops and improved cropping systems:

The introduction of crops like clover, rape, foreign high-output grain cultivars, and root and tuber crops, especially the expansion of potato cultivation, as well as the intensification of crop rotation resulted in higher yields per hectare. This was furthered by the application of new fertilisers (Guano, Chile salpeter) and drainage measures on manor lands since the 1850s. Moreover, productivity was increased by the reduction of the fallow land portion from about 25 % of the agricultural area in 1800 to 15 % in 1850 and finally to only 9–10 % in about 1870. The potato became an important crop for two reasons: Firstly, its cultivation was adequate on the sandy, dry and rather infertile soils in many places of Brandenburg. Secondly, the yield of calories earned from an area cultivated with potatoes was twice to two and a half times higher than from the same area cultivated with grain. Therefore, the farmland area necessary to sustain a family decreased from seven to eight hectares, including the fallow land which was still required at first, to two to three hectares. Consequently, a lot of small holdings were enabled again to nourish the people who lived on the farm. Potatoes with a small amount of fat, bacon or sausage advanced the basic food of the bulk of rural people.

(d) The expansion of stock farming:

On the manors, sheep husbandry was extended significantly in early nineteenth century and had highest priority in stock keeping until about 1860. This was due to an increasing demand of wool by the Prussian textile industry and due to the exports of wool to England which had been made possible by lifting the wool export ban in 1809 and by reducing the customs duty on the export of wool. On the farms, cattle breeding was extended by the expansion of forage growing and by the shortening of the access routes to the fields which allowed the replacement of draft animals with productive livestock. Cattle breeding was improved not only by providing better forage but also by keeping the cattle in the barn, even during summer. High-capacity cattle breeds were also introduced and interbred. Cattle breeding was lucrative especially since the 1830s, when the demand of animal products increased and prices rose. Because of the growing demand of milk, butter and cheese in towns and cities, particularly in Berlin and Potsdam, the husbandry of dairy cattle was extended near the towns and a ›dairy cattle zone‹ developed 20 to 30 kilometres around Berlin and Potsdam (Materna and Ribbe 1995 : 431). In the Oderbruch, that is the drained Oder marshlands, the mast of oxen was intensified. Pig breeding was of little importance in the first half of the nineteenth century even if the stock of pigs increased. Since then, pigs were almost only kept in barns, rather than pastured in fields or on common grounds, such as meadows, heath lands and woods.

The outcome of the reforms

In summary, the agricultural reforms resulted in strongly boosting labour productivity and gave rise to tremendous yield increases in both tillage and livestock farming. In Brandenburg, labour productivity around 1850 was two thirds higher than in about 1800 (Materna and Ribbe 1995 : 434). On average, the productivity per hectare increased by about 10 % between 1800 and 1830, by 50 % in the 1840s and 1850s and by another 70 % until 1870/75 whereby there were substantial variations at the local level due to the differences in soil fertility and in the expenditures of labour and capital (Materna and Ribbe 1995 : 425). As a result of the agricultural reforms, the general income level in the countryside was rising and at the end of the late 1820s, farms were prospering and even farm workers were enabled to improve their meagre income. As agriculture prospered farmers had enhanced spending power and thus, stimulated trade and handcraft in the rural areas.

The improving income conditions in the countryside, the general yield increases in agriculture and the expansion of fruit and vegetable growing allowed for comprehensive and calorie-rich nutrition. The average meat consumption increased significantly from mid-nineteenth century and doubled per capita and year in the time frame 1800 and 1900 (Teuteberg 1976 : 245). The consumption of sea fish even tripled between 1850 and 1900 due to enhanced catching and conservation techniques, and due to the construction of railways, allowing for quick and relatively cheap transportation. In general, there was a transition from the dominance of voluminous, flavourless and fibre-high nourishments, such as legumes, potatoes and rye bread, to more nutritious and flavoursome nourishments such as meat, cheese, milk and fruit. Overall, dearth was finally overcome in the mid-nineteenth century. Famines caused by crop failures, epizootics, wars, insufficient transportation and by social restraints, and which had threatened people over centuries were ultimately a thing of the past (Teuteberg 1976 : 281ff).

Presumably, the agricultural reforms and thence the improvement of mass nutrition resulted in a change of values with regard to birds:

Hypothesis I.2 *I expect that fowling became less attractive in the 1850s when in general, the nutritional problems had been solved, due to the cultivation of wasteland and the increasing productivity in agriculture, and due to new sources of income in a slowly industrialising society.*

Hypothesis II.2 *I assume that the agricultural reforms initiated in 1811 resulted in waning interest in controlling crop-devastating birds in mid-nineteenth century.*

Hypothesis III.1 *I assume that calls for the protection of insect-eating birds, which were regarded beneficial to agriculture, became successful only once birds had lost* *their significance for the masses as nourishment and as a source of income in mid-nineteenth century. This is because from my perspective, such utility thinking ranks behind resource use thinking.*

Hypothesis IV.1 *I expect that ethical and aesthetic considerations exhorting the conservation or the considerate trapping and shooting of birds came only to the fore when the significance of birds as nourishment and trading goods and as pests in agriculture was declining. I assume that in comparison to the aspects of nourishment, pest control and the protection of beneficial animals, ethical and aesthetic considerations hardly played a role before mid-nineteenth century but became more frequent afterwards. I assume that ethical considerations were predominantly anthropocentric at first, but became increasingly centred on birds themselves (non-anthropocentric) during the second half of the nineteenth century. I expect that ethical and aesthetic considerations became determinant in the bird conservation negotiations at the end of the nineteenth century.*

(3) The shortage of nourishments during World War I (1914–1918)

After the pre-industrial food shortages and famines had finally been overcome in mid-nineteenth century, World War I was the first instance to cause dearth in the developed industrial society of Germany. Dearth during the war was caused by the disruption of the international trade of food and hence the drastically receding food imports from abroad, making Germany solely dependent on German produce. In addition, farmers and farm labourers who were recruited into the army were missing as workers in agriculture and thus it was difficult for women, the youth and old people to maintain the level of agricultural production. Additionally, transportation was hampered by the lack of personnel and capacity and by destruction, making it difficult to distribute nourishments to the cities where they were needed most. Over and above that, the forcing up of prices by dealers and chafferers worsened the nutritional situation, especially of the poor people. As a matter of fact, it was easier in the countryside than in the towns and cities to compensate the lack of food in trade by additional livestock-keeping and fruit and vegetable growing and perhaps, but this question remains open, by fowling. To illustrate the nutrition problems caused by the war I am going to describe the situation in Berlin because here, like in other large cities, dearth and food scarcity were more depressing than in the countryside. At the end of the section, I will make some assumptions on the significance of small birds during the dearth in World War I.

After the beginning of the war in 1914, people feared food scarcity and hence, at the end of July 1914, there was a run into the shops and market halls of Berlin to buy and stockpile food. As a consequence, prices increased and some of the products became scarce. However, only the affluent people could afford to forage huge amounts of food, whereas the poor mainly suffered from the rising prices. On July 31st, the stores in Berlin felt impelled to limit or even cease the sale of nourishments. Despite appeals from the local authorities, stockpiling continued and resulted in a further increase of prices. Only after the first military successes had occurred, the situation normalised and prices began to level out. This was promoted by the fact that people were confident that Germany would soon emerge victorious from the war. However, food was in fact gradually becoming scarce and expensive. This started with bread and finally seized potatoes, milk and butter, meat and all other foodstuffs. On January 25th 1915, the local authorities in Germany were authorised by federal decree to ration bread and flour and on February 22nd, Berlin was the first German city to introduce a ›bread card‹ to guarantee even distribution of bread to all the citizens. Grain became even scarcer and thus, rye bread had to be supplemented with potatoes by at least 5 % in October 1914 and by 10 % in the following months. However, due to the stoppage of foreign animal feed imports and the prohibition of feeding breadstuffs to livestock, farmers started to use potatoes as fodder. This resulted in increasing potato and meat prices. To prevent further price increases the authorities of Berlin established top prices on potatoes. However, this policy referred to small enterprises only; producers and wholesale dealers avoided this policy by selling the potatoes elsewhere, especially in western parts of Germany where no top prices were introduced. As a consequence, there were hardly any potatoes on sale. This induced the Berlin authorities to cease the price policy, leading to doubling potato prices between January to April 1914 and the same time span in 1915. Due to the shortage of animal feed, milk production receded and dairy cows were slaughtered out of necessity. The reduction of milk production amounted to about 30 % between December 1st 1914 and October 1st 1915. Consequently, butter prices almost doubled between October 1914 and 1915. The city council decided to introduce ›milk cards‹ for children, breast-feeding women and ill people. Moreover, the council prohibited the trade of meat and meat-containing foods on Tuesdays and Fridays. However, agricultural produce became even scarcer. This applied to potatoes and particularly to grain since the yields of rye, barley, oat and sugar-beet receded significantly in 1916. In addition, the shortages of meat, butter and other fats were becoming more depressing, leading to the introduction of top prices on beef, veal and mutton by the Berlin authorities in spring 1916. Even sugar became a scarce resource, mainly due to the fact that farmers retrenched sugar-beet growing in favour of grain growing but also due to the rising sugar consumption. Sugar was used as a surrogate for animal feed, it was needed to produce jam which served as a surrogate for butter and finally, sugar was utilised in large quantities for the production of explosives. By spring 1916, nourishments had become twice as expensive as at the beginning of the war. Enforcedly, people in Berlin tried to improve their diet by keeping small livestock in cellars and on balconies, by growing vegetables and potatoes in flower boxes and on all possible grounds, and by purchases and barters in the countryside. In 1916, potato harvests were much lower than in previous years and amounted to only 54.2 % of the yearly average between 1906 and 1915. The potato quantities were not sufficient to supply people until the next harvest in autumn 1917 and thus, the swede was used as a surrogate to fill the gap. The swede was used to supplement bread, to make jam, soup or even ›cutlet‹. It was even used as an ingredient of coffee surrogate, pudding and beer. The peak of dearth and food scarcity was reached in winter 1916/17; as the swede became the principal foodstuff, this winter is known as the »swede winter« (Baudis 1986). After the war, the nutritional situation slowly improved. In December 1918, the amount of bread which was portioned out by the ›bread cards‹ was raised and in spring 1919, the first foreign food imports reached Berlin again. In the same year, rationing of eggs, fish, vegetables and legumes ceased, leading to a rich selection of produce but also to high prices. Apart from a shortage of potatoes in July 1920, the general dearth was overcome (Kaeber 1921 : 100f).

Against the background of the history of nutrition during World War I, I put the following hypotheses:

Hypothesis I.3 *I assume that the significance of small birds as food and trading goods increased again in World War I.*

Hypothesis II.3 *I assume that the significance of birds as vermin increased again in World War I.*

Hypothesis III.2 *I presume that in World War I, the significance of insect-eating birds as beneficial animals receded and gave way to the direct usage of birds as nourishment and trading goods.*

Hypothesis IV.2 *I expect that in World War I, ethical and aesthetic considerations with regard to birds were relegated to second place over aspects of nourishment, vermin and beneficial animal in agriculture, respectively.*

3 Birds as game, trading goods, food and delicacy

In Germany, outside the hunting grounds of the emperor and the sovereigns, the right to hunt was an accessory of freehold property until the end of the fifteenth century. Only at the beginning of the sixteenth century, jurists started to regard game as ownerless items and to consider hunting as a sovereign right, both in compliance with the comprehension of Latin law. In a decree from 1565, the Elector of Brandenburg repeated the prohibition of shooting furred and feathered game and taking the eggs of ducks and other (large) birds in the whole state of Brandenburg. However, he conceded that nobility and town citizens who had landed property in the countryside could hunt on their grounds with rifles[3]. The question of whether or not at this time rural people were allowed to catch ducks, geese and other large birds with snares, nets and traps, remains open. This was probably dependent on the degree of subjection to the landlord and the guidelines the landlord made. According to the decree of 1574, the sovereign set a high fine for the shooting of game birds on his hunting grounds and thus, was seemingly not very interested in what happened outside this area[4]. The ›edict against unauthorised hunting‹ from

[3] GStA PK, I. HA Rep. 9 Allgemeine Verwaltung, R2d, Fasc. 1; as far as I know this is the oldest existing record of hunting regulations in Brandenburg.

[4] GStA PK, I. HA Rep. 9 Allgemeine Verwaltung, R2d, Fasc. 1.

1610 changed the hunting regulations radically. The sovereign declared hunting as seigniorage and for the first time, compiled a list of furred and feathered game species he reserved for himself and his vassals. According to the edict, the shooting of swans, great bustards, capercaillies, black grouse, partridges, hazel grouse, wild geese, cranes, wild ducks and wild pigeons was prohibited and a fine was imposed[5]. This hunting regulation was affirmed by the decree from 1620 which included the egret into the list of game birds[6]. From then on, fowling of all the bird species not mentioned in the 1610 edict and the 1620 decree was free to everyone. However, as fowling and taking young birds and eggs was a common practice which might have jeopardised future trapping yields, the decree from 1615 repeated the prohibition of egg collecting and introduced a closed season for all birds from the beginning of the carnival season until Pentecost[7]. It is questionable however, how rigidly people adhered to the decree. In 1703, when the introduction of a closed season was suggested no one remembered the 1615 decree. Moreover, the sovereign deemed it necessary to release a decree in 1663 repeating the prohibition of taking the eggs of game birds. Interestingly, this decree was published again in 1664, 1670, 1677 and 1680; the 1698 and 1704 editions also restricted collecting pewit eggs to the end of March[8]. There is also other evidence that collecting pewit eggs was widespread and must have been important for the nutrition of rural people.

3.1 The significance of small birds for nutrition

Until the middle of the nineteenth century, small birds were a common part of the diet of most people. Elsholtius (1682) for example, described the quality, flavour and preparation of the meat of 51 native bird species, 30 of which can be regarded as small birds, smaller than the stock dove. Apart from the fieldfare and the field larch, all small bird species were assigned to unrestricted fowling. Most of the large birds were subject to the Brandenburg hunting regulations introduced by the decree from 1610 and thus, were reserved for the seignior and his vassals only. In order to point out the significance of small birds in everyday nutrition, I am going to reproduce a few examples of »fare birds« from Elsholtius (1682):

> »Our starlings are eaten by the ordinary man once they are young. The old ones, however, are not served on fine tables because of their abhorrent taste and the toughness of their meat.«

> »The swallow is a nourishment of the populace and the poor peasants. Their meat is very hot and thus, more convenient to utilise in medicine than for nutrition.«

> »Although titmice do not fill a hungry stomach they are easy to digest and a healthy foodstuff, however, rather for the poor than for the rich people.«

---

[5] Mylius (1736, vol. IV, sect. 1, chap. II, col. 523–526).

[6] GStA PK, I. HA Rep. 36 Hof- und Güterverwaltung, Nr. 2254, Holzordnung, pp 53–58.

[7] GStA PK, I. HA Rep. 9 Allgemeine Verwaltung, Q3a, Fasc. 5.

[8] GStA PK, II. HA Generaldirektorium, Abt. 33, Generalia Tit. XV, Nr. 2.

> »As far as the warbler is concerned neither its appearance, nor its singing, nor its meat, of which there is only a very small amount, can be praised.«

In contrast to this appraisal, the following bird species were awarded high nutritional value:

> »The siskin becomes very fat in winter and is convenient to eat. Ravenous people eat the siskins completely, with bones, meat and bowel.«

> »The linnet is used for nourishment, similar to the other small birds living in the wood.«

> »The meat of the woodpecker is not too hard but, without doubt, slightly tough; however, it is not bad in taste and especially in winter, when the woodpecker is fattest the bird is convenient to eat; this applies to the small woodpecker species in particular.«

Sparrows were not only hated as vermin (see below) but also despised in cuisine:

> »Sparrows shall not come into our bowls because their meat is not only disreputable due to its strong hotness but is also indigestive and makes incontinent.«

There is however, evidence that sparrows were eaten as well.

3.2 Some characteristics of bird markets

Until 1867, it was common use to sell and buy small birds at the weekly markets in towns. As far as birds were not subject to the hunting regulations they were assigned to unrestricted fowling. Fowling was a customary law and as such, was only codified in the ›General Prussian Land Law‹ in 1794[9]. Eventually however, fowling was confined by the bird conservation decrees of some towns in mid-nineteenth century and almost abolished by the bird conservation decrees of Berlin and the district councils of Potsdam and Frankfurt/Oder in 1867[10]. Until then, bird markets were very common in Prussia, probably however, losing significance in the 1850s as in the case of Berlin (Ruß 1882 : 8–11). The following list of the most important characteristics of bird markets illustrates the socio-economic significance of the markets and the amount and types of bird species sold:

1. In the bird markets, a large variety of bird species, ranging from small songbirds to large raptors, was offered for sale. The birds were traded for both nourishment purposes and cage-keeping. Despite the huge variety of bird species on offer, only a few species were caught in large quantities and sold for nourishment purposes. Apart from fieldfares and field larches which made up the largest portion of the sold fare birds in mid-nineteenth century, many of the bird species listed by Elsholtius (1682) were offered at the markets but

[9] *Allgemeines Landrecht für die Preußischen Staaten*; cp.: Hattenhauer and Bernert (1994).

[10] GStA PK, I. HA Rep. 87 Ministerium für Landwirtschaft, Domänen und Forsten, B Nr. 19122.

amounted to rather small quantities. Birds offered for cage-keeping were represented by a huge diversity of species but constituted only a very small portion of the overall bird quantities on sale.

2. Since the 1820s, collecting eggs, nests and clutches was a widespread hobby, especially among the youth of affluent town citizens but also among adults[11]. Collectors aimed at getting the eggs, nests and clutches of many different bird species and thus, bird traders tried to meet the demands of the collectors. The traders also benefited from the fact that many eggs broke when they were blown out.
3. Bird traders mainly belonged to the lower classes and generally stemmed from the countryside. Landless and old people as well as poor women from the countryside tried to improve their income by selling not only herbs, mushrooms and maybe home-grown fruit and vegetables, but also dead and alive birds and eggs from song-birds. There is also evidence that old invalid soldiers tried to make a living from selling nightingales which they had to catch abroad as nightingale catching was prohibited in Brandenburg since 1686.
4. Until the 1850s, hardly anyone questioned the practices of fowling and bird trading. Even if some towns in Brandenburg prohibited fowling within the town area since the 1830s because it might lead to the brutalisation of the youth, the trade of birds and eggs caught and collected in the countryside was in no way restricted. This shows that among town citizens, birds were still highly valued as nourishment and delicacies in the 1860s. Apart from that, parts of the affluent youth and some adults were interested in buying eggs, nests and clutches even if this was not accepted morally by society. In the countryside, bird fowling and trading helped to improve the nutrition of the poor and constituted a significant additional source of income, respectively. Henceforth, it seems obvious that no rural county prohibited bird fowling. Rural traders also profited from the interests of the collectors.

3.3 Highly valued in cuisine: The field larch and the fieldfare

As small birds were generally assigned to unrestricted fowling, archival tradition is very marginal. The field larch and the fieldfare are two exceptions; they were not only subject to the hunting regulations but also caught in comparatively large quantities and thus, archival records are considerable.

The *field larch* was highly valued as a delicacy. Wrapped in ham and then roasted on a spit field larches were preferred in particular if they were sapful and fatty. In the eighteenth century, the largest quantities of field larches were caught around Leipzig in Saxony. In October 1749, some 403,455 field larches were brought to the Leipzig market and mainly exported into other regions of Germany, such as Westphalia and Lower Saxony (Flemming 1749 : 199/250). The yearly catching yields may have been twice as high as most of the birds were caught in October,

[11] GStA PK, I. HA Rep. 87 Ministerium für Landwirtschaft, Domänen und Forsten, B Nr. 19998, Bd. 1.

but the months of September and November were also part of the catching season. The second largest catching yields were obtained around Halle/Saale. Every year, some 180,000 to 240,000 field larches were caught in the surroundings of Halle and then sold at the town market of Halle (Dreyhaupt 1750 : 416). In Brandenburg, field larches were caught in the surroundings of Nauen (ca. 40 km west of the centre of Berlin) every autumn to provide the royal court kitchen[12]. Field larch catching in the Nauen area demanded much manpower but amounted to relatively small catching yields. The municipality of Nauen had to provide thirty people to wander over the harvested fields and drift and catch the field larches with large nets. Despite the complaints of the two royal huntsmen and the royal hunting master, the town generally sent children, often younger than ten years, to fulfil the duty. Two neighbouring communities had to supply a cart and two menials every day to transport the catching nets and the caught birds. Another community was ordered to deliver the caught birds to the village of Wustermark from where they were brought to the royal court kitchen in Potsdam. Apart from the royal huntsmen, no one was paid for their services. Between 1767 and 1791, about 2,000 field larches on average were caught per year. Obviously, the catching quantities were very small compared to those obtained in the surroundings of Leipzig and Halle. After complaints from the citizens of Nauen about the encumbering hunting services, field larch catching finally ceased. Officially, the royal chamber argued that not only had the field larch catching in the surroundings of Nauen been too expensive, but also that the Nauen field larch was poor in quality. Even if this second argument was put forward as an excuse, the Nauen field larch stemmed from an area with infertile soils which were made responsible for inferior meat quality, whereas fertile wheat-growing areas such as the Leipzig and Halle regions were regarded ideal to make the larches fat and tasty. Most probably, the royal court kitchen was then provided exclusively with field larches caught in the surroundings of Leipzig and Halle. In these regions, field larch catching must have receded in the second half of the nineteenth century. In Brandenburg, it was banned by the bird conservation decree of the Frankfurt district in 1867. In 1904, the field larch was not regarded as a game animal in Prussia anymore and thus, protected by law.

Fieldfare trapping[13] required less equipment and only one person and therefore, the fieldfare was generally caught in much smaller quantities but at various places. Thus, it is not possible to estimate the overall trapping quantities. Whereas field larches were caught with large nets drifted over the fields by a lot of people, fieldfares were trapped in snares (*Dohnen*) furnished with rowan berries as a decoy and placed in trees in the wood. To obtain higher trapping yields the fowler used to place a number of such snares in a row (*Dohnenstieg*). However, fieldfare trapping was very time-consuming as the snares needed to be settled and equipped with decoy regularly. Apart from the royal forest personnel who had the regular right to trap fieldfares, fieldfare trapping and trading was a source of income. This was es-

[12] GStA PK, I. HA Rep. 36 Hof- und Güterverwaltung, Nr. 2276.

[13] GStA PK, I. HA Rep. 87 Ministerium für Landwirtschaft, Domänen und Forsten, B Nr. 19986.

pecially the case for the lower classes in wooded mountain areas where agriculture was hardly possible and other activities such as working in mines was badly paid. Therefore, fieldfare trapping played a significant role in the *Harz* mountains and in the Silesian *Riesengebirge* mountains but was less important in Brandenburg. In the mountain areas, fieldfare trapping had a certain economic relevance even at the end of the nineteenth century. By this time, the fieldfare had become an expensive delicacy sold in gourmet shops. The rising prices were due to the time-consuming trapping, the extension of the closed season and to the apparently receding trapping yields the contemporary witnesses bemoaned. Fieldfare trapping was finally banned by the amended German Bird Conservation Act in 1908.

4 Birds as vermin

According to archival material, the control of pests was not a matter for the government and the authorities to organise and govern before the eighteenth century. This was different in the case of predators, such as the wolf, the marten, the lynx and the otter, which were to be haunted and exterminated since the enactment of the ›timber order‹ in 1622[14]. Certainly, the predators were haunted even before this, not only by the huntsmen who worried about their game but also by farmers and stockmen who feared for their livestock. However, pest control policies by the state only started on January 26th 1701 when the ›Brandenburg fire order‹ was released which surprisingly also contained a paragraph on the extermination of the sparrow[15]. Apart from the sparrow, the crane and the great bustard became subject to the royal pest control and extermination policies of the eighteenth century. In the following paragraph, I will discuss the significance of these species as ›vermin‹ in agriculture, considering land reclamation and the worsening level of nutrition in the course of the eighteenth century. Incidentally, according to the decree of January 19th 1718, birds of prey were to be exterminated by huntsmen[16]. As this regulation is mainly interesting from the perspective of hunting and is not associated with the history of agriculture and the nutrition of the masses, the 1718 decree and its succeeding regulations shall not be discussed here.

4.1 The control of the sparrow: Sensible from an economic perspective?

According to the received opinion, the sparrow was very harmful to fields and gardens. The bird eats seeds and germ buds and especially in autumn, flocks to the grain fields, approaches the spikes and picks the grains; sometimes ripe grains fall out of spikes just when the sparrow arrives. This may result in significant harvest losses. Sparrows nested in barns and lofts where the grain was stored and thus, were a thorn in the farmers' side. They also like grapes, cherries and currants

[14] GStA PK, I. HA Rep. 36 Hof- und Güterverwaltung, Nr. 2254. Cp. also Herrmann (2003).
[15] Mylius (1737 : vol. V, sect. 1, chap. II, no. VII).
[16] GStA PK, I. HA Rep. 36 Hof- und Güterverwaltung, Nr. 2254.

(Gatterer 1782 : 414). It is self-evident that people were convinced that diminishing the sparrow population would save their produce. This was much the same everywhere else in Germany which caused the governments to introduce policies to reduce or even exterminate sparrows. According to the ›Brandenburg fire order‹ from 1701, a farmer was obliged to deliver twelve sparrow heads per year to his authority; a small peasant was urged to deliver eight heads and all the other rural residents, such as inliers, shepherds and millers who were generally landless, were expected to bring six sparrow heads to their authorities. In the *Prignitz* landscape, it appeared to the authorities in 1717 that the sparrows would proliferate again. Therefore, subjects were urged to strictly obey the 1701 order and deliver the requested amount of sparrow heads (Mylius 1736[17]). In a request to the Prussian king from September 6th 1721, the chief executive of the *Oberbarnim* district argued that people had not met their obligations to deliver the sparrow heads. He suggested the release of a revised sparrow decree containing a penalty for everyone who did not fulfil the requirements[18]. This made the government release a revised decree on December 11th 1721 which took up the penalty for the next six years[19]. On January 8th 1731, another sparrow decree was released[20]. According to this document, complaints about the damage caused by sparrows had become more frequent among rural people. Therefore, the decree requested the rural people to deliver the amount of sparrow heads as appointed in 1701, but also obliged townspeople who were owners of a garden or field to deliver two sparrow heads per year. The authorities were also asked to make compilations of the delivered heads. On June 22nd 1744, the measures were further tightened by the release of a third decree:[21] In towns, the owners of a field were requested to deliver twelve heads, and gardeners and winegrowers even fifteen sparrow heads per year. The sparrow control policy was implemented until 1767 (Herrmann 2003 : 53).

There are three possible explanations why the regulations to control the sparrow had been tightened by the decrees from 1731 and 1744:

1. The sparrows really proliferated. The reason for this is improved habitat conditions caused by the increasing population density in connection with the extension of the agricultural area over the course of the eighteenth century. Due to a habitat-driven increase in the sparrow population, the problem of vermin in agriculture might actually have gained weight in the eighteenth century.
2. Despite land reclamation and intensification, the pressure on agricultural resources, particularly on grain, increased because of population growth. Therefore, vermin could be tolerated less than before (cp. Herrmann 2003 : 52). The

[17] Mylius (1736 : vol. III, sect. 1, no. CXLII, col. 377–380).
[18] GStA PK, II. HA Generaldirektorium, Abt. 14, Kurmark Materien J-Z Tit. CCLXVIII Nr. 1, Vol. 1.
[19] GStA PK, II. HA Generaldirektorium, Abt. 14, Kurmark Materien J-Z Tit. CCLXVIII Nr. 1, Vol. 1.
[20] GStA PK, I. HA Rep. 36 Hof- und Güterverwaltung, Nr. 2257.
[21] GStA PK, II. HA Generaldirektorium, Abt. 14, Kurmark Materien J-Z Tit. CCLXVIII Nr. 1, Vol. 1.

1722 decree to control the crane, which was made responsible for crop damages, underpins the argument that the increasing shortage of produce led to the intensification of pest control. This is reinforced by the fact that the closed season was cancelled for wild ducks, wild pigeons, cranes and other migratory birds by declaration in 1728, as they may become harmful to the fields at seedtime[22].

3. The absolutistic state which aimed at the regulation of spheres of the people inter alia tightened the legal requirements for the control of the sparrow. This explanation seems plausible as at about the same time, namely in 1718, the control and extermination of raptors was regulated by law.

From my point of view, all three explanations seem plausible. However, the second explanation may be accredited the most weight since agriculture did not have the potential to raise the yields of produce as much as the growing population desired. The shortage of produce resulted in the increase of prices and in dearth, leading to the relentless fight against vermin such as the sparrow.

One question remains to be answered: Why was the sparrow control policy abandoned in 1767 when the problems of dearth were still increasing, reaching a climax in the famine of 1771/72?

People obliged to deliver a certain quantity of sparrow heads per year and the authorities responsible for the control policy may have recognised that all the time-consuming sparrow trapping and killing did not help to reduce the sparrow problem. Despite the fact that ecological knowledge was still very rudimentary, people knew that sparrows could proliferate very quickly and soon make up the original population size, particularly as they breed four times a year and on average, lay five eggs per brood. Presumably however, they did not reflect that killing a small portion of the entire population improved the nutritional conditions of the remaining sparrows, making them even fitter to compensate the losses of the meta population by rearing offspring (Herrmann 2003 : 52). In any case it is remarkable that there is no archival record of negotiations about reintroducing a sparrow control policy during the serious famine in 1771/72.

4.2 The control of the crane: Who encroached upon whose habitat?

According to the hunting decree of 1610, cranes were regarded as big game, the hunting of which the sovereign reserved for himself and some of his vassals. In 1713 however, the crane was depreciated to small game in order to increase the royal revenues by leasing the hunting right to huntsmen[23]. Shortly after that, on October 3rd 1722, King Friedrich Wilhelm released a decree allowing everyone to shoot and catch the crane because for a while, the bird had appeared frequently at various places in Brandenburg and had caused a lot of damage to the countryside,

[22] GStA PK, I. HA Rep. 36 Hof- und Güterverwaltung, Nr. 2257.

[23] GStA PK, II. HA Generaldirektorium, Abt. 33, Generalia Tit. XV Nr. 3.

especially to sown fields[24]. According to the declaration regarding the closed season from December 22nd 1728, the crane was regarded as game animal again and thus, was only to be shot by huntsmen[25]. However, shooting was allowed even in the closed season, a regulation which was valid until 1904. There seem to be no further records on the crane in the archival material. As the 1728 regulation allowed the reduction of crane numbers throughout the year wherever it was regarded necessary, there was unlikely to be any requests asking for more effective means to control the crane. This makes it difficult to estimate whether the significance of the crane as vermin actually increased in early eighteenth century, leading to the decrees from 1722 and 1728. It is also possible however, that the control of pests and hence, the reduction of the crane, was a fashionable policy issue of the absolutistic state at that time: The policies of exterminating the sparrow since 1701, of reducing raptors since 1718 and the allowance to shoot migratory birds by the decree from 1728 seem to underpin this argument. However, looking at the habitat requirements of the crane on one hand, and on the transformation of landscape since early eighteenth century on the other hand may shed light on this question.

In Central Europe, the crane settles in areas with plenty of water, such as marshlands, wide-stretching bog, moor and fen areas, and silting-up zones of lakes and ponds. When searching for a suitable breeding area the crane makes no difference between open fields and areas covered with hedges, copses or even wood but requires wetlands for its foraging. Nowadays, most of the crane hatcheries in Brandenburg are situated in the marshlands along and around the Havel river, especially in the *Havelländisches Luch* and the *Rhinluch*, in the lowlands near the towns of Friesack, Rathenow and Pritzerbe, in the marshlands of the Spree and the Warthe rivers, that is the *Spreewald* and the *Warthebruch*[26], respectively (Makatsch 1959 : 22). Crane habitats are jeopardised by drying up and meliorating the wetlands. Therefore, the crane population may have decreased due to the melioration measures of the eighteenth century, particularly due to the drainage of the Havelländisches Luch and the Rhinluch since 1718 and the melioration of the Oder marshlands, the Oderbruch, since 1747. The same holds true for the drainage and land reclamation measures during the nineteenth and twentieth centuries. However, melioration also created new crane habitats. In order to gain grasslands flat lakes were drained until about mid-nineteenth century. As a consequence, lakeshores were silting up and some of the lakes were even drying up, both leading to the creation of moors. Only the vast drainage measures since the 1960s led to a significant decrease in the crane population (Mewes 1995 : 72ff). Overall, the crane is quite flexible in adjusting to the landscape changes. According to Mewes (1995 : 85f), the crane may reach even larger populations in the intensely farmed cultural landscape than in unspoiled natural ecosystems. He assumes that in early nineteenth century, despite rather extensive land use practises, crane stocks and population density were lower than in the 1990s. Against this background, it is not

[24] GStA PK, II. HA Generaldirektorium, Abt. 33, Generalia Tit. XV Nr. 3.
[25] GStA PK, I. HA Rep. 36 Hof- und Güterverwaltung, Nr. 2257.
[26] Since 1945, the Warthebruch belongs to Poland.

possible to decide whether eighteenth and nineteenth centuries' landscape transformation resulted in an increase or decrease of the overall crane population in Brandenburg. Presumably, melioration had very different effects from region to region. Therefore, the significance of the crane as vermin can not be estimated by a suppositional rise or fall in the crane population.

However, another aspect seems more plausible to explain why since 1722, the crane was perceived mainly as vermin, rather than as a valuable game animal. Inevitably, crop damages caused by the crane increased to the same degree as people ›invaded‹ the habitats of the crane by meliorating wetlands and transforming badlands, grasslands and pastures into arable farm land. This is especially true in those areas which the cranes use as gathering places in autumn before they start their migration into the winter quarters. From this perspective, it seems coherent when the authors of the 1722 decree argue that for sometime, the bird had appeared very frequently and had caused a lot of damage to the countryside, especially to sown fields.

4.3 The control of the great bustard: What made the crop-devastator proliferate?

Similar to the crane, the perception of the great bustard changed during the eighteenth century. At first however, hunting interests dominated. According to the hunting decree of 1610, the great bustard was regarded as game bird which only the sovereign and his vassals were allowed to shoot. The decree from May 12th 1668 pointed out that the great bustard was regarded as big game, the hunting of which was a privilege for very few[27]. This was due to the fact that the bustard is a big bird but more importantly, that its meat was highly valued as a delicacy. The latter however, was only true with regard to young birds whereas old bustards were condemned because of their stiff and tough meat (Gatterer 1782 : 181).

Despite the high valuation as a delicacy, the great bustard was feared as vermin. On March 6th 1736, the Chamber of the domains in the dukedom of Magdeburg complained that the bird proliferated and caused huge damage by eating away rye and other winter seedlings in springtime. Therefore, the Chamber requested the Royal Directorate to allow royal huntsmen and the tenants of royal hunting grounds to shoot the great bustard. The Directorate was only willing to agree if the tenants felt up to paying an additional fee. However, there was a lack of interest from the tenants. Apparently, neither the harmfulness of the great bustard nor the hunting interests were strong enough at that time to pursue a new policy[28]. In 1748, the subjects, all the tenants of outlying estates and the counsel of the county of Biegen near Frankfurt/Oder complained that the great bustards had proliferated so intensely that they virtually covered the fields. The Biegeners requested the Chamber of the domains in the *Kurmark Brandenburg* to decree that the great bustards may be shot by the royal huntsmen in order to rid the arable land of this vermin. When presenting the issue to the Royal Directorate, the Chamber supported the request

[27] GStA PK, II. HA Generaldirektorium, Abt. 33, Generalia Tit. XV Nr. 3.
[28] GStA PK, II. HA Generaldirektorium, Abt. 33, Generalia Tit. XVI Nr. 15.

of the Biegeners and proposed to reduce the selling price of the shot bustards to boost sales. In turn, the Directorate suggested that this regulation should be introduced in all parts of the country and made a respective submission to the king. As Friedrich II agreed, a respective decree was sent to the Chambers of the *Kurmark*, *Neumark*, Pomeranian and Magdeburg provinces[29]. However, this solution was apparently not sufficient to reduce the great bustards. In 1751, the deputies of the Kurmark knightage complained that the bustards had proliferated in such a way that the subjects of those landlords who did not have the right to hunt big game suffered a lot as the birds caused considerable damage to the grain fields. The deputies requested the king to instruct the forestry offices to shoot the bustards more frequently. Thereupon, the Directorate asked the Kurmark Chamber to make proposals on how to prevent further proliferation of the great bustards. The Chamber proposed to exclude the bustards from the list of big game animals and consider them for small-game hunting. The Chamber argued that with this scheme further proliferation of this ›harmful game‹ could best be prevented as there would be a lot more huntsmen authorised to chase the birds. King Friedrich II agreed and thus, a respective order was sent to all the chambers of the country on October 13th 1751[30]. In 1770, the counsel of the county of Cottbus complained that the bustards not only proliferated from year to year but caused tremendous damage to the crop fields, especially in the closed season. He therefore asked the king to allow bustard shooting even in the closed season. The king complied with the request and thus, huntsmen in the county of Cottbus were conceded to diminish the bustard population even in the closed season[31].

All complaints argued that the numbers of great bustards had increased. Obviously, this statement is a convincing argument to pursue one's request to reduce a harmful vermin but perhaps, pest control in this case is rather a vehicle to put through one's nonofficial hunting interest. Therefore, having a look at the habitat requirements of the great bustard and at the outcome of land transformation during the eighteenth century may help to estimate whether the significance of the bustard as vermin really increased in the eighteenth century. The existence of clear and spacious open land represents a central habitat requirement as the great bustard loves to straggle up to 12 km a day and thereby feels hampered and threatened by earth walls, hedgerows and copses. Over and above that, the bustard prefers lush meadows (Gewalt 1959 : 20–24). The glacial valleys and the floodplains and marshlands along and around the Elbe, Havel, Oder and Warthe rivers make up excellent bustard habitats once they are drained, cleared from woods and copses and transformed into meadows, pastures and arable land. The drainage, clearing and melioration of the Havel marshlands, the Havelländisches Luch and the Rhinluch, since 1718 created spacious open land, ideal for the great bustard to settle. The few remaining single trees, small copses and diaphanous tree rows do not hamper the

[29] GStA PK, II. HA Generaldirektorium, Abt. 33, Generalia Tit. XVI Nr. 15.

[30] GStA PK, II. HA Generaldirektorium, Abt. 33, Generalia Tit. XVI Nr. 15; GStA PK, II. HA Generaldirektorium, Abt. 33, Generalia Tit. IV Nr. 14.

[31] GStA PK, II. HA Generaldirektorium, Abt. 33, Generalia Tit. XVI Nr. 15.

bustard's need to overview the countryside. Similarly, the cultivation of the lower Oder marshlands, the *Niederoderbruch*, since 1747 unintentionally created anthropogenic bustard habitats. Due to land reclamation, the overall area of grassland in Brandenburg increased significantly (Bekmann 1751 : 679). This again is evidence that spacious open land, ideal for the great bustard to settle, increased in the eighteenth century.

In summary, the complaints of contemporary writers about the increasing numbers of bustards seem justified. Hence, the vermin problem with regard to the great bustard was really aggravated during the eighteenth century. However, the depreciation of the bustard to small game in the whole country in 1751 and the exclusion from the closed season in the Cottbus county in 1770 were sufficient means to control the bird.

4.4 Concluding remarks

The examples of the sparrow, crane and great bustard show that vermin were increasingly perceived as a problem during the eighteenth century. This was partly due to the growing shortage of produce but was also a result of land cultivation. Apparently, the crane advanced as a significant vermin only once people encroached upon its habitats; the great bustard seems to have proliferated when land reclamation created anthropogenic habitats. The example of the sparrow may support best the notion that people increased their aims to control pests when agricultural produce, especially grain, became scarce and expensive.

5 The development of bird conservation for functional, ethical and aesthetic reasons

Often, bird conservation was fostered for diverse reasons, thus it is useful to discuss the historical development of bird conservation aims altogether. However, it is possible to divide the history of bird conservation in Brandenburg-Prussia into six phases. Whereas bird conservation in the first two phases was justified by aesthetics and considerations of utility, respectively, the third phase is characterised by a bundle of conservation motives. In the second half of the nineteenth century, functional reasons, that is the conservation of insect-eating birds dominate the conservation discourse (phases four and five). The last phase marks a temporary set-back in bird conservation caused by the hardships of World War I. The phases of bird conservation will now be discussed in detail.

5.1 The conservation of the nightingale and the thrush nightingale (since 1686)

The conservation of the nightingale is the earliest example of bird conservation which was exclusively justified by aesthetics. In 1685, a citizen of Halle/Saale sent a petition to the Elector of Brandenburg alleging that the bird was caught in large quantities in the surroundings of Halle. He suggested the conservation of the nightingale as her singing was »an adornment of the countryside«[32]. On August 25th 1686, the conservation of the nightingale became legally binding by decree. The decree prohibited nightingale trapping within the boundaries of the *Kurmark Brandenburg* but allowed the trade of nightingales which were caught abroad. A few years later, the thrush nightingale was included in the conservation decree. As the nightingale was highly valued as a song-bird to keep in a bird cage nightingale trapping and trading was a widespread and lucrative business. For poor people, this was a highly welcome opportunity to earn money. However, despite a lot of requests for special permits for nightingale trapping in Brandenburg in the eighteenth and nineteenth century, the royal authorities stuck to the conservation decrees and made no single exception. They did however, allow the trade of nightingales which were awarded a certificate of foreign origin. Nightingales which were imported into Brandenburg had mainly been trapped in Saxony, Bohemia and Poland.

5.2 Means to overcome a caterpillar calamity: the conservation of insect-eating birds (1792–1802)

The first aims to conserve ›beneficial‹ birds, i.e., caterpillar-eating birds, found their way into official correspondence at the end of the eighteenth century[33]. On the occasion of a pine caterpillar calamity in various areas in Prussia in 1791 and the following years, forestry officials demanded the conservation of insect-eating birds, such as starlings, woodpeckers, magpies, and especially crows and jackdaws. Some of them even fostered the conservation of all song-bird species. In order to curb the pine caterpillar calamity, crows and jackdaws were protected by decree in 1792 as they »belong to the enemies of the caterpillars, too«. Hence, the 1744 decree which requested huntsmen to shoot twenty four crows a year, as the birds are harmful to small game, was abrogated. Moreover, as the forest officials recognised that in those areas of the forests where there were ant-hills, pine trees were spared from the caterpillars and remained ›green‹, ants were also protected by decree in 1792. According to the report of a huntsman in 1794 however, the conservation of crows resulted in their proliferation and thus, small game was endangered more than before; the huntsman also stated that the caterpillar calamity had abated and therefore he demanded the conservation of the crows to be cancelled. King Friedrich II resolved that crows could be shot again in those areas where the calamity had ceased. However, the pine caterpillar calamity must have worsened in 1799, inducing the Royal Forestry Department to draw up a circular in 1800 calling for the conservation of insect-eating birds in all the forests where the pine caterpillar had become a plague. Interestingly, the circular even requested the conservation of fieldfares and other thrush species as well as the conservation of waxwings.

[32] GStA PK, II. HA Generaldirektorium, Abt. 33, Generalia Tit. XV Nr. 7. All the following incidents are accounted in the same file.

[33] GStA PK, II. HA Generaldirektorium, Abt. 33, Generalia Tit. XIV Nr. 5. All the following incidents are accounted in the same file.

This shows that forestry officials seriously feared for their pine stocks and thus, deferred fowling interests to aspects of utility. Only when the calamity had ceased in most parts of Prussia in 1802 was the prohibition of fowling fieldfares, thrush species and other song-birds revoked. Furthermore, crows and jackdaws were regarded predominantly harmful to small game and especially to small game birds and thus, were again haunted by the royal forest officials.

5.3 First approaches to permanent bird conservation: town decrees and school curricula (1813–1858)

In the first half of the nineteenth century, especially since the 1820s, collecting eggs, nests and clutches was very common among children and youth. However, the general public feared that taking eggs from the nests and destroying nests would lead to »dead heartedness and brutalisation of attitudes«. Therefore, local authorities in the district of Frankfurt/Oder were induced to avert the destruction of bird nests by assigning parents, custodians and teachers to appeal to the vandalising youth as early as 1813. Also, for mostly humane-ethical reasons, between 1843 and 1857, ten towns in Brandenburg (including Berlin) passed bird conservation decrees to prohibit taking eggs from the nests, destroying nests, and trapping and killing birds within the town area. Interestingly, no single decree prohibited the trade of those birds which were caught outside the town area and then brought into the town market. This shows that even in the 1850s, economic interests with regard to small birds had first priority and were not to be substantially confined by humane-ethical concern.

5.4 Turning the tide: the bird conservation decrees from 1867

In 1860, considerations about the utility of insect-eating birds in agriculture gave rise to a bird conservation decree in the district of Potsdam. According to the decree, some 33 bird species which were recognised as »beneficial by eating insects and other vermin« were given a closed season between December 1st and August 31st[34]. However, as October and November were the most profitable months for fowling the decree had only little effect on bird trapping and trading and thus, was not helpful for the conservation of birds. The Potsdam decree was the first attempt to govern bird conservation at the district level; this should be seen against the background that the council of the Frankfurt district regarded the town decrees sufficient and argued that there was no need for a bird conservation decree at the district level. Nevertheless, aims to manage bird conservation at the district and even at the country level (Prussia) continued. The bird conservation decrees of Berlin and the district councils of Potsdam and Frankfurt from 1867 mark a turning-point. In Berlin and the Potsdam district, and in the Frankfurt district, 34 bird species and some 37 bird species respectively, were protected all year round

[34] GStA PK, I. HA Rep. 87 Ministerium für Landwirtschaft, Domänen und Forsten, B Nr. 19121.

and banned from the weekly markets from January 1st 1868[35]. The prohibition of sale was an effective means for conservation. The 1867 decrees show that the conservation of insect-eating birds which were regarded beneficial to agriculture and forestry were clearly regarded more important than the economic interests associated with the trade of song-birds for nourishment purposes and for cage-keeping. In 1870, the province council of Brandenburg stated that once the bird conservation decrees had been passed »the fowlers who used to make the fields insecure with decoys and lime-twigs disappeared from the surroundings of Berlin«[36]. Overall, the 1867 decrees ceased the centuries-old tradition of fowling and using small birds for nourishment. Passing the decrees was only possible once the economic interest in the birds had significantly receded. However, as earning a small income from fowling was no longer an option, some lower class people, especially from the countryside, virtually had to pay for the progress in bird conservation.

5.5 The German Bird Conservation Act of 1888

Bird conservation was improved by the ›German Bird Conservation Act‹ of 1888[37] which prohibited various methods of fowling and restricted trapping and trading of small birds which were not embraced by the hunting regulations, to the winter period (after September 15th and before March 1st). Again, the conservation of insect-eating birds was the chief motive for passing the act. Its main aim was to stop the catching of birds in masses whereas the shooting and trapping of single birds remained allowed. However, non-anthropocentric, that means bird-centred considerations, were expressed more frequently in the last quarter of the nineteenth century and also played a role in the negotiations about the German Bird Conservation Act before it was finally passed in 1888. In this respect, fieldfare trapping remained a topic of very controversial debates. In well wooded German mountain areas, fieldfare trapping was economically still relevant and thus, common even in the 1880s. Snares using rowan berries as a decoy and sited in the woods were the most efficient method of fieldfare trapping. However, this did not only apply to the fieldfare but to many other song-bird species too, which the fowler was not interested in. In order to prevent or at least reduce the unattended trapping and hence, the agonising death of titmice, redbreasts and other song-birds in the snares, the German Bird Conservation Act restricted fieldfare trapping to the time span between September 21st and December 31st.

[35] GStA PK, I. HA Rep. 87 Ministerium für Landwirtschaft, Domänen und Forsten, B Nr. 19122.

[36] GStA PK, I. HA Rep. 87 Ministerium für Landwirtschaft, Domänen und Forsten, B Nr. 19122.

[37] *Gesetz, betreffend den Schutz von Vögeln, vom 22. März 1888* (Reichsgesetzblatt, p 111); cp. Grotefend and Cretschmar (1904 : III2: 1386f).

5.6 Progress and relapse: the aims of conserving the fieldfare

Repeatedly, there were demands to prohibit fieldfare trapping completely to have good reason to take action against »the senseless mass murder of the song-birds in South European countries«. Such views and conservation aims finally resulted in the release of the second ›German Bird Conservation Act‹ in 1908[38]. This act prohibited snare-catching which was the most efficient method of trapping fieldfares. Whereas before 1908, about 1,200,000 fieldfares were caught annually in Germany, the prohibition of snares resulted in almost cessation of fieldfare trapping and thus, was an effective means of conservation[39]. After 1908, only small quantities of fieldfares were trapped in nets or shot with rifles. However, during World War I the shortage of food and meat in particular, induced people to call the 1908 conservation act into question. In July 1916, the president of the Stade district, north-west Germany, argued that all possible sources of meat were to be acquired to overcome the shortage of meat. On September 21st 1916, the Federal Council of Germany decided to allow fieldfare trapping in snares until December 31st. As the amount and quality of the nutrition of the people deteriorated over the following two years, the Council decided to allow fieldfare trapping from October 1st to December 31st 1917 and from September 21st to December 31st 1918, respectively. However, only huntsmen were allowed to trap fieldfares. Several commentators complained that only rich people would profit from the trapped birds as fieldfare meat was not included in the meal voucher system which had been introduced to supply people with a minimum of bread, fat, and meat. Moreover, fieldfares which had been an expensive delicacy until 1908 anyway, were becoming even more expensive, as well as all the other nourishments. Apart from the notion of a fair supply of food to the people, the discourse with regard to the fieldfare was characterised by very controversial arguments. Some of the advocates of the re-legalisation of fieldfare trapping argued that since the trapping prohibition from 1908, the blackbird and the bullfinch had proliferated in such a way that they had become varmints to fruit-trees and berry bushes and thus, the re-legalisation would also benefit, rather than harm agriculture. The adversaries of fieldfare trapping argued that the meat quantities gained from the trapped fieldfares were »ridiculously small« and thus, were not adequate to improve the meat supply. Moreover, 80 to 90 % of the birds offered for sale as ›fieldfares‹ were actually song thrushes which cleansed the fields of snails, larvae and flying vermin. Thus, fieldfare trapping would lead to the diminishment of beneficial birds which were most important to save produce from vermin, especially whilst food was scarce during the war. Due to the slightly improved food supply in 1919, the authorities did not admit fieldfare trapping in snares in 1919, nor in the following years.

[38] *Vogelschutzgesetz für das Deutsche Reich, vom 30. Mai 1908* (Reichsgesetzblatt 1908, no. 31, p 314); cp. Sunkel (1927 : 257–260).

[39] GStA PK, I. HA Rep. 87 Ministerium für Landwirtschaft, Domänen und Forsten, B Nr. 19995. All the following incidents are accounted in the same file.

6 Discussion and summary

Despite the singularity of some historical examples, such as the conservation of the nightingale for seemingly exclusively aesthetic reasons as early as the late seventeenth century, the history of bird perception can be subdivided into certain phases depending on the dominating aspects of bird valuation. Moreover, the history of bird perception follows principles which are strongly based on the socio-economic prerequisites, particularly on the circumstances in agriculture, the level of nutrition and the sources of income in the countryside. Therefore, the general findings shall be summarised by looking at *three aspects*.

6.1 There is a *general trend* in the historical perception of birds

Edicts and decrees dealing with hunting and trapping rights of the electors and the nobility of Brandenburg start to appear in the second half of the sixteenth century and represent the oldest historical sources. These sources, as well as records on the legal proceedings against the offenders of hunting regulations which came up in late sixteenth century, show that the significance of birds as food supplement, delicacy and trading goods was not only the earliest aspect discussed in the records but also the most important until the 1850s. However, the archival material dealing with conflicting hunting interests only refers to large bird species which the sovereign reserved for himself and his vassals and thus, were subject to the hunting regulations. As fowling of small birds was a common right and, if not practised on the sovereign's hunting grounds, was free to anyone there was no reason for correspondence dealing with small birds. Hence, there is generally no archival material on small birds. It is therefore difficult or even impossible to work out the relative significance of small birds as nourishment from archival sources. Contemporary publications are more appropriate to answer this question. Therefore, hypothesis I.1 is only partly confirmed by the archival material. Whilst it is not possible to recognise an increase of small bird fowling from the archival sources of the second half of the eighteenth and the first half of the nineteenth centuries when general pauperisation took place, there is evidence that the significance of birds for nutrition decreased in mid-nineteenth century and became insignificant in the second half of the nineteenth century.

According to the stocks of the archive, the aspect of vermin control was most important in the eighteenth century and apparently, became less important in the nineteenth century. The examples of the sparrow, the crane and the great bustard have shown that the fight against vermin increased during the eighteenth century. However, the control of the sparrow was given up in 1767, probably due to the realisation that killing a small portion of the overall sparrow population would not lead to its extermination. The examples of the crane and the great bustard demonstrate another pitfall of archival research. Once legal action was sufficiently taken to diminish the two crop-devastating bird species effectively, the files were closed and hence, no further information is available on the abundance of the birds and the increasing or perhaps decreasing significance as vermin. Therefore, there is no account of the crane and the great bustard after 1722 and 1770, respectively. Overall however, hypothesis II.1 is mainly confirmed.

The idea of preserving insect-eating birds for their supposed utility to agriculture came to the fore in the 1850s and dominated the archival records in the second half of the nineteenth and early twentieth centuries. Therefore, hypothesis III.1 is confirmed.

Ethical thinking only played a secondary role until the turn of the twentieth century. First, ethical considerations were generally exclusively anthropocentric as has been shown by the bird conservation decrees of the towns. Later, from the 1880s onwards, ethical considerations were increasingly focussed on birds themselves. Aesthetic aspects led to the protection of the nightingale as early as 1686 but were generally not a topic in the archival records. Along with ethical considerations they became significant only in the beginning of the twentieth century. In conclusion, hypothesis IV.1 is approved.

6.2 There are *two turning points* in bird perception

The 1850s and 1860s mark the *first* general turn in bird perception. It seems plausible that food security since mid-nineteenth century gave rise to more sophisticated utilitarian thinking. Once malnutrition had been overcome, the significance of birds as nourishment of the lower classes vanished and the significance of insect-eating birds as ›beneficial animals‹ came to the fore. Therefore, hypotheses I.2 and III.1 are approved by the archival evidence. Simultaneously, the significance of the fieldfare and other small fare birds changed from a food supplement of the poor to a delicacy, exclusively of the affluent people; this was due to the receding trapping quantities and especially due to the bird conservation decrees of the districts in 1867, both leading to rising prices. According to the archival records, the control of crop-devastating birds only played a subordinate role in the nineteenth century. However, archival evidence is not sufficient to confirm hypothesis II.2, saying that the agricultural reforms resulted in a waning interest in controlling crop-devastating birds. Overall, the events in the 1850s and 1860s represent a qualitative change in the perception of birds and thence a transformation of values, mainly caused by the improving socio-economic conditions.

The *second* turning point in bird perception took place during World War I when the diet of most people was low in protein, fat and meat, especially in the years 1916 to 1918. Many people starved and thus, there were aims to utilise all possible sources of food. Hence, the prohibition of fieldfare trapping, part of the Bird Protection Act from 1908, was revoked in order to improve nutrition by providing an additional source of meat. This proves hypothesis I.3. At the same time, the discussion on both the necessity to control grain-eating birds as vermin and to protect insect-eating birds supposed to be beneficial to agriculture, was reviving. Conservationists who were certainly against the re-legalisation of fieldfare trapping emphasised the benefit of the fieldfare as an insect-eating bird for agriculture. Therefore, hypotheses II.3 and III.2 are confirmed. Even conservationists deferred ethical and aesthetic arguments to utilitarian aspects of bird conservation as they regarded them more convincing when arguing with the authorities. This underpins hypothesis IV.2.

6.3 The general trend in bird perception follows socio-economic development. Does this apply to the perception of biodiversity, too?

I found a general trend in bird perception over the centuries from basic utilitarian aspects, such as valuing birds for their meat and the need to control vermin, over higher utilitarian aspects such as the protection of insect-eating, and thence, beneficial birds, to ethical and aesthetic considerations. However, such development of values does not necessarily follow a one-way direction. The example of the re-legalisation of trapping the fieldfare shows that in times of malnutrition and impending famines the significance of birds as nourishment, as vermin and beneficial animal, respectively, comes to the fore again. In conclusion, the analyses of the archival sources show that the change of values with regard to birds has been strongly dependent on the socio-economic conditions and particularly, on the level of nutrition. According to the archival sources, the diversity of bird species was not worth discussing. This was probably partly due to the fact that until early nineteenth century, birds were rich not only in numbers but also in species. However, people in many parts of Germany noticed and bemoaned the receding abundance of birds in the first half of the nineteenth century. Some observers also stated the loss of certain bird species and called for bird conservation. Overall however, the discourse in mid-nineteenth century focussed on the receding abundance of birds and almost neglected the extinction of certain species. Moreover, the receding bird abundances were first a problem to the fowlers before they were regarded an economic issue of agriculture since the 1850s. In the second half of the nineteenth century, considerations about utility dominated the discourse on birds; insect-eating birds were regarded as beneficial and thus, people appreciated high abundances and fostered the proliferation of the respective species by setting up nest boxes. On the other hand, crop-devastating birds were regarded harmful and thus, there was no need to protect them by law. Legislation in early twentieth century was increasingly driven by ethical and aesthetic considerations, even giving protection to crop-devastating birds.

As has been shown, people considered the favourable and less favourable characteristics of birds very thoroughly. However, they favoured different characteristics at different stages in the past. At any time, the aspects of bird perception which were regarded as the most important determined the action people took (direct usage, vermin control and extermination, and protection for ethical and aesthetic reasons). This process was strongly influenced by socio-economic conditions. It seems very plausible that the perception and valuation of biodiversity, that is, of other wild plant and animal species, generally follows the same principle. People value species highly if they have an economic interest or if they appreciate them for aesthetic reasons. Apart from that, plant and animal species may be attributed ethical values. All this may result either in direct usage or in protection of the respective species. However, people may also call for the control and extermination of certain plant and animal species if they conflict with human interests. For obvious reasons, species are given no value if they are not known or perceived only unconsciously. This holds true for the majority of people and applies to small and unspectacular species which are particularly hard to recognise and be seen. Such species are not capable of attracting human attention or interest and they do not interfere with human needs. However, to make reliable statements on the perception of biodiversity in former periods of time, it is necessary to conduct research on plant and animal species other than birds, select other research areas, and use other written sources in addition to archival records.

Archival records

(held in the former Prussian state archive, i.e., Geheimes Staatsarchiv Preußischer Kulturbesitz Berlin-Dahlem = GStA PK)

GStA PK, I. HA Rep. 9 Allgemeine Verwaltung, Q3a, Fasc. 5
Anordnung betr. jagdbare Vögel (1615).

GStA PK, I. HA Rep. 9 Allgemeine Verwaltung, R2d, Fasc. 1
Schriften gegen die Wilddiebe (1565–1797).

GStA PK, I. HA Rep. 36 Hof- und Güterverwaltung, Nr. 2254
Sammlung von Holz-, Forst- und Jagdedikten brandenburgischer Kurfürsten und preußischer Könige aus den Jahren 1622, 1678, 1718, 1719, 1720, 1722 (2 Edikte), 1725, 1740, 1751, 1754 (3 Edikte), 1768, 1770 (2 Edikte), 1772, 1777, 1781 (2 Edikte), 1783, 1784, 1785 (2 Edikte), 1787 (2 Edikte), 1788, 1794, 1796 (2 Edikte), 1801 und 1806 (2 Edikte); Drucke (1622–1806).

GStA PK, I. HA Rep. 36 Hof- und Güterverwaltung, Nr. 2257
Emanierte Jagdedikte verschiedensten Inhalts (1676–1749).

GStA PK, I. HA Rep. 36 Hof- und Güterverwaltung, Nr. 2276
Acta des Jagdzeugmeisters Schenck betr. den Lerchenfang (1756–1796).

GStA PK, I. HA Rep. 87 Ministerium für Landwirtschaft, Domänen und Forsten, B Nr. 19112
Die polizeilichen Maßregeln zum Schutz der Feldfrüchte durch Konservierung nützlicher und Vertilgung schädlicher Tiere (Bd. 1: 1855–1867).

GStA PK, I. HA Rep. 87 Ministerium für Landwirtschaft, Domänen und Forsten, B Nr. 19121
Eingegangene Regierungsberichte auf die Zirkularverfügung vom 16. 1. 1867 betrifft den Erlaß eines Gesetzes zum Schutz nützlicher Vögel (1867).

GStA PK, I. HA Rep. 87 Ministerium für Landwirtschaft, Domänen und Forsten, B Nr. 19122
Eingegangene Regierungsberichte auf die Zirkularverfügung vom 21. 2. 1870 betrifft den Erlaß eines Gesetzes zum Schutz nützlicher Vögel (1870).

GStA PK, I. HA Rep. 87 Ministerium für Landwirtschaft, Domänen und Forsten, B Nr. 19986
Acta betreffend: die zufolge des Erlasses vom 11. August 1899 - I. B 5933 - eingegangenen Berichte der kgl. Regierungen, betr. den Krammetsvogelfang in Preussen (1899).

GStA PK, I. HA Rep. 87 Ministerium für Landwirtschaft, Domänen und Forsten, B Nr. 19995
Akten betreffend: Freigabe des Krammetsvogelfanges (1916–1932).
GStA PK, I. HA Rep. 87 Ministerium für Landwirtschaft, Domänen und Forsten, B Nr. 19998, Bd. 1
Acta betreffend den Schutz der Feldfrüchte durch Conservirung nützlicher und Abwehr schädlicher Thiere im Allgemeinen. Vogelschutz überhaupt (Bd. 1: 1817–1865).
GStA PK, II. HA Generaldirektorium, Abt. 14, Kurmark Materien J-Z Tit. CCLXVIII Nr. 1, Vol. 1
Acta Wegen Vertilgung der Sperlinge und Hamster (1721–1750).
GStA PK, II. HA Generaldirektorium, Abt. 33, Generalia Tit. IV Nr. 14
Verordnung vom 13. October 1751, dass die Trappen von der hohen Jagd ausgeschlossen und zur Niederjagd gerechnet werden sollen (1751).
GStA PK, II. HA Generaldirektorium, Abt. 33, Generalia Tit. XIV Nr. 5
Acta wegen Schonung der Krammets Vögel, der Krähen und Dohlen, zur Vertilgung der Raupen; imgleichen wegen Dultung der Ameisen (1792–1803).
GStA PK, II. HA Generaldirektorium, Abt. 33, Generalia Tit. XV Nr. 2
Acta die emanierten Edikte wegen des verbothenen Ausnehmens der Gänse, Enten, Schnepfen, auch Kiebitz-Eier betreffend (1663–1704).
GStA PK, II. HA Generaldirektorium, Abt. 33, Generalia Tit. XV Nr. 3
Acta generalia die erlassenen Edicte, wegen Schonung der Trappen, Schwäne und Kraniche betreffend (1668–1713).
GStA PK, II. HA Generaldirektorium, Abt. 33, Generalia Tit. XV Nr. 7
Acta die Schonung der Nachtigallen und die deshalb erlassenen Verordnungen betreffend (1685–1803).
GStA PK, II. HA Generaldirektorium, Abt. 33, Generalia Tit. XVI Nr. 15
Acta Wegen Schießung derer Trappen und daß ein Hahn vor 16 gr. - eine Henne aber vor 12 gr. verkauffet werden sollen (1722–1770).

References

Baudis D (1986) Vom »Schweinemord« zum »Kohlrübenwinter«. Streiflichter zur Entwicklung der Lebensverhältnisse in Berlin im ersten Weltkrieg (August 1914 bis Frühjahr 1917). In: Schultz H et al. Zur Wirtschafts- und Sozialgeschichte Berlins vom 17. Jahrhundert bis zur Gegenwart. Jahrbuch für Wirtschaftsgeschichte, Sonderband, Akademie der Wissenschaften der DDR, Institut für Wirtschaftsgeschichte, Akademie-Verlag, Berlin

Bekmann JChr (1751) Historische Beschreibung der Chur und Mark Brandenburg nach ihrem Ursprung, Einwohnern, Natürlichen Beschaffenheit, Gewässer, Landschafften, Stäten, Geistlichen Stiftern & c. Regenten, deren Staats- und Religions-Handlungen, Wapen, Siegel und Münzen, Wohlverdienten Geschlechtern Adelichen und Bürgerlichen Standes, Aufnehmen der Wissenschafften und Künste in derselben usw. Bd I. Christian Friedrich Voß, Berlin

Dreyhaupt JCv (1750) Pagus neletici et nudzici, Oder Ausführliche diplomatisch=historische Beschreibung des zum ehemaligen Primat und Ertz=Stifft, nunmehr aber durch den westphälischen Friedens=Schluß secularisirten Hertzogthum Magdeburg gehörigen Saal=Kreyses, Und aller darinnen befindlichen Städte, Schlösser, Aemter, Rittergüter, adelichen Familien, Kirchen, Clöster, Pfarren und Dörffer, Insonderheit der Städte Halle, Neumarckt, Glaucha, Wettin, Löbegün, Cönnern und Alsleben; [...] Zweyter Theil. Verlag des Waysenhauses, Halle a.S.

Elsholtius (Elßholz) JS (1682) Joan. Sig. Esholtii Diaeteticon: Das ist Newes Tisch-Buch, oder Unterricht von Erhaltung guter Gesundheit durch eine ordentliche Diät und insonderheit durch rechtmäßigen Gebrauch der Speise und des Geträncks; In sechs Büchern [...] abgefaßt, auch mit nöthigen Figuren gezieret [...]. Gedruckt durch Georg Schultze, Cölln an der Spree

Flemming JFv (1749) Der Vollkommene Teutsche Jäger. 1. Bd. Leipzig. Nachdruck der Akademischen Druck- u. Verlagsanstalt 1971, Graz

Gatterer ChrWJ (1782) Abhandlung vom Nutzen und Schaden der Thiere, nebst den vornehmsten Arten, dieselben zu fangen und die schädlichen zu vermindern. Bd II: Von den Vögeln. Weyand, Leipzig

Gewalt W (1959) Die Großtrappe (*Otis tarda L.*). Die neue Brehm-Bücherei, Heft 223. Ziemsen, Lutherstadt Wittenberg

Grotefend GA, Cretschmar C (1904) Preußisch=deutsche Gesetz=Sammlung 1806–1904. Bd III2: Landwirtschaft, 4th edn, Schwann, Düsseldorf

Hattenhauer H, Bernert G (ed) (1994) Allgemeines Landrecht für die Preußischen Staaten von 1794. Mit einer Einführung von Dr. Hans Hattenhauer und einer Bibliographie von Dr. Günther Bernert, 2nd edn, Luchterhand, Neuwied, Kriftel, Berlin

Herrmann B (1997) »Nun blüht es von End' zu End' all überall«. Die Eindeichung des Nieder-Oderbruches 1747–1753. Cottbuser Studien zur Geschichte von Technik, Arbeit und Umwelt, Bd 4. Waxmann, Münster, New York, München, Berlin

Herrmann B (2003) Die Entvölkerung der Landschaft. Der Kampf gegen »culturschädliche Thiere« in Brandenburg im 18. Jahrhundert. In: Bayerl G, Meyer T (eds) Die Veränderung der Kulturlandschaft: Nutzungen – Sichtweisen – Planungen. Cottbuser Studien zur Geschichte von Technik, Arbeit und Umwelt. Waxmann, Münster, New York, München, Berlin

Kaeber E (1921) Berlin im Weltkriege. Fünf Jahre städtischer Kriegsarbeit. Im Auftrage des Magistrats auf Grund der Berichte der städtischen Verwaltungsstellen herausgegeben. Trowitsch & Sohn, Berlin

Makatsch W (1959) Der Kranich. Die neue Brehm-Bücherei, Heft 229. Ziemsen, Lutherstadt Wittenberg

Materna I, Ribbe W (eds) (1995) Brandenburgische Geschichte. Akademie-Verlag, Berlin

Medick H (1985) »Hungerkrisen« in der historischen Forschung. Beispiele aus Mitteleuropa vom 17.–19. Jahrhundert. Sozialwissenschaftliche Informationen SOWI 14 (2) : 95–103

Mewes W (1995) Bestandsentwicklung des Kranichs (*Grus grus*) in Deutschland und deren Ursachen. Dissertation, Martin-Luther-Universität Halle-Wittenberg. Halle a.S.

Mylius ChrO (1736) Des Corporis Constitutionum Marchicarum Dritter Theil Von Kriegs-Sachen, betr. den Militair-Process, Disciplin, Straffen, Werbung, Einquartirung, March, Ausführung des Gewehrs und Pferde, Avocatorien, Aufgeboth der Ritter-Pferde und Mannschafft, Kriegs-Consistorium, Ehe-Sachen derer Officiers und Soldaten, Durch-March fremder Trouppen, Cartelle usw. In Drei Abtheilungen. Buchladen des Waysenhauses, Berlin

Mylius ChrO (1737) Des Corporis Constitutionum Marchicarum Fünffter Theil Von Policey- Hochzeit- Kindtauffen- Begräbniß- [...] und andern zur Policey gehörigen Ordnungen, [...] Manufacturen-, Commercien- [...] Dorff- und Acker- [...] Ordnungen, [...] item von Medicinal-Ordnungen [...] Scharffrichtern, Abdeckern, Schweinschneidern, Landstreichern, Zigeunern, Juden, Wirths-Häusern und dergleichen: In Fünff Abtheilungen. Buchladen des Waysenhauses, Berlin Halle a.S.

Ruß K (1882) Zum Vogelschutz. Eine Darstellung der Vogelschutzfrage in ihrer geschichtlichen Entwicklung bis zur Gegenwart nebst Besprechung aller bisherigen Maßnahmen sowie der Gesetz-Vorschläge. Voigt, Leipzig

Sunkel W (1927) Der Vogelfang für Wissenschaft und Vogelpflege. Alfred Troschütz, Hannover

Teuteberg HJ (1976) Die Nahrung der sozialen Unterschichten im späten 19. Jahrhundert. In: Heischkel-Artelt E (ed) Ernährung und Ernährungslehre im 19. Jahrhundert. Vorträge eines Symposiums am 5. und 6. Januar 1973 in Frankfurt/Main. Vandenhoeck&Ruprecht, Göttingen, pp 205–287

Wehler H-U (1996) Deutsche Gesellschaftsgeschichte. Erster Band: Vom Feudalismus des Alten Reiches bis zur Defensiven Modernisierung der Reformära 1700–1815, 3rd edn, CH Beck, München

Part III

Local, regional and nationwide perspectives on the Convention on Biological Diversity: Examples from Guatemala

Prospects and Challenges for Biodiversity Conservation in Guatemala

Regina Birner[1], Heidi Wittmer[2], Augustin Berghöfer, and Michael Mühlenberg[3]

[1] Institute of Rural Development, University of Göttingen, email to rbirner@gwdg.de
[2] Centre for Environmental Research (UFZ), Leipzig, email to heidi.wittmer@ufz.de
[3] Centre for Nature Conservation (CNC), Von-Siebold-Str. 2, 37075 Göttingen, Germany, email to mmuehle@gwdg.de

1 Introduction

Located on the land bridge between North and South America, Guatemala is characterized by exceptionally diverse ecosystems, which make the country one of the foremost repositories of biodiversity in Latin America (Detlefsen et al. 1991). Guatemala has signed the Central American Convention on Biological Diversity in 1992 and the global Convention on Biological Diversity in 1995. To support the implementation of these conventions, a National Biodiversity Strategy and Action Plan was elaborated in 1998/99 (CONAMA 1999). Various categories of protected areas cover almost 30 % of the country's surface (CONAP 2002 : 14). Under an innovative model of co-administration, civil society plays an important role in biodiversity conservation. At the same time, Guatemala is one of the poorest countries in Latin America, which suffers from slow economic development, high levels of poverty, unequal land distribution, a highly segmented society, and the effects of more than three decades of civil war which only ended in 1996. High population increase as well as commercial interests in natural resource extraction cause considerable threats to biodiversity conservation. Thus, Guatemala represents a prime example of the potentials and challenges of biodiversity conservation in developing countries. The research on biodiversity conservation presented in the contributions to Part III was conducted at the national level and in the Department of Alta Verapaz.

2 Socio-Economic and Political Situation

Guatemala covers a total area of 108,890 km^2. It is Central America's westernmost country, bordering Mexico to the North and West, Belize to the Northeast and Honduras and El Salvador to the East. The total population of Guatemala is 13,909,400 (July 2003 estimate) with a rather high population growth of 2.66 % and a fertility rate of 4.7 children born per woman (World Fact Book 2003). Information on the composition of the population differs between sources. The share of the indigenous population (mostly Mayas) is estimated to range between 43 % (World Fact Book 2003) and 60 % (Fischer 2002). Whites and others account for approximately 2 % of the population, and the remainder consists of Ladinos (mixed Indígena-Spanish or assimilated Indígena). The Indígena population comprises 23 different ethnic groups, who all speak their own languages.

Guatemala gained independence from Spanish colonial rule in 1821. During the second half of the 20th century, the country experienced a variety of military and civilian governments. In 1985/86 Guatemala witnessed a change to civilian government after various military regimes had ruled the country from 1954 onwards, with only a short democratic interlude in 1970. During the military rule, conflicts between the landless Maya majority and the landowning elite escalated into a civil war that was to last for more than three decades. The 36-year guerrilla war led to the death of more than 200,000 people (CEH 1999 : 21). Estimates of the number of displaced persons vary from 500,000 to a million and a half (CEH 1999 : 38). In 1996, the government signed a peace agreement that formally ended the conflict.

Under the Constitution of 1986, which was last changed in 1994, Guatemala is a democratic republic with a presidential form of government and a unicameral Congress. Since January 2004, the country is ruled by a conservative party alliance (GANA) led by Oscar Berger. He took over the presidential office from Alfonso Portillo Cabrera, who had led a right-wing government since 2001. Among the large number of political parties in Guatemala, there is also a small Green Party.

Guatemala's economy is characterized by a high prominence of the agricultural sector, which accounts for one-fourth of the Gross Domestic Product (GDP) and two-thirds of the export earnings, with coffee, sugar and bananas as major export products. Agriculture employs one half of the country's labour force. In 2002, the GDP per capital in purchasing power parity was estimated at 3,900 US$, and the real growth rate of the GDP was 2.2 % (World Fact Book 2003). The distribution of assets, especially land, and income is highly unequal. One percent of the population own 86 % of the cultivable territory of the country (Grochembake 2003).

According to the National Planning Agency SEGEPLAN (2001), 54 % of the population live in poverty, out of which 23 % live in extreme poverty. Especially the rural population remains highly vulnerable and food insecure (cp. Tesliuc and Lindert 2003). The combination of a drought and falling coffee prices in 2002 caused an acute crisis of malnutrition, placing thousands of children at the risk of dying from hunger (USAID 2002).

The Peace Accords of 1996 stipulated a land reform as well as investment in the physical and social infrastructure to improve the livelihoods of the indigenous and local communities in rural areas. However, the implementation of the Peace Accords has lagged behind (MINUGUA 2002). A referendum intended to change the Constitution and grant the indigenous population more rights failed. Violations of human rights have continued within the country (MINUGUA 2002). The prospects for implementing the land reform more rapidly did not increase under the new government of Oscar Berger, since he is backed by the land-owning elite (Hellweg-Larsen 2004).

The Department of Alta Verapaz, which was chosen for the research presented in Part III of this volume, is one of the 22 departments of Guatemala. Alta Verapaz covers an area of 8,686 km^2 and has a population of approximately 760,000 inhabitants. The major ethnic groups in the Departments are the Q'eqchi', Pocomchí and Achí. The Department comprises 14 municipalities and covers both mountain areas in the South and lowland areas in the North. The transition zone between the mountain areas and the lowlands, which crosses Alta Verapaz, is referred to as Franja Transversal del Norte. In the 19th century, Alta Verapaz became a major coffee producing region. German immigrants introduced cardamom production in the beginning of the 20th century, which remains a major export crop grown in the Department. The poverty level in Alta Verapaz of 76 % and the level of extreme poverty of 37 % are clearly above the average levels for Guatemala, as quoted above (SEGEPLAN 2001).

3 Biodiversity in Guatemala

3.1 Bio-Geography

Guatemala's volatile topography comprises over 30 volcanoes, which reach heights of up to 3800 m. Almost two thirds of the country's area are mountainous and of volcanic origin. The mountains stretch from West to East and divide the country into three biogeographical zones: the narrow Pacific lowland in the Southwest, which is fertile and densely populated. The mountain zone amounts to almost two thirds of the country. The mountains are of volcanic origin and covered by tropical submontane and montane forests. In the North is the depression of Petén with extensive tropical lowland forests.

It is possible to distinguish between eleven floristic regions. This diversity contributes to a particular high species richness of plants: the limestone plains of Peten, the mangrove swamps along both coasts, the rain forest of the Atlantic coast, the low savannas of Izabal and Peten, the mixed forest of the Atlantic coast, the arid desert plain-chaparral of the Oriente valleys of the Rio Motagua and Rio Negro, the mountain cloud forest of Alta and Baja Verapaz, the mixed mountain forest of the Pacific coastal region, the upland mixed forest of temperate and cold regions, the coniferous forests, and the alpine regions (Markussen 2003).

Until now more than 8,000 vascular plant species have been recorded in Guatemalan forests, and about 1,000 of them are endemic to the country. 17 conifer species and roughly 450 species of deciduous trees are found in the forests of Guatemala (Islebe and Véliz Pérez 2001).

Tropical lowland rainforest is distributed in Guatemala only in the northern lowlands (province of Peten) and the Caribbean coast near Belize. The Maya Biosphere Reserve with 3.9 million acres was established in 1990 through an agreement between three neighbouring countries, Mexico, Guatemala and Belize. The reserve contains the largest area of tropical rainforest in Guatemala and Central America, with a wide range of undisturbed natural habitats (Primack et al. 1998).

Forty percent of the migratory birds from North America stop in the forests and wetlands of this area during their seasonal migration.

In terms of zoogeographic zones the neotropics are divided into 22 zones (Stotz et al. 1996). The mountain region of Guatemala is a section of the mountains of northern Middle America and belongs to the Madrean Highlands. The degree of endemism of the mainland avifauna there exceeds 28 %, comparable to the Northern Andes with (28.7 %) and the Atlantic Forest of South America (29 %), being higher than in all other zones. The tropical montane moist forests of Guatemala are part of the Endemic Bird Areas of the North Central American highlands (Stattersfield et al. 1998). 20 bird species from the montane moist forests in the region are classified as restricted-range species because their breeding range is less than 50,000 km (Stattersfield et al. 1998).

3.2 Cloud Forests

Cloud forests are much more restricted in area than lowland forests and are therefore more vulnerable to deforestation or alteration (Labastille and Pool 1978, Hamilton et al. 1995.) This was one reason to classify cloud forests as threatened, unique in respect to biogeography and of high priority for conservation by the World Bank (Dinerstein 1995). Cloud forests are dominated by evergreen oaks and once covered more than 30,000 km^2 in Guatemala. By 1985, these forests had been reduced to less than 1,000 km^2. The Sierra de las Minas Biosphere Reserve in Alta Verapaz has been protected by decisive actions of NGOs and the Guatemalan government. This reserve covers more than 600,000 acres of cloud forest and wetlands and contains the largest cloud forest in Central America and 60 % of Guatemala's remaining cloud forest. In Guatemala cloud forests occur on the south western volcano chain as well as on the north eastern cordilleras at altitudes from about 1500 m a.s.l. on (see Hamilton et al. 1995). The ascending air mass condenses and forms fog which generates moist conditions of approximate 100 % humidity, independent of rainfall. The measurable annual rainfall is about 4,000 mm in these mountains (see the contribution by Markussen and Renner in this volume). The great humidity enables many plants to take water from the air resulting in a particular luxuriant and diverse evergreen forest vegetation. Cushion-like epiphytes of mosses, ferns and vascular plants cover the 20–25 m tall trees. Among the higher plants the families of Orchidaceae, Ericaceae and Bromeliaceae are dominant. More than half of the 800 species of orchids in Guatemala are found in the cloud forest (Behar and Tinschert 1998). The stand of trees is relatively sparse and permits a dense vegetation in the understorey with tree ferns, mosses and mountain palms (Nadkarni 1984, Catling and Lefkovitch 1989). The average low temperature results in a thick humus layer on the ground. Here and inside the cushions of epiphytes many different groups of invertebrates pass their life cycles. The abundance is comparable to the density of individuals in the forests of the temperate zone (Schulz and Menzel 1999).

One famous flagship species which depends on the cloud forest is the heraldic bird of Guatemala, the Resplendend Quetzal (Pharomacrus mocinno). In the sierra region of Aj Poop B'atz' with 55 km^2 of cloud forest area (region of Chicacnab – Chelemhá), this bird reaches its highest population density (Mandl 1989; Unger 1990). This species is classified as threatened in the world list of IUCN (Godoy 1997, see Renner and Markussen in this volume for more biological information). This bird was sacred to the Mayas and Aztecs. The Mayas worshipped a God of the name Quetzalcoatl, which was portrayed with a head ornamentation of the quetzal's tail feathers. Only noblemen or priests were allowed to wear such a head ornamentation. The birds were captured alive and deprived of their long middle-tail feathers. Afterwards they were released and the tail feathers could grow again (Janson 1993; Wheelwright 1983).

4 Policy Approaches to Conserve Biodiversity

Historically, the first natural resource legislation in Guatemala was a Forestry Law in 1921 (Detlefsen et al. 1991). A first legal provision for protected areas was made in the 1945 Forestry Law. However, only ten years later the first national parks were declared (UNEP – CEP 1996). The world renowned archaeological site of Tikal, an ensemble of Maya temples and pyramids in the rainforest, was among these first areas and its exact borderline was fixed in 1957. In 1955, 27 volcanoes had been put under legal protection as well. However, according to Castro and de León (2003 : 3) no institution had taken charge of them for more than 30 years. Until 1960 several dozens of protected areas were declared, covering about 1 % of the country's total territory. Active management was limited to a few areas, administered with differing standards, purposes and resources by several bodies, among them the University of San Carlos (USAC) and the National Institute for Anthropology and History (IDAEH) (CONAP 2002 : 14).

The first civil government of Vinicio Cerezo (1985–1990) had to cope with national and regional armed conflicts, and with a war-torn and fragmented Guatemalan society. Within this context, a low priority of environmental issues, as identified by Nations et al. (1988), seems understandable. Hence, it is astonishing that Guatemala's Constitution of 1985[4] places substantial emphasis on the protection of natural resources: Art. 64 states that the conservation of the nation's natural heritage a matter of national interest; Art. 126 declares the reforestation of the country and the conservation of the forests a matter of ›national urgency‹. In the same way, the ›rational, non-degrading use‹ of the country's flora and fauna, soil and water is called for in Art. 97. The exact consequences of these articles are to be spelled out in later legislation. In a first step, the National Commission for Environmental Affairs (CONAMA) was established in 1986. This governmental body is directly subordinate to the Presidency and in charge of assessing and coordinating at the political level those activities related to the protection and conservation of the natural environment (UNEP – CEP 1996).

[4] For the full text of the Constitution, see `www.georgetown.edu/pdba/Constitutions/Guate/guate85.html`.

As analysed in Garrelts et al. (in this volume), the Law of Protected Areas (Decreto 4–89) brought major advances in nature protection. Just as the Forestry Law (70–89), the Law of Protected Areas specified the Constitution's general demands with a set of concrete standards:[5] The laws declared national conservation objectives, created two new administrative bodies to implement them, stipulated the legal protection of a large percentage (more than 20 %) of the country's territory and prohibited certain unsustainable forms of exploitation of forest resources. Thus, they established for the first time a comprehensive regulation of the country's forestry and wildlife resources.

The Law of Protected Areas (Decreto 4–89) sets up the National Commission on Protected Areas (CONAP). This governmental body was expected to bridge the gap between the legal provisions and the actual state of in situ protection in Guatemala's protected areas[6]. CONAP has the task to run the national System of Protected Areas (SIGAP), which now comprises 120 registered areas (Castro and de León 2003 : 2). The Ley de Areas Protegidas (1989) names 17 different protection categories that are either based on international concepts, e.g., biosphere reserves, or idiosyncratic, e.g., regional parks (parque regional). In 2001, the Ministry of Agriculture presented a map with only 15 categories (MAGA 2001), a classification that is consistent with the 15 categories cited in a UNEP Technical Report, which are divided into six groups (UNEP – CEP 1996 : 10)[7]. Today, more than 3 million ha or 29 % of the national territory are under the SIGAP today (CONAP 2002 : 14). Between 1992 and 2001 a total of 45 protected areas have been declared corresponding to 5.4 % of the total national territory (Castro and de Leon 2003 : 37). The protected areas vary immensely in size, ranging from 1–2 ha to 260,000 ha. As in other countries, the degree of conservation also varies between areas: About half of the protected land is – legally – under a strict conservation regime, which represents a comparatively high percentage (around 14 %) of the total national territory.

The limited capacity of the state to manage protected areas led to the emergence of collaborative governance structures. While CONAP is the highest authority within SIGAP, it is communities, NGOs, the University of San Carlos, IDAEH, local municipalities, the National Forest Institute (INAB) and increasingly also private property owners that are administering the different protected areas. Today, 25 protected areas, corresponding to 21 % of all protected land, are under some form of co-management of the institutions named above (UNEP – CEP 1996). A prominent example is the Sierra de las Minas Biosphere Reserve: In 1989 the administration of this area that covers more than 200,000 ha was by law delegated to the NGO Defensores de la Naturaleza (Decreto No. 49–90)[8]. This NGO was the

[5] Texts of Guatemala's environmental legislation to be found under: `www.calas.org.gt/nnormativa.html`.

[6] For a detailed description of CONAP, see Castro and de León (2003).

[7] Several lists of protected areas exist with diverging data, e.g., Castro and de León (2003), www.deguate.com (2004) or UNEP-CEP (1996); the differences are due to newly declared reserves but also to different ways of counting and classifying.

[8] Guatemalan environmental law texts are published at `http://www.calas.org.gt/nnormativa.html`.

first in Latin America to receive such official authority, and to execute it with the assistance of substantial international funding (Secaira et al. 2000).

Between 1990 and 1991, a national Forest Action Plan (PAFG – Plan de Acción Forestal para Guatemala) was formulated that aimed to deal with the use patterns of forest resources in their social context (Detlefsen et al. 1991). Even though the development of the Forest Action Plan, which was supported by foreign donors, envisaged a broad stakeholder participation, the Maya communities eventually elaborated their own version of the Forest Action Plan, the PAF-Maya, articulating different, more community-oriented strategies for forest management.

The legal dimension of forest management remains conflictive: The current Forestry Law (Decreto 101–96) allows for changes in land use (Art. 46) and provides incentives for establishing monoculture plantations (Art. 80). It regulates the management of forests in more detail, but it does not recognize community forestry as a distinct form of forestry and of forest tenure. Furthermore, Guatemala lacks adequate legal measures to control illegal cutting practices (Decreto 4–89 : Art. 81), and the personnel to enforce the existing ones (Castro and de León 2003 : 39).

In reaction to the state's limited institutional capacity in forest management, an alternative instrument, the certification of sustainable forest harvesting practices, has gained ground in recent years (CONAP 2002 : 15). In December 2003, already 435,000 ha of forest were managed according to the standards of the Forest Stewardship Council (FSC), through 14 community concessions and two concessions held by logging companies. In terms of the forest area managed according to FSC standards, Guatemala occupies place 4 in Latin America, after Brazil, Mexico and Bolivia[9].

Biodiversity conservation in Guatemala benefited from regional international conservation support. As indicated in the introduction, Guatemala signed the Central American Convention on Biological Diversity in 1992 and the global Convention on Biological Diversity in 1995. Supported by international funding, a National Biodiversity Strategy and Action Plan was elaborated with the vision of a broad public consultation in 1998/99. In line with the Convention, the National Strategy and Action Plan place emphasis not only on the conservation, but also on the sustainable use of biological diversity and the creation of benefits. In spite of considerable efforts to include the indigenous population in the development of the National Biodiversity Strategy and Action Plan, this goal could not be achieved[10].

5 Prospects and Challenges for Biodiversity Conservation

As the facts presented in Sect. 2 show, the socio-economic conditions of Guatemala lead to serious challenges of biodiversity conservation. In view of the rich and

---

[9] Cp. www.certified-forests.org/data/la_table.htm.

[10] Interview with the organizers of the National Biodiversity Strategy, Guatemala City, May 2002.

unique biodiversity in Guatemala described in Sect. 3, this is a matter of global concern. Innovative policy efforts to conserve the country's rich biological diversity, as outlined in Sect. 4 constitute a potential. Against this background, the prospects and challenges for biodiversity conservation in Guatemala can be assessed as follows:

A major challenge to conservation results from the high dependence of the population on natural resources (cp. Nations et al. 1988). The fact that agricultural production employs half of the labour force illustrates this fact. The threat to biodiversity conservation is aggravated by high poverty levels and the comparatively high population growth, which leads to land scarcity. As discussed by Löning and Sautter (in this volume), the limited land availability for small-holders is one of the main reasons for deforestation, which constitutes a major challenge for nature conservation in Guatemala. Markussen and Renner (in this volume) analyse the problems of soil conservation caused by forest conversion. Using birds as an exemplary species group, Renner and Markussen (in this volume) deal with the implications of forest conversion for endangered species.

As indicated in Sect. 2, an underlying core problem of the pressure on natural resources is the unequal land distribution. The prospects to solve this problem by speeding up the land reform remain limited under the newly elected government (see Sect. 2). This problem has to be seen within the wider context of a very limited political participation of the indigenous population, a fact that is the consequence of a history of civil war, exclusion and racism, low education levels, poor infrastructural development and the lack of an effective political organization of the indigenous groups (CEH 1999; MINUGUA 2002). As outlined in Sect. 4, the indigenous groups could not be included in the development of the National Biodiversity Strategy and Action Plan, and they elaborated their own Forestry Action. Maass (in this volume) shows that the inclusion of indigenous communities requires a recognition of their cultural traditions and knowledge systems, and this is one of the major challenges of biodiversity management in Guatemala.

A high fluctuation of qualified staff within the conservation administration can be considered as an additional challenge. Many experienced professionals in the conservation sector left the public administration during the government of President Portillo, who – unlike his predecessor President Arzú – appeared to place more emphasis on party alliances than on professional qualification[11]. It remains to be seen whether this trend will continue under the new government of President Berger. Moreover, the low level of financial resources provided by the state constitute limitations to the effective protection of declared areas (CONAP 2002 : 17). CONAP's regular budget in 2003 was only about four million US$ (Castro and de León 2003 : 21).

The commercial extraction of timber, oil and other raw material constitutes another important threat to biodiversity management. Frequent disputes, for example, concerning oil extraction in protected areas, including the Maya Biosphere Reserve (see, e.g., Forests.org, 2000), underline this problem. Moreover, many

[11] Expert interviews, Guatemala City, May 2002 and August 2003.

protected areas still battle with unresolved problems concerning landownership (Detlefson et al. 1991; Castro and de León 2003).

In spite of these challenges to biodiversity conservation, there are considerable potentials to meet these challenges. The most important potential may be seen in the vibrant community of non-governmental organizations (NGOs) concerned with nature conservation. As the comparatively high percentage of protected areas under co-management shows (see Sect. 4 above), these organizations have both the interest and the professional capacity to take over tasks in nature conservation that the state is not able or willing to take up. Moreover, the NGO movement is politically active and has been able to successfully lobby for important conservation laws and policies in times when the political constellations created windows of opportunity. As a result, the percentage of Guatemala's territory that is placed under protection is comparatively high by international standards. However, as discussed by Garrelts et al. (in this volume), the exclusion of the local population in the process of declaring protected areas raises not only normative questions of participation and legitimacy, it may also have negative repercussions on the possibilities to reach conservation goals in the long run.

As the review of the conservation approaches in Sect. 4 shows, Guatemala can be considered outstanding for its innovative approaches to nature conservation. It was the first country in Latin America to formally transfer the responsibility for the management of an important protected area to an NGO (see Sects. 3 and 4). There are also initiatives, for example, by the local NGO Proyecto Eco-Quetzal, to reach conservation objectives on the basis of voluntary agreements with communities, without any state involvement (see Wittmer and Birner 2004). Private protected areas can be considered as another possibility to reach conservation goals in view of limited state capacity. As the contributions by Máñez and Zeller and by Máñez and Renner (both in this volume) show, payments for environmental services, which are discussed in Guatemala as a new policy instrument have a considerable potential to reach conservation objectives. In the forestry sector, a comparatively mission-oriented and efficient autonomous administration (INAB) has been established (Ferroukhi 2003). In spite of the limitations of acknowledging community rights to forests, community forest concessions (see Sect. 4) have been established and proved to be a particularly promising approach to meet conservation and development goals.

Another potential for biodiversity conservation can be seen in the rule of law. It was a ruling of the Human Rights Court in 2000 that restricted oil extraction in the Maya Biosphere Reserve, by invoking the human right to a clean environment (Forests.org 2000). Likewise, it was the Constitutional Court that ended the intervention of the Portillo Government in the autonomous Forest Administration (INAB) in 2002. International assistance can be considered as another potential for biodiversity conservation in Guatemala. The state agencies and the NGOs concerned with conservation were able to attract considerable international funding and expertise to support biodiversity conservation in Guatemala (Secaira et al. 2000). In summary, there are serious challenges to biodiversity conservation in Guatemala, but there is reason to hope that these challenges can be met.

References

Behar M, Tinschert O (1998) Guatemala y sus Orquideas. MayaPrin/Trade, Litho, Guatemala Ciudad, p 240

Castro F, de León F (2003) Informe Nacional de Areas Protegidas de Guatemala. CONAP,

Catling PM, Lefkovitch LP (1989) Associations of Vascular Epiphytes in an Guatemalan Cloud Forest. BIOTROPICA 21 : 35–40

CEH (1999) Guatemala – Memory of Silence, Report of the Guatemalan Commission for Historical Clarification (CEH),. Vol. 5: Conclusions and Recommendations. Guatemala City

CONAMA (1999) Estrategia Nacional para la Conservación y el Uso Sostenible de la Biodiversidad y Plan de Acción Guatemala. Informe Ejecutivo. Consejo Nacional de Medio Ambiente (CONAMA), Guatemala

CONAP (2002) Informe Institutional 2002. Consejo Nacional de Areas Protegidas (CONAP), Guatemala

Dávila J (2000) Demokratie und Korruption in Mittelamerika. INCEP, Guatemala, KAS – Auslandsinformation 3/2000, Bonn

Detlefson G, Castaneda LA, Oliva E. (eds) (1991) Plan de Acción Forestal para Guatemala (PAFG). Oficina del Plan de Acción Forestal para Guatemala, Guatemala

Dinerstein E (1995) A conservation assessment of the terrestrial ecoregions of Latin America and the Caribbean. World Bank, Washington DC, pp 129ff

Ferroukhi L (ed) (2003) Municipal Forest Management in Latin America. CIFOR, Bogor

Fischer (2002) Der Fischer Weltalmanach 2002. Fischer Taschenbuch Verlag, Frankfurt/Main

Forests.org (2000) Stop Oil Exploration in Guatemala's Maya Biosphere Reserve, `http://forests.org/archive/samerica/bifgtze.htm`

Godoy E (1997) Aves de Guatemala podrian extinguirse. Prensa Libre Guatemala 9 III, p 14

Grochembake A (2003) Guatemala's Agrarian Time Bomb. In: Tierramérca, UNEP-UNDP. `http://www.tierramerica.org/english/2003/1215/iacentos2.shtml`

Hamilton LS, Juvik JO, Scatena, FN (eds) (1995) Tropical Montane Cloud Forest. Springer, Berlin and others

Hellweg-Larsen S (2004) The Promising First Week of Guatemala's New President, Common Borders, Election News, `http://www.commonborders.org/guat_digestfirst_week.htm`

Islebe GA, Véliz Pérez ME (2001) Guatemala. In: Kapelle, M., Brown, A.D. Bosques nublados del neotrópico. Instituto Nacional de Biodiversidad, San José, pp 231–242

Janson T (1983) El Quetzal y los Mayas. Piedra Santa, Guatemala

Labastille A, Pool DJ (1978) On the need for a system of cloud-forest parks in Middle America and the Carribean. Environmental Conservation 5 : 183–190

MAGA (2001) Mapa de Áreas Protegidas. Ministerio de Agricultura, Ganaderia y Alimentación (MAGA), Guatemala

Mandl H (1989): Quetzal-Expedition. Vogelschutzreport, LBV-München 2/89 : 17–19

Markussen M (2003) Waldkonversion und Bodendegradation in Bergnebelwaldgebieten Guatemalas (Alta Verapaz) – Untersuchungen zur Biodiversitätsforschung unter besonbderer Berücksichtigung der landschaftsökologsichen Kompartimente in sensiblen Ökosystemen. Niedersächsische Staats- und Universitätsbibliothek, Göttingen

MINUGUA (2002) Report of the United Nations Verification Mission in Guatemala (MINUGUA) for the Consultative Group Meeting on Guatemala, http://www.iadb.org/regions/re2/consultative_group/gu/minugua_eng.pdf

Nadkarni NM (1984) Epiphyte biomass and nutrient capital of a neotropical elfin forest. BIOTROPICA 16 : 249–256

Nations J, Houseal B (1988) Biodiversity in Guatemala: Biological Diversity and Tropical Forests Assessment. World Directory of Country Environmental Studies. World Resources Institute / United States Agency for International Development, http://www.wri.org/data/dces.html

Primack RD, Bray H, Galletti I, Ponciano I (eds) (1998) Timber, Tourists and Temples: Conservation and Development in the Maya Forest of Belize, Guatemala, and Mexico. Island Press, Washington DC

Schulz U, Menzel F (1999) Dominante Arthropodentaxa und Sciaridae in Nebelwäldern Guatemalas – ein Inventarvergleich von hängenden Böden und Waldböden. Mitteilungsblatt der Deutschen Gesellschaft für allgemeine und angewandte Entomologie 12 : 121–127

Secaira E, Lehnhoff A, Dix A, Rojas O (2000) Delegating protected area management to an NGO: The case of Guatemala's Sierra de las Minas biosphere reserve. Biodiversity Support Program, Washington DC

SEGEPLAN (2001) Cuadros Estadisticos – Departamento Alta Verapaz: Pobreza, http://www.segeplan.gob.gt/spanish/guatemala/deptos/av/pobreza.htm

Stattersfield AJ, Crosby MJ, Long AJ, Wege DC (1998) Endemic Bird Areas of the World. Birdlife International, Cambridge

Stotz DF, Fitzpatrick JW, Parker TA III, Moskovits DK (1996) Neotropical Birds: Ecology and Conservation. Chicago University Press, Chicago

Tesliuc ED, Lindert K (2003) GUAPA – Vulnerabilidad y Crisis. En Breve, No. 22, April 2003. The World Bank, Washington DC

UNDP (2002) Human Development – Women and Education. Human Development Report Guatemala. PNUD Guatemala, http://www.pnudguatemala.org/informesdesarrollohumano/idh2002/

UNEP – CEP (1996) Technical Report No. 36: Status of Protected Area Systems in the Wider Caribbean Region: Guatemala. UNEP Caribbean Environment Programme (CEP). Regional Coordinating Unit, Kingston, Jamaica. Last Updated: August 1999, http://www.cep.unep.org/pubs/techreports/tr36cn/countries/guat.html

Unger D (1990) Proyecto Ecológico Quetzal. BIDAS-Archive, p 19
USAID (2002) Guatemala's Rural Crisis. Press Release of the U.S. Agency for International Development (USAID), March 21st, http://www.usaid.gov/press/releases/2002/fs020321_2.html
Wheelwright N (1983) Fruits and ecology of Resplendent Quetzals. Auk 100 : 286–301
Wittmer H, Birner R (forthcoming) The Role of the State in Biodiversity Governance: Theoretical Considerations and Empirical Evidence from Guatemala. Report of the Umweltforschungszentrum Leipzig Halle (UFZ), Leipzig
World Fact Book (2003) Guatemala. US Government. http://www.cia.gov/cia/publications/factbook/geos/gt.html

Calculating Incentives for Watershed Protection. A Case Study In Guatemala

María A. Máñez Costa and Manfred Zeller

Institute of Rural Development, Georg-August University of Göttingen, Waldweg 26, 37073 Göttingen, Germany, email to txonim@yahoo.com

Summary. The research is conducted in the River Basin *Mestelá* (Macizo de Cobán, *Alta Verapaz* - Guatemala), where the principal causes of threat of the biological diversity are the expansion of the exploited surfaces for agriculture and the prevalence of crops with a high dange The integration of strategies that promote watershed protection into peasants' production systems is typically associated with increased costs at the farm-household level. Against this background, the paper deals on one hand with the quantification of these costs, and on the other hand with the possibilities to compensate the peasants for these costs by introducing the policy instrument of payment for environmental services (PES). The paper focuses on payments for water supply, which is increased by environmentally friendly farming techniques that also aim to protect biodiversity. A linear programming model, which is based on empirical household data, will be used to calculate the income of farm households that differ with regard to the environmental services they generate. The income losses experienced by the farmer as a consequence of adopting systems, which provide more environmental services, will be the guideline for determining the amount of the compensation. Through the creation of different scenarios that reflect different payment schemes for environmental services and varying frame conditions, the paper shows whether the application of the PES instrument is feasible.

Key words: Payments for Environmental Services, Economic Valuation of Services, Agricultural Systems, Linear Programming

1 Introduction

1.1 Legal and policy frameworks for sustainable production systems in Guatemala

The study region, the Departamento de *Alta Verapaz*, is located in the eastern tropical and temperate highlands of Guatemala. Large-scale deforestation in Alta Verapaz began in the late 1950s when the government, in lieu of land reform, encouraged landless peasants to migrate to the department. Between 1987 and 1995, the annual deforestation rate was 1.1 percent of the total forested area, equivalent to 1860 ha per year (Jolom-Morales 1997). Degradation and loss of forest cover is caused mainly by forest clearing for cash crop production (e.g., cardamom and coffee), local demand for construction, illegal logging, rising demand for commercial forest products (particularly of the reserve's primary and old-growth forest), slash-and-burn expansion to grow subsistence crops (e.g., corn and beans), extraction of firewood (the only domestic fuel used by rural inhabitants), and finally the clearing of forest areas for the grazing of cattle (Secaira 2000).

Deforestation is a major threat not only in the study region, but in all of Guatemala. IIEE/WRI (1986) estimated the annual rate of deforestation at 2.0 %, i.e., 90,000 hectares of forest are lost each year. In Guatemala, the once-vast rainforest, dubbed by environmentalists the western hemisphere's ›second lung‹, could disappear within 20 years if deforestation practices continue at their current rate.

In Central America, a promising approach to promote forest conservation and combat land degradation is research on and promotion of sustainable agroforestry and animal husbandry systems (Malavasi and Kellenberg 2001). However, the adoption of sustainable production systems that promote environmental protection is often associated with increased costs or forgone income at the farm-household level, at least in the short-run (Máñez Costa 2003). Thus, the adoption of ecologically sustainable production systems is hampered by the fact that its costs are borne by the farmer, whereas the positive external effects of more environmentally friendly farming are predominantly treated as a by-product not to be compensated by the national or global society[1].

In recognition of the need to internalize positive externalities of sustainable agriculture and agro forestry production systems, pioneers in forest conservation, e.g., in Costa Rica and El Salvador, have implemented environmental services payment programs during the last ten years that seek to maintain socially optimal forest cover by compensating peasants for the external benefits provided by their farm households (Chomitz et al. 1999).

In Guatemala, measures to promote environmental protection have been limited so far to the establishment of protected areas reflecting different conditions and degrees of protection. Moreover, the various poverty- reduction strategies have only focused on improving the economic conditions of the rural population without explicitly considering its impact on the environment. Thus, single policy instruments pursue environmental, economic and social policy objectives and projects instead of integrated ones. There are hardly any projects in Guatemala that seek to directly address the trade-offs and synergies in smallholder agriculture between economic development in rural areas, poverty reduction and environmental protection.

The short-run trade-offs between the income needs of peasants and environmental sustainability could find a convergence in the mechanism of payments for environmental services (PES). Developing payment systems in which land users are compensated for the environmental services they deliver could be a way for society at large to compensate for the perceived benefits of agro-forestry produc-

[1] »In general, the cost of biodiversity conservation is imposed on local communities, while most of the benefits accrue to a much broader community.« (Bhattari and Hammig 1998)

tion systems. Compensatory payments for environmental services to smallholders could potentially contribute to the achievement of all three objectives of the critical triangle of rural development (Vosti and Reardon 1992): poverty reduction, economic development and environmental protection.

The legal framework enabling the possibility of payments for environmental services that we explore in this paper has already existed in Guatemala since the mid-nineties. Guatemala signed the Convenio Centroamericano de Biodiversidad (CCB) in June 1992, and the Convention on Biological Diversity (CBD) in 1995. Since then, Guatemala has compelled itself to initiate a process for developing a regulatory framework and policies for biodiversity protection. Both conventions, CCB (art. 10, 11 and 13 b) and CBD (art. 11) foresee the creation of a monetary mechanism or other adequate incentives to preserve biodiversity, as well as the incorporation of biodiversity conservation in the national jurisprudence. Both conventions focus on the integration of biodiversity concerns into policy-making by promoting the conservation and sustainable use of biological diversity of ecosystems and the fair and equitable sharing of benefits arising from their use.

The legal framework for the PES is not only provided in these two conventions, but further strengthened by the Law on Decentralisation recently enacted on 15th of April, 2002. This law makes policy measures possible that seek to empower the municipal and local governance structures for rural development and environmental protection. Yet, while Guatemala now disposes over a legal framework enabling environmental protection, the practical implementation of these laws lags behind.

1.2 Research area

Our research area is the *Mestelá* river watershed. It is located 17 kilometers south of Cobán, the capital of Alta Verapaz. The altitude of the Mestelá watershed ranges from 1400 to 2600 m ASL. The natural vegetation in the area is typical of a cloud forest. It has all but disappeared, except for areas more difficult to access for forest exploitation and farming. Yet, the remaining cloud forests hold precious biological diversity. Secondary forests under continuous extractive use can be found as ›forest islands‹ distributed over the fragmented landscape.

Next to the biological diversity in the research area, its hydrological functions (Bruijnzeel 1990) are regarded as one of the major reasons why deforestation needs to be reduced. The Mestelá river watershed provides 48.6 % of the drinking water for the city of Cobán[2]. Four hundred twenty families living in three communities located in the watershed have a direct impact on the water flow of the Mestelá River and on the city of Cobán's water catchment points. The main environmental services that the Mestelá watershed forest and agricultural ecosystems offer are the conservation of the water quality and quantity and the reduction of sediments in the water flow. Thus an important argument at the local level for forest conser-

[2] Data obtained from the municipality of Cobán.

vation and water-conserving agricultural systems is their hydrological contributions[3].

Upper watershed farmers of Mestelá applying soil conservation measures and re-vegetating the landscape are producing services for downstream people, helping to increment the water harvesting benefits, meanwhile costs remain at Mestelá. PES could manage the trade-offs among environmental options and strategies within farm households and the distribution of net benefits between Mestelá farmers and water consumers in the city of Cobán. The livelihood implications of PES are to correct the divergence between the private and social costs and benefits of ›producing and consuming‹ water and to provide the farmers of Mestelá with an economic alternative to devastating and transforming forest and perennial vegetation areas into farmland (Kahn 1998).

1.3 Research objectives

Biological diversity as a global public good as well as the provision of drinking water as a local public good are the major environmental services fulfilled by cloud forests, secondary forests and sustainable agricultural production systems in the research area. Yet peasants in the research area may incur considerable opportunity costs when conserving forests or adopting more environmentally sustainable production practices such as hedgerows, multi-cropping and tree crops.

As a prerequisite for the implementation of a mechanism for the payment of environmental services, it is important to estimate these opportunity costs for different farm-household types. Samples of the behavior in subsistence-oriented farm households in this area have been occasionally researched. Therefore, an in-depth understanding of production and consumption decisions will allow a better perception of the real patterns of resource use.

Taking the case of the Mestelá watershed as an empirical example, the paper develops a methodological approach that attempts to quantify the income losses incurred by peasants when practicing different water- conserving production measures. Second, we discuss the possibilities of compensating the peasants for these costs by introducing the policy instrument of payment for environmental services (PES) (Máñez Costa 2003).

In the following section, we present our method of data collection and conduct descriptive data analysis to categorize four different farm households[4] in the area. We further distinguish environmental measures that are already practiced by farmers in the area. In Sect. 3, a linear programming model for the four farm

[3] Environmental measures of upper watershed farmers of Mestela signify not only conserving cloud forest, but also revegetating the lanscape. Against annual crops, not only cloud forest but also perennial vegetation helps in the area to capture additional water supply from the atmosphere (cp. WRI 2002).

[4] »Careful analysis of the microeconomic behavior of smallholder farmers is thus central to understanding the roots of environmental degradation, the deepening poverty, and design of appropriate policies and strategies for reversing the problem« (Shiferaw and Holden 1997).

households is developed. The model is used to estimate the opportunity costs, i.e., the farmers' forgone private income when implementing various environmental measures. Exemplary simulation with respect to conserving secondary forests, assuming different levels of payment for environmental services will be shown. The paper concludes with a discussion of the potential arising from implementing payments for environmental services.

2 Conceptual framework

The theoretical framework for analyzing the concept of payments for environmental services (PES) as an economic instrument for environmental management draws from the utilitarian approach in economics, specifically the theory of externalities, based mostly on Pigou´s (1932) exposition of potential market failures when a externality[5] is present (Carlson 1993). From an economic point of view, the positive externalities of agroforestry farm households[6] are activities that contribute to the preservation and conservation of natural capital. Small farmers in a watershed who implement re-vegetation measures and soil conservation measures, which we assume to be positive externalities, cannot add to their income the ex-situ benefits of their practices. Yet, these measures increase water the supply so that the benefits not only accrue to the farmers but also to society. The decline of such environmental services can be directly linked to the loss of productivity and natural capital. To avoid such a situation, it is necessary to estimate the changes in the management of farm households by offering environmental services. It is also essential to calculate the opportunity costs of the preventative expenditures needed to avert natural capital loss in order to advise decision makers in environmental management (Hearne 1996).

Chomitz et al. (1999) proposed the creation of incentives and compensation mechanisms for internalizing the positive externalities of farming, making a link between theoretical economics and the real world, assuming that forested areas would improve if their owner are compensated for all services they provide. Based on this assumption, the calculation and the determination of the opportunity costs of preventative measures in the watershed is essential for the development of win-win options in Mestelá. (Klaphake, Scheumann and Schliep 2001). In our case, the economic evaluation of such opportunity costs is performed using the method of linear programming (Panell 1997). This method derives shadow values for the introduction of a new or the extension of existing farming activities. These values can be interpreted to be opportunity costs, i.e., the loss of income occurring when introducing or extending certain activities such as environmental measures into ex-

[5] Externalities are positive or negative effects of an activity of an agent upon others or upon the activities of others, which are not factored into market prices. The idea of ›internalizing externalities‹ is an attempt in economics to reduce such distortions in the price system.

[6] Refering in the case of Mestela to revegetation measures as well as the non-use of pesticides.

isting farm households. Linear programming[7] has been chosen as a proper modeling technique of farm behavior because the farm's income is linear in output prices and quantities (Cervigni 2001).

3 Data collection and the categorization of farm households

The data collection was carried out during two periods: from May to September 2001, as well as from April to July 2002. Simple random sampling was applied to select 46 out of a total of 420 families residing in the three communities in the research area. SIAS (Sistema Integrado de Asistencia Sanitaria) was the source of the sampling frame[8].

In this random sample, semi-structured interviews were carried out with the heads of the households. An interview guide was prepared beforehand as a framework for the interviews. The interview guide focused on four areas of interest: land, the demographic structure, livestock and the productive processes. The semi-structured interviews made it possible to gather in-depth information on these topics and enabled us to understand better the problems and potentials of farming communities and households in the research area better. This data was obtained in 2001. The data from the semi-structured interviews was used later to categorize households into four groups (see following section).

During the second field research phase in 2001, detailed structured questionnaires were applied to these 46 households. This data was crucial for defining costs, inputs and outputs of different production activities as well as the constraints for labor, capital and land that were fed into the construction of the linear programming models. For each of the four household types, we developed a linear programming model that aimed to represent the average household for that group. Thus, for example, the land area available for farming for a particular type of household was calculated to be the average land area possessed by all of the sample households belonging to this type.

In the second period of the data collection, the findings obtained from the first LP simulations were compared with the actual economic situation. Secondary and additional data were also gathered to further improve and validate the models.

3.1 Categorization of farm households

Following Chayanov's theory of the peasant household (Chayanov 1986), we categorized households along their major production constraints, i.e., the availability

[7] See: Anderson et al. (1977); Brandes (1970); Hazell and Norton (1986); Dabbert (1986); Bromley (1990); Dent, Harrison and Woodford (1986); Urff (1964); Binswanger (1980); Shiferaw (1997); Barbier (2001) etc. See also: Appendix: Linear Programming.

[8] SIAS is the health care system of Guatemala. It works at a local level distributing the population in 10,000 people per jurisprudence. Every single household of every jurisprudence gets registrated in a list with an individual number. We used the SIAS-Infrastructure for the data collection.

of land and forest and the use of family labor on and off- farm. In addition, we also took into account whether the household held livestock. Based on these criteria, four different groups were distinguished representing the different farm households in the research area (see Table 1).

Table 1. Farm households classification

	Aa[a]	Ab[b]	Ba[c]	Bb[d]	Units
farmed land	25	50	100	200	cuerdas
labor in own farm	3.4	8.8	14	20.1	c/w ratio per month
labor out farm	17	12.3	3.5	0	c/w ratio per month
livestock					
pigs	0	0	1	5	livestock
cattle	0	0	3	15	livestock
Possesion of primary forest	0	0	30	45	cuerdas
secondary forest	0	12	33	55	cuerdas

Source: Máñez Costa (2004)
[a] Poorest land-scarce farmers (n = 7).
[b] Land scarce spec. veg. farmers (n = 14).
[c] Vegetable farmers (n = 10).
[d] Livestock and maize farmers (n = 7).

The categorization of the farming households into four different types was motivated by the fact that household behaviour is conditioned by differences in underlying production and consumption constraints. Thus, households react differently depending on their economic constraints that are mainly land, labour and capital[9]. Given the imperfections of the food markets in rural Guatemala, households of the area initial do not behave like cash profits maximizers, but instead seek to secure their basic food needs through home production. Once they have secure subsistence households start maximizing profits[10]. In these households production and consumption decisions are non-separable in the semi-subsistence households in the research area[11], a crucial characteristic that is taken into account in the linear programming model by introducing minimum constraints for food production.

As shown in Fig. 1, in Mestelá farm households can be characterised depending on the labour input in their own household, i.e. the proportion of months per

[9] Cp. Thorner 1986.

[10] Cp. Shiferaw and Holden (1997). According to them, »this is mainly because production and resource use decisions are likely to be affected by non-profit considerations such as preference for home production of staple food, leisure consumption and other goals«.

[11] The household act as »producers and consumers, meaning that their consumption and labour supply decisions are not indepent of production and labour demand decisions« (Mlay and Nhantumbo 1999).

year working on- and off-farm. Thus the sample of households was divided in two groups, A and B:

The households included in group A worked on average 6 months off-farm, while those in the group B worked off-farm less. Furthermore, the households of Mestelá were distributed additionally along the major production constraints, i.e., the availability of land and forest, livestock tenure and capital accessibility. As reflected in Fig. 1 this additional distribution allocated the farm households in to four different types: FH *Aa* represents the poorest land-scarce farmers, FH *Ab* represents land scarce farmers specialised in market gardening, FH *Ba* represents market gardening farmers and FH *Bb* represents livestock farmers. In addition, the amount of on-farm production was important for the classification of the four groups. Fig. 1 reproduces the amount of income coming from on-farm production. It shows that for group *B*, around 100 % of their income comes from on-farm production while for group *A* only 30–70 % of their income comes from on-farm production. Therefore, farmers belonging to group *A* complement the obtained income from on-farm activities with income from wage labour off-site. For these farmers, labour off-site is the major source of income.

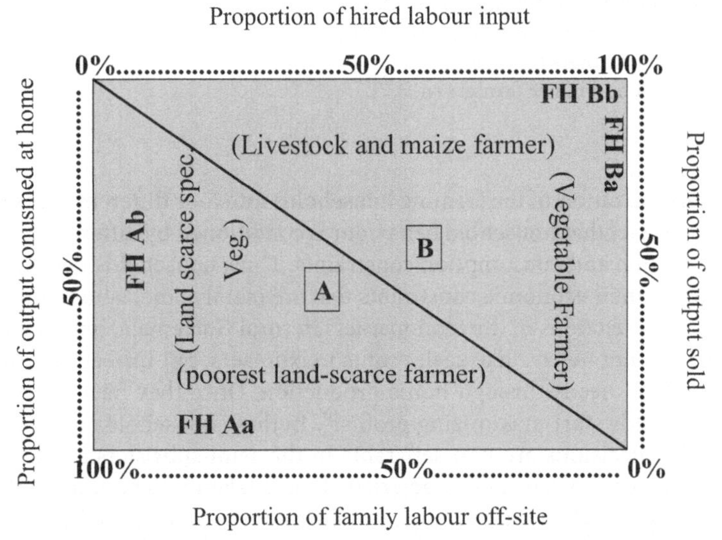

Fig. 1. Farm household classification (Source: Máñez Costa 2004, adapted from Nakajima 1986)

3.2 Environmental Services

After having categorized the farm householdss, we analysed the different environmental measures that peasants already put into practice[12] (see Table 2). These environmental measures for the peasants of Mestela are: secondary forest and natural reforestation, fruit trees intercropped with annual crops, trees in lines, living barriers, living fences, block planting, terraces and not using of Pesticides.

Table 2. Environmental measures of the FHs of Xucaneb-Mestela (Units: Cuerdas)

Environmental Measures	Farm Households			
	Aa	Ab	Ba	Bb
Natural forest	0	0	33	55
Trees in lines	0.5	2	6	10
Intercropping	1.2	6	12	20
Living fences	0.1	3	5	8
Living barriers	0.5	1.5	2.5	6.5
Block planting	6	5.5	4	3
Eco-farming	24.1	37	0	0
Terraces	0	5.5	15	20
Average of land	38.4	90	191.5	235.5

Source: Máñez Costa (2004)

All of the listed environmental measures are already implemented within the farm households of Mestelá. Almost all of the implemented environmental measures serve to conserve soil through re-vegetation. They affect the water catchments capability of the research area positively and protect the soil against soil run-off and erosion[13]. Thus we assumed that these measures contribute to the improvement of water quantity and quality[14]. Moreover, they result in more diverse agro-ecological landscapes with beneficial effects on biodiversity[15].

From Table 2 we can see that the environmental measures currently used by the farmers are correlated with the farm household's classification. This is not surprising as the classification of the farm households was based on the underlying prime

[12] »Theory suggests that, in the absence of market, policy or institutional failures, landowners would be inclined to adopt soil conservation practices that lessen the impact of the on-site costs that result from soil degradation.« (Aylward et al. 1995)

[13] »Ground cover, rather than canopy, is the chief determinant of erosion« (Wiersum 1984, cit WRI 2002).

[14] Cp. WRI (2002); Achard et al. (1998); Brown and Schreckenberg (1998); Ciesla (1995); Meyers (1992); Bruijnzeel (1991) etc.

[15] Cp. Nasi et al. 2002.

production constraints, i.e., family labor and access to land and forests. For example, farm households *Poorest land scarce farmers* represents the ›poorest peasants‹ in the area with little of arable land, low-income, lack of livestock and extreme dependency on earning wages off the farm, as well as no access to forestland. Thus, due to land scarcity, *Poorest land scarce farmer* is not likely to carry out the environmental measure ›secondary forest and natural reforestation‹. Presumably because of the high poverty level and low access to capital, *Poorest land scarce farmers* did not use terraces[16] for soil conservation, as it requires considerable labour efforts as well as capital inputs. Another example was given us as a reflectance of cash income access, which again reproduced the correlation between farm households and environmental measures, namely the use of pesticides and fertilizer (described in the table as eco-farming). Only a few farmers do not use pesticides because of the income constraint. These farmers belong to the group *A*.

3.3 Recent behaviour of the farmers after the coffee crisis

The crises on the world coffee markets in year 2000 and the corresponding decline of coffee prices have seriously worsened the wage-earning potential of the Mestelá seasonal labourers. About 68 percent of the sample households were found to engage in wage-labour activities linked to coffee farming. Thus, the coffee crisis led to a considerable decline in income in many Mestelá households. Hence, as a response to this crisis, the farmers were increasingly forced to use the remaining forestland for food crop cultivation. In this context, it is well-known facts that in such ecologically fragile areas as cloud forest, if the soils lose their vegetation cover then they are greatly exposed to the leaching and eroding effects of rain[17]. Indeed in Mestelá it was observed that farmers started to cut and burn new surfaces for staple crops production. Moreover, they sold timber to traders in order to compensate for the loss of off-farm cash income. About 76 % of the sample farmers were found to view timber as precautionary savings.

Over the years, the research area Mestelá witnessed the following:

- The forest degrading activities in the River basin Mestelá have caused forest decline and a reduction in the perennial vegetation. About 47.4 % of the total surface of the basin is categorized as being over utilized[18].
- Since 1994, the water flow experienced a reduction from an initial water flow of 70 l/sec to 32 l/sec in 2000[19]. During the dry season, Cobán now experiences frequent irregularities in its tap water supply.

[16] Due to the soil typ building terraces on this area is linked to high cash expenditure but also to high transaction cost because Q'eqchi Farmers are not familiarized with these techniques. A calculation from a local NGO gave us a price to build a cuerda of terraces of 280 quetzales a year.

[17] About 4000 mm/year.

[18] GTZ (1997) declared that agriculture and continuous extractive use of wood have provoked soil degradation.

[19] Data obtained from the Municipality of Cobán – Summer 2001.

3.4 Measuring the demand of water

As the inhabitants of Cobán are faced as water consumers with these deteriorating conditions in their water supply, we explored their willingness to pay (WTP) for environmental services.

The interview for the WTP-Questionnaire was elaborated in two steps:

- First we tried to create an instrument that could be understood by the interviewed persons and that also could give us a range of payments we could ask about. The first interview guide was done on focus groups with different social status and different areas in Coban. The interviews were »open-end«. The information about the creation of scenarios for the second interview guide was created with the obtained information from the focus groups.
- Second we elaborated a WTP-questionnaire to capture the water consumer surplus. We presented them a scenario with the actual data about water quality and quantity of Coban (see above) and explain them the role of vegetation in the water catchments area and for what the obtained money would be used. We asked them about the WTP for better quality and quantity of water[20]. The amount of the fees varied between 5, 10, 15, 20, 25, 30 and 35 quetzales over the actual water fee.

This survey found out that the population is fully aware of the environmental problems linked to the loss of vegetation cover and forest. About 96 % of the respondents believed that the degradation and deforestation of the river basin is the primary cause for the infrequent and low quality of the water supply. About 63 % of the respondents were willing to pay more for the provision of water.

These additional payments could be used for the provision of the above- mentioned environmental services provided by farmers in the Mestelá watershed. Those among the respondents who were not willing to pay more for tap water were among the poorer segments of the urban population. Using a contingent valuation method, we obtained a median WTP amount of 25.32 quetzales per month per household. This is almost double the actual median payment of 12 quetzales per month and household. The city of Cobán has 4,109 households that receive their water from the Mestelá watershed. Thus, simple calculus would yield an additional amount of 105,000 quetzales per month that could be potentially used to compensate farmers in the Mestala watershed for conserving forests and soils through adequate agro-forestry measures so as to improve the quality and quantity of water. Based on our survey results, we estimate the average household income in the Mestelá region at 1,800 quetzales per month. Thus, if the municipal government were to raise the water charges to about 25 quetzales per month, each household in the Mestelá region could obtain up to 130 quetzales per month. Of course this

[20] We defined water quality as the future possibility of drinking tap water without cooking it or filtering it (Nowaday it is impossible to drink tap water without getting sick. This is a known problem for the citizen of Coban). Water quantity was defined as a continuos and stable flowing of water during all the year as opposed to the today's circumstances of instable water flowing.

simple calculation only demonstrates that the willingness to pay on the part of urban water consumers is considerable and could contribute a considerable share of rural incomes. Compensation, however, should be linked to the opportunity costs incurred by farmers arising from environmental measures and should also be dependent on the hydrological contributions of these measures. Yet, other contributions derived from environmental measures such as biodiversity and landscape enhancement may also be figured in when deciding on which environmental measure to support through direct payment to farmers implementing those measures. In the following section, we explore the former aspect as a crucial prerequisite for the design of a decentralized payment system for environmental services.

4 Model simulations

In order to calculate the opportunity costs of implementing various environmental measures, a linear programming model (LPM) was used[21]. The LPM provides a suitable analytical framework for representing the complex input-output relationships of farm households and enabled us to model the production constraints (see Table 1) of the four different farm households.

Each environmental measure as every farming activity was considered alike, i.e., the computation for the different activities was calculated depending on the opportunity costs of labor per land unit per year.

Depending on the household type modeled, the main activities in the LPM amounted to 36, including crop and livestock production activities, miscellaneous activities for crop sales and home consumption, the environmental measures as shown in Table 2 and the seasonal sale of family labor off-farm as well as the hiring of labor for on-farm activities. We distinguished three seasons in the local labor market. The production constraints for each of the four types of farm households were modelled as observed in the survey. For example, the LP matrix for type *Poorest land scarce farmers* contains 0.45 acres of arable land (see Table 1) and 12 person-months of family labour available either for on-farm or for off-farm employment. Food subsistence constraints were modelled as a minimum requirement of 0.520 kg per person/day of maize[22] and 0.07 kg of beans per person/day. An important constraint was also the total amount of payments for environmental services that the inhabitants of Cobán were willing to pay, i.e., 105,000 quetzales.

The four models have a number of simplifying assumptions for the baseline scenario that need to be taken into account when interpreting the results. First, we

[21] The model runs in XA-Solver with Excel-Interface spreadsheet.The XA Professional Linear Programming System is copyrighted and licensed software from Sunset Software Technology.

[22] In rural Guatemala there is a calories intake of 1,148 calories per day/person through maize consumption (Total: 1,994 calories per day/person). The comsumption of maize in this area is higher than in the rest of Guatemala. The average for rural Guatemala is 318 gr pre person/day, but in poor areas men may consume about 600 gr. and women about 400 gr. (cp. FAO, Food and Nutrition Series No. 25).

assumed that the land is fixed in the short-run (i.e., no markets for land exist), although sales and rental of land were observed in the research areas. Second, we did not consider the differences in land quality mainly because of a lack of on-farm agronomic data for different soil types. Third, we assumed that farm households do not have access to formal credit that could be used for on-farm crop and livestock production or financing environmental measures. This is a realistic assumption as no credit program operated in any of the three villages. Most micro-finance programs in Guatemala cater to the urban and semi-urban non-farm enterprise sector. Fourth, we assumed certainty concerning crop prices, input-output relationships and, therefore, did not consider decision-making under risk.

We validated the models for the four household types by comparing the results obtained for the above baseline scenario with the information obtained through the primary and secondary data collection. In the baseline scenario, the models reasonably replicate the observed actual cropping pattern, amount of crop sales, seasonal income disparities and allocation of family labour between on and off-farm activities. Examples of the simulations are:

- When we simulate a higher availability of equity capital, the farm householdss tend to increase animal husbandry (mostly cattle) and basic grain production, and the optimal farm households thereby moves towards the situation of the farm households »livestock and maize farmers« that has more equity capital than the other three systems
- When we simulate a lower availability of arable land, the model reacts as farm households *Poorest land scarce farmers*, selling its family labour more off-farm.

In the following, we present the findings that simulate a policy scenario of payment to farmers for converting current crop or grazing land into secondary forests. The policy scenario simulates the impact of different annual payment levels for the environmental measure ›secondary forest or natural reforestation‹ on the different types of farm households. From the theory, we can deduce that the four types of farm households differ in their opportunity costs for environmental measures and, thereby, differ in their behaviour towards accepting a certain amount of payment. In the following, we will show the findings for type ›vegetable farmers‹.

The computation for the different activities was calculated depending on the opportunity costs. To make clear how the decision about which payments are opportune for every FH, the example of FH *Ba* will be used, i.e., the way payments were calculated for all the environmental measures will be shown. Fig. 1 reflects a scale of payments for FH *Ba* and shows the simulated land-use changes and increase in land units under PES for the maintenance of secondary forest and support of natural reforestation[23]. The first payment of 1036 quetzales replicates just the shadow cost of the measure ›secondary forest‹ for an amount of 33 cuerdas. The

[23] Not forgetting are the main constraints for FH *Ba*: an area is necessary for satisfying the subsistence demand (maize and beans) (100 cuerdas), the cultivation of vegetables, animal tenure, off-farm activities, family labour, hiring labour, capital (57,504 quetzales) and total area (163 cuerdas).

FH will not react until payments amounting to more than the shadow cost are offered. Once the LPM offers more money the FH would provide additional cuerdas of land for secondary forest (i.e., implying lower opportunity costs of secondary forest). The following payments do not significantly change the disposition of the farmer to offer more land units for this measure.

The FH reacts first when the LPM offers about 1,100 quetzales. At this time almost all possible land that the FH can provide for secondary forest is offered. This situation will place the FH in a dependent position from payments. In a country like Guatemala such conditions are not desirable due to the political and institutional instability. Therefore, the model tries to avoid these circumstances in order to recommend payments that do not place farmers in an awkward economic situation.

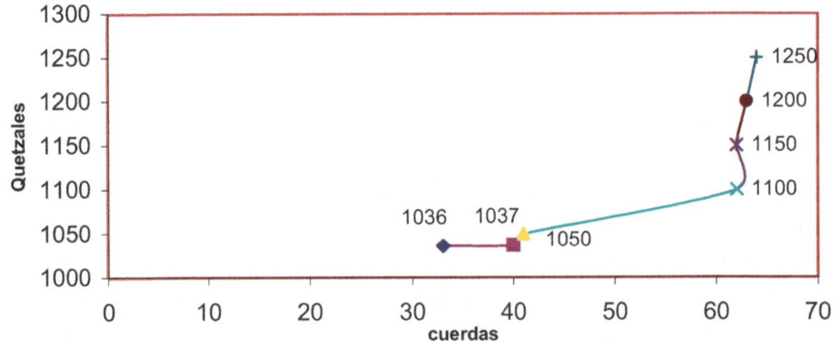

The curve does not reflect the elasticity of the supply. It simply reflects seven different payments for conserving forest and the changing in farmers' reactions to such payments. The payments are done in quetzales and the land units are cuerdas.

Fig. 2. Payments for forest conservation (FH Ba)

At this point, constraints for the payments are introduced in the LPM and payments are set to the amount shown in Table 3. Along these lines, fixed prices for environmental measures are introduced like for market products. The amounts were discussed with local experts and farmers in the area. They were also compared with ›cost and benefit data‹ from other neighbouring communities gathered by the local NGO Ecoquetzal and the European Development Programme PALTA. The payment will never be less the fixed payments of Table 3.

In this way we calculated the ›price‹ for every single environmental measure reflected in Table 2 for every single farm household type[24]. The above findings concerning the minimum level of payments for secondary forests are location-specific and would likely differ with varying biophysical, social and economic conditions

[24] See Máñez Costa (2004) for the rest of simulations.

found across different watersheds of Guatemala. With the LP, we can produce such figures for each environmental measure of every farm households.

Table 3. Payments for environmental measures (quetzales/year)

Environmental Services	Farm Households			
	Aa	Ab	Ba	Bb
Forest	0	0	455	375
Fruit trees	256	236	285	191
Trees in lines	134	135	161	173
Living barriers	206	188	242	181
Living fences	78	72	97	81
Blockplanting	53	55	68	33
Ecofarming	29	75	35	162
Terraces	0	435	504	373

Source: Máñez Costa (2004)

5 Findings and Conclusions

The LPM provides a tool for simulating the possible impacts of various payment levels for different environmental services. The LPM gives us estimation about the farmers' reactions to the policy instrument and could help to find a satisfactory mixture of options.

The calculation enables us to provide a more substantiated estimation of the required level of payments so that a specific environmental measure will be increasingly adopted in the farmers' production systems.

We discussed different scenarios with urban water consumers, local stakeholders and farmers. The groups are aware of the fact that economic scarcity in the area will drive farmers to continue using more land covered by forest for staple food cultivation, mainly maize and beans, to reduce the fallow periods and to abandon conservation techniques such as re- vegetation measures. They also agree and react positively to the ›possible‹ introduction of this policy instrument. Strengthening that financial compensation would significantly enhance the likelihood of more environmentally friendly production systems.

On these grounds, the creation of such an incentive mechanism for conservation is a first step that can transform the environmental measures into an attractive production factor and in that way slow down the loss of vegetation cover and forest in the area as well as provide farmers an alternative income source. A second step could be a cost-efficiency calculation for the total amount of money that we ›could have‹ yearly from the water payments made by the citizens of Cobán for environmental measures in Mestelá. However questions about the implementation of cost-efficient measures are difficult to analyse and answer because in order to do so, we have to generate a specific order of preferences. These preferences change depending on the goal, i.e., either water catchments protection, or biodiversity protection and conservation, or landscape conservation. In the case of watershed protection, the creation of such a preferences-list is linked to the in-depth study of the hydrological function of re-vegetation measures and that would go beyond the scope of this study.

Acknowledgments

This contribution profits from the comments of the participants of the International Conference on Rural Livelihoods, Forests and Biodiversity, held between the 19th and 23rd of May, 2003, in Bonn, Germany.

Appendix: Linear Programming

Linear programs are characterized by the following properties (cp. Footnote 7):

- Optimization of a problem/mathematical program: The problem involves finding the best value for a given objective, subject to a given set of constraints.
- Deterministic: The behavior of the model is completely determined, that is, there is no probability contributing to the objective or constraints. Proportionality: Each variable contributes in direct proportion to its value.
- Additivity: The variables in the objective and each constraint contribute the sum of the contributions of each variable. Divisibility: The variables can take on continuous values subject to the constraints.
- Any function satisfying the proportionality and additivity properties is called a linear function, and will always have the form
  $f(x_1, \ldots, x_n) = a_1 x_1 + \ldots + a_n x_n (+d)$

References

Achard F, Eva H, Glinni A, Mayaux P, Richards T, Stibig HJ (1998) Identification of Deforestation Hot Spot Areas in the Humid Tropics. TREES Publication Series B No.4, European Commission, Luxembourg

Anderson JR, Dillon JL, Hardaker JB (1977) Agricultural Decision Analysis. Iowa State University Press, Iowa

Aylward et al. (1995) Economic Incentives for Watershed Protection: A Report on an Ongoing Study of Arenal, Costa Rica. CREED Working Paper Series No 3. The International Institute for Environment and Development, London and the Institute for Environmental Studies, Amsterdam

Barbier B, Bergeron G (2001) Natural Resource Management in the Hillsides of Honduras. Bioeconomic Modelling at the Micro watershed Level. Research Report, IFPRI, Washington

Binswanger HP (1986) Attitudes toward Risk: Experimental Measurement in Rural India. In: American Journal of Development Studies 20 : 5–21

Brandes W, Wörmann E (1970) Landwirtschaftliche Betriebslehre. Hamburg

Bromley DW, Larson BA (1990) Property Rights, Externalities, and Resource Degradation. In: Journal of Development Economics 33 : 235–262

Brown D, Schreckenberg K (1998) Shifting Cultivators as Agents of Deforestation: Assessing the Evidence. In: Natural Resource Perspective Number 29, April 1998, London

Bruinjzeel (1990) Hydrology of Moist Tropical Forests and Effects of Conversion: A State of Knowledge Review. UNESCO IHP, Humide Tropics Programme, Paris

Carlson GA, Zilberman D, Miranowski JA (1993) Agricultural and Environmental Resource Economics. Oxford University Press, New York

Ciesla WM (1995) Climate change, forests and forest management: an overview, FAO Forestry Paper 126, Forest Resources Division, FAO, Rome

Clark KL, Lawton RO, Butler PR (2000) The Physical Environment. In: Nalini MN, Nathaniel T (eds) Oxford University Press

Chayanov AV (1986) Theory of Peasants Economy. Manchester

Chomitz (1999) Financing environmental services: The Costa Rican experience and its implications. In: Science of the Total Environment 240 : 157–169

Dent JB, Harrison SR, Woodford KB (1986) Farm planning with linear programming : concept and practice. Butterworths, Sydney

Ellis F (1988) Peasant Economics. Farm Households and Agrarian Development. Cambridge University Press, Cambridge

FAO (1992) Maize in Human Nutrition. Food and Nutrition Series No. 25. Rome

Haber W (2000) Plants and Vegetation. In: Nalini M. Nadkarni and Nathaniel T. Monteverde: Ecology and Conservation of a Cloud Forest. Oxford University Press. Oxford

Hazell PBR, Norton RD (1986) Mathematical Programming for Economich Analysis of Agriculture. MacMillan Publishers. New York

Hearne R (1996) Economic valuation of use and non-use values of environmental goods and services in developing countries. Project Appraisal 11(4) : 255–260

Jolom-Morales (1997) Caracterización de la Actividad de Cacería en la Reserva de la Biósfera Sierra de las Minas y Diseño de un Plan de Monitoreo. Fundación Defensores de la Naturaleza. Guatemala

Klaphake A, Scheumann W, Schliep R (2001) Biodivesity and International Water Policy. International Agreements and Experiences Related to the Protection of Freshwater Ecosystems. Research Report commissioned by the Federal Ministry for the Environment, Nature Conservation and Nuclear Safety, Berlin

Malavasi EO, Kellenberg J (2002) Program of Payments for Ecological Services in Costa Rica. Paper presented on the International Expert Meeting on Forest Landscape Restoration in Heredia, Costa Rica

Máñez Costa (2003) Strategien und Optionen für den Natur- und Umweltschutz in agrarischen Produktionssysteme – Ein Fallbeispiel aus Guatemala. BfN, Bonn
Meyers N (1992) The Primary Source: Tropical Forests and Our Future. Norton Publishers. New York
Mlay GI, Falção M, Nhantumbo I, Kowero G (2003) Policy impact on woodland resource management, use and conservation in Mozambique: Case study of selected sites in Dondo. Nhamatanda. Gondola and Manica District. In: Kowero G, Campbell BM, Sumaila R (eds) Policies and governance structures in woodlans of Southern Africa. CIFOR, Bogor
Nasi et al. (2002) Forest Ecosystem Services: can they pay our way out of deforestation. GEF. San José. Costa Rica Ortiz Malavasi E, Kellenberg J (1999) Program of Payments For Ecological Services in Costa Rica. Costa Rica – United Kingdom
Panell D (1997) Introduction to practical linear programming. New York
Secaira E (2000) La Conservación de la Naturaleza, el Pueblo y Movimiento Maya, y la Espiritualidad en Guatemala: Implicaciones para Conservacionistas. Proarca - Capas. Guatemala
Singh I, Squire L, Strauss J (1986) Agricultural Household Models. Johns Hopkins University Press. Baltimore. USA
Shiferaw B, Holden ST (1997) A farm household analysis of resource use and conservation decisions of smallholders: an application to highland farmers in ethiopia. Discussion Paper No D-03/1997. University of Norway
Thorner D (1986) Chayanov and The Theory of Peasants Economy. Madison University Press, Wisconsin
von Urff W (1964) Productionsplanung in der Landwirtschaft. Unter besonderer Berücksichtigung der Methode des Linear Programming. Berlin
Vosti S, Reardon T (1992) Agricultural Development, Sustainability, and Alleviating Poverty: The Critical Triangle. IFPRI, Washington

The cultural context of biodiversity conservation

Petra Maass

Institute of Cultural and Social Anthropology, University of Göttingen, Theaterplatz 15, 37073 Göttingen, Germany, email to pmaass@gwdg.de

Summary. Due to the recognition that many areas of the world that contain high levels of biodiversity are cultural landscapes inhabited by indigenous and local communities, the significant role such communities play in preserving natural resources has been underlined in the Convention on Biological Diversity (CBD). In particular, the Convention calls for the acknowledgement and wider application of local knowledge systems being embedded in traditional lifestyles as they can contribute to the *in situ* conservation of biodiversity. The purpose of this contribution is to analyse the role of indigenous communities and local knowledge systems in the global environmental debate. It draws on an ethnographic case study of *Maya-Q'eqchi'* communities living adjacent to protected areas in Guatemala. The operative paradigm, that underlies the anthropological perspective, indicates that an understanding of the cultural context is essential to the success of any initiative designed for the sustainable conservation of natural resources. Accordingly, the applied approach assumes that indigenous environmental knowledge has to be encountered as a social product integral to the respective cultural system it has been generated in. Equally, human cognitive understandings of nature are culturally embedded, bounded to locality and intertwined with the broader context. This implies a multidimensional reality in which diverse economic, social, political and historical aspects intersect. The field-based research is concerned with these contextual dimensions of indigenous knowledge, whereas the particular purpose aims to explore the significance of cultural values such as social identities related to the local landscape and beliefs in the intimate attachment of humans to nature that are closely tied to natural resource use patterns, subsistence activities and ritual practices that define indigenous perceptions of the natural environment.

Key words: CBD, indigenous communities, local environmental knowledge, protected area management, Q'eqchi', cultural landscapes, social and spiritual values of biodiversity

1 The global context

> »The core idea (...) is that the existence and the future course of biodiversity is dependent upon both biological and sociocultural processes (...) it is essential to see biodiversity in its multilevel process of conditions and impacts.«[1]

[1] Kamppinen and Walls 1999 : 14f.

At the end of March 2003, the author of this contribution while participating at a harvest with a *Q'eqchi'* farmer in a remote village in *Alta Verapaz*, observed clouds of smoke in the sky. She asked about their origin and supposed the informant would attribute the smoke to extensive forest burnings in the northern lowlands. Instead he expressed his concern about the increased conflict between the United States and Iraq. Having followed the events on the radio news, he interpreted the smoke as an indication of the recent attacks on Baghdad. As the worldview of the farmer does not imply continental geographies, transatlantic distances and country boundaries, the war of global range could accordingly even reach the local peasant's life in the countryside in Guatemala. Before this conversation between rows of maize, we had walked to the field passing areas at different stages of fallow and forest. On the way the farmer shared his knowledge on the use of almost 100 plant and tree species. Although the conflict in Iraq indeed had consequences for the local economy, it is beyond the intention of this contribution to discuss its impacts on living conditions of rural communities in Guatemala. Neither does it concentrate on the role of modern media in traditional cultures, nor does it intend to analyse the influence of forest-fires on the deforestation process in the study area. The intention of this contribution is to question the significance of people's spatial concepts and interpretations of phenomena occurring in their environments. The overall frame given to the anthropological considerations on this topic is defined by the interface of *global* environmental policies and their implications on cultural dynamics at the *local* level. In the context of the research project on ›Valuation and Conservation of Biodiversity‹, the present portrayed study initially questioned how to protect biodiversity in accordance with community-based natural resource management that is basically built upon local environmental knowledge. Thereby, the primary aim was to move towards an understanding of notions in the more encompassing sense of knowledge associated with social mechanisms, historical and political experiences, cultural identities and interpretations of people through which they structure and comprehend their environment. During the field-based research, the focus altered from contextual aspects of knowledge generation, transmission and transformation towards internal dynamics such as inherent values and implicit symbolisms of human-nature relations. These tend to be underestimated issues in the international environmental debate on biodiversity conservation policies and equally are often dismissed in the frame of implemented initiatives in protected area management.

1.1 Indigenous communities and biodiversity conservation

The significant correlation of regions inhabited by indigenous[2] communities and areas of highest remaining biodiversity has given evidence to the close inter-

[2] Having been discussed over decades, the question of how to define *indigenous* has become a highly politicised matter and is still far from being clarified. The UN has defined the term according to the widely accepted definition elaborated by José Martínez-Cobo: »Indigenous communities, peoples and nations are those which, having a historical continuity with pre-invasion and pre-colonial societies that have developed on their terri-

dependence of natural resources and human ecology. Indigenous resource management systems have evolved over time as adaptive responses to the specific natural conditions of particular local environments. In many areas of the world, ecosystems in regions of high conservation priority have been shaped by indigenous resource use related to subsistence patterns like agriculture, forest extraction, hunting and gathering practices. Living in close relation to complex ecosystems, indigenous people have developed an understanding of the diverse reciprocal interactions of such systems within their environment. Thus, extensive knowledge related to local environments often has conserved and even enhanced biodiversity (Laird 2000 : 348). Although many governments have refused to recognise this connectedness and the legitimacy of indigenous systems of resource management and property in the past, the important role indigenous and local communities play as decisive stakeholders for initiatives concerning the protection and sustainable use of biodiversity has been increasingly recognised in the global environmental discourse. In particular, the contribution that local knowledge systems can make to conservation efforts has been widely acknowledged and has become an evolving subject of national and international law.

1.2 The Convention on Biological Diversity

Among other agreements, the most important global policy framework for conservation strategies was negotiated at the United Nations Conference on Environment and Development held in Rio de Janeiro in 1992. One of the major outcomes in underlining the role of indigenous and local communities in *in situ* conservation is the Convention on Biological Diversity (CBD). Official commitments embodied in article 8(j) of the CBD mark the starting point of the present study. In this provision clear reference is made to the significance of local and indigenous communities in environmental protection. The section requires the contracting parties to take measures ›as far as possible‹, and ›subject to their national legislations‹ to »respect, preserve and maintain knowledge, innovations and practices of indigenous and local communities embodying traditional lifestyles relevant for the conservation and sustainable use of biological diversity and promote their wider application with the approval and involvement of the holders of such knowledge, innovations and practices and encourage the equitable sharing of the benefits arising from the utilization of such knowledge, innovations and practices« (Gündling 2000 : 8)[3].

tories, consider themselves distinct from other sectors of the societies now prevailing in those territories, or parts of them. They form at present non-dominant sectors of society and are determined to preserve, develop and transmit to future generations their ancestral territories, and their ethnic identity, as the basis of their continued existence as peoples, in accordance with their own cultural patterns, social institutions and legal systems« (Posey 2002 : 26). The criterion *self-identification* tends to be regarded as the central aspect of the concept *indigenous* that also has been emphasised in other international agreements like the ILO-Convention 169 (Chatty and Colchester 2002 : 14).

[3] Similarly, with reference to *in situ* conservation, article 10(c) calls upon the signing countries to »protect and encourage customary use of biological resources in accordance with

2 The discursive context

»... nature is always constructed by our meaning-giving and discursive processes, so that what we perceive as natural is also cultural and social ...«[4]

Since the CBD entered into force, scientific attention has been drawn to cross-disciplinary approaches to comprehend the driving forces underlying environmental and social changes that are associated with global phenomena. The challenge lies in moving beyond disciplinary boundaries in order to encounter the complexity of environmental, social, political and economic issues relating to the sustainable use and conservation of biodiversity. To capture these interconnected phenomena more holistically conceived frameworks have been emphasised. Focusing on the interactions of local and global patterns of resource management, a growing body of contributions in social sciences emerged, examining the dynamic linkages of human societies with their natural environments. In this way, the implications of protected area management as a significant instrument in achieving the objectives of the CBD concerning *in situ* conservation came up as a field of study. Much of the academic work reveals the social dimension of linking conservation efforts through protected areas with local level development. By doing so, Furze et al. suggest that the imposition of protected areas, »while obviously concerned with managing ecosystems, is fundamentally about the management of people, their aspirations and their relationship with nature« (1996 : 36). The necessity to expand perspectives towards integrated conservation strategies that would combine nature protection objectives with human development issues has been increasingly featured, as the conventional practice of protected area management generally overlooked the presence of local communities and their cultural dependence upon local environments. The living conditions and concerns of local residents frequently remained disregarded and their specific knowledge systems underestimated. In most cases the fundamental operational principle even assumed that natural resources had to be protected against local people who would affect the physical environment negatively (Chatty and Colchester 2002 : 3). In this way of thinking, causes of environmental degradation like deforestation, erosion, degeneration of soils and loss of habitats of wildlife had been traced back to rates of population growth, pressure on scarce resources, extractive activities and ignorant attitudes of rural peasants being incapable of coping with external changes (Banuri and Apffel Marglin 1993 : 4). Nevertheless, findings of field-based research of social scientists gave evidence that local customary practices of resource extraction in many cases were sustainable, continuously adapting to changing conditions and compatible with the aims of prevailing conservation approaches that focused on endangered species (Orlove and Brush 1996 : 339). Most indigenous ecological principles are largely attached

traditional cultural practices that are compatible with conservation or sustainable use requirements« (Gündling 2000 : 8). Although the CBD explicitly delivers the application of the requirements to national legislations, it has become an important platform upon which national policies, regulations and legislations have been designed.

[4] Escobar 1999 : 2.

to experience and prediction in relation to human subsistence. These principles also include values, norms and beliefs regarding the maintenance of the »balance of nature«, which imply specific conservation methods and practices (Slikkerveer 2000 : 174). Local people who depend essentially on natural resources for their nutritional, economic, social and cultural needs, have developed social mechanisms and regulations which had conservational impacts on natural resources, despite the fact that they originally might have been motivated by distinct considerations (Kalland 2000 : 329).

2.1 Anthropological perspectives

Although Cultural Anthropology is foremost associated with small-scale societies, it equally dedicates its attention to the impacts of broader social and global phenomena on cultural systems at the local level. Employing comparative and cross cultural approaches, anthropologists are concerned with the ways in which human societies interact with their natural environments and the modes in which the natural world is conceptualised and classified. A major paradigm underlying anthropological research refers to the assumption of ›cultural relativism‹. The concept emphasises that cultures can only be understood in their own terms. This means that we cannot study other societies applying the values and norms that regulate social institutions and human behaviour in our own society. The imposition of values on other cultures has been described in terms of ›ethnocentrism‹ which implies the belief in the superiority of one's own culture (Furze et al. 1996 : 43). From this perspective, a primary aim in distinct sub-fields of the profession has been to investigate how different societies have acquired knowledge about their specific natural environments and how such knowledge is generated, communicated and represented by cultural means within specific epistemic, historical or political settings. Theoretical considerations developed within the disciplines have been particularly important to the contemporary debate on indigenous knowledge. By analysing the complex interrelationship of people and culture to natural environments studies in Cultural Ecology have stressed the role of knowledge in adapting to specific physical conditions. Building on Ethnobiology, contributions in Cognitive Anthropology have examined the systematic structure of knowledge systems and organising principles that underlie culture and behaviour. Scholars in the field of applied Development Anthropology widely discussed the role of local knowledge for rural development strategies. Anthropological research on protected areas emerged in the 1970s and draws on a diversity of perspectives. Current studies in Environmental Anthropology aim to move towards an understanding of how natural resources are used, encountered, perceived and endowed with significance. Similarly, research in Ethnoecology approaches primarily perspectives and behavioural patterns and includes classifications of the environmental relationships of particular communities or cultures (Slikkerveer 2000 : 170). In discussing the political dimension of indigenous knowledge, authors like Pottier (2003) conceptualise knowledge as *power*. Accordingly, any analysis must include an appreciation of social relations imbued with aspects of power and control being central to the production and articulation of knowledge. Such advancing contributions emerging from the field of Political Ecology analyse the impacts of the global expansion of policies and economics at local level interrelationships of resource users with regional, national and international actors. This approach places the concept of »interaction« above that of »intervention« (Slikkerveer 2000 : 172) and devolves more responsibility and power in decision-making to local communities. It implies the adoption of new participatory concepts and methods in conservation practice. Thereby, it has been claimed that participation is not to be seen as a means to achieve externally-desirable goals, but rather must be based on an interactive long-term process of dialogue, negotiation and conflict resolution to understand how local livelihoods are constructed by people's own meanings and priorities.

In terms of applied methodology for an understanding and explanation of the distinct levels of interaction, a wide range of methods has been developed in Anthropology. The ethnographic method serves as a major framework for qualitative fieldwork and offers a means for in-depth and site specific understandings. As Anthropology is an empirical science based on observable data, ethnography is a process of recording, describing and interpreting indigenous ways of knowing, valuing and organising the world through participatory observation[5].

By drawing on ethnographic case studies on human interaction with nature, environmental anthropologists have placed special emphasis on the significance of symbolically motivated criteria within knowledge systems. They underlined the crucial role of cultural values that are highly variable, difficult to quantify and often contrast with values underlying ›western‹ scientific approaches, that dominated conservation thinking until recently (Laird 2000 : 356). In this discursive frame a number of critiques challenged. These questioned the inherent ethnocentric assumptions that had been transferred to local settings with no regard to the complexity of human-nature interactions and without concessions to experiences and knowledge of local people. Despite calls from anthropologists for the scientific credibility of such local knowledge to be recognised, conventional principles commonly remained the foundation of conservation management (Howitt 2001 : 45). The »managerial ethic of control« has been critically analysed by Banuri and Apffel Marglin (1993 : 17). It is denying the complexity of interfering ecological, ideological, social, political, and economic factors that contribute to any given environmental situation. Similarly, scholars like Berkes outline the limitations of scientific paradigms, which are based on a »instrumental attitude« towards nature (1999 : 10). In »pursuing simplicity and clarity«, scientific thinking »minimises the extension of interaction to a relatively narrow range of direct causes and effects« (Howitt 2001 : 69). Similarly, Chatty and Colchester refer to this reductionist principle that implies »the existence of only one reality« (2002 : 7). Applying an approach that reduces complex aspects into discrete parts that can be analysed, the target of science is to discover, predict, and control this reality and to summarise the obtained knowledge into universal laws or generalisations. These critical ex-

[5] For specifics on ethnographic methods in indigenous knowledge research see Grenier (1998); Antweiler (1998 : 478ff); Slikkerveer (2000 : 172ff).

pressions concerning the inadequacies of the predominant ›Cartesian‹ worldview, in which an intensified dichotomy of reality separates not only culture from nature, but likewise subject from object, mind from body and cultural sciences from natural sciences, came along with the quest for an »entirely new ecological paradigm« (Kalland 2000 : 321).

2.2 Cultural concepts of nature

As a contribution to this paradigmatic shift, a basic assumption in Cultural Anthropology that refers to the cultural ›construction‹ of nature may offer conceptual considerations. In Anthropology it is commonly assumed that human perceptions of nature do not simply reflect an essential reality of a biophysical environment, but comprise a much broader »social reality« (Townsend 2000 : 24). From this perspective, the human relationship to nature is constructed by social interaction and symbolic means. Ethnographic studies unveil significantly different ways for thinking about, relating to, constructing and experiencing the natural (Escobar 2001 : 151). Although there is no unified view on what characterises »cultural models« (Kempton 2001 : 59) of nature that are constituted by ensembles of meanings, the notion is commonly shared that many indigenous concepts do not rely on the ›occidental‹ worldview that separates the *natural* from the *cultural* (Escobar 2001 : 151). Accordingly, the word *nature* as conceptualised in ›western‹ cultures in a dichotomous relation to *culture* or *society* does not correspond with indigenous worldviews. As the close attachment to land is a defining characteristic of indigenous cultures, the natural environment is not viewed as a ›resource‹, but rather as living space even being endowed with ›sacred‹ meanings, which define the existence and identity of people. Referring to this holistically perceived relation, Banuri and Apffel Marglin conclude briefly: »there is no separation between nature and culture« (1993 : 15). This integral concept of social reality implies, that the material and the spiritual dimensions of culture are not considered as opposed spheres, but rather as non-fragmental extensions of human society. Similarly, knowledge is not perceived as a resource that can be extracted from its origin, as it is inextricably embedded in the respective cultural context. Taking these considerations into account, a concept for bridging the gap between the *natural* and the *cultural* has been developed aiming at the conservation of ›bio-cultural‹ diversity. With an emphasis on *in situ* conservation of biological *and* cultural resources as »interdependent phenomena« (Zent 1999 : 93), it indicates »the crucial complementarity for achieving an alternative, less exploitive philosophy of nature« (Slikkerveer 2000 : 174).

2.3 Indigenous knowledge

> »Just as there can be no ›nature‹, there can never be just one ›knowledge‹ either.«[6]

Since the early 1990s, *indigenous knowledge* (IK) has become a major theme in the global environmental discourse. The growing appreciation that primarily emerged

[6] Gray 2000 : 62.

in the field of rural development can be interpreted as a reaction to the previous failure of the grand development theories. Scholars engaged in this field had criticised externally introduced ›top-down‹ policies and uni-directional initiatives that were based on modernisation and technology transfer and argued in favour of more participatory and decentralised ›bottom-up‹ initiatives. In search of locally grounded approaches to global issues, knowledge being acquired in local settings has been discovered as a meaningful alternative to conventional development strategies (Agrawal 1995 : 415f). The current debate on the issue tends to pertain mostly to the domain of natural resource management and has shifted from initially questioning *what* is the usefulness of such local knowledge systems for sustainable development strategies to *how* can it be used to ensure equitable sharing of benefits arising from the utilisation of knowledge, innovations and practices as claimed in the CBD (Slikkerveer 2000 : 169). This crucial aspect has been framed in terms of *intellectual property rights*. The field of knowledge has not just emerged as a subject of attention among anthropologists, environmental researchers or development agents, but also biodiversity prospectors increasingly have become aware of the previously underestimated value of IK. Therefore, it has been argued by anthropologists in the debate on the relevance of local environmental knowledge for biodiversity conservation, that it should not be labelled as an under-used resource of empirical information concerning ecological relationships »to be shared for the benefit of humanity at large«, or as »culturally and socially free ›human capital‹ to be harnessed in the service of biobusiness« (Nygren 1999 : 273)[7].

Although the relevance of IK for conservational issues has been widely recognised, defining the term still remains difficult. What is meant by *indigenous knowledge* in the discursive context relating to biodiversity conservation? A broad range of terms can be found in current writings, which are mostly used interchangeably[8]. Grenier, for instance, defines indigenous knowledge as »the unique, traditional, local knowledge existing within and developed around the specific conditions of women and men indigenous to a particular geographic area« (1998 : 1). The term *local knowledge* equally has been widely applied and characterised by Escobar in a broader sense as »a mode of place-based consciousness, a place-specific ... way of endowing the world with meaning« (2001 : 153). Purcell defines it as »the body of historically constituted (emic)[9] knowledge instrumental in the long-term adaptation of human groups to the biophysical environment« (1998 : 260). Similarly, Kimmerer describes the term as a »diachronic database« of collective intellectual contributions of peoples that are the result of observations over a long time period (2002 : 433)[10]. Despite the absence of a widely accepted terminological and con-

[7] It is beyond the scope of this contribution to discuss the issue on intellectual ownership of knowledge. For details on this theme see the contributions in Laird (2002).

[8] For an overview to the terminological and conceptual diversity within the academic debate see Antweiler (1998).

[9] ›Emic‹ indicates *from within* a culture.

[10] The term *traditional knowledge* is often used synonymously in contemporary writings. Referring to knowledge in the present context ›traditional‹ is avoided as it is often conceived as a counterpoint to ›modern‹ and implies a rather static rather than dynamic and

ceptual framework, it has been largely asserted that IK needs to be understood as an integral part of the respective cultural system and cannot be disembedded from the local context in which it has been developed (Ellen 2002 : 249). Referring to the scientific documentation, Antweiler notes that it is »not just a question of recording the knowledge contents as a product« or a set of information, but rather it is necessary to describe the relevant framing »social, economic, cultural, and ecological processes« (1998 : 478). As it is woven into social structures of a specific culture and »bound by time and space, by contextual and moral factors«, IK contrasts according to Banuri and Apffel Marglin with ›modern‹ knowledge, »which bases its claim to superiority on the basis of universal validity« (1993 : 13)[11]. Although rooted in particular local settings, and in this sense to be understood as »situated knowledge« (Nygren 1999 : 267), IK is not to be seen as exclusively *local*, but rather as a result of complex negotiations linked to knowledge interfaces. Therefore, it must be approached as a dynamic and fragmented »plurality of local knowledges«, rather than a unitary concept of knowledge, as mentioned by Pottier (2003 : 1). Likewise, Nakashima and Roué employ the term »indigenous knowledge systems« as »the complex arrays of knowledge, know-how, practices and representations that guide human societies in their innumerable interactions with the natural milieu: agriculture and animal husbandry; hunting, fishing and gathering; struggles against disease and injury; naming and explaining natural phenomena; and strategies for coping with changing environments« (2002 : 315). As comprehensively outlined, indigenous knowledge particularly pertaining to the field of natural resource management is expressed in multiple ways and first of all closely related to subsistence patterns. It is the result of a continuous process of generation, experimentation, transmission, innovation, and adaptation to changing conditions. Thus, being constantly transformed by humans it must be approached as a dynamic entity that permanently undergoes modification and negotiation between people and their environments. Knowledge repertoires are a result of encounters in which local and global, traditional and modern elements are intricately intermingled. Nygren refers to this aspect by underlining the »contested and hybrid character« of knowledge (1999 : 268). This process of knowledge accumulation and transformation is likewise influenced by broader socio-economic, political and historical factors and involves the conjunctive adoption of external notions.

The present study concentrates on those aspects that apply directly to indigenous peoples' knowledge about natural resource management. The preferential term ›indigenous environmental knowledge‹ as used by Ellen and Harris (2000) is understood in accordance with the operational definition provided by Berkes who describes ›traditional ecological knowledge‹ as »a cumulative body of knowledge, practice, and belief, evolving by adaptive processes and handed down through gen-

---

evolving perception (Kalland 2000 : 322). Otherwise ›traditional‹ refers to »cultural continuity transmitted in the form of social attitudes, beliefs, principles, and conventions of behaviour and practice derived from historical experience« (Berkes 1999 : 5).

[11] This contribution does not intend to interrogate the relation between ›modern‹ or ›western‹ science and indigenous knowledge. For this see Agrawal (1995); Ellen (2002); Kimmerer (2002).

erations by cultural transmission, about the relationship of living beings (including humans) with one another and with their environment« (1999 : 8). Berkes has undertaken to conceptualise indigenous knowledge as a »knowledge-practice-belief complex« (1999 : 13). This implies a primary level of ecological knowledge regarding animals, plants, soils, and landscapes being framed by a second level of knowledge on land and resource management systems. A third level refers to social institutions and relations, whereas the fourth level is given by the worldview, which includes belief systems and cultural ethics that shape environmental perceptions.

2.4 The landscape approach

As concluded by Ellen and Harris (2000 : 26), the failure to take into account the co-existence of the various levels of this knowledge-practice-belief complex and the interconnections between both empirically *and* symbolically motivated criteria within any knowledge system inevitably leads to limited understandings about how knowledge operates and how it is situationally successful. To capture this interconnectedness, more holistic approaches have been employed to study the nature-society interface. In discussing the role of Anthropology in research on biodiversity conservation, Myer (1998) proposes a ›landscape‹ orientation as conceptual framework for case studies focusing on indigenous knowledge at the local level.

In Anthropology landscapes are regarded as socially constructed places in the sense of natural spaces manipulated by human communities with interactive political, economic, social and biological aspects. According to this understanding, landscapes are formed by human use and likewise characterised by belief systems as culturally conditioned experiences and understandings. By analysing the relationship between identity and space Hernando Gonzalo argues that landscape refers to symbolically constituted spaces that are conceived only through personal experience that constitutes the basis of identity. In this sense, all space is experience, and therefore all ›nature‹ and ›land‹ is ›landscape‹ (1999 : 256).

As the concept of ›sacred land‹ can be found in many indigenous societies, Howitt concludes that the geomorphic landscape reflects the same cosmological truths that shape the relationships within the currently living community of people. Thus, »much is inscribed into and recorded upon the landscape – either physically or symbolically – which affects resource management practice« (2001 : 172). In ›reading‹ such landscapes, even the most learned outsider is reduced to ›illiteracy‹ and the absence of ›literacy‹ in complex multicultural environments succeeds as a common source of misunderstanding and conflict in many settings of conservational management (Howitt 2001 : 174).

3 The local context

»When we deal with culture we are dealing with symbols that communicate the soul of a host community.«[12]

The finding that areas with high biological diversity tend to have high cultural and linguistic diversity (Kempton 2001 : 58) is also evident in the case of Guatemala. The Central American country features substantial biological and cultural diversity[13]. Within the multi-ethnic society, around half of the twelve million inhabitants belong to 22 indigenous groups of Mayan descent. The *Q'eqchi'* represent the fourth largest of these distinct ethnic groups. The majority of the 360,000 people live as subsistence peasants in relatively dispersed villages in rural areas of the department *Alta Verapaz*. Due to the remoteness of their settlements, a large number of elder people in particular, are illiterate and monolingual speakers of *Q'eqchi'*. Exposure to the formal education system as well as more frequent interaction with the urban centres and regional markets led to a process of acculturation towards the predominantly urban *Ladino* culture among younger people who are more likely to speak Spanish. Apart from the language, the religious realm is of important significance in the cultural frame. Since the Spanish invasion in the 16th century, the religious belief system has been composed of a syncretic combination of catholic elements and components of the traditional *Mayan* belief system. Since the 1970s, protestant denominations were increasingly brought into the area, being less tolerant regarding the incorporation of autochthonous beliefs and customs. The growing influence of these churches had transforming effects upon the social structures within the communities and led to severe disruptions from the Mayan *cosmovision*[14].

As Wilson notes, the term *Q'eqchi'* originally referred to the common language rather than to the entire ethnic group with shared cultural attributes. The people refer to themselves as *ral ch'och*, which means »sons and daughters of the earth«

---

[12] Furze et al. 1996 : 162.

[13] Since Guatemala signed the CBD, a national biodiversity strategy has been ratified that includes considerations relating to article 8(j). In addition, the need to protect indigenous knowledge has been politically acknowledged in the national *Agreement on Identity and Rights of Indigenous Peoples* that was established in 1997. In this document attention is drawn to the management of biological resources, and it guarantees the rights of indigenous communities to participate in the use, administration and conservation of the natural resources existing in their lands. Academic institutions and agencies working in the field of development and environment are urged to promote the contributions of indigenous people and the dissemination of their knowledge.

[14] Cosmovision refers to »the way in which the members of a particular culture perceive their world, cosmos or universe. It represents a view of the world as a living being, its totality including not only natural elements such as plants, animals and humans, but also spiritual elements such as spirits, ancestors and future generations« (Slikkerveer 2000 : 171). For an analysis on the encounter of the traditional religion and Christianity see Wilson (1995).

or they call themselves *aj k'aleb'aal*, translated as »people of the cornfield« (Wilson 1995 : 309). As these expressions indicate, the cultural identity basically relates to location and points to the cultural significance of maize and associated agricultural activities. Most people identify themselves as being from a certain community. Thus, not meaning a particular exclusive village, but rather the larger surrounding social and natural environment of a specific place. As the mainstay of livelihood is based on agriculture, the majority of the *Q'eqchi'* is largely dependent on the local natural resources. The entire economic and social life revolves around the *milpa* system. This consists basically of the cultivation of maize being produced through swidden-farming and generally carried out by single household units on smallholdings. Other crops such as beans, chillies, and tomatoes are cultivated on the *milpa* in diverse forms of multi-cropping. Also, home gardens provide a wide range of crops raised for consumption and for sale at local markets. The livelihood is supplemented with small-scale forest extraction such as logging and gathering of non-timber products. Since the 1970s, coffee as the major cash crop has increasingly been replaced by cardamom as a mainstay of the regional economy. Gender is a major organising principle in the local production system. Although there is a certain degree of flexibility within the division along gender lines, most agricultural production is related to the labour of men, whereas household maintenance and food processing are completed by women who also cultivate home gardens and tend domestic animals. Parts of the produced items are taken by women to sell at the local markets. Communal labour and co-operative ventures are an integral part of community life, which in terms of socio-political organisation is traditionally characterised by a high degree of egalitarianism. Traditionally, land tenure had been governed by a communal management system. Due to the privatisation of communal territories since colonial times, which led to extremely unequal land distribution, temporal wage and migrant labour has become a common source of secondary monetary income. Limited access to sufficient quantities of arable land, the need for immediate external income and a growing demographic pressure entailed continued clearing of primary forest for agricultural use and shorter rotation cycles. Especially in the highlands of *Alta Verapaz*, where villagers need three times as much land for subsistence than in the lowlands with richer soils and a more suitable climate, the ideal fallow time of four to six years declined to shorter periods of restoration (Wilson 1995 : 42). Consequently, erosion and the decline of soil fertility and crop yields became a common environmental problem[15].

The present conditions of the *Q'eqchi'* communities are interwoven with the history of conquest and colonisation and the economic, political and social marginalisation of the indigenous population throughout the past centuries. The resistance to the terms of their incorporation into colonial and post-colonial nation state has produced a struggle for legal recognition. Accordingly, there has been much political unrest in the area until the 20th century. The policies of successive governments historically sought to advance the process of national integration and to promote cultural assimilation. This strategy towards the homogenisation of

---

[15] For details on this theme see Markussen and Renner (in this volume).

the national society began throughout the Guatemalan highlands during the late 1970s. The growing oppositional activism and an arising guerrilla movement were encountered by massive military aggression. In the early 1980s, the discriminating policies against the indigenous population involved extreme violence. The army and paramilitary forces sought to control the situation by means of threats, torture, community massacres and killing of individuals who were regarded to be linked to the insurgents. In 1996, by the signing of the Peace Accords between the government and the guerrilla organisation, democratic structures were formally re-established. Nevertheless, much distrust and social imbalance remains within all domains of the national society and are notably encountered in any intercultural setting.

Fig. 1. House of a *Q'eqchi'* family in the lowlands of *Alta Verapaz*

3.1 The ethnographic experience

The field research took place within four farmers' communities located in the margins of private protected areas in the eastern highlands as well as in two communities settled in the surroundings of the National Park *Laguna Lachuá* in the northern lowlands of *Alta Verapaz*[16]. As the constitutive idea in the ethnographic case study is that social conditions and impacts relating to biodiversity conservation are processes, the research was conducted by qualitative methods like participant observation and semi-structured interviews to identify and understand these processes. Other than secondary data review, the methodology applied in the field has been extended by community surveys and participatory methods such as transect walks with key informants. It also included interviews with staff members and field workers of governmental conservation agencies and private development organisations, local village representatives, teachers, and scientists. However, the essential insights gained in the process of fieldwork emerged from personal relations with women and men, youth and elders in the local communities. To collect the aspired ethnographic data a structured conversational approach was the primary applied method[17]. As it has been difficult to approach ›embodied‹ knowledge and ›emic‹ perspectives by inquiries, the interviews were supplemented by observation and active participation. As indigenous knowledge to a large extent is intuitively practiced and seldom explicitly articulated by informants, the sharing of daily activities proved to be the best method to gain insights and to elicit, describe and interpret different knowledge expressions. In the frame of guided field walks, key informants were primarily men, but many interviews were also conducted with women in their domestic environment. Although their knowledge reflects different sets of labour responsibilities as they are not directly involved in the agricultural processes of production, they very often reported to the same degree as men detailed information on the subsistence activities and agricultural practices of the family households.

3.2 Protected area management and local livelihoods

In discussing social impacts of protected area management Furze et al. (1996) reveal, that one of the most common problems encountered by local communities when protected areas are established, is the restriction of access to resources. Consequently, there is likely to be a conflict between the necessity for rural communities to supply their living and the management of nearby protected areas. This also applies to the National Park *Laguna Lachuá* where the establishment of the protected area had many implications for the livelihood opportunities of the people living in the surrounding forest-edge villages. At the beginning of 1970s, the colonisation of the region in the northern lowlands of *Alta Verapaz* was officially initiated. Formerly the area was covered widely with tropical rainforest and inhabited by few scattered settlements. The National Institute of Agrarian Transforma-

[16] Although the general findings of the study are based on the fieldwork conducted in both areas, the contribution highlights the investigations carried out in the surroundings of the National Park *Laguna Lachuá*.

[17] A semi-structured interviewing technique uses predetermined questions and topics but also allows new topics to be pursued as the informal interview develops. The important feature of conversational interviews is that they attempt to approach the informants' ›emic‹ perceptions of their own reality.

tion (INTA) distributed land to farmers many of whom came from the highlands of *Alta Verapaz.* In 1975, as the pressure on the land constantly intensified, the National Park *Laguna Lachuá* was officially declared in order to conserve the biological diversity of the forest resources. The area now covers 14,500 hectares and is administered by the National Forest Institute INAB. Additionally the management has been promoted by the World Conservation Union (IUCN)[18]. Nowadays, 45 communities are located in the wider surroundings of the park, being inhabited by a total number of 10,000 people of whom 90 percent belong to the *Q'eqchi'*. Succeeding the immigration of displaced farmers who had left their homelands due to the military repression in the 1980–90s, the natural resources increasingly deteriorated due to the ongoing deforestation process in the area. Many of the peasants settled without legal land titles, so that differentiating claims caused a high level of land-use conflicts among the communities. Consequently, although forbidden by the authorities, cases of logging, hunting and gathering occurred repeatedly within the limits of the park. These illegal activities led to a series of conflicts between the local residents and the park agents.

Based on the notion that there is little chance of protecting the resources within the park boundaries from extractive activities without the involvement of the local people, a central concern has been to encourage effective conservation with active community participation. This purpose is achieved by a *co-management* framework, which has been described as a »key concept« in resource management (Kalland 2000 : 329). The main principle of the concept aims to link environmental conservation with human development issues. This management plan has been implemented in 36 of the communities and includes initiatives like diversification of agricultural products and methods, promotion of marketing and income-generating alternatives, support for the legalisation of land tenure, strengthening of local institutional capacities and also involved programmes on environmental education. Although many of these projects have been widely appreciated by the population, reservation was frequently expressed in interviews concerning measures that would not be accompanied by corresponding instructions to capacitate the people for the practical application of the proposed initiatives. Apart from such technical questions, informants indicated concerns regarding certain paternalistic attitudes of park agents who are predominantly *Ladinos.* In an interview one of the park managers stressed the necessity to ›educate‹ the local residents, as they are ›unaware‹ of the ecological diversity in their habitats and ›ignorant‹ of conservational initiatives.

The *co-management* arrangement also includes a branch of research carried out by the *Escuela de Biologia* of the State University San Carlos. A scientific research programme has been employed which is based on an intersectoral ecosystem approach and actively seeks to integrate local knowledge with scientific studies. A process has been initiated of concluding a comprehensive assessment of the parks' biophysical resources and their reciprocal relations with the economic and

[18] For an analysis of the Guatemalan conservation policy and the implementation practice of corresponding programmes see Birner et al. (in this volume).

social dimensions of human resource use activities. Investigations in cooperation with 17 communities have been undertaken so far in the fields of biology, agronomy and sociology. Integrating socio-economic with ecological information by combining quantitative and qualitative data may provide new perspectives on nature conservation and sustainable resource use. Data collection on flora, fauna, soil characteristics and water quality was conducted in the park itself and in the adjacent zones. In this frame, local taxonomies were investigated with local specialists. A study in ethno-botany for instance, explored 209 plant species being used for medical purposes. In the long term it is planned to further incorporate local and scientific information for the design of educational components as a contribution to resolving the conflicts between conservation efforts and the needs of the *Q'eqchi'* people. Another recent study aimed at finding agreements with community members to reduce the hunting of wildlife animals to a sustainable level in the area. The investigation was based on the monitoring carried out by the local peasants themselves, as they have detailed knowledge of the habitat, behaviour and migration patterns of the animals. The outcome of the joint evaluation consisted in the elaboration of a calendar indicating temporal restrictions to the hunting practices. The hunters adopted these restraints, as the principle of seasonal and rotational use of resources is an integral part of their traditional resource use patterns.

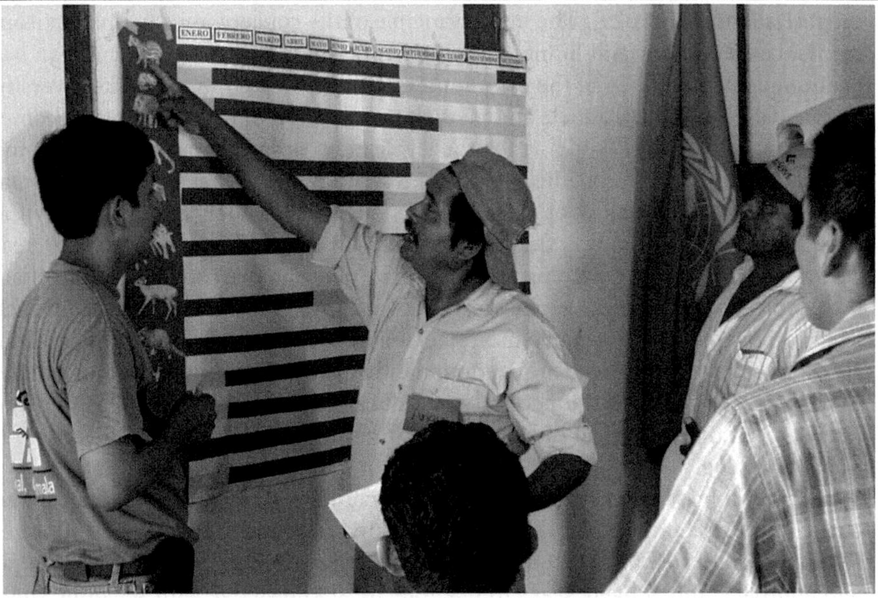

Fig. 2. Discussion on the hunting calendar

Taking into account the history of conflicts and lack of communication between local people and state authorities, it was particularly interesting to participate in an open discussion on this hunting calendar between villagers, park guards and administrative agents, initiated for the first time in September 2002 by a group of biologists who became intermediaries in this communication process. The encounter between the population and the park management did not simply reflect competing interests, but also gave evidence of underlying multiple realities and ways of endowing the world with meaning that met in this interface situation.

3.3 Expressions of indigenous knowledge

The present investigation is associated with a current research project that employs an ethno-ecological approach to document indigenous environmental knowledge of plants, soils and other natural components. The objective is to identify locally evolved knowledge and practices that can contribute to responding to environmental changes and may be better adapted to the local conditions than externally introduced technologies. The joint field research took place in two communities at the margins of the National Park. The obtained insights were predominantly gained through case studies conducted with five *Q'eqchi'* farmers in each village.

In general, it can be concluded that most of the farmers in the two communities under study have established a highly differentiated system of integrated forest and agricultural management. This involves a wide range of managed vegetation zones from permanently cultivated to swidden lands at different stages of recovery and areas of primary forest. Within these diverse zones high densities of plants and animals exist. Although the slash and burn practices had negative impacts on the forest, it did allow for regeneration of secondary forest cover. While crops are growing, the vegetation is renewing the site for the next cultivation cycle. In addition, a complex system of agroforestry combining food crops and trees could be identified. The forest resources are widely used for food gathering, harvesting of medical plants, grazing, fishing, hunting and the collection of wood. The role trees play in soil protection is widely recognised. Seeds are collected, and multiple tree species are planted and maintained along the edges of the *milpa*, which can contain up to forty different crops besides maize. Combined with notions on climate conditions and observations of the lunar cycle, the majority of the farmers have acquired sophisticated knowledge on species and crop diversities, which they use for livelihood security to encounter uncertainty and minimise risk. Many indigenous crops fit the different ecological niches of their lands, thereby reinforcing the agro-biodiversity. Within the larger farming system, home gardens and fruit orchards close to the domestic area play an important role in the livelihood of the *Q'eqchi'*[19].

The experience-based process of knowledge accumulation includes the constant absorption of notions being introduced from outside. The agricultural knowledge embedded in the traditional practices of slash and burn cultivation has been

[19] As home gardens contain high levels of species, the value of the associated knowledge for *in situ* conservation of plant genetic resources has been recognised widely. For a detailed discussion on home gardens see Holl (in this volume).

mixed with elements of modern technologies. Development agencies have introduced new crops and organic measures to maintain soil fertility, which have been widely integrated into the local production system. At the same time, the common application of chemical products has been observed.

Fig. 3. Children at the harvest of cardamom

The patterns of knowledge distribution are configured by learning situations. The process of intergenerational knowledge transmission is evident, as children at the age of six to eight commonly join their parents in their specific labour domains. In this way they acquire knowledge tacitly, unobtrusively and unreflectively as part of the process of socialisation. To a considerable extent indigenous knowledge is transferred through non-verbal demonstration, imitation and repetitive practice. Therefore, it has to be approached as ›embodied‹ practice that is coded as part of doing and recognising in practical contexts, by experience and informal apprenticeship but which is not explicitly formulated into a set of rules. Among the *Q'eqchi'* most knowledge particularly of natural processes is only partially consciously known and often remains implicit. If explicitly transmitted, it is orally communicated and thereby encoded by language. Being stored in people's minds it is not solely expressed through subsistence activities, but rather in narratives and myths, values and beliefs, rituals and laws. While there is common knowledge shared by the majority of the communities, a considerable part of knowledge is asymmetrically distributed and only held by sections of the population according to gender, age, occupation and experience. In the field of ritual practice for instance, knowledge is exclusively held by the elders of the communities.

It has to be kept in mind that knowledge is not a static set of information, but rather a process that also includes the transformation and erosion of notions. A major finding of the investigation led to the conclusion that traditional land use practices and accumulated knowledge of domestic plants have been eroding during the past decades. In many interviews, mostly elder respondents, men and women expressed their preoccupation concerning the perceived erosion of knowledge. In particular, the loss of knowledge on the use of medical plants led to an impoverishment of local medicine and caused an increased dependence upon pharmaceutical products. The loss of adaptive capacities in the field of agricultural patterns occurs in the frame of an overall acculturation process through modernisation of farming practices and the incorporation into market economy. Due to the increased use of chemical products as fertilisers and pesticides following the process of land use intensification and monocropping, traditional techniques to maintain soil fertility or to avoid plant diseases have widely been displaced. Despite contemporary patterns that increasingly undermine local systems of knowledge and resource use, there remains a great variety of local knowledge systems that can contribute to community-based conservation efforts.

As Zent (1999 : 91) states, few scientific studies on indigenous ethno-ecologies referring to local environmental knowledge and resource management systems have focused explicitly on knowledge loss. He argues that it makes little sense to attempt to conserve knowledge without first attempting to analyse the local patterns of knowledge loss itself. Accordingly, there is a need to identify specific causal variables underlying the transformation and fragmentation in order to develop strategies to achieve the protection of such knowledge systems as claimed in article 8(j) of the CBD. The erosion of knowledge is not a natural and inevitable process, since it has a largely social and political origin. If indigenous knowledge is disappearing, it is, according to Agrawal (1995 : 431), primarily because processes of modernisation and cultural homogenisation threaten the lifestyles of indigenous people.

In the present case there are numerous socio-economic factors across dynamic spatial and temporal scales, which had transforming effects on the modes of livelihood of the *Q'eqchi'*. The inequality of land distribution and the frequent lack of secure legal land rights and self-determination interfere with a long history of traditional management of natural resources. Of major significance have been the implications of the long lasting civil war that has been accompanied by significant political and socio-cultural changes within the communities. In order to reinstate control over rural communities being accused of supporting the guerrilla movement, entire villages were destroyed by the militaries in the 1980s. Thousands of *Q'eqchi'*, mainly elders, died in the course of the armed conflict and with them valuable knowledge disappeared. The army attacks led to displacements from original lands, mass migrations, and the abandoning of traditional practices. As massacres became routine, many people fled their original homelands and lost a large percentage of their seed stocks, especially for garden crops (Wilson 1995 : 42). A strat-

Fig. 4. Although much knowledge on organic methods to maintain soil fertility has been lost, many farmers plant a bean species (*Mucuna pruriensis*) that is known to enrich the soil

egy of forced migration was implemented by the military clustering members of different ethnic groups into new settlements, affecting the social cohesion within the newly established villages. Consequently, networks among peasants, through which farmers formerly obtained and exchanged their seed stocks, were disrupted or even destroyed. In the slow process of rebuilding the village economy, the displaced communities concentrated foremost on re-establishing maize crops as a central link of social life. The experience of violence originating from political authorities and highly asymmetrical power relations created a situation of fear and silence among the villages. According to the dominant ›emic‹ perspective, the decades of structural disempowerment had a devastating impact on community life, traditional beliefs and practices and led to a continuous distrust in any state agency and institutions associated with the state. The long period of repression and the historic neglect of the indigenous population in all fields of the national society are also reflected in the educational policy. Until recently, indigenous knowledge and languages have been excluded from what equally contributed to a severe decline of cultural self-reliance. Related to the national peace process, the political impact of indigenous issues has increased and alternative bilingual and multicultural teaching models and efforts to incorporate local knowledge systems into the formal education system are now on the way to being implemented.

3.4 The sacred landscape

When talking about subsistence patterns, economics and ecology, informants recurrently used religious symbolism to describe their experiences and referred to spiritual features inherent in their way of life. This led to the question of how nature is socially constructed and how cosmological principles reflect the environment and connect with production methods. An example of cultural frameworks for resource management is the *milpa* cycle that carries symbolic significance for the *Q'eqchi'*. Building on the myth that the *Maya* originally were made of maize, the people have a strong sense of identification with their *milpa*. As Wilson puts it, it is of great value to have own land on which to cultivate maize, because doing so is a »vital aspect« (1995 : 49) of cultural identity, which provides economic safety and maintains the relationship with the local landscape. The participation in the reciprocal process of planting and harvesting is an »archetypical feature of membership« (1995 : 47) in *Q'eqchi'* communities. As has been stated formerly, identity is not primarily linked to language or ethnicity, but built around the local community and the landscape it is placed in. The influence of the specific local landscape upon the culture and the intimate attachment of people to their land, finds corresponding expressions in ritual performances. Equilibrium and reciprocity are central concepts in the *Mayan* cosmovision. The relationship between humans and the universe is characterised by interdependence. »Responsible human action is carried out with consciousness« of this reciprocal linkage and successful farming depends on »maintaining a harmonious relation with cosmic forces through a series of rituals« associated with agrarian activities (Nigh 2002 : 455). The *Q'eqchi'* communities have developed a set of specific rituals that answer fundamental questions about how humans should behave towards the environment. Knowledge, values, and identity are transferred intergenerationally through the annual, cyclical repetition of livelihood activities that are associated with rituals stored in the memory of the elders. Rituals are symbols for the lifecycle of the communities. Foremost, agrarian rituals practiced at the time of burning, seeding, weeding and harvesting are integral to the interaction between humans and the natural world. These expressions of the indigenous cosmovision are also interlinked with other subsistence activities such as hunting and fishing and they help to maintain and reinforce social and spiritual cohesion within the community. By rituals which are inseparable from everyday knowledge of fertility and health, the social, economic, spiritual, and cosmological spheres of the *Q'eqchi'* are intimately related. As Gray notes, rituals are particularly important, whereby »the forces of the invisible world are drawn into the world in order to ensure growth and fertility« (2000 : 64). Among others, the main ritual practiced by the *Q'eqchi'* peasants is the so-called *Mayehak* directed at the earth deity *tzuul taq'a*[20] which has been described as a »core symbol« of the

[20] Literally *tzuul* means mountain, *taq'a* means valley (Wilson 1995 : 51).

indigenous culture (Wilson 1995). This ritual related to the agrarian calendar is performed by farmers and their families in the centre of their *milpa* but has also been adapted to externally introduced crops such as cardamom. By performing the ritual, the farmers request permission to engage in an activity that will affect the environment. As elder informants explained, they are carried out to ensure good harvests and protection from the *tzuul taq'a*. Combined are behavioural regulations and taboos like abstinence, feasting and avoiding certain food items. If these codes are ignored, the harvest will be poor or damaged by animals. Likewise, misbehaviour can also cause human illness as the life cycle of crops and humans are believed to be intertwined. There are other rituals collectively practiced by the eldest men and women of the communities at particular sites such as groves. These are considered to be sacred, as it is believed that the *tzuul taq'a* reside in caves of mountains. As they provide human health and crop fertility, the *Q'eqchi'* create a reciprocal relationship with the tellurian deities through sacrifice. Sacrifice establishes and maintains a relationship of balance between the humans and the local landscape. As offerings, different items such as animal blood or cacao beans are provided and candles and incense obtained from the *Copal* tree are burned. This principle of reciprocity has long been recognised as fundamental to indigenous economic systems (Gray 2000 : 64) and is equally an essential concept in social and political relations within the *Q'eqchi'* communities. Sacrifice symbolically »inscribes« the land tenure of community and single households onto the landscape (Wilson 1995 : 88). Through the symbolic relationship of reciprocity, the *Q'eqchi'* establish what is essentially a »social relation« (Nigh 2002 : 455) with the elements of the inhabited landscape. Howitt compares the intense and intimate relationship between indigenous people and their country to a »kinship relationship« that includes responsibilities and »obligations of nurturance« (2001 : 54). This close relationship between villagers and the landscape anchors community identity. In view of the multitude of hills and valleys as characteristic physical features of the topography of *Alta Verapaz*, the mountain orientation in the cosmology of the *Q'eqchi'* is not surprising. Although the catholic concept of god may have determined aspects of institutional and ritual life, it is not essential to the everyday life of the people (Hernando Gonzalo 1999 : 259). This is governed by *tzuul taq'a* as »god's sentinels here on earth, guarding the fruits of creation from human abuse« (Wilson 1995 : 68). The mountain-valley deity is seen as the land itself, as well as a spirit inhabiting it. Or as Hernando Gonzalo puts it, *tzuul taq'a* »is not just the god to whom everything belongs, the source of all order ... not a category separate from the manifestations of nature, but nature itself« (1999 : 260).

In interviews the forested and mountainous areas have been described as central points of reference, proving the *Q'eqchi'* with a sense of security, orientation and identification. Such beliefs associated with mountains grant protection of a wide range of species of animals and especially medical plants. Mountains are seen as sources of life, water, health and well-being and practices such as rituals intended to maintain these physical and cultural sanctuaries can work to preserve the biological integrity. In many cases, taboos apply to particular vulnerable sets of natural resources. Breaking a taboo can result in sanctions from the natural world such as illness, loss of crops or drought. These regulations can be interpreted as social mechanisms for local conservation practices. Taboo associations attached to particular species of trees, forest groves, mountains and temple sites have played an important role in the protection of particular ecosystems by the local communities.

Fig. 5. Landscape in northern *Alta Verapaz*

There exist numerous holy caves amongst the mountains and hills in *Alta Verapaz* that draw pilgrims from other regions. In this way, the identity based on the collectively constructed landscape is reassured and spiritual conceptions that promote biodiversity and environmental awareness grounded in cultural traditions are disseminated. Nevertheless, the belief in the sacred power associated with mountains has undergone cultural transformation. As the »sacredness« among the *Q'eqchi'* has a strong situational aspect, displacement from the original land has created a chasm between the people and the mountain deities. But despite profound dislocations and social disruptions the *Q'eqchi'* have experienced in the past, worship of the territorial guardian spirits is still carried out and community rituals connected with agricultural activities are practised to strengthen social relations and restore productivity and fertility. In terms of the ›mountain cult‹, Wilson (1995 : 58) affirms that the *tzuul taq'a* should be seen as a fluid and continually redefined figure, rather than a legacy of the ancient pre-Columbian past. Cultural beliefs and customs persist and still form the base upon which traditional agriculture is widely practiced and can be seen as functioning to reinforce a community's set of rules for managing common resources. The established sacred sites, as sanctuaries for biodiversity and locations for spiritual practices, have survived substantial cultural changes over the past decades. Socio-cultural life, spirituality and biological diversity combine to provide the context for indigenous knowledge and sustainable production methods. Thus, the local conception of the natural environment contains mechanisms for maintaining balance in the biophysical, human and spiritual circuits.

Fig. 6. Since ancient times the *Ceiba pentandra*, which is the tallest of the trees in the area, has been considered by the *Q'eqchi'* as a holy tree to be worshiped as a symbol linking earth and heaven and as a physical expression of life, growth and historical continuity

4 Concluding remarks

»We should not forget that management of natural resources is foremost a question about *social* relations, a means of regulating *people's* access to these resources.«[21]

4.1 From local to global

This contribution attempts to exemplify the diverse connections between nature and culture, which adopt specific shapes according to the social, historical, political and geographic contexts in which they occur. In consequence, strategies and policies oriented at indigenous community development and conservation issues, which tend to be designed by external institutions, should take into account these diverse contexts. In the case of Guatemala, the historical developments and implications especially of the war have to be taken into account as they affect social and political spheres of the country long after the formal re-establishment of democratic structures. *In situ* conservation of biodiversity as conceptualised in the provisions of the CBD cannot succeed without the participation of indigenous populations gaining control over the use of territories and the resources on which they rely (Agrawal 1995 : 432). Even though the recognition of indigenous knowledge is a necessary condition for participatory processes into contemporary resource management, it is often not a sufficient condition to achieve long-term sustainability. A fundamental issue to provide long-term protection for local conservation in the area is to legalise communal resource-use rights, which would also entail the reaffirmation of cultural identity. The lack of secure land tenure illustrates that environmental conflicts at the local level often do not only reflect different values and senses of place, but are linked to conflicts of interest that originate far from park boundaries and require consideration of much larger political and legal issues. To avoid these conflicts, there is a need to assume that it is in the common interest to preserve natural *and* cultural diversity. Despite the global recognition of indigenous rights, the formal acknowledgment often remains nominally. Although the national policy has undergone revisions with the change of administration since the end of the war, its general purpose has not been significantly altered. The indigenous population still is widely marginalised in the social, economic, political and environmental domains of national society. Beyond the relationships of power and control, competing land uses and conflicting resource management do not simply reflect competing interests. Such conflicts reflect much deeper »ontological schisms between worldviews – ways of seeing the world and ways of thinking about peoples' places within the world« (Howitt 2001 : 59). As the conservation of biodiversity, in the sense of science which separates the phenomenon of non-human diversity from local knowledge and livelihoods, is not a value of the *Q'eqchi'* farmers, the relatedness of people, culture, land and knowledge through which they conceptualise their world might serve as a basis for a meaningful consideration of biodiversity conservation. It has been assumed by Kalland that »infusing nature with spirits is apparently no guarantee for the well-being of the environment«

[21] Kalland 2000 : 328.

(2000 : 323) and beliefs may have little to do with actual behaviour towards the natural environment as often economic needs are more decisive for human behaviour than religious beliefs. Nevertheless, the indigenous cosmovision implies manifold conservational aspects as it provides philosophical principles of ethical responsibility and social norms of reciprocity and respect for ecosystem integrity that promote ecologically sustainable behaviour. For this, an adaptive ›co-management‹ needs to be modified by realising the implications of a multicultural definition on environment. As often proposed by informants, a community based conservation approach should encounter the erosion of local knowledge and practices as emanations of the cosmovision by putting in place mechanisms for their revitalisation. The incorporation into the formal education systems has been considered as an initial step to address local concerns. As knowledge is linked to language, its protection might be achieved through conserving the local language ensuring the transmission within the societies themselves, empowering indigenous people »so as to increase their control over processes of change, and ensuring continued access to the environments upon which their ways of life depend« (Nakashima and Roué 2002 : 323).

4.2 Rethinking scientific assumptions

The experience of multi-disciplinary research as well as field-based encounters between environmentalists, scientists and local residents led to the notion that science itself takes place in a cultural context of which scientists in many cases are unaware. The exercise of examining environmental relations and constructions of nature from a cross-cultural perspective not only deepened understandings of indigenous perceptions but also provided insights into cultural implications underlying the ontology and epistemology of ›western‹ science. Scientists should not only widen their scope towards paradigms of other disciplines, but should likewise reflect on the inherent values and ethics of their own analytical approaches and consider alternative ways of defining reality that include pathways between the natural and the cultural realms. Such a paradigmatic shift needs to emerge if the complexity of ecological and social relationships underlying processes of biodiversity loss is to be approached and solutions to environmental problems according to the provisions of the CBD are to be found. Global environmental policy frameworks and local initiatives only can be successfully negotiated if differences of perceptions and assessments of such problems are understood. In the quest for a global solution to the conservation of biodiversity, local knowledge, customary practices, social values and religious beliefs provide the foundation upon which practical solutions responsive to the needs of indigenous communities involved should be based on.

> »...western analysis would say that environmental degradation is caused by relations between people: unequal access to the land and extension of the agroexport economy. For the elders, however, ecological crisis results form the breakdown in relations between people and the conscious environment, specifically the mountain spirits that inhabit and constitute the landscape. Science is not seen as a valid source of explanation for ecological problems; instead, it is part of those problems.« Wilson (1995 : 290f)

References

Agrawal A (1995) Dismantling the Divide Between Indigenous and Scientific Knowledge. Development and Change 26(3) : 413–439

Antweiler C (1998) Local Knowledge and Local Knowing. An Anthropological Analysis of Contested »Cultural Products« in the Context of Development. Anthropos 93(4–6) : 469–494

Banuri T, Apffel Marglin F (1993) A Systems-of-Knowledge Analysis of Deforestation, Participation and Management. In: Banuri T, Apffel Marglin F (eds) Who Will Save the Forests? Knowledge, Power and Environmental Destruction. Zed Books, London New Jersey, pp 1–23

Berkes F (1999) Sacred Ecology. Traditional Ecological Knowledge and Resource Management. Taylor and Francis, Philadelphia

Chatty D, Colchester M (2002) Conservation and Mobile Indigenous Peoples. In: Chatty D, Colchester M (eds) Displacement, Forced Settlement, and Sustainable Development. Berghahn Books, New York Oxford, pp 1–20

Ellen R (2002) ›Déjà vu, all over again‹, again. Reinvention and Progress in Applying Local Knowledge to Development. In: Sillitoe P et al. (eds) Participating in Development. Approaches to Indigenous Knowledge. Routledge, London New York, pp 235–258

Ellen R, Harris H (2000) Introduction. In: Ellen R et al. (eds) Indigenous Environmental Knowledge and its Transformations. Critical Anthropological Perspectives. Harwood, Amsterdam, pp 1–33

Escobar A (1999) After Nature. Steps to an Antiessentialist Political Ecology. Current Anthropology 40(1) : 1–30

Escobar A (2001) Culture Sits in Places. Reflections on Globalism and Subaltern Strategies of Localization. Political Geography 20(2) : 139–174

Furze B, De Lacy T, Birckhead J (1996) Culture, Conservation and Biodiversity. The Social Dimension of Linking Local Level Development and Conservation through Protected Areas. Wiley, Chichester

Gray A (2000) Indigenous Peoples, their Environments and Territories. In: Posey DA (ed) Cultural and Spiritual Values of Biodiversity. Intermediate Technology Publications UNEP, London, pp 61–66

Grenier L (1998) Working with Indigenous Knowledge. A Guide for Researchers. International Development Research Centre, Ottawa

Gündling L (2000) Implementing Article 8(j) and other Provisions of the Convention of Biological Diversity. COICA Climate Alliance, Quito

Hernando Gonzalo A (1999) The Perception of Landscape amongst the Q'eqchi', a Group of Slash-and-Burn Farmers in the Alta Verapaz (Guatemala). In: Ucko P, Layton R (eds) The Archaeology and Anthropology of Landscape. Shaping your Landscape. Routledge, London New York, pp 254–263

Howitt R (2001) Rethinking Resource Management. Justice, Sustainability and Indigenous Peoples. Routledge, London New York

Kalland A (2000) Indigenous Knowledge. Prospects and Limitations. In: Ellen R et al. (eds) Indigenous Environmental Knowledge and its Transformations. Critical Anthropological Perspectives. Harwood, Amsterdam, pp 319–335
Kamppinen M, Walls M (1999) Integrating Biodiversity into Decision Making. Biodiversity and Conservation 8(1) : 7–16
Kempton W (2001) Cognitive Anthropology and the Environment. In: Crumley CL (ed) New Directions in Anthropology and Environment. Intersections. Altamira Press, Walnut Creek, pp 49–71
Kimmerer RW (2002) Weaving Traditional Ecological Knowledge into Biological Education. A Call to Action. BioScience 52(5) : 432–438
Laird S (2000) Forests, Culture and Conservation. In: Posey D (ed) Cultural and Spiritual Values of Biodiversity. Intermediate Technology Publications UNEP, London, pp 347–358
Laird S (ed) (2002) Biodiversity and Traditional Knowledge. Equitable Partnerships in Practice. Earthscan, London
Myer L (1998) Biodiversity Conservation and Indigenous Knowledge. Rethinking the Role of Anthropology. Indigenous Knowledge and Development Monitor 6(1) `http://www.nuffic.nl/ciran/ikdm/6-1/myer.html`
Nakashima D, Roué M (2002) Indigenous Knowledge, Peoples and Sustainable Practice. In: Munn T (ed) Encyclopedia of Global Environmental Change, vol. 5, Social and Economic Dimensions of Global Environmental Change. Wiley, Chichester, pp 314–324
Nigh R (2002) Maya Medicine in the Biological Gaze. Bioprospecting Research as Herbal Fetishism. Current Anthropology 43(3) : 451–477
Nygren A (1999) Local Knowledge in the Environment-Development Discourse. From Dichotomies to Situated Knowledges. Critique of Anthropology 19(3) : 267–288
Orlove BS, Brush SB (1996) Anthropology and the Conservation of Biodiversity. Annual Review of Anthropology 25 : 329–352
Posey DA (2002) Upsetting the Sacred Balance. Can the Study of Indigenous Knowledge Reflect Cosmic Connectedness? In: Sillitoe P et al. (eds), Participating in Development. Approaches to Indigenous Knowledge. Routledge, London New York, pp 24–42
Pottier J (2003) Negotiating Local Knowledge. An Introduction. In: Pottier J et al. (eds) Negotiating Local Knowledge. Power and Identity in Development. Pluto Press, London, pp 1–29
Purcell T (1998) Indigenous Knowledge and Applied Anthropology. Questions of Definition and Direction. Human Organization 57(3) : 258–272
Slikkerveer J (2000) Ethnoscience, ›TEK‹ and its Application to Conservation. In: Posey DA (ed) Cultural and Spiritual Values of Biodiversity. Intermediate Technology Publications UNEP, London, pp 169–177
Townsend PK (2000) Environmental Anthropology. From Pigs to Policies. Waveland Press, Prospect Heights
Wilson R (1995) Maya Resurgence in Guatemala. Q'eqchi' Experiences. University of Oklahoma Press, Norman
Zent S (1999) The Quandary of Conserving Ethnoecological Knowledge. A Piaroa Example. In: Gragson T, Blount B (eds) Ethnoecology. Knowledge, Resources, and Rights. University of Georgia Press, Athens, pp 90–124

Direct payments for conservation – the importance of environmental measures in farming systems for bird populations in a fragmented landscape. A case study in Guatemala

María A. Máñez Costa[1] and Swen C. Renner[2]

[1] Institute of Rural Development, Georg-August University of Göttingen, Waldweg 26, 37073 Göttingen, Germany, email to `txonim@yahoo.com`

[2] Conservation and Research Center, Smithsonian Institution, 1500 Remount Rd, Front Royal 22630 VA, USA, email to `renners@crc.si.edu`

Summary. The name Guatemala is derived from the Nahuatl *guauhtemallan*, which means ›land of trees‹. Guatemala is home to a unique mix of plants and animals. The country's biological diversity is among the highest in Central America and it ranks among the 25 most plant-rich countries in the world. Increasing conservation efforts in this region are crucial to preserving Guatemala's wealth of ecosystems. As a case study, the region of the *Macizo de Cobán* has been chosen. The area contains one of biodiversity's hotspots, consisting of large areas of cloud forest. Degradation and loss of forest cover is caused mainly by slash-and-burn increased agricultural expansion to grow subsistence crops (e.g., corn and beans), forest clearing for cash crops (e.g., cardamom and coffee), extraction of firewood (the only domestic fuel of rural inhabitants), local demand for construction, illegal logging increasing demand for commercial forest products (particularly of the reserve's primary and old-growth forest) and finally, the clearing of forest for cattle grazing. Slash-and-burn is becoming a pushing method that is destroying the most intricate part of the tropical nutrient cycle and continues to degrade and destroy this biologically rich region. Due to the socio-economic conditions and the pressure on resources in the Macizo de Cobán, it is almost impossible to conserve forest as a pristine area. Therefore, environmental measures must be implemented to mitigate the loss of biologically diverse forests. Revisions from Ferraro and Kiss (2002), Nasi et al. (2002), Vosti et al. (2002), Chomnitz (1998), Espinoza and Smyle (1999), about incentives for conservation clarify that direct payment to land owners could pay the way out of forest decline. To elucidate the problem at the local level, interdisciplinary data from bird populations and environmental measures in the Macizo de Cobán will be used. Based on this data policy recommendations for forest conservation will be made.

Key words: interdisciplinary approach, environmental measures, ecological services, species extinction, conservation, PES

1 Introduction

Environmental degradation and unfavourable environmental conditions disproportionately affect the health, livelihood and security of poor people (World Bank 2000). In the study region degradation of soil and vegetative resources already threatens agricultural productivity, biodiversity and water quality and availability[3]. Therefore, to create incentives and to compensate the poor for conserving or managing resources of value to others and themselves is the main objective of this contribution.

In the long-term, the goals of poverty alleviation, agricultural growth and environmental protection should be complementary, as human development depends on the ability of the environment to provide a variety of goods and services and to sustain these into the future (Vosti and Reardon 1997; Terborgh 1999).

The first step to address this critical triangle is to put rural households and communities first (Zeller et al. 2000) and to have an in-depth understanding of the forces that drive farmers to deforest.

Marginal hillsides in Guatemala suffer from oblivion. Local policies that link the outcomes of the critical triangle of development have until now seldom been implemented. Some international development agencies have developed agendas for a broader development under sustainable conditions but their measures have been limited so far to the establishment of protected areas reflecting different conditions and degrees of protection (WWF and IUCN). Moreover, various poverty-reduction strategies have only focused on improving the economic conditions of the rural population without explicitly considering its impact on the environment. Thus, single policy instruments pursue environmental, economic and social policy objectives and projects instead of integrated ones.

Therefore, the design of policies that focus on development as an overall description of environmental sustainability and poverty alleviation are needed for the area and also for other river basins in Guatemala in order to reach a level of sustainability that allows the poor to develop in the long term. Direct payments for conservation to smallholders could potentially contribute to the achievment of all three objectives of the critical triangle of rural development: poverty reduction, economic development and environmental protection (Vosti and Reardon 1997).

Here we present joint data, ecological and socio-economic, evaluating the state of the problem. This approach is innovative because both studies were interdisciplinary and have data generated from the same region.

2 Background of the problem

The research area is located in the centre of Guatemala in the region of Alta Verapaz. Thee study area is referred to as Macizo de Cobán, and is located in the south-east of the city of Cobán. El Macizo de Cobán lies in the north on the eastern

[3] Compare Scherr and Yadav (1995), cited by Scherr and White (2002).

tropical and temperate highland of Guatemala. The area belongs to the humid corridor formed by the Sierras de Lacandón to the West and the Sierra de Chamá and Santa Cruz to the East (Méndez et al. 1995). The area is categorised through annual precipitations of between 1600 mm and 2500 mm (classification by Köppen), while the chosen study area in contrast has rainfalls of about 4000 mm yearly (own data). The majority of the vegetation pertains to the biome mountainous tropical and cloud forest. Montane cloud forests are a highly endangered ecosystem. They hold higher numbers and significantly higher abundances of sensitive and indicator species than temperate forests. This area is also notorious for migrant species (Kapelle and Brown 2001, Doumenge et al. 1995) and for species endemic to the Central American highlands (Stattersfield et al. 1998).

Endemics – or restricted range species – are mostly are dependent on restricted habitat use in special areas. Centres of such areas with a higher degree of endemism are named Endemic Bird Areas (EBA from Stattersfield et al. 1998). Within Central America there are 30 EBAs. The Macizo de Cobán is located within the EBA 018, ›North Central American highlands‹, which is 15,000,000 ha large, 500–3500 m in elevation, key habitats are montane forest, pine-oak forest, and deciduous forest which are threatened by moderate habitat loss. EBA 018 is classified as ›Priority – Urgent‹. Two of the 20 restricted range species of EBA 018 are threatened (*Oreophasis derbianus*, *Tangara cabanisi*), neither were present at Macizo de Cobán or perhaps are already extinct there. Nevertheless, 13 of the 20 restricted range species were recorded. Out of the 13, at least three were exclusively montane evergreen and pine-oak species (*Strix fulvescens*, *Lampornis viridipallens*, *Troglodytes rufociliatus*). Data from Renner (2003) about endemic bird species diversity of the area are reflected in Table 1.

Possingham et al. (2002) stated that Red Lists are not designed (i) to set priorities for resource allocation for species recovery, (ii) to inform reserve system design, (iii) to constrain development and exploration and (iv) to report the state of the environment. On the other hand, Lamoreux et al. (2003) stated that the misuse is indeed given and agreed with Possingham (2002) »(i)t is naïve and counterproductive from all point of views to use threatened species lists alone to allocate resources for recovery, ... «. Therefore, several more aspects were integrated to evaluate the research area. On one hand, biological data about habitat comparison, measure of influences, body mass, etc and on the other hand socio-economic data to fulfil the gap of information.

The region has one of the highest proportions of Mayan inhabitants (89.0 %) in Guatemala. The distribution of rural and urban population in Alta Verapaz is also disproportionate: 15.8 % urban population and 84.2 % of rural population[4]. The *Q'eqchi'*, the local Mayan population of Macizo de Cobán, are victims of high incidences of poverty, limited infra-structural development, low market integration and restrained off-farm income opportunities. The subsistence economy of the Q'eqchi', based on traditional slash and burn farming, has led not only to well-

[4] United Nations Development Program. Guatemala: los contrastes del desarrollo humano (Guatemala: UNDP 1998 : 217).

documented loss of biodiversity but also to economic loss due to soil degradation (INAB-MAGA-PAFG-MARN-CONAP 2002).

Table 1. Endemic and specialised birds of the Sierra Yalijux and the Chelemhá Plot. Listed are all species classified in the Endemic Bird Area # 018 ›North Central American highlands‹ (Stattersfield et al. 1998 compiled with data from Renner 2003).

Species	Global status[a]	Altitude (m)	Present in study area[c]	Habitat[b]
Podilymbus gigas	extinct (1987)	1,500	-	water
Oreophasis derbianus	vulnerable	2,000–3,000	extinct (1990)	NF
Cyrtonyx ocellatus	near threatened	1,000–3,000	*	PO
Otus barbatus	near threatened	1,800–2,500	not rec.	NF, PO
Strix fulvescens	least concern	1,200–3,000	*	NF, PO
Campylopterus rufus	least concern	900–2,000	not rec.	NF, PO, SF, e, a
Lampornis viridipallens	least concern	1,400–2,200	*	NF, PO, e
Lampornis sybillae	least concern	1,400–2,200	not rec.	NF, PO, e
Doricha enicura	least concern	1,000–2,200	*	NF, PO, SF
Atthis ellioti	least concern	1,500–3,500	*	NF, PO, SF
Asphata gularis	least concern	1,500–3,000	*	NF, PO, SF
Xenotriccus callizonus	near threatened	1,200–2,000	not rec.	deciduous forest
Notiochelidon pileata	least concern	1,000–3,000	*	NF, PO, SF
Troglodytes rufociliatus	least concern	1,700–3,500	*	NF, PO, e
Melanotis hypoleucus	least concern	1,000–3,000	*	NF, PO, SF
Turdus rufitorques	least concern	1,500–3,350	*	NF, PO, e
Tangara cabanisi	least concern	1,000–1,700	not rec.	NF, PO
Ergaticus versicolor	near threatened	1,800–3,500	*	NF, PO, SF
Icterus maculialatus	least concern	500–1,800	not rec.	PO, SF
Carduelis atriceps	near threatened	2,000–3,500	*	PO, e, SF
Cyanocorax melanocyaneus	least concern	600–2,450	*	e, PO, SF

[a] adapted from Stattersfield et al. (1998).
[b] NF: montane evergreen forest, PO: pine-oak forest, SF: secondary growth, e: forest edge, clearings; a: agricultural areas.
[c] Renner (2003): * observed in Chelemhá.

The Macizo de Cobán belongs to one of the poorest areas in Mesoamerica with a GDP of € 264 per capita[5]. It has also has one of the lowest Human Development

[5] 1,852.20 Quetzales/Data Source Segeplan 2003 (Secretaría de Planificación y Programación de la Presidencia); 1€ = US$ 1.19 in November 2003.

Index (HDI)[6] scores (0.355) in Guatemala. Its HDI rank is 32 places below its real GDP per capita rank, indicating comparatively poor performance on human development in relation to the material resources at its disposal[7]. This is another example of an area where the poorest people inhabit one of the biologically richest areas.

Between 1987 and 1995 the annual deforestation rate for Guatemala was 1.1 % of the total forested area, equivalent to 80,000 ha per year (Jolom-Morales 1997), while worldwide deforestation data from the FAO (2001) suggest that in the tropics, the annual rate of deforestation is about 0.8 %. Table 2 reflects the actual causes for deforestation[8]

Table 2. Actual causes of deforestation[a]

Causes of deforestation	%
Shifting Agriculture	79
Livestock	10
Illegal logging	5
Uncontrolled burning	3
Plagues and plant diseases	2
Export agriculture	1

[a] Source: Gálvez (2000)

3 Predicting the future of the region

Spatial diversity (species-area relationship) patterns have important implications for conservation of biodiversity, and understanding these patterns contributes to our knowledge of community structure.

Species and area are related with $S = cA^z$, where S number of species, A area, z and c constants (e.g., MacArthur and Wilson 1967, Rosenzweig 1995, Krebs 1999, Waltert et al. 2003).

The inverse species-area relationship (Preston 1962) might assess the number of extinct forest dependent bird species threatened from deforestation. This procedure has been proven to be valid since it produced results that were concor-

[6] The HDI measures a country's achievements in terms of life expectancy, educational attainment and adjusted real income. UNO. »Human Development Index Report.« United Nations, 02 February 2000).

[7] United Nations Development Program. Human Development Report 1999 (New York: UNDP and Oxford University Press, 1999), pp. 134-137.

[8] For more data about deforestation in the Macizo de Cobán see Markussen and Renner (in this volume).

dant with assessments of mammal and bird threat status made by conservationists. The observations were consistent both in Neotropical and south-east Asian tropical rainforest (Brooks and Balmford 1996, Brooks et al. 1999a, 1999b, 1999c, Waltert et al. 2003). Rearranging the relationship $S = cA^z$ one can calculate the number of species most likely to become extinct in a fragment (or nested subset) of a given size by dividing $S_{surviving} = cA^z_{surviving}$ by $S_{original} = cA^z_{original}$. This results in $S_{surviving}/S_{original} = (A_{surviving}/A_{original})^z$ and one can estimate the extinct species $S_{extinct} = S_{original} - (S_{original} - (A_{surviving}/A_{original})^z)$, because $S_{extinct} = S_{original} - S_{surviving}$.

Assuming the regional meta-population of the Macizo de Cobán has no or limited individuals' or genetic exchange (the next natural montane cloud forest is located in the Sierra de las Minas, 20 km southwards, separated by the valley of the Río Polochic), one can set $A_{surviving}$ to 55,000 ha as the remaining natural montane cloud forest with $A_{original} = 165,000$ ha for the Macizo de Cobán (Mühlenberg et al. 1989). The original area of EBA # 018 Central American highlands is $A_{original} = 15,000,000$ ha.

Focusing on the 21 natural montane cloud forest specialists endemic to the Central American highlands (Table 1; Stattersfield et al. 1998), ten should be present in natural forest and oak-pine forests. Six out of the 21 were observed in the study plot (Table 1). Fig. 1 illustrates the relationship between the 21 mentioned endemics and area.

According to the 21 endemics from EBA # 018 »Central American highlands« (Stattersfield et al. 1998) there will be 2.68 endemics remaining ($S_{surviving}$) for the 55,000 ha surviving natural forest in the Macizo de Cobán. For extreme z-values, 1.4 and 5.47 endemics will remain, respectively. Contrastingly, 13 endemics were observed.

Therefore, a lot of more species are still present than might be carried by natural forest and species-area equilibrium. Even when an area is not decreasing and the deforestation rate will immediately be pushed to zero, four to five of the observed ten natural forest highland endemics will not survive because the area-species equilibrium is not reached. It must be highlighted, that the background is theoretic and the natural situation might be different.

Results from the socio-economic study in the same area show that development under the actual conditions will drive to further conversion of forest and forested land into agricultural land. The current crises in the world coffee markets and the corresponding decline of coffee prices have seriously worsened the wage-earning potential of the area's seasonal labourers. About 68 % of the sample households were found to engage in wage-labour activities linked to coffee farming. Thus, the coffee crisis led to a considerable decline in income in many households in the Macizo de Cobán. Hence, as a response to this crisis, the farmers were increasingly forced to use the remaining forestland for food crop cultivation. In this context, it is a well-known fact that in such ecologically fragile areas, if the soils lose their vegetation cover they are greatly exposed to the leaching and eroding effects of rain. Indeed, in Macizo de Cobán it was observed that farmers started to cut and burn new surfaces for staple crops production. Moreover, they sold timber to traders in order to compensate for the loss of off-farm cash income.

Currently, farmers are a threat to biodiversity, especially concerning endemic birds. The aim should be to motivate farmers to use the forests in such a way that both their income improves and conservation is guaranteed.

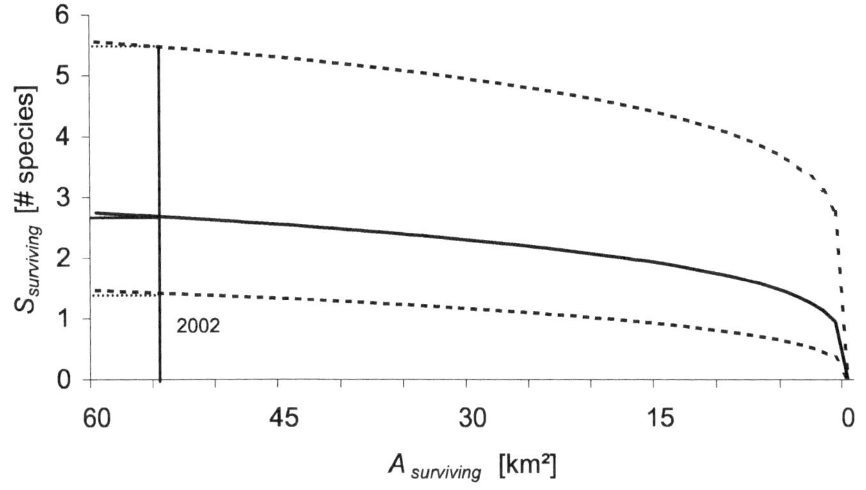

Fig. 1. Relationship between 21 highland bird species of EBA # 018 remaining ($S_{surviving}$) and remaining closed forest cover within areas of the Macizo de Cobán. Dotted curves represent calculations using extreme $z = 0.17$ and $z = 0.34$, respectively.

About 76 % of the sample farmers viewed timber as precautionary savings. Furthermore, forest and vegetation conservation also makes »sense from the viewpoint of minimising high environmental risk under considerable uncertainty« (Nasi et al. 2002). This coincides with the safety-first criterion of Low (1974) ensuring minimum conditions to meet farmers' necessities by conserving environmental measures within farming systems. In the opposite situation, when trees are harvested and vegetation eradicated, farmers are in a way ›destroying‹ their own minimum needs and staying empty handed.

Fig. 2 gives an overview of the development of land use for the next few years if the actual socio-economic situation continues.

Comparing the trend of both Figs. (1 and 2), we can conclude that land use patterns will drive the loss of forest and the intrinsic loss of species in Macizo de Cobán. As indicated in Fig. 2, further natural forests will vanish and thus, as indicated in Fig. 1, the tendency of endemic species extinction will be aggravated. The services of the ecosystem »cloud and montane forest« will not only be affected by

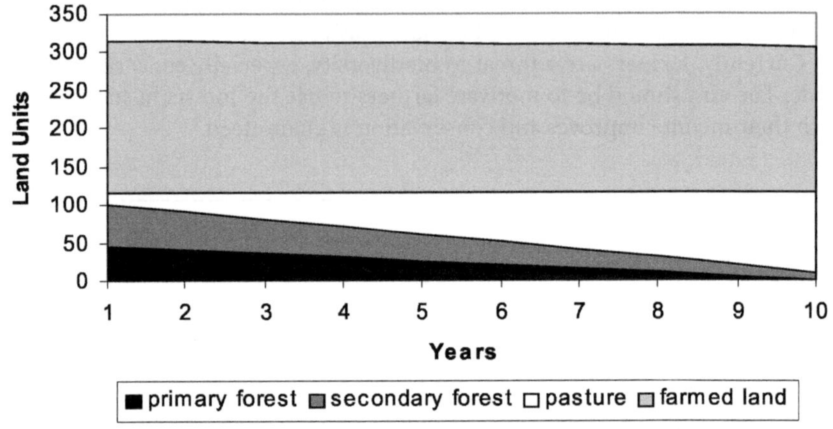

Fig. 2. Prediction of land use in the Macizo de Cobán (source: Máñez Costa forthcoming).

the loss of biological diversity but also by the depletion of soil productivity[9] and consequently, to the endangerment of the basic developmental conditions for the inhabitants of the Macizo de Cobán.

4 The needs of environmental conservation

In Guatemala now, different efforts to reach conservation in a broader sense have been established[10]. With all these measures, reforested areas and other revegetation initiatives have recovered about 50,000 ha per year. Nevertheless, deforestation continues at the rate of about 80,000 ha per year (Jolom-Morales 1997). Consequently despite all efforts, forest policies have barely been successful so far.

Cervigni (2001) exposed that recent theories of extinctions and the basic postulate of economic theory are consistent with each other. Both are based on the fact that »resources are allocated among competing uses« (Cervingi 2001). The economic theory further reveals that farmers allocate resources among competing uses in such a manner that monetary returns could be maximised (Cervigni 2001 : 39). Therefore, it could be assumed that land uses with less monetary return would disappear in favour of those with more monetary returns.

It is therefore a fact that forests and habitat conservation needs new economic and policy instruments to be implemented. Thinking of the sustainable triangle of development, as reflected in Fig. 3, there is a field where goods and payments for these goods have still not found a market or policy instrument to regulate them. On one hand, we have farmers producing different goods and on the other hand

[9] Cp. Markussen (2003).
[10] Cp. Birner et al. (in this volume).

we have society consuming such goods but not paying for them. Such goods, still without markets, are landscape conservation, habitat and environmental protection and biodiversity conservation. The only market relation between society and farmers is established through agricultural market products. For this reason, environmental policy instruments for these ›goods without monetary returns‹ should be developed to compensate farmers and to incite them.

The proposed solution of the problem might be the instalment of direct payments to local farmers. These payments can incite the farmers to preserve either natural forest or old secondary forest, or both.

Direct payments have so far been implemented in Mesoamerica, mostly in Costa Rica. Ferraro and Kiss (2002) describe a project where Costa Rica pays rural residents about € 29.17 annually per hectare forest protected, and access demand for conservation contracts suggests that these payments are higher than necessary (Ortiz 2002, cit. Ferraro and Kiss 2002). Espinoza and Smyle (1999) presented different direct payments to forest owners for biodiversity protection in Costa Rica in the range of € 4.12 to € 9.3 per ha and year. Due to these payments, and also to the forest conservation policy of Costa Rica, Malavasi and Kellenberg (2003) reported a deceleration of deforestation.

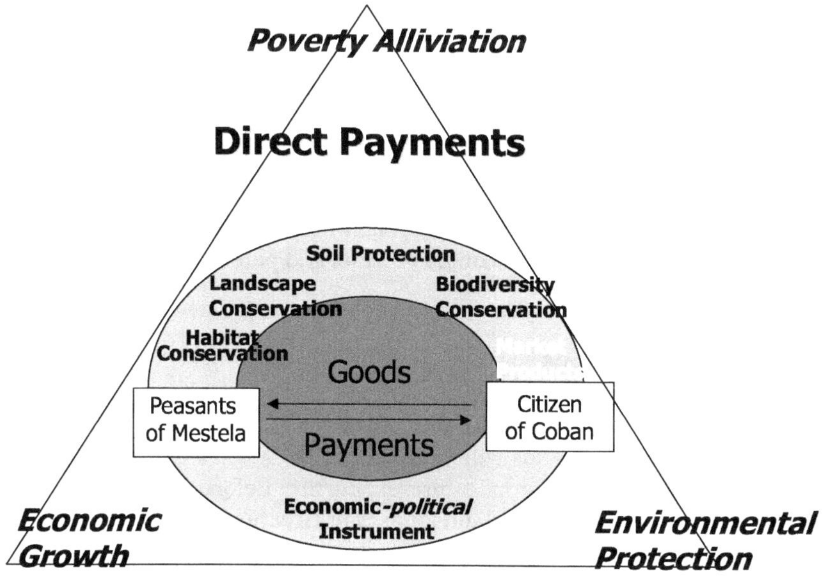

Fig. 3. Conceptual framework for direct payments (Source: Máñez Costa 2003, adapted from Vosti and Reardon (1997)

Although in Costa Rica direct payments were implemented and an initial success in reduction of deforestation is recognisable, further data or success stories are lacking. Forest still decreases in Guatemala but the decrease of forest cover might be reduced (Jolom-Morales 1997; not necessarily ›untouched‹ forests, Nasi et al. 2002, Jolom-Morales 1997). However, the decrease might not be necessarily based on ›untouched‹ forests (Nasi et al. 2002), but on all types of forest in general.

Aylward (2002) summarised studies about economic evaluation of environmental services in Guatemala (Table 3). There is a wide range of data obtained through these studies but so far there is no implementation of the results. Certainly, Aylward (ibid) argues that the »results – such as they are – for caution before extolling the value of environmental services«. Thus, he suggests examining smaller scale environmental services supply issues in rural settings, particularly in relation to healthy ecosystem concerns.

Data for Guatemala was calculated until the 2003 payments were made to farmers in the Macizo de Cobán. The baseline data was collected in the context of household interviews. Máñez Costa evaluated the environmental services that farmers are already offering and that contribute somehow to preserving the contemporary environmental conditions of the study area.

Fig. 4 reflects that farmers react positively to direct incentives. Under these calculations, farmers include direct payments for forest conservation in their production function as a remunerated activity. Based on this assumption, we can also derive that direct payments for natural forest conservation will serve their purpose.

The calculations show that direct payments will/could positively affect the environment[11]. As reflected in Fig. 4, direct payments could achieve the maintenance of at least secondary forests for the following 10 years.

Summarising and including the interdisciplinary data already presented about bird populations and forest conservation under direct payments, we can conclude that at least ten endemic bird species might be preserved from extinction, when direct payments to preserve secondary forests and perennial vegetation occur.

5 Policy Implications

Even if deforestation comes to an immediate standstill, 10 bird endemics of EBA Central American highlands will vanish from the Macizo de Cobán. Direct payments might be a suitable tool to prevent further deforestation or generate new natural-like reforestation areas and landscape structures promoting biological diversity in areas like old secondary forest.

In Guatemala, policy instruments for environmental protection should definitively include the three goals of the critical triangle of development. Without direct payments forests remnants will disappear completely and with it the habitat for the remaining bird species (as shown in Figs. 1 and 2). If forest owners of the Macizo de Cobán do not see forest and perennial vegetation as a valuable commodity and

[11] Cp. Máñez Costa and Zeller (in this volume).

Table 3. Summary of Studies of Environmental Services in Guatemala (source: Aylward 2002).

Study	Location or Ecosystem of Study	Environmental Service			Type of Analysis		Comment
		Hydro Services	Carbon	Tourism	Bio-physical	Economic	
Brown et al. (1996)	Sierra de Las Minas Biosphere Reserve	Yes			Yes	Yes	valuation in the case of cloud forests
Vel squez, et al. (no date)	Xayá-Pixcayá watersheds	Yes				Yes	cost estimates rather than values per se
Pape and Ixcot (1998)	Lale Amatitlán.					Yes	valuation of water and water quality in the lake not of hydrological services
Velasquez and Montalvo (1998)	Chixoy River					Yes	valuation of water - unclear if of hydrological services
Márquez (no date)	various		Yes		Yes		Summary of other studies
OCIC Consulting Group (2000)	Mayan Biosphere Reserve, Peten		Yes		Yes		Cost estimates rather than values per se
Márquez (1997)	Agroforestry systems.		Yes		Yes		primary data
Winrock International (1998)	coffee, Sololá.		Yes		Yes		primary data
Medina et al. (2000)			Yes		Yes		primary data
Solano Alvarez (2000)	coffee plantation		Yes		Yes		primary data
Fundación Solar/ Gremial de Huleros (1999)	rubber plantation		Yes		Yes		primary data
Fundación Solar (1999)	rubber plantations.		Yes		Yes		primary data
Morales Ralda (2000)	rubber plantation - S Guatemala.		Yes		Yes		primary data
Fundación Solar (1999)	hardwood forests - Lake Atitlán.		Yes		Yes		primary data
Castellanos (2000)	evergreen and cloud forest in Chiquimula		Yes		Yes		primary data
López (1998)	countrywide		Yes		Yes		secondary data
López (1998)	Volcán San Pedro			Yes		Yes	contingent valuation and also valuation of forestry and agriculture alternatives
Ortiz (2000)	Guatemalan Protected Areas	Yes	Yes	Yes	Yes	Yes	secondary data
Martínez (2000)	forests-Sololá, Baja Verapaz.	Yes	Yes		Yes		primary data
Castanier (1999)	Punta de Manabique						varios services using technical coef cients

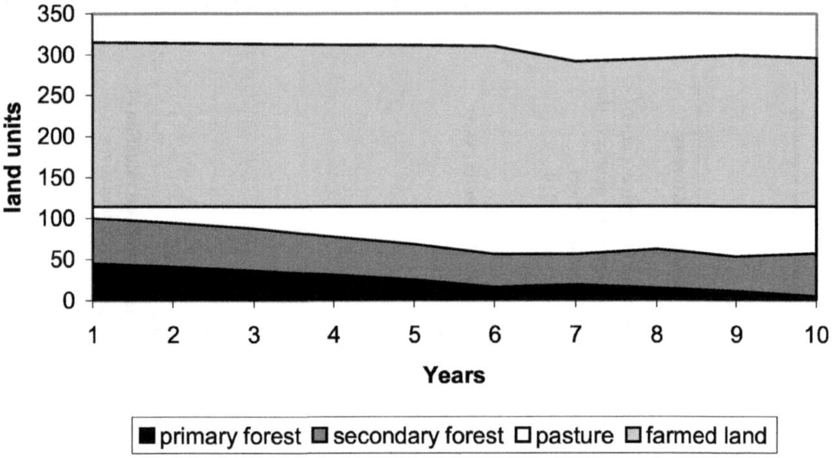

Fig. 4. Modelling land use change for the Macizo de Cobán with direct payments (source: Máñez Costa, forthcoming).

in forest preservation an alternative land use, the equilibrium of the critical triangle will be displaced. Consequently, neither conservation nor development aims will/could be reached.

References

Aylward B (2002) Mecanismos de mercado y Servicios Ambientales. Planteamiento Conceptual y Revisión de Experiencias Internacionales. Guatemala, Proyecto Mirna : 114

Brooks TM, Balmford A (1996) Atlantic forest extinctions. Nature 380 : 115

Brooks TM, Pimm SL, Oyugi JO (1999a) Time lag between deforestation and bird extinction in tropical forest fragments. Conservation Biology 13 : 1140–1150

Brooks TM, Pimm SL, Kapos V, Ravilious C (1999b) Threat from deforestation to montane and lowland forest birds and mammals in insular south-east Asia. Journal of Animal Ecology 68 : 1061–1078

Brooks TM, Tobias J, Balmford A (1999c) Deforestation and bird extinctions in the Atlantic forest. Animal Conservation 2 : 211–222

Cervigni R (2001) Biodiversity in the balance: land use, national development, and global welfare.Edward Elgar, Cheltenham

Chomnitz KM, Kumari K (1998) The domestic benefits of tropical forest: A critical review. The World Bank Research Observer 13 : 13–35

Doumenge C, Gilmour D, Ruíz Pérez M, and Blockhus J (1995) Tropical montane cloud forests: conservation status and management issues. In: Hamilton L, Juvik J, Scatena F (eds) Tropical montane cloud forests. Springer, Berlin, pp 24–37

Espinoza NG, Smyle J (1999) El pago de servicios ambientales y el desarrollo sostenible en el medio rural. Turrialba, RUTA

FAO (2001) State of the worlds forests. FAO, Roma

Ferraro P, Kiss A (2002) Direct Payments to Conserve Biodiversity. Science 298 : 1718–1719

Gálvez JJA, Rodas OA (2000) Potencial de producción forestal de Guatemala. Guatemala : 28

INAB-MAGA-PAFG-MARN-CONAP (2002) Programa de Forestería Social. Guatemala, MAGA : 49

Jolom-Morales M (1997) Caracterización de la actividad de cacería en la reserva de la biosfera sierra de las minas y diseño de un plan de monitoreo. Guatemala

Kappelle M, Brown AD (2001) Bosques nublados del neotrópico. Instituto Nacional de Biodiversidad, San José

Krebs CJ (1999) Ecological methodology. Addison-Welsey Educational Publisher, Menlo Park

Lamoreux J, Akcakaya HR, Bennun L, Collar NJ, Boitani L, Brackett D, Bräutigam A, Brooks T, da Fonseca GAB, Mittermeier RA, Rylands AB, Gärdenfors U, Hilton-Taylor G, Mace G, Stein BA, Stuart S (2003) Value of the IUCN Red List. Trends in Ecology and Evolution 18 : 214–215

Low ARC (1974) Decision taking under uncertainty: a linear programming model of peasant farm behaviour. Journal of Agricultural Economics 62 : 311–320

MacArthur RH, Wilson EO (1967) The theory of island biogeography. Princeton University Press, Princeton

Malavasi EO, Kellenberg J (2002) Program of Payments for Ecological Services in Costa Rica. International Expert Meeting on Forest Landscape Restoration, Heredia, Costa Rica, CATIE

Máñez Costa MA (2003) Strategien und Optionen für den Natur und Umweltschutz in agrarischen Produktionssystema - Ein Fallbeispiel aus Guatemala. Treffpunkt Biologische Vielfalt III : 213–219

Mendez CA, Sisk TD, Haddad NM (1995) Beyond birds: multitaxonomic monitoring programs provide a broad measure of tropical diversity. In: Bissonette JA, Krausman PR (eds) Integrating people and wildlife for a sustainable future. Proceedings of the First International Wildlife Management Congress. The Wildlife Society, Bethesda, MD, pp 451–456

Mühlenberg M, Hovestadt T, Unger D (1989) Schutz des Nebelwaldes in Guatemala. Ökologische Station der Universität Würzburg, Würzburg

Nasi R, Wunder S, Campos JJ (2002) Forest Ecosystem Services: Can they pay our way out of deforestation? Forestry Roundtable, Costa Rica, GEF

Possingham HP, Andelman SJ, Burgman MA, Medellín RA, Master LL, Keith DA (2002) Limits to the use of threatened species lists. Trends in Ecology and Evolution 17 : 503–507

Preston FW (1962) The canonical distribution of commonness and rarity. Ecology 43 : 185–215

Renner SC (2003) Structure and diversity of cloud forest bird communities in Alta Verapaz, Guatemala, and implications for conservation. Niedersächsische Staats-und Universitätsbibliothek, Göttingen

Rosenzweig ML (1995) Species diversity in space and time. Cambridge University Press, Cambridge

Scherr SJ, White A (2002) A new agenda for forest conservation and poverty reduction. Washington, Forest Ternds : 99

Stattersfield AJ, Crosby MJ, Long AJ, Wege DC (1998). Endemic bird areas of the world. Priorities for biodiversity and conservation. BirdLife International, London

Terborgh J (1999) Requiem for nature. Island Press, Washington DC

UNO (2000) Human Development Index Report. New York, United Nations

Vosti S, Reardon T (1997) Agricultural Sustainability, Growth, and Poverty Alleviation: A Policy and Agroecological Perspective. Baltimore, Johns Hopkins University Press

Vosti S, Witcover J, Carpentier CL (2002) Agricultural Intensification by Smallholders in the Western Brazilian Amazon: From Deforestation to Sustainable Land Use. Washington, IFPRI : 30

Waltert M, Langkau M, Maertens M, Härtel M, Erasmi S, Mühlenberg M (2003) Predicting losses of bird species from deforestation in Central Sulawesi. submitted

Winker K, Rappole JH, Ramos MA (1995) The use of movement data as an assay of habitat quality. Oecologia 101 : 211–216

World Bank (2001) World development indicators 2001. World Bank, Washington DC

Zeller M, Meyer RL (eds) (2002) The Triangle of Microfinance. Financial Sustainability, Outreach and Impact. IFPRI, Washington DC

Land use changes and abiotic aspects as basic conditions for conservation of biodiversity in a tropical montane cloud forest (Guatemala)

Michael Markussen[1] and Swen C. Renner[2]

[1] Geographical Institute, Department of Landscape Ecology, Georg-August University of Göttingen, Goldschmidtstraße 5, 37077 Göttingen, Germany, email to mmarkus@gwdg.de
[2] Conservation and Research Center, Smithsonian Institution, 1500 Remount Rd, Front Royal 22630 VA, USA, email to renners@crc.si.edu

Summary. The tropical montane cloud forests of Guatemala need to be protected for their high number of ecological functions and endemic flora and fauna. For that, a complete land use potential and mosaic landscape evaluation was carried out with special consideration of the resources ›forest‹ and ›soil‹. Processes regarding the forest conversion were investigated, as well as studies with respect to the soil resource potential. The main objectives were the analysis of mosaic landscapes changes. Satellite imaging supported analysis of the land use changes, as well as the determination of the potential of use within the context of the degradation processes, soil evaluation and soil hazards. This has contributed to the achievement of the following aims: (1) to characterise the chemical, physical and biological changes of the soil by traditional milpa land use system in the highlands of Guatemala (Sierra Yalijux); (2) spatial evaluation of the forest cover changes in the same area; (3) illustration of the interrelationship between forest conversion, soil degradation and changes of the micro climate and their theoretically effects on biodiversity. The inclusion of the investigated soil quality and soil differentiation within the nutrient turnover in the tropical mountainous cloud forests might be used as a basis for selection of protected areas as well as for recommendations for reforestation. At the investigation site in Guatemala, many changes occur concerning forest conversion, soil quality and micro climate. Soil quality declines with increasing land use intensity. The soil degradation as a result of agricultural use is high. For example, carbon and nitrogen are reduced in the following consecutive habitats: primary forest > milpa 15 years > milpa 25 years > milpa 60 years > secondary scrub > fallow land. An increase of soil temperature is identified after logging, determined by a false-time-series, which leads to effects on the composition of soil organisms.

Key words: soil fertility, Guatemala, micro climate, tropical montane cloud forest, habitat fragmentation, deforestation, nutrient cycle

1 Introduction

Due to their high number of ecological functions and endemic flora and fauna, the tropical montane cloud forests of Guatemala need to be protected (Budowski 1965;

Veblen 1976; LaBastille and Pool 1978; Stadtmüller 1987; Catling and Lefkovitch 1989; Hamilton et al. 1995; Aldrich et al. 1997; Kapelle and Brown 2001; Menchú Tum and Budowski 1997). These forests were classified by the World Bank as vulnerable and biogeographically unique (Dinerstein et al. 1995). Therefore, they are considered to be of high priority for conservation. To guarantee sustainable development of tropical montane cloud forests in Guatemala it is necessary to identify sustainable land use systems. Sustainable land use systems might be achieved applying soil management strategies like prevention of erosion (UPROBON 2000; Schulz and Unger 2000). The integration of protection of natural resources and the sustainable land use in the surrounding area is a practicable way to protect the forests and to assure the alimentation of the population.

Both the Convention on Biological Diversity (agreed in Rio de Janeiro 1992; Bundesminister für Umwelt, Naturschutz und Reaktorsicherheit, n.d.) and the Agenda 21 (Bundesministerium für Umwelt, Naturschutz und Reaktorsicherheit, 1997) require the determination and monitoring of ecosystems and habitats. Agenda 21 further demands the implementation of a continuous monitoring system for soil degradation (e.g., Chapter 12.6).

2 Study area

The study area is located in the central highlands of Guatemala, in the region of Alta Verapaz. The research area is located in the Sierra Yalijux in Alta Verapaz with the geographical coordinates 15°21' N to 15°27' N altitude and 90°0' W to 90°7' W longitude. The study sites were selected at the Private Reserve Chelemhá, managed by the non-governmental organization *Unión para Proteger el Bosque Nuboso* (UPROBON), and in the surrounding communities of Chelemá and San Antonio Las Puertas as well as Chicacnab and Mestelá (Fig. 1).

Mean annual rainfall is about 5,100 mm and potential evapotranspiration is 1,700 mm (MAGA 2001). Mean annual temperature in the primary forest under natural conditions is 11.2°C, and is 15.3°C at the traditional cultivation sites. Mean annual relative humidity in the natural ecosystem is 95.9 % and 87.4 % at the milpa-system. The milpa-system is the traditional form of slash-and-burn agriculture by Maya-Q'eqchi' in Central America with cultivation of corn and bean. Soil temperature in 0.1 m below surface is 11.0°C (forest) up to 16.6°C (milpa-system).

Parent materials of the soils are Permian Limestones and are about 246 to 287 million years old (Donelly et al. 1990). The soils are deeply weathered and acid Cambisols with pH 4 to 6 (Markussen 2004). Predominant soil-types are Ferralic, Eutric and Dystric Cambisol (FAO/UNESCO), respectively Oxic Humitropepts, Typic Eutropepts, Oxic Dystropepts (USDA).

Mean vegetation genera are *Pinus*, *Quercus*, *Alnus*, Bromeliaceae, Solanaceae and Orchidaceae.

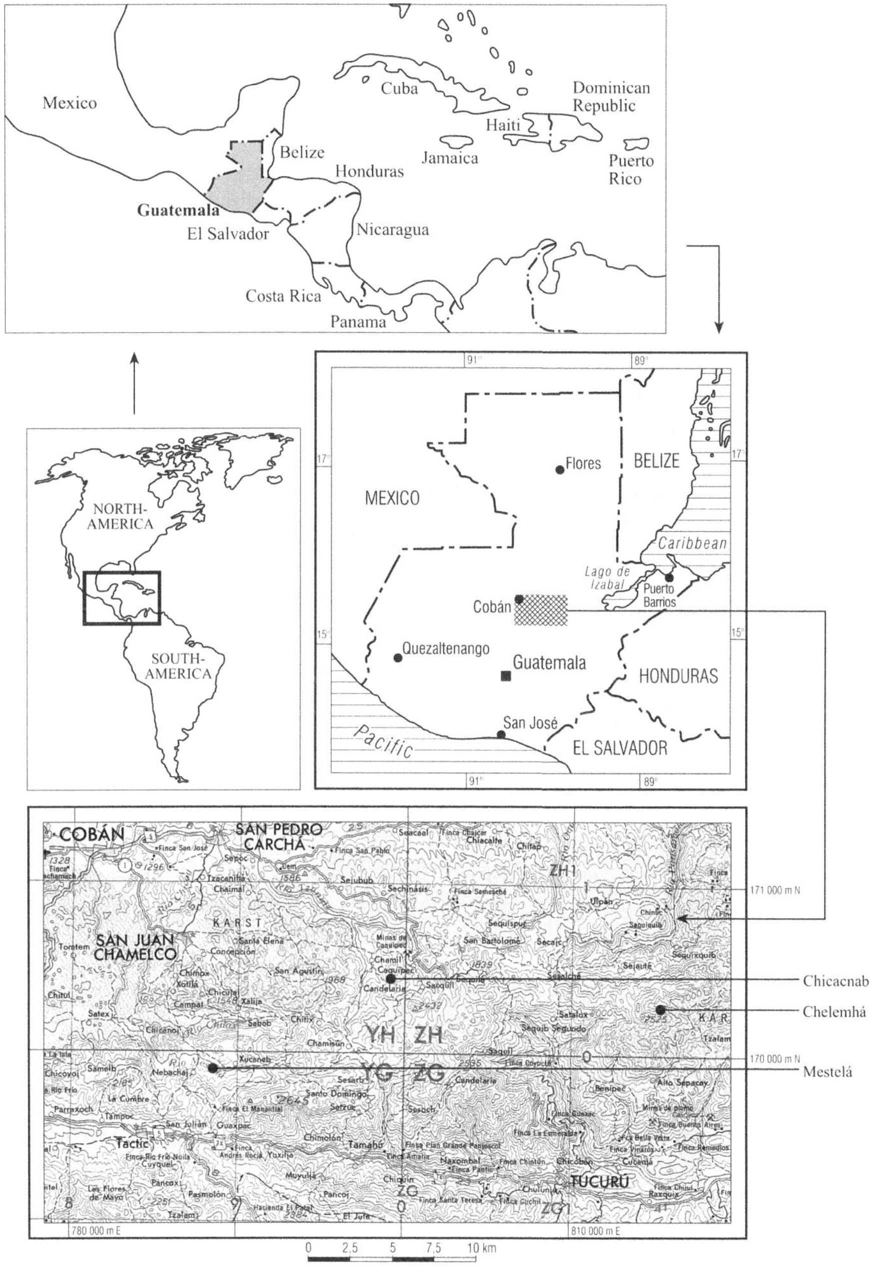

Fig. 1. Location of the study sites

3 Methods and Material

Forest distribution and spatiotemporal changes are determined by studying and evaluating satellite images (LANDSAT-TM 5 and -ETM 7) dated April 14th 1986 and January 23rd 2000, as well as aerial views from January 23rd 1964, February 14th 1991 and January 20th 2000.

Changes of chemical, physical and biological soil parameters depending on deforestation and intensive land use were determined as follows: C_t, C_{org}, N_t, pH (H_2O, KCl and $CaCl_2$), conductivity, P, effective cation exchange capacity (CEC-eff.: Na, K, Mg, Ca, Mn, Fe, Al, H), total nutrient contents (Na, K, Ca, Al, Cu, Mg, Mn, Zn, Fe, Cr, Co, Ni, P, S, Mo, Cd, Tl, Pb), particle size distribution, bulk density, water content. For more details about methods and material see Markussen (2004). Each parameter was measured according land use system: primary forest, secondary shrub vegetation and secondary forest, milpa-system of different land use temporal intensity (used for 15, 25, 60, and more than 100 years).

Three climate data logger were installed to collect data to determine climatological differences between habitats (primary forest, secondary forest, milpa-system).

4 Results

Fig. 2 illustrates the general properties, how deforestation, soil degradation and changes in micro climate affect biodiversity. Increasing human population in rural areas leads to a lack of land. Thus, fallow periods are shorter or left out and incipient soil degradation is the result. Agricultural productivity decreases and the need in further area for land use increases. The ensuing deforestation leads to habitat fragmentation, habitat loss and a disturbed water cycle, which consequently influences flora and fauna. With deforestation, conditional soil degradation yields changes in soil functionality. Additional changes in the micro climate have an impact on the diversity of life in the ecosystems. And last but not least, deforestation and soil degradation effects the agricultural productivity of the peasants.

4.1 Forest cover and forest conversion

The natural forest cover changed in a conspicuous way in Guatemala between 1950 and 2000. The loss of primary forest due to deforestation is visible considering the forest cover changes: 65 % of the total area of Guatemala was forested in 1950 and decreased to 26 % in 2000 (Fig. 3). This requires proactive decisions in order to stop further decline of forested areas.

Information and data about forest conversion in Guatemala at the regional level is rare. From the lowland areas in the region Petén, data are available (Hayes et al. 2002). For the highlands, this study is the first investigation regarding deforestation at the regional level. Fig. 4 shows the forest cover and land use in the Sierra Yalijux in 2000. Primary forest and pine covers 25.3 %, secondary forest, shrub land, coffee

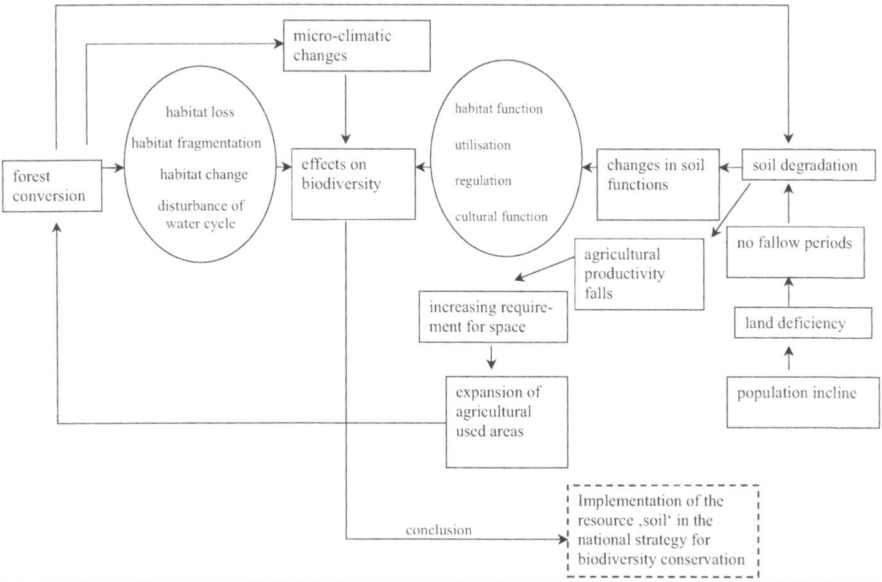

Fig. 2. Interrelationship between forest conversion, soil degradation and changes of the micro climate (Markussen 2004)

and cardamom 60.8 % and milpa-system with vegetation free area 14.0 %. There is no visible change of forest cover in the montane cloud forest of the Sierra Yalijux between 1986 and 2000 (Markussen 2004), contrastingly to the national mean.

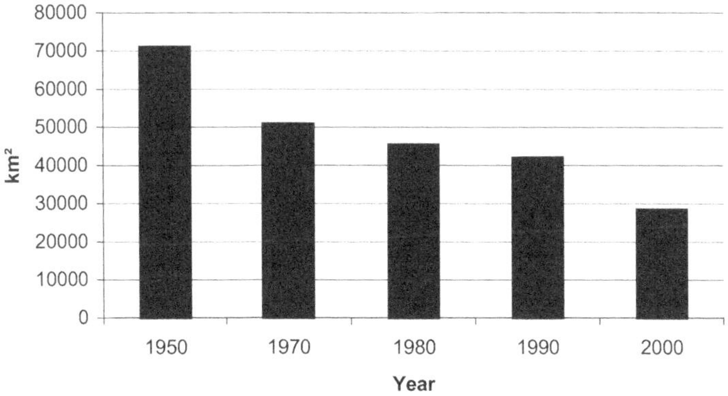

Fig. 3. Development of forest cover in Guatemala from 1950 to 2000. Adapted from Loening and Markussen (2003)

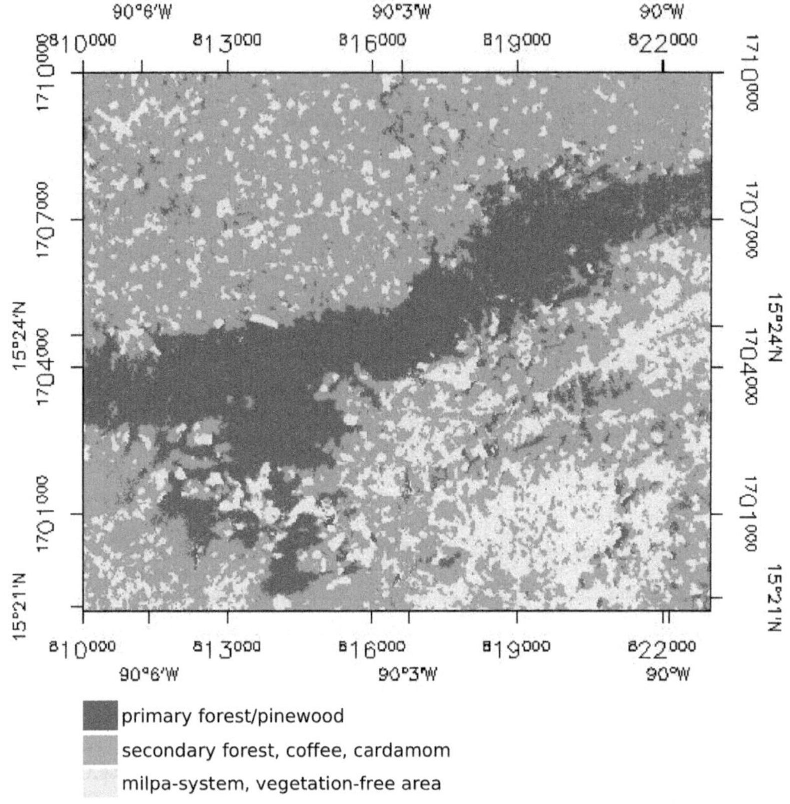

Fig. 4. Forest cover and land use in the Sierra Yalijux, 2000 (Markussen 2004)

4.2 Soil properties and its degradation

As stated above, both the Convention on Biological Diversity and Agenda 21, claim the determination and monitoring of ecosystems and habitats. Agenda 21 further demands the implementation of a monitoring system for the continuous control of the soil degradation with the objective of improving living conditions in the adjacent areas. At the investigation site in Guatemala, many changes have taken place concerning forest conversion, soil quality and micro climate. One intention of the present study was the agrarian-ecological valuation of the soil quality and the land use potential, as well as the registration of the soil degradation from surfaces which are agricultural intensively used, by means of a *false-time-series*. One results of the present investigation is the decline of soil quality due to the increase of land use intensity and temporal duration of land use, respectively. Soil degradation as a result of agricultural use is high. The contents of carbon and nitrogen are reduced significantly. According to these results a decreasing gradient in temporal progression can be determined as follows: primary forest > milpa 15 years > milpa 25 years > milpa 60 years > secondary scrub > fallow land. The total carbon content in the upper soil (0–30cm) of Chelemhá is 18.0 % in primary forest and 19.1 % in secondary forest. The contents decrease to 12.8 % (with burning), 9.0 % (without burning) and 7.5 % (without burning and with soil protection strategies) after 100 years of land use. It appears that the total nitrogen reacts faster to conservation strategies than the total carbon (Fig. 5). With the growing of legumes in the form of hedgerows, the accumulation of nitrogen is analytically verifiable. Implemented soil protection strategies increase the N contents from 0.5 to 0.7 % according to traditional land use systems. Thus, the applied soil management should be implemented more rigorously.

The distinct relief leads to steep slopes, where despite natural and anthropogenic hazards, no one quits land use.

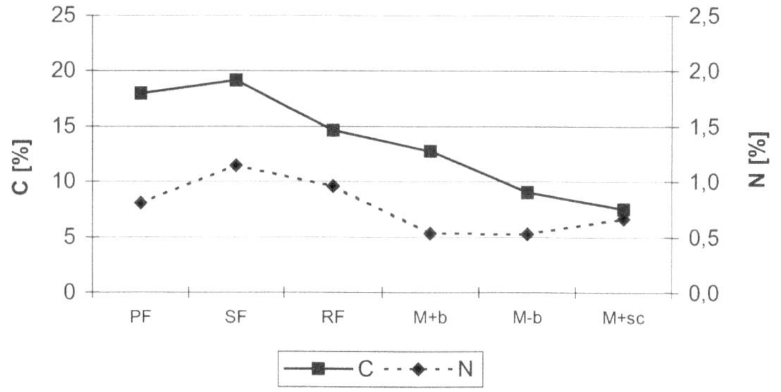

Fig. 5. Contents of carbon (C) and nitrogen (N) in Chelemhá (0–30 cm) (Markussen 2004). PF (primary forest), SF (secondary forest), RF (reforestation), M+b (milpa 100 years with burning), M-b (milpa 100 years without burning), M+sc (milpa 100 years with soil conservation)

The decline of carbon and nitrogen through land use is more distinct in Chicacnab. In the upper horizon (Ah), these content changes are particularly serious (Fig. 6). The sequence in terms of carbon and nitrogen is: primary forest > 15 years old milpa > 25 years old milpa > 60 years old milpa > secondary shrub > fallow land. The upper soil in primary forest is rich in humus with 35.2 %. With anthropogenic impact these contents decrease drastically after 15 years to 25.7 %, after 25 years to 15.5 % and after 60 years to 10.0 %. Secondary shrub and fallow land show carbon contents of 6.3 % and 7.1 %.

The total carbon pool in the upper soil in Chelemhá (0–30 cm) shows remarkably high contents, whereby a decrease in dependence of the land use is detectable (Fig. 7). The primary forest contains 324 t/ha total carbon in the upper soil, the sec-

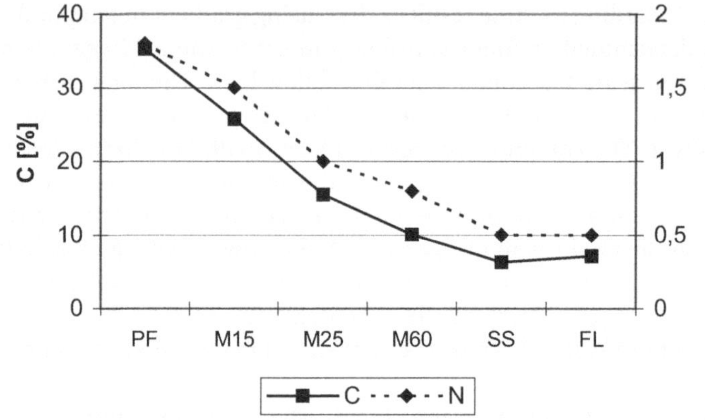

Fig. 6. Contents of carbon (C) and nitrogen (N) in Chicacnab (upper layer [Ah]) (Markussen 2004). PF (primary forest), M15 (milpa over 15 years), M25 (milpa over 25 years), M60 (milpa over 60 years), SS (secondary shrub), FL (fallow land)

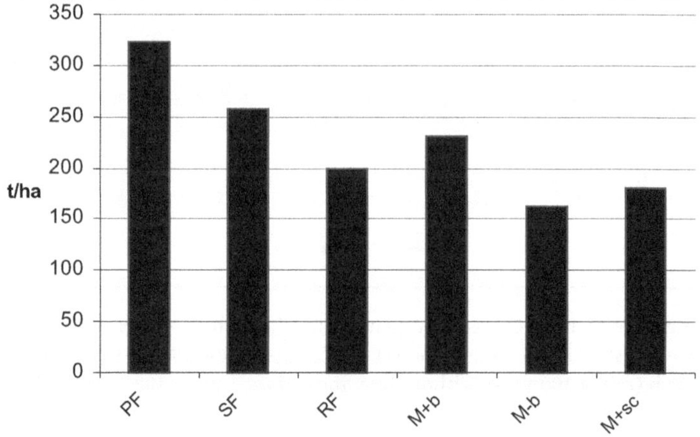

Fig. 7. Total carbon pool in upper soil in Chelemhá (0–30 cm) [t/ha] (Markussen 2004), PF (primary forest), SF (secondary forest), RF (reforestation), M+b (milpa 100 years with burning), M-b (milpa 100 years without burning), M+sc (milpa 100 years with soil conservation)

ondary forest 258 t/ha, land use without burning 162 t/ha. The difference between undisturbed forests and anthropogenic affected land is visible. It is interesting that at the milpa with soil conservation strategies a lower total carbon content was detected (M+sc, Fig. 5). However, taking into account the bulk density, the total carbon pool is higher within milpa with soil conservation (M+sc) than milpa without burning (M-b). The accumulation of carbon is a positive result of the implementation of soil conservation strategies.

The total carbon and nitrogen pools are very high at the investigation site, but at the same time there is a high influence on the dynamics of total carbon and nitrogen contents in dependence on agricultural activities.

4.3 Micro climate changes

Air and soil temperature

One further objective of this investigation was to determine the extent of the traditional land use-system and the effects on micro climate. Data logger were used to collect soil and air temperature, relative and absolute humidity, dew point and precipitation with high temporal resolution.

Soil and air temperature are important for the bio-ecological stocktaking. The soil temperature is a steering factor for processes in the soil. Changes affect processes of decomposition and weathering. Increasing temperatures intensify the normal decomposition of humus (Scheffer and Schachtschabel 1998).

Montane cloud forests show differences in micro climatic parameters in comparison to different land use systems (Figs. 8 to 11). The monthly mean air temperature in primary forest lies between 8°C and 12°C annually. Temperature peaks in June and September are the consequence of seasonal variation. The anthropogenic influence is obvious after taking into consideration temperature conditions in secondary forest and milpa-system. Likewise, the peaks are in June and September, but the temperature in secondary forest is 2°C to 3°C higher than in primary forest, and in the milpa-system 4°C to 5°C higher than in primary forest (Fig. 8). This leads to different habitat conditions caused by land use changes.

Notable are the differences in maximum air temperatures. As a result of direct solarisation at the milpa-system, maximum temperatures of 32.7°C are achieved. The maximum air temperature in primary forest is below 20°C.

The monthly mean soil temperature in profundity of 0.1 m provides analogical results. This parameter has the same sequence: primary forest < secondary forest < milpa-system. Soil temperature increase in logged over areas and affects the composition of soil organisms. The differences between the habitats are similar to air temperature. Because of the direct solarisation, the soil temperature in the milpa-system rises up to 18°C in September (Fig. 9). In primary forest mean air temperature in 0.5m and mean soil temperature in 0.1m are similar. Soil temperature peaks at 21.3°C. In September 2002 for example, the maximum air temperature in primary forest was 16°C, whereas in the milpa-system it was 28.3°C.

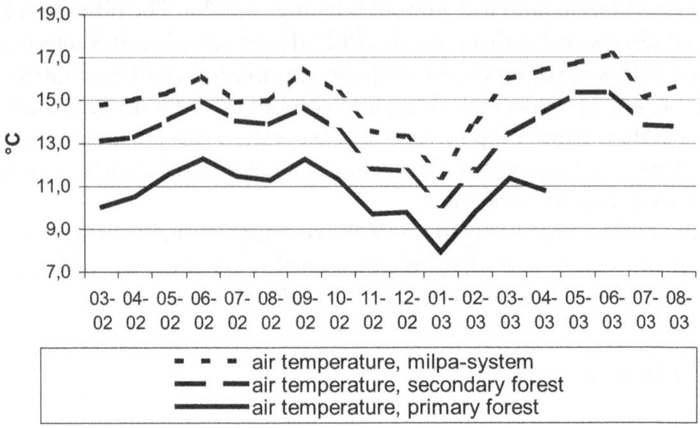

Fig. 8. Mean monthly air temperature in different habitats 0.5 m above ground, March 2002 to August 2003 (Markussen 2004)

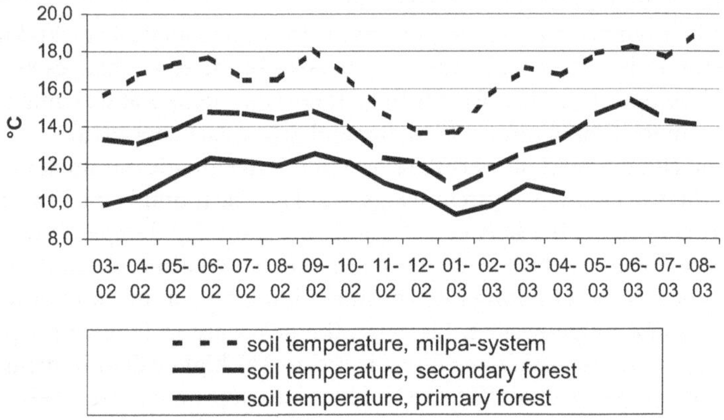

Fig. 9. Mean monthly soil temperature in profundity of 0.1 m in different habitats, March 2002 to August 2003 (Markussen 2004)

The minima and maxima (temperatures) also have the following order: primary forest < secondary forest < milpa-system. It is conspicuous that maxima values are relatively constant during an annual run, whereas minimum values fluctuate seasonally.

Air and soil temperatures in primary forest are steady during the day. For example, on August 1st 2003 the air temperature oscillated within 3°C. In milpa-system the temperatures oscillated on the same day, within 9°C.

Recapitulating, we resume that the diurnal climate within primary forest does not vary strongly. In the milpa-system the climate depends on the time of day, with a range of 11°C. In primary forest, the micro climate is more stable than in agricultural used areas as a consequence of the clearance of forest cover.

Relative and absolute humidity

Zahorka (2002) characterises the maximum air humidity as an important ecological function. Natural forests have two fundamental advantages. First of all, the forest protects the soil against direct solar radiation and prevents wind induced dehydration. Secondly, soil erosion decreases by interception.

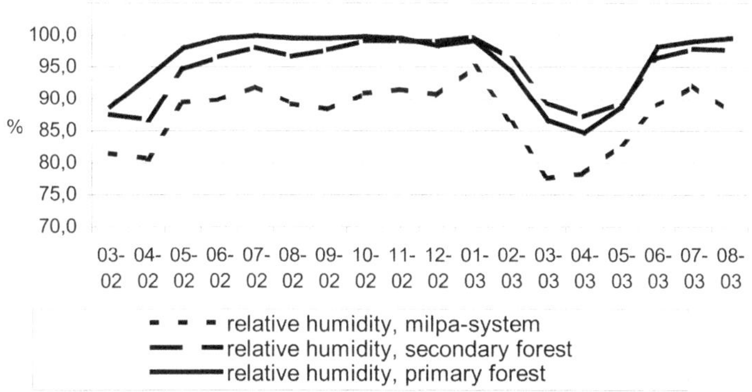

Fig. 10. Mean monthly relative humidity in 2.0 m in different habitats, March 2002 to August 2003 (Markussen 2004)

In the present case, it is obvious how the anthropogenic influence changes the humidity parameters. The monthly mean relative humidity in 2.0 m above the ground is, on average, 10 % lower in the milpa-system than in primary forest (Fig. 10). The tendency of humidity in different habitats over the course of the day is particularly responsible for this. The relative humidity in primary forest during the day is continuously 100 % (Fig. 11). At the agricultural sites the humidity decreases to 40 % between 1300 and 1400 (Fig. 11). These micro climatic conditions lead to different basic conditions for flora and fauna.

The reason for the difference of relative humidity between the investigated habitats is the loss of forest cover and the changes in evapotranspiration and interception.

Precipitation

The total rainfall between 19 March 2002 and 14 August 2003 was 6,343.2 mm (Fig. 12), that means an average rainfall per year of 5,100 mm. The dry season

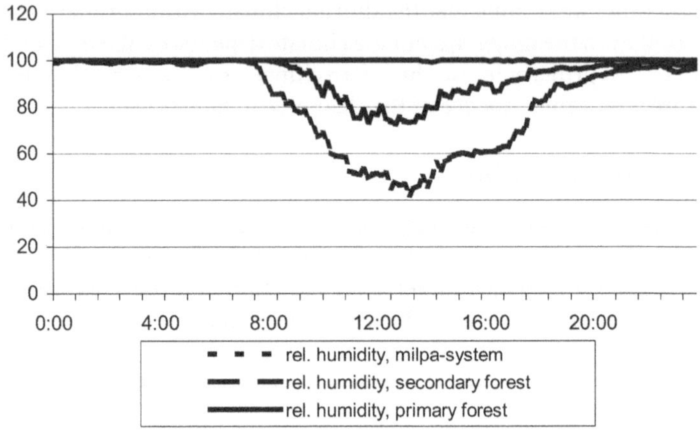

Fig. 11. Relative humidity 2.0 m above ground on 01 August 2002, in different habitats (Markussen 2004)

is between February and April, maxima are in July and December. The stock precipitation in the secondary forest for the same period is 4,226 mm, equal to 65 % of the total rainfall. 35 % is kept back by interception at the cover of vegetation. A part of this evaporates relatively fast without reaching the grassroots. Only a small part arrives through stemflow in the soilwater cycle.

As a result of very high contents of relative humidity, fog in this region plays an important role. Data indicates that there is an additional entry of ›horizontal rain‹ (i.e., precipitation caused by fog) amounting to 30 % (Markussen 2004). Thus, tropical montane cloud forests have important ecological functions. For detailed investigations about the fog entry see Brown et al. 1996, Bruijnzeel and Proctor 1995, Bruijnzeel and Hamilton 2000.

5 Discussion

The results illustrate that deforestation of the tropical montane cloud forests in Guatemala have significant effects on habitat quality, soil processes and micro climate. The situation in the investigation area is positive because deforestation is currently not increasing in contrast to the national level. The comparison between the satellite images of the Sierra Yalijux from 1986 and 2000 shows that distribution of forest cover is constant. Nevertheless, the cloud forests are still threatened because of population increase and rural development. Mean factors for deforestation in Guatemala are roadmaking, colonisation and soil quality (Grunberg et al. 2000). Sader et al. (2001) gathered data at two investigation sites in the lowlands of Guatemala during 1986 to 1990. They found that more than 90 % of the deforestation is located within a 3 km distance of streets and rivers. Sader et al. (1994)

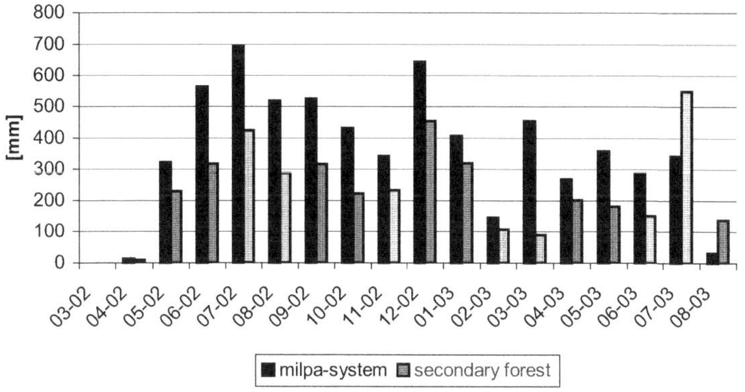

Fig. 12. Precipitation in the milpa-system and secondary forest, March 2002 to August 2003 (Markussen 2004)

calculate for the northern Petén region, a deforestation rate of 0.4 % (1986–1990) and a growth of 0.03 %. That means a disparity of 12:1 regarding the relationship between deforestation and growth.

The transformation of montane cloud forests into agricultural areas leads to pedological changes in the soil. This phenomenon exists throughout the Central American region (Beach 1998). Fischer and Vasseur (2000) show the deforestation problem using the example of Panama. Similar to Guatemala, there is a decline of forest cover from 70 % (1947) to 44 % (1995). They point out the advantage of agroforestry systems for the conservation of soil and biodiversity. In Panama a lot of project activities exist regarding alley cropping, home gardens, live barriers, shade trees with perennial crops, windbreaks etc. Agroforestry could also be an alternative for the highlands of Guatemala. However, there are not many attempts to adopt such alternatives for the Sierra Yalijux today. Experiments are being tried by the non-governmental organisations UPROBON (*Unión para Proteger el Bosque Nubosos*, `http://www.chelemha.org`), PEQ (*Proyecto Ecológico Quetzal*, `http://www.ecoquetzal.org`) and PROEVAL-RAXMU (*Proyectos Evaluados para Conservación de la Naturaleza*, `http://www.proeval-raxmu.org`). An actual overview about the problems of neotropical montane cloud forests is given in Kapelle and Brown (2001), Aldrich et al. (1997), Hamilton et al. (1995), Brown and Kapelle (2001), Hamilton (2001) and Bubb (2001). Islebe and Pérez (2001), Brown et al. (1996), Schuster et al. (2000) refer especially to the cloud forests in Guatemala.

6 Conclusions

The resource ›soil‹ is of major importance, especially regarding discussions about biodiversity. However, it has had insufficient attention. Forest conversion and soil degradation have high effects on the biodiversity of flora and fauna. Forest conversion is leading to habitat loss, changes and fragmentation. Soil degradation will lead to changes of the soil functions, shifting of biological diversity in the soil and lower harvest, and consequently, to an expanded demand of agricultural areas (i.e., forest conversion). Taking these effects into account, it is necessary to implement the resource ›soil‹ into the national biodiversity strategy of Guatemala (cp. Fig. 2).

The integration of preservation of natural resources and sustainable land use in the environment of protected areas in the Sierra Yalijux is a useful way to protect the forests and to assure the livelihoods of the local population. The development or realisation of guidelines may contribute to the maintenance or even growth of the food production, and above all, a sustainable land use in the cloud forest covered highlands of Guatemala. The implementation of soil management and the reduction of human population pressure are important assumptions to conserve natural habitats in the cloud forest of the Sierra Yalijux.

Acknowledgments

We thank Prof. Dr. Gerhard Gerold and Prof. Dr. Michael Mühlenberg who supervised these research investigations. We are grateful to Malte Voigt and Luis Alejandro Cacao Bartolón for their research assistance. We acknowledge the help of Dr. Jürgen Grotheer, Petra Voigt and Anja Södje with their analyses of the physical and soil parameters in the laboratory.

References

Aldrich M, Billington C, Edwards M, Laidlaw R (1997) Tropical Montane Cloud Forests: An Urgent Priority for Conservation. WCMC Biodiversity Bulletin No. 2

Beach T (1998) Soil catenas, tropical deforestation, and ancient and contemporary soil erosion in the Petén, Guatemala. Physical Geography 19(5):378–405

Brown AD, Kappelle M (2001) Introducción a los bosques nublados del neotrópico: una síntesis regional. In: Kappelle M, Brown AD (eds) Bosques nublados del neotrópico, pp 25–40

Brown MB, de la Roca I, Vallejo A, Ford G, Casey J, Aguilar B, Haacker R (1996) Un Análisis del Valor de Bosque Nuboso en la Protección de Cuencas. Reserva de Biósfera Sierra de las Minas, Guatemala y Parque Nacional Cusuco, Honduras

Bruijnzeel LA, Proctor J (1995) Hydrology and biogeochemistry of tropical montane cloud forests: what do we really know? In: Hamilton LS, Juvik JO, Scatena FN (eds) Tropical Montane Cloud Forests, Ecological Studies 110, Springer-Verlag, New York, pp 38–78

Bruijnzeel LA, Hamilton LS (2000) Decision time for cloud forests. Water-Related issues and problems of the humid tropics and other warm humid regions. IHP Humid tropics programme series No. 13. UNESCO

Bubb P (2001) Desarrollo de una base de datos para los bosques nublados del neotrópico. In: Kappelle M, Brown AD (eds) Bosques nublados del neotrópico, pp 51–62

Budowski G (1965) The choice and classification of natural habitats in need of preservation in Central America. Turrialba 15(3):238–246

Bundesminister für Umwelt, Naturschutz und Reaktorsicherheit (n.d.) Umweltpolitik. Konferenz der Vereinten Nationen für Umwelt und Entwicklung im Juni 1992 in Rio de Janeiro. Dokumente. Klimakonvention, Konvention über die biologische Vielfalt, Rio-Deklaration, Walderklärung, a copy is available online at `http://www.biodiv.org/convention/articles.asp`

Bundesministerium für Umwelt, Naturschutz und Reaktorsicherheit (1997) Umweltpolitik. Agenda 21. Konferenz der Vereinten Nationen für Umwelt und Entwicklung im Juni 1992 in Rio de Janeiro. Dokumente, a copy is available online at `http://www.un.org/esa/sustdev/documents/agenda21/english/agenda21toc.htm`

Catling PM, Lefkovitch LP (1989) Associations of Vascular Epiphytes in a Guatemalan Cloud Forest. Biotropica 21 (1):35–40

Dinerstein E, Olson DM, Graham DJ, Webster AL, Primm SA, Bookbinder MP, Ledec G (1995) A Conservation Assessment of the Terrestrial Ecoregions of Latin America and the Caribbean. World Bank, Washington DC

Donnelly TW, Horne GS, Finch G, López-Ramos E (1990) Los Bloques Maya y Chortí, Centro America Norte. In: The Geology of North America, Vol. H. The Caribbean Region. The Geological Society of America

Fischer A, Vasseur L (2000) The crisis in shifting cultivation practices and the promise of agroforestry: a review of the Panamanian experience. Biodiversity and Conservation 9:739–756

Grunberg W, Guertin DP, Shaw WW (2000) Modeling Deforestation Risks for the Maya Bioshere Reserve, Guatemala. Twentieth Annual ESRI International User Conference. San Diego, California. June 26–30, 2000, Unpublished script

Hamilton LS (2001) Una campaña por los bosques nublados: ecosistemas únicos y valiosos en peligro. In: Kappelle M, Brown AD (eds) Bosques nublados del neotrópico, pp 41–49

Hamilton LS, Juvik JO, Scatena FN (eds) (1995) Tropical Montane Cloud Forests. Ecological Studies 110. Springer

Hayes DJ, Sader SA, Schwartz NB (2002) Analyzing a forest conversion history database to explore the spatial and temporal characteristics of land cover change in Guatemala's Maya Biosphere Reserve. Landscape Ecology 17:299–314

Islebe GA, Véliz Pérez ME (2001) Guatemala. In: Kappelle M, Brown AD (eds) Bosques nublados del neotrópico, pp 231–241

LaBastille A, Pool DJ (1978) On the Need for a System of Cloud-forest Parks in Middle America and the Caribbean. Environmental Conservation, vol 5, no 3: 183–190

Loening L, Markussen M (2003) Pobreza, deforestación y sus eventuales implicaciones para la biodiversidad en Guatemala. In: Economía, Sociedad y Territorio, vol IV, no 14, Juli-Dezember 2003, El Colegio Mexiquense, Zinacantepec, México, pp 279–315

MAGA (Ministerio de Agricultura, Ganadería y Alimentación) (2001) Mapas digitales a escala 1:250.000 elaborados por el Sistema de Información Geográfica (SIG-MAGA), CD-ROM, Guatemala

Markussen M (2004) Waldkonversion und Bodendegradation in Bergnebelwaldgebieten Guatemalas (Alta Verapaz). Ein Beitrag zur Biodiversitätsforschung in sensiblen tropischen Ökosystemen. EcoRegio 11, Shaker Verlag, Aachen, available online at `http://webdoc.sub.gwdg.de/diss/2003/markussen/index.html`

Menchú Tum R, Budowski G (1997) Cloud forest in the mountains of Guatemala: support needed for providing protected status. Mountain Research and Development 17, 1:91

Sader SA, Hayes DJ, Hepinstall JA, Coan M, Soza C (2001) Forest Change Monitoring of a Remote Biosphere Reserve. International Journal of Remote Sensing 22, 10:1937–1950

Sader SA, Sever T, Smoot JC, Richards M (1994) Forest Change Estimates for the Northern Petén Region of Guatemala - 1986-1990. Human Ecology 22, 3:317–332

Scheffer F, Schachtschabel P (1998) Lehrbuch der Bodenkunde. Stuttgart

Schulz U, Unger D (2000) Integration von Landnutzung von Regenwaldschutz – eine Fallstudie aus Guatemala. In: Landnutzungsplanung und Naturschutz. Wissenschaft und Technik-Verlag, pp 94–105

Schuster JC, Cano EB, Cardona C (2000) Un metodo sencillo para priorizar la conservacion de los bosques nubosos de Guatemala, usando passalidae (coleoptera) como organismos indicadores. Acta Zoologica Mexicana, nueva serie, núm. 80, Instituto de Ecología A.C., Mexico, pp 197–209

Stadtmüller T (1987) Cloud Forests in the Humid Tropics. A Bibliographic Review. The United Nations University. Centro Agronómico Tropical de Investigación y Enseñanza

UPROBON (2000) Schutzgemeinschaft Nebelwald: 1. Vereinsbericht. Union para Proteger el Bosque Nuboso. Cobán, Guatemala. Unpublished report

Veblen TT (1976) The urgent need for forest conservation in highland Guatemala. Biol. Conserv. (9):141–154

Zahorka H (2002) Tropischer Regenwald – Ökologische und soziale Funktion. Das Beispiel Südostasien. Libertas Verlag, Sindelfingen

Human impact on bird diversity and community structure in a tropical montane cloud forest in Alta Verapaz, Guatemala, with special reference to the Quetzal (*Pharomachrus mocinno*)

Swen C. Renner[1] and Michael Markussen[2]

[1] Conservation and Research Center, Smithsonian Institution, 1500 Remount Rd, Front Royal 22630 VA, USA, email to `renners@crc.si.edu`

[2] Geographical Institute, Department of Landscape Ecology, Georg-August University of Göttingen, Goldschmidtstraße 5, 37077 Göttingen, Germany, email to `mmarkus@gwdg.de`

Summary. Using birds as an exemplary animal group, the human influence on diversity, community structure and nutrition guild composition is examined in tropical montane cloud forest in Guatemala. Human impact is measured as differences in diversity and body mass in both natural forest and secondary forest (as a consequence of slash-and-burn agriculture). A comparison of these measures is made between the two different habitat types. In terms of species richness, α-diversity (punctual diversity) is higher in used habitats than in natural forest. However, the mean body mass per species is lower in secondary growth than in natural forest. The latter implies that the nutritional conditions of birds in natural forest are qualitatively better than in secondary vegetation. The remaining natural forests in the study area in the region of Alta Verapaz are highly fragmented and have been reduced to less than 50 % of its original cover. Some species are considered to be extinct (e.g. *Oreophasis derbianus* Mountain Guan) or are threatened to vanish from the study area like, e.g., *Penelopina nigra* Highland Guan, *Pharomachrus mocinno* Resplendent Quetzal. Both of these species are dependent on natural forest, the latter because of breeding holes. Both species will become extinct when the last natural forest is gone, which with the current mean national deforestation rate in Guatemala, is within 120 years. The study area is located in the mountain cloud forest zone of the Sierra Yalijux, Alta Verapaz, Guatemala (15°28'N, 90°20'W) between 1,900 and 2,550 m. The investigations were conducted within a 102 ha study plot, including both habitat types with equal spatial dimensions. Birds were censused with several standardised methods. The main factor for fragmentation and deforestation – i.e., habitat loss – is slash-and-burn agriculture by peasants. The Resplendent Quetzal *Pharomachrus mocinno* was used as an example to explain the limiting factors of the regional populations and estimate their viability.

Key words: birds, Guatemala, fragmentation, tropical montane cloud forest, Pharomachrus mocinno, conservation

1 Introduction

The (bio-)diversity in Guatemala is rarely discovered (World Conservation Monitoring Centre 1992). The real distribution of the avian fauna remains uncertain and is based mainly on assumptions (Howell and Webb 1995). In particular, the community composition of highly endangered habitats like montane cloud forest between 1,500 and 3,000 m above sea level is critically reduced and affected (FAO 2001). The aim of this study was to increase the knowledge and understanding of the bird community and their function, and to yield strategies for the conservation of the forest remnants or propose changes from a biological point of view.

The human impact on bird communities in temperate landscapes is frequently investigated compared with, e.g., tropical landscapes. Evidence from these studies show that human used landscapes might also support a degree of diversity that is comparable with natural habitats (e.g., Karr 1971; Benton et al. 2003). But is this applicable for subtropical or tropical landscapes, too? Hughes et al. (2002) and Schulz et al. (in press) investigated the influence of human activities in the tropics for different kinds of organisms and concluded that often diversity is higher in secondary habitats than in natural environs. However, Hughes et al. stated that a higher degree of diversity might not be due to the secondary habitat but to frequency of disturbance and that reproduction and fitness in secondary habitats is lower than in natural habitats (2002).

Here, the difference of diversity and bird community structure – as a measure of human impact – between natural and secondary habitat is analysed with special emphasis on the Resplendent Quetzal *Pharomachrus mocinno*, a species dependent on natural forest because of breeding holes and feeding trees. The population viability in the region for the Quetzal is estimated (Gilpin and Soulé 1986; Soulé 1986, 1987).

2 Study area

The study area is located near the community of Chelemhá in the Sierra Yalijux. This mountain ridge belongs to the north-eastern most slopes of the northern Cordillera of Guatemala. The 102-ha study plot of Chelemhá belongs to the Municipio Tucurú in the region of Alta Verapaz (90° 04' W, 15° 23' N, 1,980 m to 2,550 m above sea level). The plot is comprised of approximately 50 % natural oak-pine forest. It belongs to the largest continuous fragment of oak-pine montane cloud forest (approximately 25 km^2) at the Sierra Yalijux with a remaining total of 55 km^2 (Markussen 2003).

Near-natural vegetation is found at several locations in the Sierra Yalijux including the Chelemhá Plot. The study area is used by locals, who sometimes extract timber (maximum diameter of stems is 10 cm) and carry out limited hunting (larger birds, mainly *Penelopina nigra* for food or *Dives dives* due to competition for nutrition from corn, as well as larger ground living mammals).

The natural forest overstory reaches a height between 25 and 35 m. Secondary growth areas are a patchwork of different habitats including pine reforestation (*Pinus maximinoii*), fallow land, old (10 years, approximately 7 m in height) and young patches of secondary forest (up to 4 years, approximately 3 m in height). Secondary vegetation is comparatively dense. Visibility at eye level is below 2 m. The Chelemhá Plot for bird census consists of secondary vegetation of different age stages between 4 and 15 years, but no agriculture.

From 1950 to the present day, the natural forest cover in Guatemala has decreased from 65.0 % to 26.6 %, with a contemporary mean deforestation rate of 1.7 % (FAO 2001; World Bank 2001). Deforestation in the highlands and the study plot is believed to be less than that of national lowland areas. Further details of the study area like soil conditions, vegetation, climate etc. are discussed by Markussen and Renner in this volume.

3 Methods

Between March and September in 2001 and 2002, a combination of mark-recapture and transect census methods to determine the avifauna of the study plot were used (cp. Terborgh et al. 1990). A total of 5,304 net hours (i.e., hours per 12 m net) were conducted. Birds were captured between 0600 and 1430, at 12 sites each with 96 metres of mist nets (12 x 2.5 m, mesh-width: 12 mm). Six mist net lines were placed in natural forest with a distance of approximately 150 m between them, and six were placed in young secondary forest. Because the latter forest was very dense, the nets were established at tracks and the distance between net lines varied between 75 and 150 m. After determination of species, sex, age and body mass, all individuals were marked with an aluminium leg band and released. Morphometrics were gathered to compare the populations between the two forests regarding body mass, size etc. and recaptures are commonly used as a measure of habitat quality (Winker et al. 1995; Dranzoa 1998; Waltert and Mühlenberg 2001; Renner 2003).

Capturing procedures at each net line were repeated, twice in 2001 and four times in 2002. The minimal temporal distance between capturing at one net line was three weeks. In addition, birds were censused three times using a transect system located within the plot with a total length of 3,825 m. This was visited between 0500 and 0900 to record resident bird populations during their daily main activity in 2002. The transect censuses and mist net methods were used in both primary forest and secondary growth. In total, three repetitions were made at each transect section during the main breeding season between March and July 2002.

For bird field identification, Howell and Webb (1995), Edwards (1998), Land (1970), and Peterson and Chalif (1973) were used. For nomenclature, the American Ornithologists' Union checklist (1998) was followed including the relevant Supplements (American Ornithologists' Union 2000, 2002).

To estimate the presumed diversity – not all species and individuals can be observed diversity statistics were applied using Bootstrap and Abundance-based Coverage Estimator of species richness (ACE) (Colwell 2000). ACE is a comparably new estimator established by Colwell and Coddington (1994) because their studies using the Chao method (Chao and Lee 1992) steadily overestimated the samples. Bootstrap was used because Jackknife is more suitable for small sample sizes (Magurran 1988; Krebs 1999).

Diversity was compared between both habitat types using the Wilcoxon-Test applying Statistica 5.0 and the Sørensen-Quantitative method (Sørensen 1948; Magurran 1988; Krebs 1999). Based on species richness, abundance and diversity, the impact of human land use on the cloud forest bird community is evaluated.

Sørensen-Quantitative values were proceeded with a Multi Dimensional Scaling (MDS) to illustrate differences between the diversity of mist net lines in natural forest and young secondary forest (MANOVA, independent variable: Habitat, $p = 0.05$). Bootstrap, ACE, and Sørensen were determined applying EstimateS 6.0 (Colwell 2000).

Population model affinities were calculated following the instructions of Magurran (1988).

All levels of significance were set to $p = 0.05$, unless otherwise indicated.

4 Diversity pattern and community structure of the birds in the Sierra Yalijux

4.1 Species set and composition

In total, 99 species were observed at the study plot; 64 were recorded by mist-netting and 75 with transect census techniques. Seventy five of the 99 were presumed to be resident breeding species. This is the same as the total number of species from the transect census, including species that were mainly captured and believed to be breeding and incidental non-standardized recordings which were nevertheless believed to be resident.

Nearctic migratory species were excluded except for Wilson's Warbler *Willsonia pusilla* and MacGillivary's Warbler *Oporornis tolimiei*, because both were present during the early study (2001) and main breeding period. The maximum estimated number of species was 67 using Bootstrap estimator for mist nets. The inclusion of standardised transects compensates for the fact that mist netting is not the best method for a complete inventory of a bird community (e.g., Terborgh et al. 1990; Remsen 1994; Remsen and Good 1996; Poulsen 1994).

The community of the natural forest plot follows a log-norm distribution. This indicates an undisturbed area. The complete study plot also follows log-normal distributions. The secondary forest compartment is indifferent and exact classification would yield log-normal distribution, but with $p = 0.12$ near significant. This indicates no clear classification to one out of the four population models.

Geometric series and broken stick would indicate an area highly influenced by humans (Whittaker 1965, 1970, 1972; Magurran 1988). The species individual relationship for the entire plot and the two compartments are illustrated in Fig. 1 and summarised in Table 1.

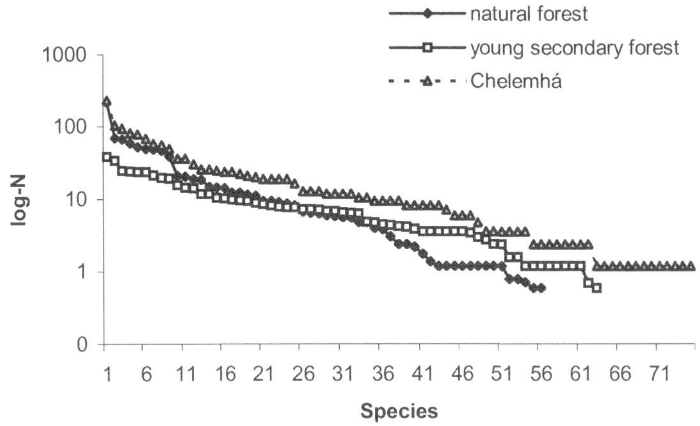

Fig. 1. Species rank distribution in the study plot of Chelemhá independent of each habitat type using transect census data from 2002

4.2 Estimation of species richness

For estimating unweighted species richness, ACE and Bootstrap are suitable estimators. Because it is hardly possible to observe all species and all individuals in a given area (cp. Magurran 1988; Rosenzweig 1995; Krebs 1999), extrapolation of the observed species number is frequently used.

As indicated in Table 1, Bootstrap and ACE are relatively close to the observed species number (S_{obs}). Bootstrap estimates species numbers especially well. Differences between methods are obvious: while transect census has a low bias, mist netting is unreasonably highly over or underestimated by both Bootstrap and ACE but more over estimated by ACE (Table 1). The second index newly established by Colwell and Coddington (1994), Incidents-based Coverage Estimator of species richness (ICE), estimates even more deviation than Bootstrap or ACE.

Differences are given by method: transect census is generally underestimated by both Bootstrap and ACE in the two habitat types compared to mist netting. This indicates that the lower species-individuals relationship might influence the two estimators or that there are more species within young secondary forest than this habitat might be able to carry. The latter might be explained by the diversity-stability hypothesis discussion (cp. Pimm 1986; McCann 2000; and others), which is not discussed here in more detail because it is not yet generally accepted.

4.3 Habitat selection

Considering each of the two major vegetation types of the study plot, differences concerning diversity and species richness appear obvious. In natural forest, the

Table 1. Diversity and species richness in Chelemhá. Singletons, Doubletons, Jackknife, Bootstrap and ACE after Colwell (2000). The lower part of the table indicates the population model affinities for transect census from 2002

	total		natural forest		secondary vegetation	
Index	N	TC	N	TC	N	TC
Individuals (*N*)	865	1493	368	955	497	538
Species (S_{obs})	64	75	31	56	55	63
Singletons	19	10.79	8	0	16	6
Doubletons	13	6.61	8	3	13	4
Bootstrap	72.62	78.09	36.35	50.15	65.17	62.60
± SD	± 2.73	± 1.54	± 0.77	± 0.00	± 0.00	± 0.00
ACE	84.19	82.11	40. 56	46.00	72.43	61.86
± SD	± 8.50	± 2.58	± 4.61	± 1.47	± 6.48	± 2.13
log-norm		yes		yes		(yes)
log-series		no		no		no
geometric series		no		no		no
broken stick		no		no		no

N: mist netting, TC: transect census

observed species number (S_{obs}) was 32 (Bootstrap: 33), while in secondary vegetation it was higher with 57 (60) species. The diversity statistics are summarised in Table 1. Focusing on 75 residents, most species have no clear preference for the two habitat types. From transect census data, Common Bush Tanager *Chlorospingus ophthalmicus* was most abundant (91.6 % of observed individuals were in natural forest), followed by Grey-Breasted Wood-Wren *Henicorhina leucophrys* (85.1 %), Slate-Coloured Solitaire *Myadestes unicolor* (84.0 %), Tufted Flycatcher *Mitrephanes phaeocercus* (81.2 %), and Mountain Thrush *Turdus plebejus* (81.2 %) with relatively higher abundance in natural forest. Higher singing and display activity or territorial behaviour is not the explanation for the higher abundance of species with more than 80 % observed in natural forest of the Chelemhá Plot. It is more likely that the greater amount of vertical vegetation strata is responsible for a higher number of possible territories in natural forest which therefore might increase the number of individuals. However, this is only valid for species with equal frequency in all vertical strata of the natural forest.

Two species accounted for more than 80 % of recordings in secondary habitats (Great-Tailed Grackle *Quiscalus mexicanus* and Black-Capped Swallow *Notiochelidon pileata* with 100.0 %). The medium relative distribution between the habitats of all species was 48.5 % in natural forest (N $\geqslant$ 15 individuals per 100-ha in total).

Diversity settings were significantly different between the two habitats ($p < 0.01$, Wilcoxon-Test). Using a Multi Dimensional Scaling (MDS, Fig. 2) of mist net data, differences between both habitats are significant (Rao's $R_{2,9} = 8.27$, $p < 0.01$). This also implies the differences between both habitats.

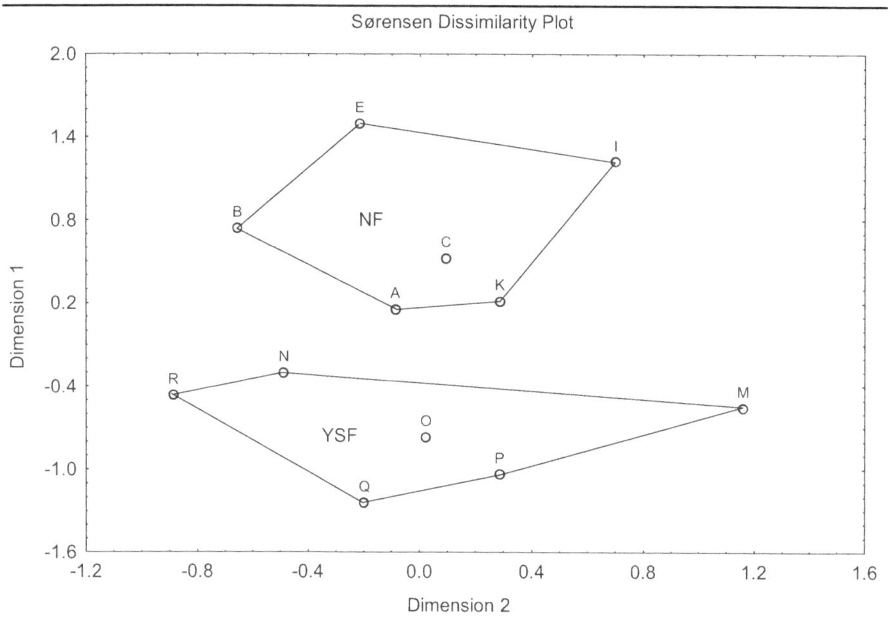

Fig. 2. Dissimilarity plot of the twelve mist net lines in the Chelemhá Plot. NF: natural forest, YSF: young secondary forest. Each letter represents a mist net line. Rao's $R_{2,9} = 8.27$, $p < 0.01$

4.4 Recaptures

Recapture rates are generally regarded as a good method of evaluating habitat quality (e.g., Lambert 1992; Johns 1992; Holbech 1996; Dranzoa 1998; Waltert and Mühlenberg 2001; for exceptions see Winker et al. 1995).

The 25 species recaptured in the Chelemhá Plot are listed in Table 2. Six were recaptured exclusively in natural forest, 12 exclusively in secondary forest and seven in both habitat types. Individuals with more than one recapture were limited to eight species. Recaptures of second or third order were always in natural vegetation. This indicates that individual turnover and hence territoriality is higher in natural forest than in secondary vegetation (cp. Waltert and Mühlenberg 2001; Renner 2003). Here, territoriality refers to the number of territories rather than the area.

In summary, recaptures indicate – especially in second order recaptures – a ›better‹ habitat quality of natural forest. Individuals were recaptured more than once and mainly at the same net line. Second order recaptures were observed exclusively in natural forest.

Immature individuals are most frequently found in secondary habitats in the Chelemhá Plot (this is not significant due to too low N for tests). According to the model of Winker et al. (1995), this indicates that immature birds first use secondary

Table 2. Captures, recaptures and recapture rate per habitat in the Chelemhá Plot. Excluded are all recaptures on the same day (cp. Renner 2003)

Species	Recaptures								Recapture rate		
	01	r1	r2	r3	r4	Total	NF	SF	Total	NF	SF
Chlorospingus ophthalmicus	1	12	0	1	0	23	9	14	0.16	0.16	0.15
Lampornis amethystinus	1	2	1	0	0	7	6	1	0.06	0.08	0.02
Catharus frantzii	9	7	3	4	0	35	29	6	0.56	0.71	0.27
Lamprolaima rhami	1	10	0	0	0	12	12	0	0.19	0.23	0.00
Diglossa baritula	7	8	4	0	0	21	0	21	0.47	0.00	0.57
Basileuterus belli	4	5	3	0	1	19	11	8	0.49	0.46	0.53
Buarremon brunneinucha	2	5	1	0	0	7	1	6	0.23	0.17	0.24
Henicorhina leucophrys	6	3	3	1	0	12	10	2	0.43	0.53	0.22
Wilsonia pusilla	1	2	1	0	0	5	0	5	0.20	0.00	0.21
Atlapetes gutteralis	1	2	2	0	0	8	0	8	0.35	0.00	0.38
Myadestes occidentalis	1	2	0	0	0	4	1	3	0.18	1.00	0.14
Lampornis viridipallens	0	1	0	0	0	1	1	0	0.05	0.06	0.00
Empidonax flavescens	1	1	0	0	0	2	2	0	0.11	0.20	0.00
Myadestes unicolor	0	1	0	0	0	1	1	0	0.08	0.13	0.00
Troglodytes musculus	1	1	0	0	0	1	0	1	0.17	-	0.17
Troglodytes ruficiliatus	0	1	0	0	0	1	0	1	0.17	0.00	0.33
Asphata gularis	1	2	0	0	0	2	0	2	0.40	0.00	0.50
Basileuterus rufifrons	0	1	0	0	0	1	0	1	0.33	-	0.33
Oporornis tolmiei	1	1	0	0	0	1	0	1	0.33	-	0.33
Catharus aurantiirostris	0	1	0	0	0	1	1	0	0.50	0.50	-
Melanotis hypoleucus	0	1	0	0	0	1	0	1	0.50	-	0.50
Myioborus miniatus	0	2	0	0	0	2	0	2	1.00	-	1.00
Thryothorus modestus	1	1	0	0	0	2	0	2	1.00	-	1.00
Abeillia abeillei	0	1	0	0	0	1	0	1	1.00	-	1.00
Xiphorhynchus erythropygius	0	1	0	0	0	1	1	0	1.00	1.00	-
Total:	38	60	18	5	1	171	85	86			

01: first capture in 2001, at least one recapture in 2002; r1: one recapture per a definite individual in total, r2: two recaptures per individual in total, and so on; Habitat: recapture in both habitat (Both) or exclusively in one habitat; Recapture rate (R^t) is recaptures per captures. NF: natural forest, SF: secondary forest.

habitats (here: suboptimal or pessimal) because optimal habitats/territories in natural forest are occupied by adults.

4.5 Body mass distribution

Body mass distribution (Fig. 3) was used to indicate the nutritional situation of the bird community. It is assumed that individuals with higher body mass are feeding in habitats with better nutrition sources. Single captured individuals in natural forest had a higher mean body mass.

Table 3. Mean body mass in g (BM) of captured birds and detected individuals in Chelemhá. N: mist net captures (without recaptures), TC: number of individuals recorded with transect census techniques. Missing body mass data was compiled by data from Dunning (1993). NF: natural forest, YSF: young secondary forest

		Body mass					
		Total per species		NF		YSF	
Species	BM	N	TC	N	TC	N	TC
Chondrohierax uncinatus	278.0		324.4				324.4
Ortalis vetula	563.0		1972.6		328.6		1,644.0
Penelopina nigra	890.0		16625.2		8,108.5		8,516.6
Dendrortyx leucophrys	350.0		1226.3				1,226.3
Cyrtonyx ocellatus	200.0		233.6				233.6
Columba fasciata	342.3		6794.6		1,966.0		4,828.5
Claravis mondetoura	73.0	4,599.0		2993.0		1,606.0	
Geotrygon albifacies	304.0	2,432.0		608.0		1,824.0	
Bolborhynchus lineola	53.6		312.8		250.3		62.6
Ciccaba virgata	250.0		291.8		291.8		
Strix fulvescens supersp. *varia*	716.5		836.3		418.2		418.1
Caprimulgus arizonae	53.0		247.2				247.2
Cypseloides niger	45.6		159.8				159.8
Aeronautes saxatalis	32.1		37.5		37.5		
Campylopterus hemileucurus	10.6	328.6		63.6		265.0	
Colibri thalassinus	5.4	5.4	63.1	0.0		5.4	63.1
Abeillia abeillei	3.4	130.7	7.8	80.4	3.9	50.3	3.9
Hylocharis leucotis	3.4	3.4		3.4		0.0	
Amazilia cyanocephala	5.9	146.3		5.9		140.4	
Lampornis viridipallens	5.6	705.6	13.1	425.6	6.5	280.0	6.5
Lampornis amethystinus	5.7	176.7	532.3	0.0	398.5	176.7	133.9
Lamprolaima rhami	6.9	138.0	128.9	110.4	98.2	27.6	30.7
Eugenes fulgens	7.7	139.1	9.0	77.3		61.8	9.0
Doricha enicura	3.1	3.1		0.0		3.1	
Tilmatura dupontii	2.2		2.6				2.6
Atthis ellioti	2.4	12.2	19.9	4.9	11.4	7.3	8.5
Trogon collaris	64.4		150.3		150.3		
Trogon mexicanus	69.3		647.2		380.7		266.6
Pharomachrus mocinno	220.0	220.0	5,650.3	220.0	4,063.9	0.0	1,586.4
Asphata gularis	55.4	55.4	452.8	0.0	75.4	55.4	377.4
Aulacorhynchus prasinus	154.5	772.5	1262.7	154.5	841.6	618.0	421.1
Picoides villosus	35.0	35.0	449.5	35.0	329.6	0.0	119.9
Colaptes auratus	124.5	18,426.0	726.8	7,096.5	218.0	11,329.5	508.8
Automolus rubiginosus	47.8	1,099.4	446.6	95.6	103.0	1,003.8	343.6
Sclerurus mexicanus	25.0		233.5		233.5		
Xiphorhynchus erythropygius	41.1	41.1	95.9	41.1	32.0	0.0	64.0
Lepidocolaptes affinis	27.5	1,705.0	32.1	1,430.0	32.1	275.0	
Grallaria guatimalensis	88.3	88.3		0.0		88.3	
Camptostoma imberbe	13.3	39.9		0.0		39.9	
Zimmerius vilissimus	11.3	11.3	554.1	0.0	437.0	11.3	117.1
Mitrephanes phaeocercus	8.6		220.9		179.4		41.4
Cantopus pertinax	27.2		31.8		31.8		
Empidonax affinis	17.0	765.0		136.0		629.0	
Empidonax flavescens	12.2	12.2	14.2	0.0	14.2	12.2	
Vireo plumbeus	20.3	20.3	165.5	0.0	165.5	20.3	
Cyanocorax melanocyaneus	106.0	212.0	866.5	0.0		212.0	866.5
Cyanolyca pumilo	49.3	1528.3	921.0	98.6	571.5	1429.7	349.4
Corvus corax	1,200.0		1400.8		1,400.8		
Tachycineta thalassina	14.2		16.6				16.6
Notiochelidon pileata	12.2		284.8				284.8
Thryothorus modestus	17.3	34.6	525.1	0.0	350.0	34.6	175.1
Troglodytes musculus	11.3	67.8	237.5	0.0	75.0	67.8	162.5
Troglodytes rufociliatus	11.3	67.8	250.6	0.0	163.4	67.8	87.2
Henicorhina leucophrys	14.5	14.5	829.4	0.0	706.0	14.5	123.4
Myadestes occidentalis	36.1	36.1	1,306.8	0.0	443.7	36.1	863.1
Myadestes unicolor	38.6	849.2	2,117.8	38.6	1,778.9	810.6	338.9
Catharus aurantiirostris	28.8	57.6	369.8	0.0	184.9	57.6	184.9
Catharus frantzii	27.3	54.6	1,848.4	0.0	1,418.5	54.6	429.9
Turdus infuscatus	73.4	220.2	1,371.1	220.2	891.1	0.0	480.0
Turdus plebejus	79.5	1,828.5	1,856.0	1351.5	1,508.0	477.0	348.0
Turdus grayi	69.0	414.0	5,396.8	207.0	4,066.9	207.0	1,329.9
Turdus rufitorques	72.0		672.5				672.5
Melanotis hypoleucus	63.8	63.8	1,042.8	63.8	548.8	0.0	494.1
Vermivora chrysoptera	8.8	79.2		44.0		35.2	
Parula superciliosa	8.5	25.5	89.3	0.0	53.6	25.5	35.7
Dendroica fusca	7.3	7.3	8.5	0.0	8.5	7.3	
Oporornis tolmiei	11.4	22.8	66.5	0.0	66.5	22.8	
Geothlypis poliocephala	11.0	11.0		11.0		0.0	
Wilsonia pusilla	7.9	15.8	18.4	0.0	6.1	15.8	12.3
Myioborus miniatus	8.8	114.4	102.7	70.4	34.2	44.0	68.5
Basileuterus rufifrons	11.3	11.3		0.0		11.3	
Basileuterus belli	10.7	32.1	1,124.3	0.0	713.6	32.1	410.6
Chlorospingus ophthalmicus	17.8	35.6	4,134.9	35.6	3,786.0	0.0	348.9
Euphonia elegantissima	15.0		175.1		164.8		10.3
Chlorophonia occipitalis	25.8		60.2		60.2		
Volatinia jacarina	8.5	17.0		0.0		17.0	
Sporophila torqueola	8.0	16.0	18.7	0.0		16.0	18.7
Tiarias olivacea	9.0	18.0	31.5	0.0		18.0	31.5
Diglossa baritula	9.0	27.0	94.6	27.0	33.8	0.0	60.8
Haplospiza rustica	15.6		54.7				54.7
Atlapetes gutteralis	34.8	34.8	1,259.6	0.0	514.4	34.8	745.3
Buarremon brunneinucha	44.3	88.6	3,620.1	44.3	2,127.9	44.3	1,492.2
Melozone biarcuatum	30.8	61.6	107.9	0.0		61.6	107.9
Aimophila rufescens	36.5	1,022.0		693.5		328.5	
Zonotrichia capensis	17.2	34.3	200.3	0.0		34.3	200.3
Ergaticus versicolor	10.0	70.0		0.0		70.0	
Spizella passerina	20.3	40.6		0.0		40.6	
Saltator atriceps	79.7		558.5		55.8		502.7
Dives dives	96.2		224.7				224.7
Quiscalus mexicanus	110.0		3,652.1				3,652.1
Carduelis notata	11.0	33.0	141.3	0.0	33.2	33.0	108.1
Carduelis atriceps	19.8	19.8		0.0		19.8	
Total:		39,398.0	78,031.0	16,486.6	40,973.4	22,911.4	37,057.6

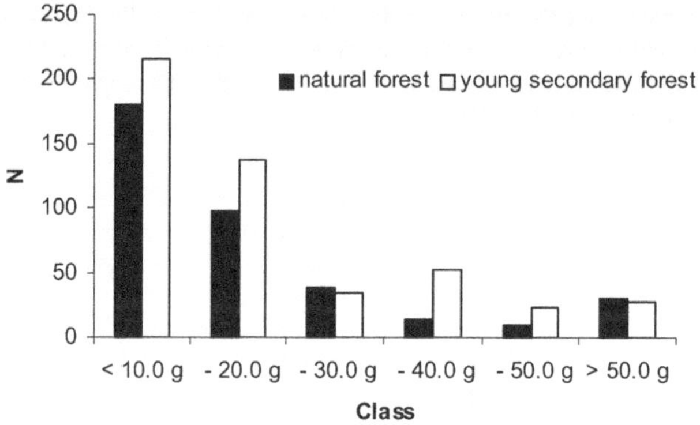

Fig. 3. Body mass distribution in the Chelemhá Plot for all detected individuals

The total body mass (bird biomass) in natural forest for detected resident birds was higher in natural forest than in secondary forest, with 40,973.4 g and 37,057.6 g (cp. Table 3) respectively (data was gathered from standardised transect census and dedicated to area). The difference is based mainly on a few observed larger animals like raptors observed above or within natural forest.

Captures indicate that individual body mass is generally higher in natural forest than in young secondary forest.

Analysing the mass distribution in relation to individual numbers, small birds are more frequent than larger birds (Fig. 3).

4.6 Pharomachrus mocinno and its regional population

One prominent species still frequently observed at the study area in the Sierra Yalijux is *Pharomachrus mocinno*, the Resplendent Quetzal. The Quetzal, the national bird of Guatemala, is used here to explain differences and habitat affinities in more detail as an example which is valid for many species. The general patterns affect species like jays and raptors present in the study area or even mammals like *Alouatta pigra*, the Yucatán Black-Howler Monkey. Direct influences like poaching of Quetzals by humans is not recorded but the indirect influence is set by the limiting factors (see below) and habitat use.

Mühlenberg et al. (1989) studied *P. mocinno* at several locations in Guatemala in 1988 to analyse the population viability of the meta-populations. First, the results from 1988 will be discussed, and then the results from the study in 2001/2002. Finally, both sets of results will be compared.

Mühlenberg et al. (1989) recorded at least three viable Quetzal populations in Guatemala (Sierra Cuchumatanes in the western highlands [effective population of 7,200 individuals], Chelemhá [5,700 individuals], and Sierra de las Minas [27,000 individuals]). Two more which might support viable populations were recorded

(Chamá East [4,900 individuals] and Chamá West [4,500 individuals]) but were not investigated sufficiently due to inaccessibility. The theoretical basis for the classification of the minimal viable population was 5,000 individuals, presuming 500 active reproducing individuals as the minimum necessary (Hovestadt et al. 1991). In addition, several correctional factors were added like breeding success per male (2.5x), breeding success per breeding pair (2x) and variance in breeding pair success (2x) in order to assess the individual number of the minimal viable population. The factors were investigated by Unger (1988).

Mühlenberg et al. concluded that the minimum area for *P. mocinno* is 111 km^2 for 15 males/km^2. Considering the 33 males/km^2 determined in the study, the minimum area needed for viable meta-populations with no presumed exchange of individuals is 50.5 km^2 (but cp. Powell and Bjork 1994, 1995 for immature individuals). The latter value indicates (i) the presumption of 15 males/km^2 is too low (ii) males are crowding within the forest remnants to find suitable habitats for feeding and reproduction in conjunction with the female density (iii) the local Quetzal population of Chelemhá is condemned to extinction.

Unfortunately, the remaining natural forest is less than half of the estimated minimal area in Chelemhá and its surroundings. Mühlenberg et al. estimated the male population as 33 individuals/km^2 in Chelemhá in 1988. By 2002, the individuals were approximately 18 males/km^2 – approximately half of the population of 1988. The differences can be explained by methodological differences and natural population dynamics. Mühlenberg et al. used triangulation as a method but the temporal effort was less than in 2002. On the other hand, natural population dynamics in birds are reported to be high (e.g., Marchant et al. 1992). Differences might be explained by both, but differentiation is hardly possible. The decline should not be overstated because missing data during 1988 and 2002 might indicate a comparably high flux of population dynamics.

Due to insufficient information about living and behaviour it is difficult to evaluate the viability of populations (cp. Terborgh 1999). Even today the situation has not improved and knowledge about species like Quetzals is poor. Habitat loss will overrun conservationists and many species will vanish without sufficient knowledge or not even having been described by the scientific community (e.g., Terborgh 1999).

In 1973, a sanctuary to preserve the Quetzal was established at the southern slopes of the Volcano Atitlan (LaBastille 1973). Mühlenberg et al. (1989) concluded from their data that the south-eastern slopes of the volcano were not suitable for *P. mocinno* and none were detected there. This discrepancy might be explained by the differing areas (south versus south-eastern) or that the populations are already extinct. When establishing reserves for Quetzals it is necessary to include the seasonal movements, particularly of immature individuals, otherwise protection will be insufficient (Loiselle et al. 1989; Powell and Bjork 1994, 1995).

Population decline of *P. mocinno* is forced by habitat destruction by humans (agricultural expansion, slash-and-burn, erosion, etc.) but also by natural impacts like mudslides, fires, etc. (LaBastille 1974). Because breeding holes are the limiting factor for the Quetzal population, studies were conducted about artificial nests

(LaBastille 1974). These nests might be a good alternative to increase the number of breeding holes. In 50 % of the installations, artificial breeding holes were accepted by female *P. mocinno*. Unfortunately, breeding success was not proven (LaBastille 1974).

The analyses of *P. mocinno* might be used as part of a Population Viability Analyses (PVA) and basis for Minimal Viable Population estimation (MVP, see above). Soulé introduced the procedure of a PVA in 1986. The PVA is a method to determine the likelihood of survival of a local population within a previously defined period (e.g., 100 years). Depending on species and local populations, the procedure also has to incorporate different parameters. Applied to *P. mocinno*, estimation of survival for the next 100 years is directly linked to the presence of the habitat where reproduction occurs.

As illustrated in Fig. 4, the persistence of natural forest is essential for the Quetzal, since reproduction is strongly influenced by the presence of breeding trees.

The human impact on the population is obvious. While *P. mocinno* can survive individually in secondary vegetation and feed on the necessary feeding trees (food supplying trees are abundant and used by Quetzals frequently in old secondary forests, own data), the natural forest with a maximum of low human impact is essential. This is still present in the Sierra Yalijux. When the natural forest vanishes, Quetzals will still find suitable breeding tree holes for a limited time period. On the breeding trees have tumbled over, the breeding success of the population is terminated. Without natural forest, the rotten tree stumps are unlikely to be ›reproduced‹.

Assuming that the deforestation rate in the Sierra Yalijux is the same as in Guatemala, 1.7 % annually (currently resembles approximately 1,000 ha annually of the study area in the Sierra Yalijux), natural forest will vanish within the next 60 years. The breeding hole tree stumps – if not cut down by locals – will rest for maybe a further 60 years surrounded by secondary vegetation. After this, the Quetzal will no longer exist in the Sierra Yalijux.

Current investigations on regional deforestation in Guatemala show that the situation is less drastic than the national mean (Markussen 2003; Markussen and Renner in this volume). The deforestation between 1986 and 2000 was generally lower (near zero) in the Sierra Yalijux. Natural forest and oak-pine forests continued at a similar rate of nearly 25 % of the complete study region (based on 150 km^2). Non-governmental actions and state-aided conservation projects have partially yielded a decrease of deforestation of the tropical montane cloud forests in the Sierra Yalijux over the past few years.

5 Discussion

Advantages and disadvantages of the diversity indices are discussed in Magurran (1988), Krebs (1999), and specific traits on the bird community of the Sierra Yalijux in Renner (2003). The best index for larger samples is still Bootstrap. The comparable overestimation is unreasonably high with ACE and other estimators.

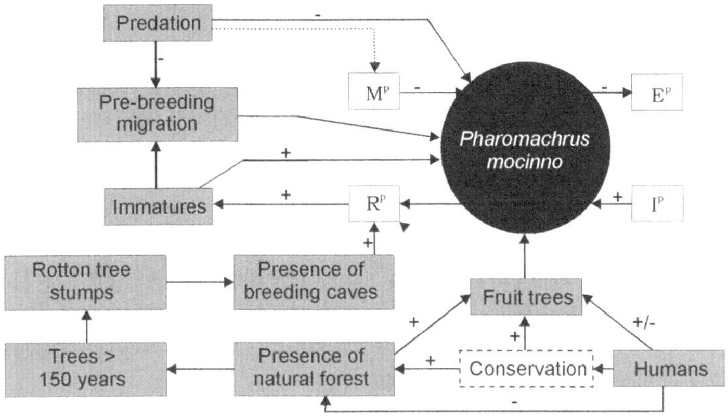

Fig. 4. Factors influencing the population of *Pharomachrus mocinno* in Chelemhá. +: positive/increasing effect, -: negative/decreasing effect. M^P: mortality, E^P: emigration, I^P: immigration, R^P: reproduction success

While diversity is generally lower in natural forest than in secondary forest in the Chelemhá Plot, body mass distribution and recapture rates indicate that natural forest is still of better quality. Reproductive success as a measure of fitness might give an answer if tropical secondary landscape habitats can preserve a comparably high degree of diversity, at least in the Chelemhá Plot. During the study period the major bulk of the bird community species were recorded but several migrating species could not be recorded due to the relatively narrow time period of the study during the main breeding season from March to September. At the end of the breeding season, the number of species increased in both breeding seasons 2001 and 2002 due to vagrants and dispersing immature individuals. This caused an untypical and relatively high overestimation of the species numbers and avoided the saturation of the species-individuals curves. Without vagrants and post-breeding dispersers, the estimation of the species numbers is more likely to be accurate, fit better with the observed species numbers S_{obs} (cp. Fig. 1) and saturation is almost achieved. The rainforest remnants of Chelemhá are an important refuge (cp. Haffer 1974) for rare species like *Pharomachrus mocinno*, *Penelopina nigra* and Blue-Throated Motmot *Asphata gularis*. All three species need primary vegetation to breed successfully. *P. mocinno* for instance needs the 4 - 10 m high and 1 - 2 m in diameter old rotten tree stumps to excavate the breeding holes. Those tree holes are dependent on the old oaks which can only grow up within primary vegetation. Our own estimations yield that with deforestation ongoing at the national rate, the oak-pine forest will be gone in approximately 60 years (cp. Markussen and Renner in this volume). After a reprieve of 50 more years – the time a dead breeding hole tree stump for *P. mocinno* will tumble over – the effective breeding time of the Quetzal will be strongly reduced. Therefore, conservation of the forest for the Quetzal is essential.

The Quetzal could serve as a good target species for the rest of the (bird) community. E.g. the Mountain Pygmy-Owl *Glaucidium gnoma* also uses the same tree stumps in a more rotten state and exclusively hunts in primary vegetation. *P. nigra* is exclusively abundant in natural forest or old secondary forest and reproduces there. The aim of conservation strategies is to preserve the cloud forest remnants of the Sierra Yalijux (Schulz and Unger 2000; UPROBON 2000). With *P. mocinno* as a target species this aim could be fully reached.

Extending to another animal group, *Alouatta pigra*, the Yucatán Black Howler Monkey, should also be included in the conservation efforts regarding the Quetzal. Like the Quetzal, the monkey species is dependent on natural forests. In Chelemhá, it was observed exclusively in natural vegetation or secondary vegetation older than 15 years and trees taller than 10 m.

While the mentioned species are still present, there are others which probably are extinct. For example, the Horned Guan (*Oreophasis derbianus*) was last reported in the Sierra Yalijux by local Q'eqchi in 1990. In 2002 it was not observed.

A visit in August 2003 to the cloud forests of the Sierra de las Minas, 20 km southwards on the other side of the valley of the Río Polochic, indicated several individuals of Horned Guan at the Sierra de las Minas. Further results and comparison of the two sites will follow.

When analysing the mark-recapture data, the most striking aspect is that the secondary growth has twice as many species as the primary forest, i.e., 57 and 33 observed species, respectively. The most likely explanation is that the secondary growth is a habitat with many dispersing individuals which are not permanent residents. This assumption is proved by the fact that some species like Sparkling-Tailed Woodstar *Tilmatura dupontii* and others were only recorded as immature and only during the late breeding season in August. Additionally, many Neotropical migrants like *Wilsonia pusilla*, prefer to forage in secondary vegetation in March and April. This species was recorded exclusively in secondary growth (30 to 1 captures with the same relation in transect census). Last but not least, an estimation of theoretical extinction rates and analyses of not yet reached species-area equilibrium are useful to evaluate a fragmented natural forest. For further aspects of extinctions and endemics cp. Máñez Costa and Renner in this volume.

6 Conclusions

For several reasons, young secondary forest in tropical landscape is not suitable to preserve a high degree of bird diversity, even if species richness appears to be higher than in natural forest. Body mass distribution and recapture rates (presumably reproductive success also) indicate the advantages of natural forest over young secondary forests. Nevertheless, reproductive success must be evaluated in respect to natural forest in comparison with different landscape habitat mosaics.

The Resplendent Quetzal *Pharomachrus mocinno* shows a clear dependence on natural habitats and will vanish without preservation of natural forests.

Acknowledgments

Michael Mühlenberg in particular and Bruce McKenzie gave valuable comments on an earlier version of the script. We especially thank the people in Chelemhá and all participants in the project.

References

American Ornithologists' Union (1998) The check-list of north American birds. American Ornithologists' Union, Washington DC

American Ornithologists' Union (2000) 42nd supplement to the American Ornithologists' Union check-list of north American birds. Auk 117:847–858

American Ornithologists' Union (2002) 43rd supplement to the American Ornithologists' Union check-list of north American birds. Auk 119:897–906

Benton TG, Vickery JA, Wilson JD (2003) Farmland biodiversity: is habitat heterogeneity the key? Trends in Ecology and Evolution 18:182–188

Chao A, Lee SM (1992) Estimating the number of classes via sample coverage. Journal of the American Statistical Association 87:210–217

Colwell RK, Coddington JA (1994) Estimating terrestrial biodiversity through extrapolation. Philosophical Transactions of the Royal Society of London. Series B: Biological Sciences 345:101–118

Colwell RK (2000) EstimateS 6.0: Statistical estimation of species richness and shared species from samples. Version 6.0, `http://viceroy.eeb.uconn.edu/estimates`

Dranzoa C (1998) The avifauna 23 years after logging in Kibale National Park, Uganda. Biodiversity and Conservation 7:777–797

Dunning JB (1993) CRC handbook of avian body masses. CRC Smithsonian Institution, Front Royal

Edwards EP (1998) A field guide to the birds of Mexico and adjacent areas. Belize, Guatemala, and El Salvador. University of Texas Press, Austin

FAO (2001) State of the worlds forests. FAO, Roma

Gilpin ME, Soulé ME (1986) Minimum viable populations: processes of species extinction. In: Soulé M (1986) Conservation biology. Sinauer, Sunderland, pp 19-34

Haffer J (1974) Avian speciation in tropical South America. With a survey of the Toucans (Ramphastidae) and Jacamars (Galbulidae). Nuttall Ornithological Club, Cambridge

Holbech LH (1996) Faunistic diversity and game production contra human activities in the Ghana high forest zone with references to the western region. Dissertation, University of Copenhagen, Copenhagen

Hovestadt T, Roeser J, Mühlenberg M (1991) Flächenbedarf von Tierpopulationen als Kriterien für Maßnahmen des Biotopschutzes und als Datenbasis zur Beurteilung von Eingriffen in Natur und Landschaft. Forschungszentrum Jülich GmbH, Jülich

Howell SNG, Webb S (1995) A guide to the birds of Mexico and northern Central America. Oxford University Press, Oxford

Hughes JB, Daily GC, Ehrlich PR (2002) Conservation of tropical forest birds in countryside habitats. Ecology Letters 5:121–129

Johns AD (1992) Vertebrate responses to selective logging: implications for the design of logging systems. Philosophical Transactions of the Royal Society of London. Series B: Biological Sciences 335:437–442

Karr JR (1971) Structure of avian communities in selected Panama and Illinois habitats. Ecological Monographs 41:207–233

Krebs CJ (1999) Ecological methodology. Addison-Welsey Educational Publisher, Menlo Park

LaBastille A (1973) Establishment of a Quetzal cloud-forest reserve in Guatemala. Biological Conservation 5:60–62

LaBastille A (1974) Use of artificial nest-boxes by Quetzals in Guatemala. Biological Conservation 6:64–65

Lambert FR (1992) The consequences of selective logging for Bornean lowland birds. Philosophical Transactions of the Royal Society of London. Series B: Biological Sciences 335:443–457

Land HC (1970) Birds of Guatemala. Livingston Publishing Company, Wynnewood

Loiselle BA, Blake JG, Moermond TC, Mason DJ (1989) Low elevation records of Resplendent Quetzals in Costa Rica. Journal of Field Ornithology 60:86–88

Magurran AE (1988) Ecological diversity and its measurement. Princeton University Press, Princeton, New Jersey

Marchant JH, Hudson R, Carter SP, Whittington P (1992) Population trends in British breeding birds. British Trust for Ornithology, Norfolk

Markussen M (2003) Waldkonversion und Bodendegradation in Bergnebelwaldgebieten Guatemalas (Alta Verapaz) – Ein Beitrag zur Biodiversitätsforschung in sensiblen Ökosystemen. EcoRegio 11, Shaker, Aachen

McCann KS (2000) The diversity-stability debate. Nature 405:228–233

Mühlenberg M, Hovestadt T, Unger D (1989) Conservation of the cloud forest in Guatemala [in German: Schutz des Nebelwaldes in Guatemala]. Ökologische Station der Universität Würzburg, Würzburg

Peterson RT, Chalif EL (1973) A field guide to Mexican birds. Mexico, Guatemala, Belize, El Salvador. Houghton Mifflin Company, Boston

Pimm SL (1986) Community stability and structure. In: Soulé M (1986) Conservation Biology. Sinauer, Sunderland, p 309

Poulsen BO (1994) Mist-netting as a census method for determining species richness and abundances in an Andean cloud forest bird community. Le Gerfaut 84:39–49

Powell GVN, Bjork RD (1994) Implications of altitudinal migration for conservation strategies to protect tropical biodiversity: a case study of the Resplendent Quetzal *Pharomachrus mocinno* at Monteverde, Costa Rica. Bird Conservation International 4:161–174

Powell GVN, Bjork RD (1995) Implications of intratropical migration on reserve design: a case study using *Pharomachrus mocinno*. Conservation Biology 9: 354–362

Remsen JV, Good DA (1996) Misuse of data from mist-net captures to assess abundance in bird populations. Auk 113:381–398

Remsen JV (1994) Use and misuse of bird lists in community ecology and conservation. Auk 111:225–227

Renner SC (2003) Structure and diversity of cloud forest bird communities in Alta Verapaz, Guatemala, and implications for conservation. Niedersächsische Staats- und Universitätsbibliothek, Göttingen, published online at `http://webdoc.gwdg.de/diss/`

Rosenzweig ML (1995) Species diversity in space and time. Cambridge University Press, Cambridge

Schulz CH, Waltert M, Kessler PJA, Pitopang R, Shahabudin E, Veddeler D, Steffan-Dewenter I, Mühlenberg M, Gradstein SR, Tscharntke T (in press) Biodiversity indicator taxa of tropical land-use systems: comparing plants, birds and insects. Ecological Applications

Schulz U, Unger D (2000) Integration von Landnutzung und Regenwaldschutz – eine Fallstudie aus Guatemala. In: Landnutzungsplanung und Naturschutz, pp 94–105

Soulé ME (1986) The fitness and viability of populations. In: Soulé ME (1986) Conservation biology. Sinauer, Sunderland, pp 13–18

Soulé ME (1987) Viable populations for conservation. Cambridge University Press, Cambridge

Sørensen T (1948) A method of establishing groups of equal amplitude in plant sociology based on similarity of species content. Det Kongelige Danske Videnskabernes Selskab Biologiske Skrifter 5:1–34

Terborgh J (1999) Requiem for nature. Island Press, Washington DC

Terborgh J, Robinson SK, Parker TA, Munn C, Pierpont N (1990) Structure and organization of an Amazon forest bird community. Ecological Monographs 60: 213–238

Unger D (1988) Welche Funktion hat der extreme Sexuladimorphismus des Quetzal (*Pharomachrus mocinno*)? Universität Tübingen, Tübingen

UPROBON (2000) Schutzgemeinschaft Nebelwald: 1. Vereinsbericht. Union para Proteger el Bosque Nuboso. Cobán, Guatemala

Waltert M, Mühlenberg M (2001) Zur Beziehung zwischen Abundanz und Habitatqualität: Fangraten territorialer Gelbbartbülbüls Andropadus latirostris. In: Gottschalk E, Barkow A, Mühlenberg M, Settele J (2001) Naturschutz und Verhalten. Umweltforschungszentrum Halle Leipzig, Leipzig, pp 103–110

Whittaker RH (1965) Dominance and diversity in land plant communities. Science 147:250–260

Whittaker RH (1970) Communities and ecosystems. Macmillan, New York

Whittaker RH (1972) Evolution and measurement of species diversity. Taxon 21: 213–251

Winker K, Rappole JH, Ramos MA (1995) The use of movement data as an assay of habitat quality. Oecologia 101:211–216

World Conservation Monitoring Centre (1992) Global Biodiversity. Status. Chapman and Hall, London

World Bank (2001) World development indicators 2001. World Bank, Washington DC

What drives biodiversity loss in the land of trees? A review of the economic and historical parameters causing deforestation in Guatemala

Ludger-Josef Loening[1] and Hermann Sautter[2]

[1] Ibero-America Institute for Economic Research, University of Goettingen and The World Bank, 1818 H Street NW, Washington DC 20433, USA, email to jloening@worldbank.org
[2] Department of Economics, University of Goettingen, Platz der Goettinger Sieben 3, 37073 Goettingen, Germany, email to hsautte@gwdg.de

Summary. This paper analyses some of the fundamental causes of deforestation and habitat loss in Guatemala. A literature review indicates that agricultural expansion during the past was highly correlated with discriminatory labour and educational politics. These past politics have repercussions until today, not only on the persistence of rural poverty but also on forest cover loss. After a period of large-scale agricultural and pasture expansion, it appears that spontaneous colonisation has become the single most important direct source of deforestation. Rural non-farm employment, schooling, communal tenure regimes and ethnicity are among the most important parameters of land use decisions. However, contrary to common believes, there is evidence that deforestation is not dominated by subsistence-driven patterns, and that agricultural intensification techniques promoted by several environmental movements increase the demand for additional land.

Key words: Biodiversity loss, deforestation, economic determinants of land use, rural poverty, education, rural non-farm employment, Guatemala

1 Introduction

Over the past decades, decline and extinction of species have merged as a major environmental issue[3]. Given the importance commonly attached to habitat modifications as a threat to biodiverse natural resources, analysing the processes that

[3] We have greatly benefited from interviews and discussions with Margret Dix, Silvel Elías, Maynor Estrada, Juventino Gálvez, Gisela Gellert, Juan Carlos Godoy, David Kaimowitz, Celia Marcos, Odgen Rodas, Augusto Rosales and Norman Schwartz. The results and opinions presented here are the authors' only. They do not necessarily reflect the points of view of the people above, and should not be attributed to the institutions the authors are affiliated with.

determine land use decisions is essential[4]. Like many other countries, Guatemala has experienced a dramatic loss of forest cover during the past decades. In fact, according to the Food and Agricultural Organization (FAO 2001), deforestation in Central America is higher than elsewhere in the world, and the annual percent loss of forest cover in Guatemala is higher than in countries predominantly in the public eye, such as Brazil or Indonesia. Within this context, the aim of this paper is to analyse the socio-economic factors that determine the deforestation process in Guatemala. Clarifying the role of the economic parameters of deforestation, in turn, makes it possible to propose policy interventions that can mitigate an important part of biodiversity loss.

According to Southgate and Basterrechea (1993), the deforestation process in Guatemala has much in common with its Latin American counterparts. As such, the country provides an interesting case study. Generally, the analysis of deforestation in the developing world has been concentrated on medium and large countries, which are politically stable, and of comparatively higher levels of economic wealth. However, this kind of ›selection bias‹ may impact the results of deforestation studies, and influence the conclusions and policy decisions drawn from them. Thus, a closer look at a small, poor, and unstable country such as Guatemala is inherently valuable. Sad to say, the Guatemalan case offers nearly a textbook example for exploring some of the fundamental factors of deforestation. Given the small size of the open economy and considering its heavy social contrast, much appears to be clearer here than in other settings.

The reader should be aware that most evidence for Guatemala comes from a scattered literature, unpublished reports or booklets, and disperse individual files, which are usually focused on a particular issue, location or time period. Many of these sources have never been used to generate a consistent picture. Available information is sometimes contradictory and varies in accuracy. Nevertheless, the following sections will piece together the available evidence. It will be argued that to some degree or another, coherent trends can be identified. To examine the causal structure of deforestation, this paper is organised as follows: Sect. 2 presents the historical and socio-economic context. Sect. 3 refers to the magnitude and spatial deforestation patterns. Sect. 4 addresses the question of the agents of deforestation. Sect. 5 explores some elements of the underlying causal structure of land use decisions, as identified by the literature. Sect. 6 concludes.

[4] The rational to see deforestation as a major cause of biodiversity loss is provided by the theory of island biography. The species-area relationship, upon which island biography is based, is often used to make *pragmatic* estimates of the extinction of species resulting from deforestation. Empirical examinations of this relationship generally demonstrate that it is a good approximation at broad scales. It is less accurate, however, as a rule of thumb at finer resolutions. For a detailed discussion of recent evidence, see Loening (2004).

2 From past to present

While deforestation in Guatemala is a contemporary problem, land use changes in Guatemala *cannot* be understood without first looking at the historical and socio-economic context. In the next paragraphs it will be argued that agricultural expansion is correlated with Guatemala's past exclusionary labour politics, since the elite believe their wealth to depend upon a low-wage and uneducated labour force. From the 19th century until the early 1970s, the expansion of agricultural land was mainly due to the production of export crops. However, deforestation patterns have changed dramatically during the last decades. Since then, most of the forest cover loss can be attributed to managed and, more recently, to spontaneous colonisation. These patterns, in turn, have their roots in the exclusionary policies of the past, which have caused rural poverty and underdevelopment.

The next paragraphs distinguish four different phases of deforestation. These can be related to the Mayan civilisation and the Spanish colonial period, the production of export crops after independence from Spain, the massive spontaneous colonisation processes since the 1960s, and the declaration of protected areas in the light of increased migration up until today.

2.1 Mayan civilisation and Spanish colonial period

The historical deforestation patterns in Guatemala show similarities to current processes. Environmental degradation already occurred during the time of the Mayan civilisation. It was concentrated, ironically, in the Petén region and in the lowlands of northern Guatemala – the same place where the overwhelming part of deforestation occurs today[5].

The Mayan empire reached its peak in the period from 250 AD until 900, and then suffered a rapid decline in both population and cultural sophistication. The civilisation was partially rebuilt in 1200, but this latter Mayan civilisation again went into decline. Due to the soil characteristics in that region, which are not suitable for long-term agriculture, Cabrera (1995) claims that the Mayas were forced to practice shifting cultivation. There are various hypotheses regarding the civilisation's collapse, including political reasons. Culbert (1988) and O'Hara et al. (1993) argue that environmental decline played a key role. Much of the evidence is from carbon-dated core samples showing deforestation, drying, soil erosion and reduced crop yields in major agricultural areas. By the early 9th century, agricultural output could no longer support the dense population in the region. This lead to migration out of the Petén, a decline in population, and the final collapse of the civilisation.

[5] Interestingly, the fact that deforestation already occurred in the Petén during that time implies that the threat of deforestation in terms of biodiversity loss may be less severe than commonly believed, i.e., after fleeing, species can return and invade sites through successful natural mechanisms.

The main unresolved question concerns the extent to which drying and soil erosion were the endogenous result of human activity, or if they arose from exogenous climate change. The evidence suggests that both factors were important. Most likely the agricultural shortfall was partly determined by the Maya as a result of deforestation and soil erosion. At the same time, the high population density they had achieved through their intensive agriculture left no margin of safety when climate conditions changed, even if just marginally. Shriar (2001) presents evidence suggesting that the population density in the Petén at the height of the Maya civilisation was higher than that prevailing today.

At the time of the Spanish conquest in 1524, the Petén region had already been abandoned, and much of the population was concentrated in the highlands[6]. Because of the relatively fertile grounds, the indigenous population likely depended on less extensive forms of agriculture. Due to the lack of precious metals, the Spanish rapidly directed their attention to agricultural production. Both the Spanish and the Mayan culture had a strong focus on agriculture, but the motivation for agricultural production was not the same. While the Mayan practised agriculture to cover the living needs of the community and relied on a communal structure of property, the Spanish exploited the land to export a small number of export crops, such as cacao and indigo. They based their land use on a structure of private property.

Initially, the process of colonisation meant changes in the structure of property rights. According to Luján (2000), these changes consisted of the appropriation of agricultural areas and the consequent expulsion of the Maya people to less fertile areas, particularly to the highlands. However, the Spanish allowed the indigenous to rely on communal forms of land tenure. Cabrera (1995) argues that moderate deforestation during this period was principally due to two reasons. First, immigration from Europe and the need to create a surplus from production for the Spanish crown resulted in a moderate increase in farm land. Second, the expropriation of indigenous agricultural land may already have encouraged shifting cultivation, because the indigenous were expulsed to areas less appropriate for agriculture.

2.2 Production of export crops

The massive expansion of agricultural land began after independence from Spain in 1821. Forests were considered useless and opened to distribution. ›Using‹ the land was the mechanism to assure property rights. Consequently, privileged landowners allowed the indigenous and campesinos to clear forest and plant their own crops for a limited period of time. After a while, the landowners introduced commercial crops. The peasants moved on and repeated the process. Until the

[6] Some authors speculate that the origin of the word Guatemala comes from *Quauhtemalan*, which in Nathuatl means land of trees. *Quauhtemalan* is a translation of the word *K'ichee'* (*ki* = much, *chee'* = trees) by the indigenous Nathual accompanying the Spanish conquistadors (see Cabrera 1995 and Maya' 1995). *K'ichee'* is also one of the 21 indigenous languages spoken in Guatemala today. The etymological narrative suggests that forests covered the overwhelming part of the country when the Spanish arrived.

1970s, the massive expansion of agricultural land was due to the production of relatively few export crops, such as coffee, and later also sugar cane, banana, cotton and cattle. Deforestation during this period was concentrated in southern Guatemala, and in the parts of the highlands suitable for coffee production. Historical data showing the rapid increase of major export crops can be found in Luján (2000) as well as SEGEPLAN and MAGA (1984).

Cabrera (1995) discusses several factors that favoured the loss of forest cover during that period. First, governments provided cheap credit for agricultural expansion and gave incentives for European immigration. The credit advantage and availability of cheap land was the origin of the German coffee producers, particularly in the highlands of the Department Alta Verapaz. Second, deforestation was initially associated with the expansion of the coffee sector. However, the diversification of export crops beginning in the 1940s additionally spurred the expansion of agricultural land.

Most qualitative research agrees that increased agricultural production due to the conversion of forest areas contributed to Guatemala's high agricultural growth rates until the mid 1970s (Elías et al. 1997; Schweigert 1993; Valenzuela de Pisano 1996; World Bank 1996). However, according to the World Bank (2003), exclusionary patterns of development can also be attributed to past agricultural policies. These policies, in turn, not only had important historical consequences for the distribution of incomes, but they are still relevant today. Key elements, as summarised below, refer to mass land expropriations as well as the discrimination of the indigenous and campesinos regarding labour and education.

Luján (2000) remarks that the historical practice of expropriating land from the indigenous gained momentum with the *Reforma Liberal* in 1871. Legal instruments encouraged the conversion of communally held indigenous lands into individually titled holdings[7]. A central aim was the formation of large plantations, in particular for coffee. Since the ideal terrain for coffee occurs between 800 and 1,500 meters of altitude, the indigenous peoples who had been cultivating this land were once again compelled to locate to less fertile grounds. Diversification of export crops brought additional expropriations for the peasants.

In 1950, according to the World Bank (2003), communal lands accounted for 12 percent of total agricultural land. This share dropped to 4.8 percent in 1964 and only about 1 percent in 1979. However, the number of farm families between 1950 and 1979 possessing parcels of land too small to provide subsistence incomes increased by 37 percent, and the number of landless peasants increased to about 1/4 of the rural workforce. Estimates from 1979, the last published agricultural census, indicate that less than 2 percent of the population owned at least 65 percent of the land. With an estimated Gini coefficient of 0.85 for the distribution of land in 1979,

[7] Today, Southgate and Basterrechea (1993) argue that large-scale land redistribution – an issue of faith for many observers – cannot be regarded as panacea for Guatemala. That is, there is simply not enough land available to give all rural households a farm large enough to support an entire family.

the World Bank (1996) places Guatemala highest on the list for land inequality in Latin America[8].

The exclusionary labour and educational politics are closely connected with land expansion. The World Bank (2003) and the United Nations Development Programme (UNDP 2000) claim that land politics in the past were designed to create a low-wage labour force by reducing the agricultural land available to the indigenous. Because insufficient cheap labour was considered to be the main barrier to an expansion of export crops, the elite actively sought to create and maintain poverty during that period. The expropriation of indigenous communal lands helped to create rural underemployment by forcing families into marginal areas, or leaving them without access to sufficient land. Insofar, Guatemala's agricultural sector developed essentially on the backs of indigenous and campesino workers. According to UNDP (2000), in conjunction with mass land expropriations, an increasing share of the indigenous and campesino population was forced into numerous forms of mandatory labour.

In 1873, the *Contribución de Caminos* decreed that all able male citizens were obliged to provide free labour for public projects to build roads or to pay a commutation fee. While this free work applied to all male citizens, in practice only the indigenous and campesino population was forced to perform it. In 1877, Guatemala instituted its *Mandamiento* forced-labour system in which villages were required to supply crews of up to 60 people for periods of 15 to 30 days to coffee plantations. In 1934, the *Ley Contra Vagancia* obliged landless peasants to work at least 150 days per year on plantations. Proof of service was required in the workers' personal workbooks.

Moreover, indentured labour was also common. Under this system, advances were given to workers in anticipation of a certain amount of work in the future. Debts were then deducted from the worker's harvest or required in cash. Such debts commonly built up to such high levels that the plantation owners essentially controlled the workers. Debts were monitored by local public authorities who were authorised to arrest any defaulters, as evidenced by debt recordings in the personal workbooks that the indigenous peasants were required to keep. Lovell (1988) finds that these coercive labour laws did not register in a uniform way across the country. However, many of the indigenous and campesinos did loose their lands and were forced into mandatory forms of labour.

These forced labour laws remained in effect until the middle of the 20th century. After the democratic revolution in 1944, the Agrarian Reform Law of 1952 finally prohibited all forms of servitude and slavery. Regarding agricultural land, according to Luján (2000), the reform had a moderate redistributional character. However, the agrarian reform was abandoned with a military and US Central In-

[8] According to the World Bank (2003), this compares with 0.82 for El Salvador, 0.81 for Panama, 0.80 for Costa Rica and Nicaragua, 0.77 for Honduras and Bolivia, and 0.61 for Mexico. Since there have been no significant distributional changes during the past 50 years, UNDP (2002) reports increased land fragmentation.

telligence Agency-backed *coup d'état* in 1954[9]. At that time a decree was issued that permitted a reintroduction of a semi-feudal system under which landowners could avail themselves of a cheap labour force by providing subsistence plots on their plantations in exchange for labour during the harvest period. In some parts of the country, this practice still continues today, and minimum labour standards are often not enforced. In consequence, the economy that resulted has provided little incentives for workers or firms to accumulate human capital, and to engage in non-farm activities.

2.3 Directed and spontaneous colonisation

Colonisation policies had the most severe impact on forest cover loss. Land colonisation was concentrated in the northern lowlands, particularly in the Petén[10]. It began shortly after the military coup in 1954, which both dismantled the previous agrarian reform and initiated government-promoted land colonisation. In early 1960, within the context of a growing agro-export economy, high population growth rates and increasing land scarcity, the government decided to open the northern lowlands to colonisation and development. Katz (2000) documents that the opening of the agricultural frontier, particularly in the Petén, was an easy way for the military governments to increase agricultural production, and to provide an escape for land-hungry peasants, while not compromising the government's anti-resdistributional policies.

The Petén is Guatemala's northernmost department and compromises 1/3 of the national land mass or about 36,000 km^2 of originally low-altitude moist tropical forest ecosystems. The Petén ecosystem is characterised by the fact that most nutrients are stored in the vegetation, and these nutrients are continually recycled between the soil and the biomass. Because of these characteristics, the elimination of the vegetative cover decreases soil fertility and the productive potential of the land. Nevertheless, the Petén still contains one of the largest remaining areas of tropical rainforest in Mesoamerica with high levels of biodiversity, as Méndez (1999) documents.

Until the 1960s, Guatemalans widely neglected the distant hinterland, considering it a dangerous and unsafe place. Population density and immigration was negligible[11]. Schwartz (1990) illustrates that during the Spanish colonial period the

[9] In a visit to Guatemala, former President Bill Clinton made an unusual statement in 1999. Reading from hand-written notes, he issued an ›apology‹ for the US role in Guatemala: »For the United States, it is important that I state clearly that support for military forces or intelligence units which engaged in violent and widespread repression [...] was wrong. The United States must not repeat that mistake.« The Washington Post, March 11, 1999.

[10] Land colonisation was also important in the *Franja Transversal del Norte*, the region south of the Petén. Despite its geographic proximity to the Petén, colonisation patterns were different here. In particular, a focus was laid on small and medium farm settlement. Unfortunately, colonisation and deforestation patterns in this region are poorly documented.

[11] Based on census data, Spielmann (1973) reports that during 1950–1964 there was no significant migration to the Petén. Moreover, only 30 percent of all migrants moved to ur-

small economy of the Petén was initially based on cattle and horse ranching. The traditional cattle raising systems were based on open cattle grazing in the large areas of natural pasture. After a decline in ranching in the middle of the 19th century, the economy was based on agriculture and the extraction of non-timber products. There was no year-round road access from the Petén to the rest of Guatemala. The majority of the Petén's habitants lived in the department capital or small towns.

In 1959, the Guatemalan government made a serious effort to colonise and ›develop‹ the department, and created the *Empresa Nacional de Fomento y Desarrollo de El Petén* (FYDEP). Kaimowitz (1995) describes how this state enterprise behaved as a largely autonomous regional government with nearly unlimited authority. Throughout most of its history, military officers ran the National Enterprise for Economic Development of the Petén. A lack of continuity at the administrative level, pressure from metropolitan politicians, and mismanagement hindered the effective use of financial resources. Because of its historic association with previous military governments and its reputation for corruption, FYDEP was dissolved in the mid-1980s-Guatemala's transition to civilian rule.

According to Schwartz (1990), FYDEP began to promote colonisation into the Petén by large landholders. The decision to colonise was based at least in part on the conclusions of studies by Latinconsult, a consulting firm. It argued that cattle ranching would be the most appropriate land use for much of the Petén. FYDEP employed four criteria to set land prices: access to roads and markets, potential for commercial timber exploitation, soil quality, and access to water resources[12]. FYDEP imposed regulations on buyers to prevent land speculations and took measures to conserve the forests. However, due to its limited administrative capacity, infraction was the rule. FYDEP also built a basic infrastructure of roads, small bridges, a modern airport, grain storage facilities, and brought electricity to most towns. In 1970, FYDEP opened an all-weather dirt road from central Petén to the central highlands. Farmers in the Petén began sending maize to the south, and to a lesser extent also beans, rice and cattle.

Three major types of people applied to purchase land made available: entrepreneurs and professionals from the large cities, politicians and military officers from various regions, and medium size ranchers from eastern Guatemala. Because there were no immediate prospects for labour-intensive industries, and agricultural intensification could not be sustained, FYDEP wanted only gradual migration flows. Therefore, FYDEP was never eager to sell land to *milperos* (maize farmers) or indigenous migrants but preferred large-scale and middle-class Ladino settlers.

However, starting in 1966, politicians frequently pressured the enterprise to sell land to thousands of land-poor settlers from the south. Ever since then, FYDEP

ban centres: most people preferred to migrate to the southern coast plantations. Based on unpublished 1994 census data, Gellert (2000) finds that migration to metropolitan Guatemala and, in particular the Petén, has sharply increased.

[12] Land prices in the Petén rose rapidly. Kaimowitz (1995) reports that initially the FYDEP sold land ranging from US$ 2–9 per hectare. A down payment was required, with the remainder to be paid over 10–20 years at zero interest. Today, depending on the land characteristics and tenure form, a hectare values approximately US$ 60–600.

has been unable to cope with uncontrolled spontaneous colonisation. According to Schwartz (1995a, 1995b), immigration finally ran out of control when the level of violent military conflict decreased in the mid-1980s. Ever since that time, anarchic colonisation processes have been taking place. Grandia (2000) suggest that population growth rates may have exceeded 9 percent per year. Total population may have risen from approximately 25,000 in the 1960s to over 500,000 for the early 1990s.

The government decision to sell extensive areas of land in the Petén to major ranchers at below market prices was clearly a key factor in the rapid deforestation that occurred between the 1960s and the 1980s. Road construction together with considerations related to cattle, spurred deforestation. Expectations from rising land prices and the need to clear forestland as a way to enhance tenure security also drove deforestation. However, Kaimowitz (1995) finds that subsidised livestock credit was of marginal importance. Only a small percent of the Petén ranchers received credit. Moreover, livestock credit in the department has steadily declined as a result of both increasing real interest rates and a general decline in public agricultural loans. Since there appears to have been little private lending for livestock in the Petén, most of the investment by large and medium-size landowners was probably self-financed.

The Petén has evidently served as the country's main agricultural frontier, and as an important site for large-scale ranching. Government policies have permitted large landholders with privileged land access to reap substantial rents, which were made possible by public investment in infrastructure. It is less clear, however, to what degree logging, timber extraction operations, shifting cultivation and oil exploitation were responsible for deforestation, and how all these activities are linked to one another[13]. Due to the anarchic settlement process and scarcity of data, much remains speculative.

Colonisation patterns changed, however, for the period of the late 1980s. In fact, most qualitative research agrees that since Guatemala's transition to civilian rule in 1985, and the ceasing of the military conflict, one of the greatest threats to the Petén is the migration of thousands of land-poor peasants into the region (Elías et al. 1997; FLACSO 2000; Katz 2000; Schwartz 1995b; Valenzuela de Pisano 1996; World Bank 1995). Rural Guatemalans fleeing from land scarcity and with limited employment opportunity come to the Petén in search of agricultural production. In spite of the lack of reliable data on forest cover change in the department over the last decades, virtually all observers agree that the changes in land use and forest cover have been dramatic.

[13] Oil exploitation in the Petén started in the 1980s but companies withdrew their activities during the peak of the civil war. Rosenfeld (1999) claims that with the end of guerilla activity, the development of the oil industry in the Petén is proceeding. Until today, however, only few contracts have been granted. The Xan Oil Field in the Laguna del Tigre National Park is the most important.

2.4 Protected areas and civil war refugees

The last phase of deforestation in Guatemala can be associated with the return of the civil war refugees during the mid-1990s, and the massive expansion of protected areas by environmental groups. According to the final report of the Commission for Historical Clarification (1999), the internal military conflict – lasting from 1960–1996 – left 1 million persons displaced in a total population of 10 million. Reportedly, a substantial number of civil war peasants that fled to Mexico are now returning. Cabrera (1995) argues that extensive rural migration combined with land-scarcity exaggerates violent conflict over land and deforestation.

Given the increased threat of invasion into the forest margins by smallholders, environmental movements together with governmental institutions have responded with the massive establishment of protected areas[14]. According to the National Commission for Environment (CONAMA et al. 1999a), the majority of protected areas have been set up with increased international financing and technical support between 1986–1995. UNDP (2002) states there were 123 protected areas in 2001, covering approximately 28.6 percent of the country's total landmass. However, CONAMA et al. (1999a) reports that the protected area in Guatemala is not representative regarding ecosystem variability. This frequently leads to environmentalists calling for further areas. Only 13 percent of Guatemala's parks had an approved and executed administrative framework in 1997. However, this percentage has recently increased and suggests a gradual improvement in the management of the formerly ›paper parks.‹

The majority of the protected areas are administrated by the *Consejo Nacional de Areas Protegidas* (CONAP), a reportedly weak public institution, which is subordinated to the *Presidencia de la República*. In some cases, non-governmental organisations (NGOs) or other institutions are accepted as co-administrators. According to the area protected by parks and their financial revenues, environmental NGOs play de facto the predominant role in Guatemala. In principle, most protected areas seek to combine the goals of nature conservation with improving rural livelihoods, and they are run as an Integrated Conservation and Development Project (ICDP). Despite the magnitude of the protected areas and the supposedly heavy impact they have on the land use decisions of the peasants, there is no systematic information regarding the protected areas' contribution to forest conservation and rural development. However, an extremely weak executive and local administrative capacity (Godoy 1998), a high administrative and wage overhead (Galindo 1999), and the disappointing results of many ICDPs in developing countries, all suggest that the conservation impact might be modest in magnitude (for a summary see Loening and Markussen 2003).

The most ambitious conservation project, created in 1990, is the Maya Biosphere Reserve in the Petén. With 16,000 km^2 it corresponds to the size of the

[14] In practice, not on paper, Guatemala's 1999 National Biodiversity Conservation Strategy focuses nearly exclusively on protected area issues. It was made without proper consultation of the indigenous groups, who represent the overwhelming share of the rural population.

country El Salvador and covers approximately 1/2 of the department, or about 1/3 of the entire Guatemalan System of Protected Areas (SIGAP). The reserve contains three different types of management units. A core area of national parks and biological reserves has the highest level of protection. A large multiple-zone allows some forms of extractive harvesting, ranching, farming, hunting and commercial logging. A buffer zone south of the reserve has no restrictions. The reserve is run by a unique blend of NGOs, international development institutions, and national authorities. According to Valenzuela de Pisano (1996) and Shriar (2001), agencies have a strong focus on nature conservation and place little or no emphasis on the underlying socio-economic causes that lead to the immigration of peasants. A common feature of conservationists is the belief that deforestation follows subsistence-driven patterns (see for example Nations et al. 1999 and Defensores de la Naturaleza 2001). Consequently, extension agencies promote ›sustainable‹ agriculture techniques that aim to intensify agricultural production[15].

3 Magnitude and location of forest cover loss

Having presented the main trends of deforestation, this section analyses the magnitude and location of forest cover change in Guatemala. In addition, it provides a brief account of the country's environmental diversity. Overall, given the historical and socio-economic context, there can be little doubt that deforestation has been rapid. Despite considerable data gaps, this is confirmed by virtually all deforestation studies that have been carried out for Guatemala.

3.1 Magnitude of deforestation

Fig. 1 scatters data on forest cover change as a percentage of total land area from 1950–2000. The data is taken from Loening and Markussen (2003) who synthesise the results from more than 24 studies on deforestation and land use in Guatemala. Most studies are unpublished consultant or government reports. According to local experts, national forestry inventories were carried out in 1950 by the *Instituto Nacional Forestal* (INAFOR), 1975–1976 by Mittak (1977), 1987–1988 by Sagastume (1992), and 1998–1999 by INAB (2000). The original inventories prior to Sagastume (1992) are reported to be lost. The most reliable estimate probably comes from the FAO (2001). It compares data from satellite imagery for 1987–1988 and 1998–1999.

The data is typically based on aerial photography and, for later periods, on satellite imagery. Sometimes agricultural census, socio-economic surveys or other estimates complement it. Since each author employs different methodologies and much of the data is of dubious quality, forest cover estimates vary greatly. However,

[15] Conservation organisations employ a number of such techniques. In the Petén, the use of green manures is often promoted. This system most commonly involves the development of an *abonera*, a plot on which velvet beans (*frijol abono*) are planted, and which later in the year is used as a plot to grow maize.

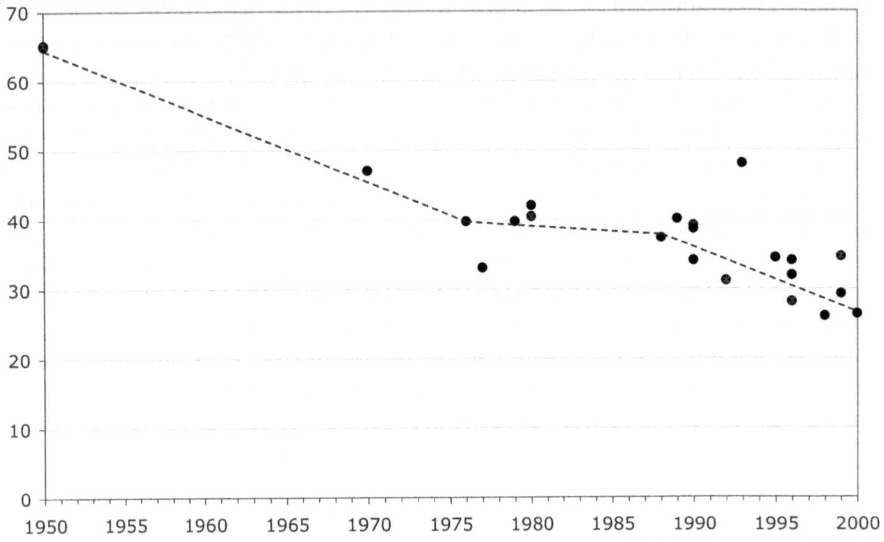

Source: Author's elaboration, based on data from Loening and Markussen (2003).
The dotted line is speculative and suggests a decreasing trend in deforestation patterns during a period of the civil war.

Fig. 1. Guatemala: Forest cover estimates, 1950–2000 (in percent of total land area)

the overall trend clearly suggests that forest cover change has been dramatic. In 1950 forests covered nearly 65 percent of Guatemala's total land area. Only about 26 percent remained forested in 2000. This implies an estimated loss in forest resources of approximately 60 percent for the entire period (70,451 km^2 in 1950 compared to 28,497 km^2 in 2000).

According to Fig. 1, there may have been changing patterns over time in deforestation. A speculative interpretation, indicated by the dotted line, would be to assume that deforestation slowed down in the 1980s. Most likely this decrease is associated with the peak of Guatemala's internal military conflict during that period. Kaimowitz (1995, 1996) argues that the instability of the civil war discouraged deforestation activities. In particular, it reduced land prices and the number of farmers interested in purchasing land from the government. In addition, farmland was abandoned when there was violent conflict. Especially in regions where violence was heavy, the rate of deforestation was reduced.

Some studies present estimates on annual forest cover change. The estimates analysed in Loening and Markussen (2003) suggest annual forest cover change lies between 54,000 and 90,000 hectares. However, the data does not allow an inference about whether deforestation has increased over time. In general, environmental movements and conservation agencies tend to report higher figures than forestry institutions or the FAO. Kaimowitz (1996) claims that most estimates are too high.

Annual deforestation has rather been between 50,000 and 60,000 hectares than the 80,000 or 90,000 hectares mentioned in most studies. This is empirically confirmed by the FAO (2001) using satellite imagery, and by Baumeister (2001) analysing land use changes over the past decades. According to the FAO (2001), annual deforestation was approximately 1.7 percent between 1990 and 2000. This is comparatively high and places Guatemala among the top 1/3 of all countries worldwide that are affected by deforestation.

3.2 Location of deforestation

For the period 1975–1988, the latest information available at the department level, CONAMA et al. (1999a) and Elías et al. (1997) report deforestation as being a countrywide phenomenon. In absolute terms, however, deforestation is located in the northern lowlands, particularly in the Petén. In fact, more than 1/2 of the country's deforestation is reported to be in the Petén. A back of the envelope calculation from Cabrera (1995) suggests annual deforestation at approximately 39,000 hectares. Based on comparatively good quality data, Sader et al. (2001) report that in the buffer zones of the Maya Biosphere Reserve annual deforestation even exceeded 3 percent during the 1990s (see Table 1). Note that overall deforestation trends show increasing patterns, despite much of the environmentalist's conservation efforts. Moreover, deforestation seems to have increased significantly in 1990, the year when the reserve was founded[16].

Table 1. The Petén: Annual forest clearing rates in Maya Biosphere Reserve, 1986–1997 (in percent)

Area	1986–1990	1990–1993	1993–1995	1995–1997
Multiple Use Zone	0.05	0.16	0.25	0.25
Buffer Zone	0.74	2.71	3.76	3.28
Total reserve	0.04	0.23	0.33	0.36

Source: Sader et al. (2000, 2001).

Other areas threatened heavily by deforestation are the highlands of the eastern regions but also the central regions. Less critical are the Western Highlands. Deforestation in all parts of the highlands is usually associated with high levels of soil erosion. In fact, many of the cleared lands have been so degraded that they are no longer useful for agricultural production.

[16] In a preliminary analysis for the period 1997–1999, unfortunately plagued by data deficiencies, Sader et al. (2000) find that deforestation might be lower than in previous periods. However, total destruction of the reserve is probably higher than ever due to numerous forest fires. UNDP (2002) reports that forest fires were particularly high in 1998 but eventually have decreased over time.

3.3 Environmental diversity

Given the poor quality of most estimates regarding forest cover change, it does not come as a big surprise that much less is known about the environmental diversity of the country. There is neither proper measurement for the conservation of biodiversity nor information about the evolution of species loss over time. Nations et al. (1989), UNDP (1998) and CONAMA et al. (1999b, 1999c) summarise the available evidence. Based on incomplete inventories, they claim Guatemala to be ›the country with one of the highest levels in species diversity‹ in Central America. However, an objective framework to assess the biological importance of geographic areas is missing in all these estimates. UNDP (2002) lists 1170 endemic species for Guatemala and ranks the country 24th out of 25 countries with respect to forest biodiversity. Due to its natural location and its biogeographic situation, Myers et al. (2000) classify the country equally as a ›hotspot‹ regarding species diversity. The reasoning is essentially based on Guatemala's pronounced climatic conditions and its varied landscape. De la Cruz (1982) suggests that Guatemala has 14 Holdridge Life Zones.

Other studies are more cautious. For example, in a World Bank priority setting report for Latin America, Dinerstein et al. (1995) identify ecoregions of comparatively high biodiversity conservation value based on final conservation status and biological distinctiveness. The regions are then classified into four broad categories. According to this scheme, experts define ecoregions in descending order as globally, regionally, bioregionally or locally outstanding. For the case of Guatemala, the report claims that the tropical forests in the Petén and cloud montane forests in the highlands are seriously threatened and of great biodiversity value, although these forests are only categorised here as bioregionally outstanding. Insofar they are classified behind such regions as the Amazon tropical moist forests, which are considered to be of global importance.

4 Direct sources of deforestation

Table 2 presents data for direct sources of resource degradation in the period 1993–1997. It is based on estimates from national forest experts. The estimate suggests shifting farming systems are the main deforestation factor, followed by pasture expansion, illegal logging, and firewood consumption. Commercial agriculture for export crops and other factors, such as logging activities or natural disasters, appear to be of minor importance. According to Banco de Guatemala, the forest sector accounts for approximately 2.5 to 2.6 percent in GDP during the 1990s[17]. UNDP (2002) claims that about 97 percent of the country's forest production during that period was used for firewood production. This leaves only around 3 percent for

[17] Direct information from Banco de Guatemala. Given the lack of precise baseline information, in particular for firewood energy consumption, the responsible statistician suggested that the reliability of this number might be weak.

industrial wood production and other purposes. In fact, Guatemala has a negative trade balance for wood products, principally due to the importation of paper products.

Table 2. Guatemala: Direct sources of deforestation, 1993–1997 (in 1,000 hectares and percent)

Direct source	Annual deforestation	Percent
Migratory agriculture	64.37	78.5
Pasture expansion	8.20	10.0
Illegal logging	4.10	5.0
Firewood consumption	2.46	3.0
Forest fires	1.64	2.0
Natural factors	0.82	1.0
Commercial agriculture	0.41	0.5

Source: MAGA, PAFG and INAB (1998).

Somewhat surprising then is the low share of forest degradation attributed to firewood collection. According to the World Bank and UNDP (2003), 80.3 percent of all Guatemalans use firewood as their main source of energy. In particular, the rural poor continue to rely almost entirely on firewood for cooking. Income, education, access to basic social infrastructure, and ethnicity are important determinants of fuel choice. While firewood collection is often thought to be an important source of deforestation, this issue is little studied in Guatemala. The estimate in Table 2 suggests the impact of firewood collection on forest degradation to be low[18]. However, research for other countries has ascertained equally that most deforestation is caused by clearing for agriculture or logging, and not by wood collection. For example, Heltberg (2001) finds that the impact of firewood collection for the case of rural India is highly localised. Much wood is not collected from forested land. Firewood collection causes forest degradation only in certain places, particularly in areas of high population density, around cities, and on fragile and sloping lands.

Commercial agriculture and licensed logging, though occasionally opening up penetration roads and enabling farmers to move deeper into the woods, seem to be of marginal importance. However, illegal and selective logging, especially in the Petén, has been reported to be a direct source of deforestation. Due to the secrecy involved, evidence on this topic is merely narrative. Schwartz (1990, 1995a) reports

[18] See Schwartz (1995a) for a contrary argument. Katz (2000) reports that the average Guatemalan rural household may use the equivalent of 2 or 3 trees per month for firewood purposes, or approximately 25–30 trees per year. Of that, approximately 1/2 come from felled trees.

clandestine sawmills operating in Petén and Mexico that use Petén wood and employ workers from Guatemala, Mexico and Belize. In addition, there is a substantial number of independent small-scale loggers reported who combine woodcutting with other income-producing activities, such as the extraction of shate, chicle latex and *pimienta gorda*, or hunting. According to the FAO (1999), radical environmental movements may have unintentionally spurred these clandestine activities. Because of the their emphasis on strict preservation, conservation groups have made it difficult for companies to obtain licenses to harvest trees and non-timber products.

Table 3. The Petén: Land use change, 1987–1993 (in 1,000 hectares and percent)

Category	Land use change	Percent
Pasture	36	8.5
Grasslands and natural pasture	10	2.4
Mixed use (agriculture and pasture)	127	30.0
Abandoned agricultural land	111	26.2
Agricultural land with residual forests	140	33.1
Inundated land areas	−1	−0.2
Forests	−423	−100.0

Source: World Bank (1995) and own calculations.

The main controversy on the direct sources of deforestation in Guatemala concerns whether agricultural expansion comes mainly from migratory agriculture farming systems or large-scale pasture expansion. Since pasture expansion has been particularly high in the Petén, in two influential studies Kaimowitz (1995, 1996) argues that large-scale livestock expansion has favoured the conversion of forest cover here rather than land clearing by small farmers. Moreover, Kaimowitz (1995) claims that large-scale pasture expansion may even have protected certain forests from land clearing by small farmers. Other studies are more cautious and simply refer to pasture expansion and small farmer's land clearing without quantifying effects (see, e.g., Schwartz 1990, 1995a). Much of the uncertainty is due to the lack of reliable data on land use changes over time, the anarchic situation in the Petén, and changing deforestation patterns over time. The issue is further complicated by the fact that agricultural encroachment and pasture expansion often go hand in hand. Forest areas are typically converted by farmers and then later used for cattle systems. In addition, with the exception perhaps of farmers with prime agricultural land, the first thing that almost any small farmer in Guatemala does when accumulating money is to purchase cattle.

As pointed out in Table 3, pure pasture expansion in the Petén accounted for less than 10 percent, and mixed crops and pasture were 30 percent of total forest

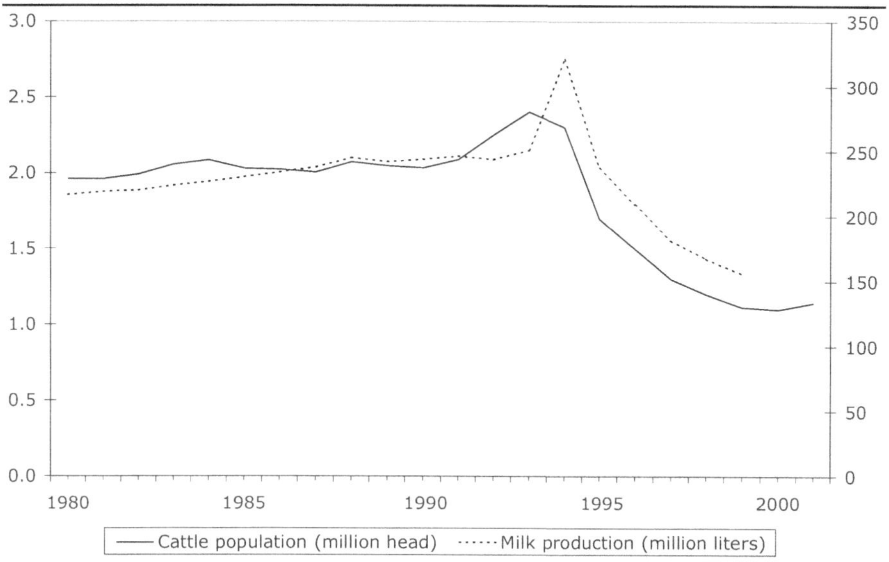

Source: CEPAL (2003), based on data from Banco de Guatemala and the National Statistics Institute. The left axis refers to the amount of cattle population.

Fig. 2. Guatemala: Cattle population and milk production, 1980–2001 (in million head and million litres)

cover loss. If none of this mixed-used category were pasture, livestock expansion would have presented less than 10 percent of total forest cover loss. If all of it were pasture, livestock expansion would have presented more than 38 percent. However, the available evidence gradually suggests that large-scale pasture expansion is becoming less important than it was in earlier periods. Already in 1981, a study from the Central Bank reported that many ranches in the Petén had failed or were not profitable. Banco de Guatemala (1981) argues this was due to inadequate road access, water supply problems, unacceptable low stocking rates, inexperienced management, inadequate knowledge of cattle ranching and the use of methods that work in the south coast of Guatemala but not in the Petén. Moreover, the study claimed the returns to ranching per unit of hectare were lower than agriculture and, hence, not particularly attractive.

Fig. 2, based on information from the Economic Commission of Latin America (CEPAL) Macroeconomics Database, shows the evolution of cattle population in Guatemala for the period 1980–2001. About 10 percent of total livestock is located in the Petén. Since no disaggregated time series data for the Petén is available, some care should be taken when interpreting Fig. 2. However, the figure clearly suggests that cattle population was almost stagnant during the 1980s, and has continually decreased since the mid-1990s. Therefore, livestock expansion and associated forest clearing may have become much less important during the past decades than in earlier periods. In addition, Kaimowitz (1996) argues that after the mid 1970s international conditions for beef exports continually worsened due to the restrictive European Community, U.S. and Mexican import regulations. As expected for a low-income country, the domestic demand for beef in Guatemala is traditionally low.

Table 4. Guatemala: Land use change, 1979–1999 (in 1,000 hectares and percent)

Category	1979	1999	Change	Change in percent
Agricultural land	1,310	1,490	180	15.7
Other agricultural land area combined with open forests	3,766	4,638	872	76.0
Pasture	1,334	1,400	66	5.8
Other land	156	185	29	2.5
Forests	*4,323*	*3,176*	−1, 147	−100.0

Source: Baumeister (2001) and own calculations.

This finding is consistent with the data on land use change for Guatemala presented in Table 4. The estimate suggests that deforestation is almost entirely associated with the expansion of other land area combined with open forests. Baumeister (2001) claims this category to be associated with agricultural expansion and shifting cultivation[19]. Finally, empirical case studies from Grunberg et al. (2000) as well as Sader et al. (2001) support the view that smallholders encroachment along river and road corridors has become the single most important factor associated with forest cover loss in northern Guatemala.

To summarise, the decline of large-scale pasture expansion, the negligible importance of commercial farming and logging, and the unsuitability for cultivation of the remaining land area (CONAMA et al. 1999a) within the context of continuing high deforestation, indicate that smallholders' migratory agriculture is the main direct source of deforestation in Guatemala during the 1990s.

5 Underlying determinants of deforestation

This section analyses some of the underlying economic parameters of land use decisions in Guatemala for the period 1990–2004 as they are identified in the lit-

[19] The estimate from Baumeister (2001) is based on satellite imaginary, but partly also on socio-economic data. Regarding the measurement of pasture expansion, survey and census data can lead to biases if ranchers have incentives to report less cattle.

erature. The *number* of deforestation studies and consulting reports sharply contrasts with the *analytical* work on this subject. For example, in co-operation with national forest institutions, the FAO (1999) synthesised more than 35 studies and reports on deforestation for Guatemala, and conducted targeted interviews to help overcome information constraints. Overall, the report remains inconclusive. The FAO asserts that there is no general agreement on the causal structure of deforestation in Guatemala.

However, a closer look here at selected deforestation studies does reveal some important findings. The following paragraphs present the findings on the following parameters of deforestation: (i) poverty, population growth and migration, (ii) road construction and soil conditions, (iii) agricultural intensification, (iv) property rights, (v) ethnicity, and (vi) rural non-farm employment and education.

Most of the reviewed studies concentrate on a particular issue, location or time period. Much of the analysis focuses on the Maya Biosphere Reserve in the Petén. Research has a strong focus on anthropological, sociological and nature conservation issues, and is almost entirely qualitative. In fact, Grunberg et al. (2000), Defensores de la Naturaleza (2001) and Loening (2004) are the only papers for Guatemala that empirically address some of the underlying factors of deforestation. In general, most of the research implicitly refers to a subsistence-based explanation of land use. Ironically, in particular regarding the effects of agricultural intensification, property rights and non-farm employment, the available evidence strongly supports the analytical relevance of a market-based explanation, i.e., deforestation occurs because it is relatively profitable[20].

5.1 Poverty, population growth and migration

The relationship between poverty and deforestation is rarely addressed explicitly. However, some studies do refer to the country's broader socio-economic context. In line with the previous historical overview, it is argued that the root causes of high migration are inequality of land distribution and the unfavourable socio-economic conditions that prevail in most parts of the country (Elías et al. 1997; Valenzuela de Pisano 1996; Schwartz 1990, 1995a, 1995b; World Bank 1995). In a descriptive analysis, based on unpublished census data, Gellert (2000) essentially qualifies rural migration as a survival strategy of the poor. Broadly speaking, despite much data uncertainty, the single greatest consistent finding of qualitative research is that natural population growth and rural migration flows, in particular to the Petén, are key parameters for environmental degradation and forest decline (Bilsborrow 1992; Elías et al. 1997; FLACSO 2000; Katz 2000; Nations et al. 1999; Schwartz 1990; Southgate and Basterrechea 1993; Valenzuela de Pisano 1996; World Bank 1995).

[20] For a theoretical discussion of subsistence and market-based deforestation models, see Angelsen (1999). An early advocate of seeing rural habitants as integrated into a larger market economy was the anthropologist Tax (1953). His ethnographic research documents that the socio-economic life of the Guatemalan indigenous population is deeply imbedded in a competitive market, and not dominated by irrational beliefs, cultural or religious influences.

Equally, a consultant report from Chemonics International et al. (2000) finds deforestation on the Maya Biosphere Reserve in the Petén essentially due to a high natural population growth and migrants from other areas of the country. However, Bilsborrow (1992) identifies severe research gaps regarding the exact underlying causes of rural migration, the origin, and the socio-economic characteristics of migrants.

Recently, for the Petén, some of these issues have been empirically documented by Grandia et al. (2001). One of the most interesting findings of the analysed household survey data refers to immigration. Many of the migrants are relatively young families. Migration increased sharply in the 1970s and reached its height during the early 1990s. While Ladino migration decreased over time, indigenous migration increased. The overwhelming part of the population in the Petén, however, is Ladino and not indigenous. Migrants come predominately from the north and northeast of Guatemala – part of the so-called ›poverty-belt‹ – and to a lesser extent also from the southwest. The main reason to come to the Petén is land-scarcity. Other striking results from Grandia et al. (2001) indicate that agricultural production is generally *not* subsistence-oriented, and that the overwhelming part of the migrants feels to be better off in the Petén.

Empirically, only Grunberg et al. (2000) and Loening (2004) have addressed the deforestation-population nexus. In a spatial regression for the buffer zone of the Maya Biosphere Reserve in the Petén, Grunberg et al. (2000) find that population growth and deforestation show a significant relationship for the period 1986–1999. However, the analysis is plagued by serious deficiencies. In particular, spatial and temporal auto-correlation problems and the low quality of the population data[21] in the Petén limit the reliability of the analysis. Moreover, the regression makes no reference to the potential endogeneity of the variable with respect to infrastructure, soil conditions or socio-economic factors. Nevertheless, the effect of population growth appears to be quite important. Using household survey data for 1999 and 2000, and dealing with endogeneity, the results from Loening (2004) indicate that 1 additional household member increases agricultural land use by about 10 percent.

5.2 Soil conditions and road construction

In line with spatial regression analysis for many other countries, Grunberg et al. (2000) find that well draining soil types are more likely to be deforested than poorly draining soil types in the Maya Biosphere Reserve of the Petén. In addition, over time, the deforestation rate increases on well draining soils while remaining close to zero on poorly draining soils. As expected, perennial roads are good indicators for deforestation threats in Grunberg's et al. (2000) simulation model.

[21] Population estimates for the Petén vary greatly. At the low end, the National Statistics Institute (INE) reports a population growth rate of 4.1 percent. Grandia (2000) argues that this figure does not account adequately for migration. Also, natural population growth is higher in the Petén than elsewhere. At the high end, according to a survey conducted by a consulting firm, there is an estimated annual growth of 9.5 percent. Virtually every institution seems to manage differing statistics falling in between these two estimates.

Equally, Sader et al. (2001) report that forest cover change in the Maya Biosphere Reserve essentially occurs along river and road corridors. Also, Rosenfeld (1999) finds that spontaneous colonisation processes are indirectly promoted by the road and pipeline construction undertaken by oil exploitation activities. Finally, in his voluminous study of the historical colonisation processes for the Petén, Schwartz (1990) remarks that with the opening of an all-weather road from the Petén to the rest of the country in 1970, deforestation and spontaneous encroachment increased.

5.3 Agricultural productivity

Agricultural intensification is actively promoted by conservation agencies as an effective tool to reduce deforestation in the Petén and elsewhere in the country. However, the evidence for Guatemala does not support this as an effective strategy. In a qualitative analysis, Shriar (2001) finds that extension agencies are frustrated by the lack of interest smallholders have for such practices in the Maya Biosphere Reserve. He argues that agricultural intensification involves high opportunity costs of labour, and this seems to make such techniques unattractive to most peasants. Moreover, unfavourable market conditions – except for maize – and other factors, such as fear of snakes and vandalism, limit farmers' interest in applying agricultural intensification techniques. Equally, Chemonics International et al. (2000) report no evidence that the promotion of such techniques does contribute to conserving the forest cover in the Petén.

In an empirical case study for the Sierra de la Minas Biosphere Reserve, Defensores de la Naturaleza (2001) find that the adoption of agricultural intensification techniques significantly increases farmers' demand for agricultural land use. The study is located in core areas of the second largest biosphere reserve in Guatemala (2,400 km^2) and uses survey data from 1998 for Q'eqchi' households. Overall, the study finds that farmers who use ›sustainable‹ agriculture techniques put their saved labour to use in ways that increase deforestation, for example by increasing the amount of area for planting maize or by establishing cash crops in forest areas[22]. Defensores de la Naturaleza (2001) suggests that the effect is due to the open access situation, where farmers lack the incentive to be efficient in land use. Loening (2004) using both countrywide and Petén-specific household survey data confirms these results. That is, the availability of technical assistance or the use of green manures exhibits a significant deforestation-increasing effect. In other words, at present, the promotion of ›sustainable‹ agriculture techniques seems to have the *exact* opposite effect on agricultural land use practices than the one desired by extension agencies.

[22] Defensores de la Naturaleza (2001) argue that agricultural intensification techniques are still a good instrument for forest conservation because of a number of positive externalities, such as improvements of rural livelihoods, local participation and the reduction of forest fires. Given that such techniques proved to have the opposite effect on forest conservation, this argumentation seems puzzling.

5.4 Property rights

Regarding the effects of land tenure and property rights on natural resource exploitation, the available evidence is entirely qualitative. The most comprehensive analysis comes from Katz (2000) and is based on a study by the World Bank (1995). For the Petén, a key finding is that insecure private land tenure regimes have encouraged resource-mining activities because ›using‹ the land is the necessary mechanism for establishing property rights. Moreover, in accordance with the observations from the previous paragraph, uncertainty over property rights frequently lead farmers to produce crops with short-growing cycles. Consequently, conservation NGO movement activities geared towards the adoption of ›sustainable‹ agriculture are frustrated by smallholders' lack of interest. Because of their limited size, communal land tenure regimes only play a minor role in the Petén.

In stark contrast to the Petén are the Western Highlands of Guatemala. They cover approximately 18 percent of Guatemala's territory and are home for more than 30 percent of the national population. Just as the Petén, the Western Highlands are ecologically diverse. They also constitute the principal watershed for most rivers in Guatemala. The fact that deforestation has been relatively low despite the high level of poverty among the predominantly indigenous smallholders, the high population density, and the lack of property rights in terms of legally registered titles has stimulated numerous debates. The World Bank (1995), Elías et al. (1997) and more pronounced Katz (2000) argue that the ›social capital‹ of the indigenous fosters a sense of ownership and respect for boundaries that can partially substitute for legal property rights[23]. Hence, deforestation is reduced because smallholders have incentives for conserving their resources. In the Western Highlands, the social capital base has allowed a series of common property regimes and communal owned forests to be established. There is no ›race for property rights‹ like in the Petén, where the migratory nature of population has prevented the creation of social capital.

5.5 Ethnicity

The empirical results from Loening (2004) indicate that indigenous households have a reduced demand for agricultural land area. The effect is very strong. Since in the popular debate in Guatemala the indigenous are often blamed for being one of the main deforestation agents, the negative correlation with land use and its high significance are enlightening. An interpretation, unfortunately, is less straightforward. In line with Katz (2000), one could assume that the indigenous social capital base and hence, informal tenure regimes, imply a deforestation-reducing effect. In

[23] The basic idea behind the notion of ›social capital‹ is that relationships among individuals give rise to something valuable. This intangible value can then be drawn upon to improve individual and collective well-being. Social capital may have its foundations in ethnicity, religion, shared history or other group membership. It is manifested in collective knowledge, including environmental knowledge, respect for group rules and norms, and the creation and maintenance of self-governing institutions (see Katz 2000).

other parts of the country, however, the indigenous could show a reduced demand for agricultural land area because of their small-scale farm practices, or because they are expelled to marginal agricultural areas with nothing left to exploit. Another reason for the reduced demand of the indigenous for agricultural land could be due to cultural attitudes, i.e., in the sense of a higher appreciation of natural resources.

5.6 Rural non-farm employment and education

Employment in the non-agricultural sectors of the rural economy and temporary farm wage employment exhibits a strong deforestation-decreasing effect, as evidenced by Loening (2004). In fact, non-farm employment opportunities appear to be the single most important factor among the various households' decision parameters that affect deforestation. Moreover, the empirical evidence reveals that poverty and deforestation may be linked via the availability of non-farm employment. This is because households can diversify their sources of income, which allows them to augment their incomes and to minimise the effect of exogenous shocks on farm activities. The empirical results also indicate that access to piped water, sanitation and connection to electricity – virtually unavailable in many rural areas – are important complementary factors for a policy aimed to reduce deforestation. Access to basic infrastructure significantly increases the probability of being employed in the non-farm sector.

An issue widely ignored in the deforestation literature is the role of education[24]. For the case of Guatemala, the results presented in Loening (2004) suggest that schooling exhibit a significant deforestation-reducing effect. Overall, 1 additional year of schooling implies a reduction of about 7 percent in cultivated land area. In other words, a household head with complete primary education (6 years) exhibits a reduced demand for agricultural land of more than 1/3 compared to his counterpart with no formal education. For the Petén, the outcome is non-linear and more pronounced for basic education up to 3 years. The main channel of a deforestation-reducing effect of schooling operates by enhancing the probability of being employed in the non-farm sectors that require skills obtained by formal education. However, other factors, such as a higher appreciation of biodiverse natural resources, lower fertility and increased rural-to-urban migration, equally play a role.

6 Concluding summary

This paper has addressed some of the fundamental causes of deforestation and habitat loss in Guatemala. The factors, which are associated with deforestation in

[24] Godoy and Contreras (2001) point out that this is surprising, but may reflect disciplinary parochialism, since researchers of education seem to be not concerned about environmental issues, and investigators working on deforestation seem to know little about the effects of human capital.

Guatemala, seem to have much in common with other countries in Latin America. This makes Guatemala an interesting case study and, to some degree, the summarised findings may be important for other settings as well.

The socio-economic causes that spurred habitat fragmentation are multifaceted and complex. Agricultural expansion during the 19th and early 20th century was correlated with discriminatory educational and labour polices. These past policies, in turn, have repercussions until today, not only in terms of the persistence of rural poverty but also on deforestation. At the risk of oversimplification, the available evidence supports a view that after a period of large-scale agricultural and pasture expansion, smallholder's encroachment has become the single most important direct source of deforestation. A major pull factor was the government-sponsored opening up of previously inaccessible forest areas in northern Guatemala and the Petén. For Guatemala's main agricultural frontier, with high levels of biodiversity, the findings of this review suggest that land conversion and degradation is associated with persistent rural poverty and land scarcity in immigrants' areas of origin. In addition, there is evidence that most of the Petén's migrants do not follow subsistence-oriented agricultural patterns, and that resource mining in the agricultural frontier improves individuals' livelihoods. Insofar, it is fair to argue that rural poverty alleviation is *the* key issue for tackling the overwhelming majority of the deforestation and biodiversity loss in Guatemala. Much, if not most, of habitat loss is a side effect of non-agricultural policies.

The underlying determinants of land use decisions show several interesting findings. First, rural non-farm employment and increased schooling has the potential to reduce pressure on the forests. Since educational reforms in Guatemala are already in place, schooling may reduce forest clearing in a potentially more progressive way than via the financing of new programs to conserve tropical forests. Second, other important parameters are related to the land tenure regime and ethnicity. The fact that indigenous households show a reduced demand for agricultural land could be due to the indigenous peoples' social capital base in the sense of communal tenure regimes, as well as a higher appreciation of natural resources, or other factors. Third, contrary to common beliefs, agricultural intensification programs in Guatemala seem to increase the demand for additional land. These results call into question some of the current conservation practices promoted by NGOs and bilateral donors. Altogether, Guatemala is proving a showcase, suggesting that policies focussing on natural resource conservation without being concerned about the socio-economic well-being of the affected population – in terms of non-farm employment opportunities and human capital policies – may be seriously misguided.

References

Angelsen A (1999) Agricultural Expansion and Deforestation: Modelling the Impact of Population, Market Forces and Property Rights. Journal of Development Economics 58 : 85–218

Banco de Guatemala (1981), Establecimiento de una empresa ganadera en Petén. Informe Económico 28 : 21–70

Baumeister E (2001) Formas de acceso a la tierra y al agua en Guatemala. Cuadernos del Desarrollo Humano 4, Sistema de las Naciones Unidas en Guatemala, Guatemala

Bilsborrow RE (1992) Rural Poverty, Migration, and the Environment in Developing Countries: Three Case Studies. World Bank Policy Research Working Paper 1017

Cabrera C (1995) Síntesis histórica de la deforestación en Guatemala: perspectivas para un nuevo siglo. Tikalia 15 : 83–101

CEPAL (Comisión Económica para América Latina) (2003) Macroeconomic Database. http://www.cepal.org/, October 2003

Chemonics International, BIOFOR Consortium, and International Resource Group EPIQ ICQ (2000) Guatemala: Assessment and Analysis of Progress toward SO5 Goals in the Maya Biosphere Reserve. Consultant report submitted to USAID/G-CAP, Guatemala

Commission for Historical Clarification (Comisión para el Esclarecimiento Histórico) (1999) Guatemala: memoria del silencio. United Nations Office for Project Services, Guatemala

CONAMA et al. (Comisión Nacional de Medio Ambiente) (1999a) Conociendo el Sistema Guatemalteco de Areas Protegidas. Estrategia Nacional para la Conservación y Uso Sostenible de la Biodiversidad, Guatemala

CONAMA et al. (1999b) Las áreas silvestres de Guatemala: ¿tienen amenazas? Estrategia Nacional para la Conservación y Uso Sostenible de la Biodiversidad, Guatemala

CONAMA et al. (1999c) La vida silvestre: uso y conservación. Estrategia Nacional para la Conservación y Uso Sostenible de la biodiversidad, Guatemala

Culbert TP (1988) The Collapse of Ancient States and Civilizations. University of Arizona Press, Tucson

de La Cruz JR (1982) Clasificación de zonas de vida de Guatemala a nivel de reconocimiento. Ministerio de Agricultura, Ganadería y Alimentación & Instituto Nacional Forestal, Guatemala

Defensores de la Naturaleza (2001) Impacto de la agricultura sostenible sobre la conservación de la biodiversidad, Reserva de la Biósfera Sierra de las Minas, Guatemala. Biodiversity Support Program, Washington DC

Dinerstein E, Olson DM, Graham DJ et al. (1995) A Conservation Assessment of the Terrestrial Ecoregions of Latin America and the Caribbean. World Bank, Washington DC

Elías S, Gellert G, Pape E et al. (1997) Evaluación de la sostenibilidad: el caso de Guatemala. Facultad Latinoamericana de Ciencias Sociales & World Wildlife Fund, Guatemala

FAO (Food and Agriculture Organization of the United Nations) (1999) Annotated Bibliography Forest Cover Change Guatemala. Forest Resources Assessment Programme Working Paper 13

FAO (2001) State of the World's Forests. Rome

FLACSO (Facultad Latinoamericana de Ciencias Sociales) (2000) Nuevas perspectivas de desarrollo sostenible en Petén. Guatemala

Galindo JL (1999) Propuesta para una estrategia financiera del Sistema Guatemalteco de Areas Protegidas. Documento técnico 10, Consejo Nacional de Áreas Protegidas, Guatemala

Gellert G (2000) Las migraciones como estrategias de sobrevivencia de los excluidos y sus determinantes territoriales. In: Gálvez V (ed) Guatemala: exclusión social y estrategias para enfrentarla. Facultad Latinoamericana de Ciencias Sociales, Guatemala, pp 173–343

Godoy J (1998) Matrices de proyectos en ejecución y un perfil conceptual: diagnóstico resumido del SIGAP. Consultant report submitted to USAID, Guatemala, pp 25–27

Godoy R, Contreras M (2001) A Comparative Study of Education and Tropical Deforestation among Lowland Bolivian Amerindians: Forest Values, Environmental Externality, and School Subsidies. Economic Development and Cultural Change 49(3) : 555–574

Grandia L (2000) ¿Cuántas personas quiere usted que vivan en el Petén? In: FLACSO (2000) Nuevas perspectivas de desarrollo sostenible en Petén. Guatemala, pp 137–151

Grandia L, Schwartz N, Corzo A et al. (2001) Guatemala: salud, migración y recursos naturales en Petén, resultados del módulo ambiental en la Encuesta de Salud Materno Infantil 1999. Instituto Nacional de Estadística, US Agency for International Development, Measure/DHS+ and Macro International, Calverton

Grunberg W, Guertin D, Shaw W (2000) Modeling Deforestation Risk for the Maya Biosphere Reserve, Guatemala. Paper presented at the 20th annual ESRI International User Conference, San Diego, California

Heltberg R (2001) Determinants and Impact of Local Institutions for Common Resource Management. Environment and Development Economics 6(2) : 183–208

INAB (Instituto Nacional de Bosques) (2000) Informe final, fase I: mapa de cobertura forestal para 1999. Guatemala (mimeo)

Kaimowitz D (1995) Land Tenure, Land Markets, and Natural Resource Management in the Petén and Northern Transversal of Guatemala. Center for International Policy Research, Bogor (mimeo)

Kaimowitz D (1996) Livestock and Deforestation in Central America in the 1980s and 1990s: A Policy Perspective. Center for International Policy Research, Bogor (mimeo)

Katz E (2000) Social Capital and Natural Capital: A Comparative Analysis of Land Tenure and Natural Resource Management in Guatemala. Land Economics 76(1) : 114–132

Loening LJ (2004) Economic Growth, Biodiversity Conservation, and the Formation of Human Capital in a Developing Country: The Case of Guatemala. Goettingen Studies in Development Economics 13, Peter Lang, Frankfurt

Loening LJ, Markussen M (2003) Pobreza, deforestación y sus eventuales implicaciones para la biodiversidad en Guatemala. Economía, Sociedad y Territorio 4(14) : 279–315

Lovell GW (1988) Resisting Conquest: Development and Guatemalan Indians. In: Kirk JM and Schuyler GW (eds) Democracy, Development and Change. Praeger, New York, pp 101–107

Luján J (2000) Breve historia contemporánea de Guatemala. Colección popular 552, Fondo de la Cultura Económica, Mexico

MAGA (Ministerio de Agricultura, Ganadería y Alimentación), PAFG (Plan de Acción Forestal), INAB (Instituto Nacional de Bosques) (1998) Proyecciones del Programa de Incentivos Forestales y necesidades de inversiones adicionales. Guatemala (mimeo)

Maya' QK (1995) Comentarios sobre algunos nombres de comunidades e idiomas mayas, Boletín de Lingüística Universidad Rafael Landívar 9(51) : 1–5

Méndez C (1999) How Old is the Petén Tropical Forest? In: Nations JD, Rader CJ, Neubauer IQ (1999) Thirteen Ways of Looking at a Tropical Forest: Guatemala's Maya Biosphere Reserve. Conservation International, Washington DC, pp 31–38

Mittak WL (1977) Guatemala: estudios para la reforestación nacional, fortalecimiento al sector forestal. Documento de Trabajo 25, UNDP/FAO, Guatemala (mimeo)

Myers N, Mittermeier RA, Mittermeier CG et al. (2000) Biodiversity Hotspots for Conservation Priorities. Nature 403 : 853–858

Nations JD, Houseal B, Ponciano I et al. (1989) Biodiversidad en Guatemala: evaluación de la diversidad biológica y los bosques tropicales. Centro para el Desarrollo Internacional y el Medio Ambiente & World Resource Institute, Washington DC

Nations JD, Rader CJ, Neubauer IQ (1999) Thirteen Ways of Looking at a Tropical Forest: Guatemala's Maya Biosphere Reserve. Conservation International, Washington DC

O'Hara SL, Street-Perrott FA, Burt TP (1993) Accelerated Soil Erosion Around a Mexican Highland Lake Caused by Prehispanic Agriculture. Nature 362 : 48–50

Rosenfeld A (1999) Oil Exploitation in the Forest. In: JD Nations, Rader CJ, Neubauer IQ (eds) Thirteen Ways of Looking at a Tropical Forest: Guatemala's Maya Biosphere Reserve. Conservation International, Washington DC, pp 68–76

Sader SA, Hayes D, Coan M et al. (2001) Forest Monitoring of a Remote Biosphere Reserve. International Journal of Remote Sensing 22(10) : 1937–1950

Sader SA, Martínez EB, Irwin DE et al. (2000) Estimación de la deforestación en la Reserva Biósfera Maya 1997–1999. In: FLACSO (ed) Nuevas perspectivas de desarrollo sostenible en Petén. Guatemala, pp 321–324

Sagastume R (1992) Informe mapa de la cubierta forestal de la República de Guatemala. Plan de Acción Forestal, Guatemala (mimeo)

Schwartz NB (1990) Forest Society: A Social History of Petén, Guatemala. 3rd ed, University of Pennsylvania Press, Philadelphia

Schwartz NB (1995a) Colonization, Development and Deforestation in Petén, Northern Guatemala. In: Painter M, Durham WH (eds) The Social Causes of Environmental Destruction in Latin America. University of Michigan Press, Ann Arbor, pp 101–130

Schwartz NB (1995b) Reprivatización y privación: sistema tradicional y contemporáneo de tenencia de la tierra en el Petén, Guatemala. Mesoamerica 29 : 215–232

Schweigert TE (1993) Commercial Sector Wages and Subsistence Sector Labor Productivity in Guatemalan Agriculture. World Development 21(1) : 81–91

SEGEPLAN (Secretaría General del Consejo Nacional de Planificación Económica), MAGA (Ministerio de Agricultura, Ganadería y Alimentación) (1984) Recopilación de información comparativa de los censos agropecuarios 1950, 1964 y 1979 y series de superficie, producción y disponibilidad por rubros agrícolas, 1950–1982. Guatemala (mimeo)

Shriar AJ (2001) The Dynamics of Agricultural Intensification and Resource Conservation in the Buffer Zone of the Maya Biosphere Reserve, Petén, Guatemala. Human Ecology 29(1) : 27–48

Southgate D, Basterrechea M (1993) Population Growth, Public Policy and Resource Degradation: The Case of Guatemala. Ambio 21(7) : 460–464

Spielmann HO (1973) Ursachen, Merkmale und Bedeutung der Bevoelkerungsverschiebungen in Guatemala. Institut für Geographie und Wirtschaftsgeographie, Hamburger Geographische Studien 30, Verlag Ferdinand Hirt, Universitaet Hamburg

Tax S (1953) Penny Capitalism: A Guatemalan Indian Economy. Smithsonian Institution, Institute of Social Anthropology, Publication 16, US Government Printing Office, Washington DC

UNDP (United Nations Development Programme) (1998) Guatemala: los contrastes del desarrollo humano. Informe Nacional de Desarrollo Humano 1998, Sistema de las Naciones Unidades en Guatemala, Guatemala

UNDP (2000) Guatemala: la fuerza incluyente del desarrollo humano. Informe Nacional de Desarrollo Humano 2000, Sistema de las Naciones Unidades en Guatemala, Guatemala

UNDP (2002) Guatemala: desarrollo humano, mujeres y salud. Informe Nacional de Desarrollo Humano 2002, Sistema de las Naciones Unidades en Guatemala, Guatemala

Valenzuela de Pisano I (1996) Agricultura y bosque en Guatemala: estudio de caso en Petén y Sierra de las Minas. Universidad Rafael Landívar & World Wildlife Fund, Guatemala

World Bank (1995) Guatemala: tenencia agraria y manejo de los recursos naturales. Report 14553-GU, Washington DC (mimeo)

World Bank (1996) Guatemala: Building Peace with Rapid and Equitable Growth. Country Economic Memorandum, Report 15352-GU, Washington DC

World Bank (2003) Guatemala: Poverty in Guatemala. Report 24221-GU, Washington DC

World Bank, UNDP (2003) Household Fuel Use and Fuel Switching in Guatemala. Joint World Bank & UNDP Energy Sector Management Assistance Programme, Washington DC (mimeo)

Part IV

Epilogue: A general perspective on the evaluation of interdisciplinary research

Permanent Evaluation: An important tool for a quality assurance in interdisciplinary research

Philip W. Balsiger and Rudolf Kötter

Interdisciplinary Institute for Philosophy and History of Science, University of Erlangen-Nuremberg, Germany, email to pebalsig@phil.uni-erlangen.de

Summary. If scientists plan to solve a given problem which has arisen in the real world they are often forced to collaborate with scientists stemming from various disciplines. From a methodological point of view this actually implies that, while seeking the best solution, several disciplinary research programmes have to be interconnected. This is less to achieve goals like ›understanding biodiversity‹; rather it is to describe the scientific problem of biodiversity in an adequate extension and also to consider the inherent complexity. This is what a simple disciplinary approach can no longer achieve. Hence, it can be taken for granted that research on biodiversity definitely challenges science because the impacts of several scientific disciplines are needed. To grant the quality of such research, a form of evaluation is needed which takes into consideration all different disciplinary and supradisciplinary aspects. Today, generally most of the treatments of evaluation in use are related to a disciplinary basis. However, those disciplinary patterns of assessment are only adequate to some extent for the specific aspects of interdisciplinary or transdisciplinary research. Nevertheless, evaluation should contribute to the success of a research project and it should also be an instrument to manage and assure the quality of a project. To maintain this task for supradisciplinary projects as well, this contribution suggests a form of evaluation which is based on full transparency among the various partners during the research process concerning the criteria and a discursive manner of actualising.

1 Introduction

There appears to be a broad agreement among biologists in the field of biodiversity that research is strongly related to supradisciplinary approaches[1]. In most of the cases this approach is simply called interdisciplinary (Weber and Schmid 1995), but an alternative sometimes used is transdisciplinary. When examining various papers claiming such a close relationship, it is easy to find that in most cases there lacks a clear definition of what exactly is meant when using a term like interdisci-

[1] Here the term of ›supradisciplinarity‹ is used as a collective term, unifying the terms of ›interdisciplinarity‹ and ›transdisciplinarity‹. This is necessary in order to use both these terms in terminologically well defined ways.

plinarity. In respect to the philosophy of science, it is often even quite difficult to find an explanation for this stated need.

Hence, in this contribution we will first discuss the basis on which a strong terminological solution may be settled. This suggestion should help to avoid misunderstandings. Following this clarification, we have to figure out which of these various forms of supradisciplinary approaches fits best with the form of research done in the field of biodiversity. Subsequently, we have to gain insight into how research in biodiversity is actually done. In addition, we have to clarify whether or not the arguments used by biologists in biodiversity research, explaining the need of supradisciplinary approach, is persuasive or not. Based on sound arguments, this will help to decide which form of supradisciplinary research may really be considered a suitable form of research in the field of biodiversity. Upon understanding the terminology, we are talking about having good arguments for the use of one, or the other term. Once we understand what is typical for supradisciplinary research, we will then analyse what sort of consequences arise for a goal-leading evaluation of that kind of research project. We then will focus on the problem of evaluation of research projects, and its need in the field of biodiversity. Finally, we will suggest how supradisciplinary research projects in general and research projects in the field of biodiversity in particular, can be evaluated in an adequate manner.

2 On terminological inconveniences

2.1 Problem-oriented vs. topic-oriented research

One reason that so-called ›interdisciplinary‹ research often ends in an unsatisfying manner is due to the fact that such projects often start without knowing their precise goal. In reaction, scientists start to behave in an opportunistic way or they renounce further involvement in such projects. In fact, scientists often believe that a scientific problem has been labelled, when at the very most only the outline of a topic has been formulated. The most important difference between problem-oriented research (which leads to interdisciplinarity) and theme or topic-oriented research (which leads to multidisciplinarity) will be discussed in the following section.

The term of problem-oriented research was introduced by the Belgian philosopher of science, Pierre de Bie in 1970. However, de Bie's terminological solution focused on a type of practice which will, in the following text, be described as transdisciplinary (see de Bie 1973 : 9). The essential differences are a close form of collaboration defined by the relationship of a problem and an open form which is guided by a topic. By formulating a topic the conceptual frame is set, inside of which scientists dealing with the topic are free to formulate their own concrete scientific problems. All contributions will then be taken as elements of a set, which is bounded by the topic, but they don't need to show any closer relations between each other. One can work on various topics but they will not be solved. A problem needs a solution and by posing a problem the expectations and criteria are delivered, which a good solution has to meet. Every step in a problem oriented research regarding its content or its organisational framework has to contribute to the solution of the problem.

Such confusion about problem-oriented or topic-oriented work may happen when researchers start to realise that the multidisciplinary approach they have chosen, intending to solve a scientific problem, does not really work out the way they expected when starting the project. To figure out whether they are working on a problem or on a topic one can ask whether they are looking for a solution or whether they are contributing to, e.g., an explanation for a given topic. If they work on a solution, their supradisciplinary collaboration needs to be carried out in a specific manner. In contrast, if they treat a topic, scientists will choose a manner of collaboration which does not *apriori*, ask for a strong coherence among participating disciplines. Here is an example of what is meant by this differentiation. If scientists start to think about the sense and the need for a Natural Protection Area they are treating a topic. As soon as they start to establish a Natural Protection Area, they are treating a problem. The first activity needs an explanation, the second a solution!

With respect to the difference between problem-oriented and topic-oriented research we should now be able to state that interdisciplinary research projects will only be accepted as independent projects if they are based on a well defined problem. Unfortunately, in research practice it is heavily underestimated that outstanding and successfully driven interdisciplinary research should be exclusively based on a concrete problem-orientation.

2.2 Interdisciplinarity and the question of how to guide the research process

For all scientists involved, establishing a Natural Protection Area is a concrete and challenging problem to solve – the basis for successful supradisciplinary collaboration is given. It no longer makes any sense to develop individual disciplinary suggestions. Rather, it becomes necessary to collaborate in a particular manner. Hence, we can say that it is a necessary condition for a problem-oriented supradisciplinary collaboration to show every participating scientific discipline exactly where and how it will contribute to the solution of the given problem. As it demands a special methodological and organisational input to initiate a problem-oriented supradisciplinary research, the term ›interdisciplinary‹ should only be dedicated to those cases where various disciplines jointly contribute to a solution while keeping their autonomy. In particular, this means that they may not mutually be reduced to service providers. Concerning the methodological requirements for a truly interdisciplinary collaboration, in a paper of Henk Zandvoort (Zandvoort 1995) we find a so-called »interactive model«. It describes how in an interdisciplinary research project participating disciplines will knot together. Zandvoort is not talking about ›scientific disciplines‹; rather he prefers to use the term »research programme« which was first introduced by Imre Lakatos (Lakatos 1970). Thus, for Zandvoort there exists a guide-supply relationship among the various research programmes involved in the research process in question. Zandvoort uses the following formulation:

> »The cooperative development of research programmes comes about in the following way. Some of the research programmes do not define their own primary problems. Instead, they aim at solving problems arising in and defined by other research programmes. The latter programmes may not themselves have the (efficient) means to solve those problems. The programmes generating the problems I have called ›guide programmes‹, because they act as guides for the programmes that aim at solving those problems. The latter programmes I have called ›supply programmes‹, because, when successful, they satisfy the needs of the other programme«. (Zandvoort 1995 : 53)

This means in particular, that abstract descriptions which arise from the perspective of a discipline that is actually in the ›guide-mode‹ have to be cemented in such a way that new tasks are created for disciplines in the ›supply-mode.‹ It is then decisive that the direction in which the abstraction is to be put into concrete form as well as the other disciplines can only be gained by means of the initial nature of the problem, as has been shown by the above example. A serving relationship differs from a guide-supply-relationship. In the latter, the discipline is not and does not remain solely responsible for the treatment of a given problem but merely takes over the additional task of distributing complementary research projects. Even if one discipline is especially recommended as the first partner in the treatment of a problem such as climatology on the carbon dioxide problem, that problem will be developed by consulting further disciplines. This implies that the ›guide-supply-relationship‹ among the participating research programmes can by all means switch throughout the research process. This would not be the case in a relationship of services.

For this reason the initial and most important step towards a successful interdisciplinary collaboration has to be made by formulating a structural description of the given problem. This formulation should precisely point out (i) that the problem in question cannot to be handled by a single discipline, and (ii) what exactly a research programme expects from other research programmes in respect of their contributions to the intended solution of the given problem, i.e., what kind of ›guide-supply-relationship‹ should be established.

To support a joint structural description it must be ensured that all expectations are perceived in an undistorted manner. In particular, this implies that scientists have to agree on the common language as well as the disciplinary heuristics. These are both necessary conditions to ensure that a scientific discipline will perceive the problem-situation in question.

2.3 Transdisciplinarity

In many cases, for example in the so-called research on sustainability, problems have a particular quality. They emerge in our everyday life by any implementation of scientific knowledge. But those disciplines which have developed the basis of knowledge cannot solve those problems which arise from any implementation as negative side-effects, alone. Therefore, in the frame of research we suggest using the term ›transdisciplinarity‹ (Kötter and Balsiger 1999). This covers all those problems which cannot simply be described and evaluated scientifically in an internal manner. It is obvious that the science-transgressing nature of this kind of problem strongly requires an interdisciplinary collaboration to solve the given problem. Hence, the specific character needed to mark a science-transgressing problem and therefore to be permitted to call it a transdisciplinary problem is given:

(a) if the problem is emerging from a non-scientific field (e.g., economics, politics, everyday life)
(b) if its solution is urgently requested in this specific field
(c) if it is publicly classified as a relevant problem
(d) if it is implemented in science by research contracts or by specific funding of a project.

In addition to these criteria, we further suggest ›transdisciplinarity‹ if scientists recognise that some special developments have the potential to become socially relevant, or if society is not yet aware of their capability or if scientists would like to take on the task of elucidating. It should be emphasised that this terminological suggestion envisages a specific relationship between science and society. In our opinion, the term transdisciplinarity is not suitable for the characterisation of this interplay in a more general way. Therefore in (d) we only relate in a pragmatic manner to the most usual way research projects are awarded.

The distinction between interdisciplinary and transdisciplinary research suggested above, first of all only describes pure types of research. Often one can spot research projects which represent hybrid forms of transdisciplinary, disciplinary and even applied research[2]. This was observed in the large Priority Programme Environment, installed and run by the Swiss National Research Foundation from 1992 to 2001.

2.4 Biodiversity and its real relationship with supradisciplinary research

After all these rather large terminological differentiations and explanations, the following questions have to be posed: (i) what are the characteristics of biodiversity research? (ii) based on these characteristics, is there a true and exclusive need in biodiversity research to explore the given problems by using a supradisciplinary approach? If the answer to this question is positive, which is the most adequate form?

It is hard to argue against a definition of biodiversity research which focuses on both the quantitative and the qualitative aspect of the appearance of species in a defined geographical space and the genetic variety in a specific species. Based

[2] Applied research means that the result of a research process (usually a concrete product) is to be used immediately in, e.g., industrial production processes.

on such a definition, one could easily claim that the key issue of biodiversity research is simply a biological one. At a first glance this seems to be correct, because at the first stage of biodiversity research the given problem situation of diversity is always judged and evaluated in respect of the idea of evolution. However, in addition to this, further questions are arising which are fairly transgressing the field of biological disciplines. Thus it should be mentioned that the problem situation of diversity is also evaluated in respect to the situation of human beings, especially concerning human needs. Hence, this activity of scientific evaluation and judgement definitely does not belong to the biological issue mentioned above. If we ask about the impacts of economy, tourism and landscape architecture on biodiversity, or, from another perspective we ask what the consequences of a switch in the spectrum of species or even the disappearance of species for human beings are and whether these processes can be stopped and at what price, then we have to deal with problems which can only, if ever, be solved by transdisciplinary research.

3 Evaluation

Each form of evaluation of research projects can be divided into two different parts: (i) the so-called evaluation ex ante is only related to research proposals. In the first instance, it has to decide whether a proposal is, with regards to any formal prerequisites, capable of being funded or not. In the second instance, it has to establish an order in relation to the dignity of funding among those proposals which have passed this first barrier. In this situation, aspects like the relevance of a research question and its fecundity as well as the prospect of success become decisive.

From these results, we have to differentiate the so-called evaluation ex post. With this we can now decide in a conclusive manner whether the goals stated in the proposal have been achieved or not. If the project has been successful, using the evaluation ex post, we can then rate to what extent the goals have been achieved and to what limit reasonable differences may be accepted.

The key problem when evaluating scientific projects and performances is the large number of criteria involved. These criteria have to be interconnected in such a manner that the on-based system becomes unambiguous and reproducible. Finally, it has to be accepted by all participants. In the past, standards have been developed to evaluate the practice in disciplinary research. It remains undecided whether these standards really express scientific rationality and whether or not they are accepted (Beywl 1988).

Evaluating transdisciplinary research projects is even more complicated and complex. In respect to the content and the organisation, the evaluation itself has to follow the complicated structures of such projects and it is not allowed to just evaluate the respective disciplinary outcome. Because the success of transdisciplinary research projects is also dependent on its perception by the external partners, the evaluation of such research projects also has to consider the way scientists and external partners collaborated.

Although the projects in the Swiss Priority Programme Environment were rather heterogeneous in respect of their topics, as well as in respect of their organisation, there were a number of specific difficulties which were the same for all participating scientists. Here we will just discuss three of them which seem to be symptomatic for projects in the research field of sustainability.

(1) As mentioned above, one source for quite a lot of difficulties in supradisciplinary research is the missing consciousness of a precise differentiation between theme or topic-oriented research and problem-oriented research. As observed in the case of the Swiss Priority Programme Environment (Balsiger and Kötter 2000), research grant foundations often call for the treatment of topics. In addition, they expect the projects, in a more or less clearly expressed manner, to join up with problem-oriented research compounds. This vagueness is also found in both evaluation processes. Due to the character of supradisciplinary research projects, a standardised evaluation procedure does not exist. Peer reviewers prefer to use the well-known disciplinary criteria, but this is where they often miss the specific aspect of supradisciplinary research projects (see Grossenbacher 1996). To evaluate a supradisciplinary research project there are three guiding-questions which may help to produce an adequate assessment.

First, has the problem-situation of the project been developed in cooperation with partners from outside of science and has it been taken into consideration that the solution of the given problem must also be accepted by these partners?

A positive answer to this question is a necessary condition in order to talk about a transdisciplinary project. Subsequently, the question arises whether or not each of the disciplinary partial projects also achieves a scientific contribution for its self and whether or not these projects could also be considered single projects in a normal evaluation procedure.

If this second question is answered positively, we can definitely say that we are not dealing with an interdisciplinary project. In contrast, if the answer is negative we can add the third question. Do the partial projects, to reach their own goals, have to get specific support from other partial projects? A positive assessment of whether partial projects are mainly dependent on each other is not decided in respect to the description of the given problem and the design of the project as mentioned above. Both have to immediately make clear what kind of character a potential solution will have, e.g., does the project look for explanations or for technical or social strategies on how to treat the given problem, and what kind of organisational development will the project have. In transdisciplinary research it is important that all partners involved (those who are asking for the project, those who are treating the project scientifically and the peer reviewers) agree right from the start whether they want to deal with a problem-oriented or a topic-oriented project.

(2) Because of their problem-orientation, transdisciplinary projects often have a temporal structure. Hence, it would be extremely coincidental if all partial projects could start their research at the same time. At the academic level, a research paper is often work towards a qualification, e.g., a doctoral thesis, and therefore such work has to be adapted to a specific manner of training and additional difficulties such as temporal and financial ones will emerge. Another problem is

that what the professional profile collaborators can gain in a supradisciplinary research project might not necessarily fit with the structures of the specific disciplinary field in which they want to become qualified. At present, it is an open question whether the professional skills acquired in the frame of a transdisciplinary project, are favourable for the start of a professional career.

(3) The education and the training of skills for transdisciplinary work is hard fit with the structures of formation today[3]. The only thing that is clear is that simply accumulating non-disciplinary offers in a disciplinary curriculum does not really work. In fact it would be necessary to practise various forms of problem-oriented working in the frame of small projects. Thereby, students should learn how to formulate problems in a scientifically adequate way and to recognise where, and to what extent various disciplines can contribute. Students should also learn about the various ways of thinking that is practiced in different disciplines (*Denkstil*). Nevertheless, it is also important that scientists planning to research in the field of sustainability are able to communicate with partners from outside science, at the same level. As observed in the Swiss Priority Programme Environment, it is less the problem of communicating among scientists than it is the problem of communication between scientists and non-scientists. Therefore, it is not only a problem of a lingual mediation but also of how to perform and how maintain the respect and the competence of the counter-part.

References

Balsiger PW, Kötter R (2000) Transdisziplinäre Forschung—Ein Erfahrungsbericht zum Schwerpunktprogramm Umwelt (SPPU) des Schweizerischen Nationalfonds. In: Brand KW (ed) Nachhaltige Entwicklung und Transdisziplinarität: Besonderheiten, Probleme und Erfordernisse der Nachhaltigkeitsforschung. Analytica, Berlin, pp 181–194

Beywl W (1988) Zur Weiterentwicklung der Evaluationsmethodologie. Grundlegung, Konzeption und Anwendung eines Modells der responsiven Evaluation. Lang, Frankfurt a.M. and others

de Bie P (1973) Problemorientierte Forschung. Hauptströmungen der sozialwissenschaftlichen Forschung. Ullstein, Frankfurt a.M. and others

[3] Often, in the frame of supradisciplinary research projects scientists are talking about ›interdisciplinarity‹. In this case, the term is seldom taken in its strongest interpretation. If this was the case, disciplinary contributions might have built on each other in respect of the contents as well as in respect of the temporal development of the project. Moreover, they would also be determined by the way the problem is given, and not by the demand of the profile of qualification of particular disciplinary fields. In other words, it could not be guaranteed that the necessary papers would always have the level of a dissertation. For each Ph.D student, the mutual dependencies could become high risk and could endanger the success of the whole project as well. In its strongest interpretation, interdisciplinary research in general is not a good panel for qualification works unless such work can be integrated into corresponding structures of a project.

Grossenbacher-Mansuy W (1996) Projektauswahl und Evaluation im SPP Umwelt. Panorama 6 : 46–50

Kötter R, Balsiger PW (1999) Interdisciplinarity and transdisciplinarity—a constant challenge to the sciences. Issues in Integrative Studies 17 : 87–120

Lakatos I (1970) Falsification and the Methodology of Scientific Research Programmes. In: Lakatos I, Musgrave A (eds) International Colloquium in the Philosophy of Science, Cambridge University Press, London, pp 91–195

Weber M, Schmid B (1995) Reductionism, holism and integrated approaches in biodiversity research. Interdisciplinary Science Reviews 20(1) : 49–60

Zandvoort H (1995) Concepts of Interdisciplinarity and Environmental Science. Cognitive Patterns in Science and Common Sense. In: Kuipers TAF, Mackor AR (eds) Groningen Studies in philosophy of science, logic, and epistemology (= Poznan studies in the philosophy of the science and the humanities, No 45), Amsterdam Atlanta GA. (USA), Rodopi, No 45, pp 45–68

GPSR Compliance

The European Union's (EU) General Product Safety Regulation (GPSR) is a set of rules that requires consumer products to be safe and our obligations to ensure this.

If you have any concerns about our products, you can contact us on ProductSafety@springernature.com

In case Publisher is established outside the EU, the EU authorized representative is:

Springer Nature Customer Service Center GmbH
Europaplatz 3
69115 Heidelberg, Germany

Batch number: 10755329

Printed by Printforce, the Netherlands